Advanced Agro-Engineering Technologies for Rural Business Development

Valeriy Kharchenko
Federal Scientific Agroengineering Center VIM, Russia

Pandian Vasant
Universiti Teknologi PETRONAS, Malaysia

A volume in the Advances in Environmental Engineering and Green Technologies (AEEGT) Book Series

IGI Global
DISSEMINATOR OF KNOWLEDGE

Published in the United States of America by
IGI Global
Engineering Science Reference (an imprint of IGI Global)
701 E. Chocolate Avenue
Hershey PA, USA 17033
Tel: 717-533-8845
Fax: 717-533-8661
E-mail: cust@igi-global.com
Web site: http://www.igi-global.com

Library of Congress Cataloging-in-Publication Data

Names: Kharchenko, Valeriy, 1938- editor. | Vasant, Pandian, editor.
Title: Advanced agro-engineering technologies for rural business development / Valeriy Kharchenko and Pandian Vasant, editors.
Description: Hershey, PA : Engineering Science Reference, an imprint of IGI Global, [2019] | Includes bibliographical references and index.
Identifiers: LCCN 2018030370| ISBN 9781522575733 (hardcover) | ISBN 9781522575740 (ebook)
Subjects: LCSH: Agricultural development projects. | Rural development projects. | Agricultural engineering. | Agriculture--Economic aspects.
Classification: LCC S675 .A29 2019 | DDC 631.3--dc23 LC record available at https://lccn.loc.gov/2018030370

This book is published in the IGI Global book series Advances in Environmental Engineering and Green Technologies (AEEGT) (ISSN: 2326-9162; eISSN: 2326-9170)

British Cataloguing in Publication Data
A Cataloguing in Publication record for this book is available from the British Library.

Advances in Environmental Engineering and Green Technologies (AEEGT) Book Series

Sang-Bing Tsai, University of Electronic Science and Technology of China Zhongshan Institute, China & Ming-Lang Tseng, Lunghwa University of Science and Technology, Taiwan & Yuchi Wang, University of Electronic Science and Technology of China Zhongshan Institute, China

ISSN:2326-9162
EISSN:2326-9170

Mission

Growing awareness and an increased focus on environmental issues such as climate change, energy use, and loss of non-renewable resources have brought about a greater need for research that provides potential solutions to these problems. Research in environmental science and engineering continues to play a vital role in uncovering new opportunities for a "green" future.

The **Advances in Environmental Engineering and Green Technologies (AEEGT)** book series is a mouthpiece for research in all aspects of environmental science, earth science, and green initiatives. This series supports the ongoing research in this field through publishing books that discuss topics within environmental engineering or that deal with the interdisciplinary field of green technologies.

Coverage

- Sustainable Communities
- Cleantech
- Waste Management
- Contaminated Site Remediation
- Radioactive Waste Treatment
- Water Supply and Treatment
- Green Transportation
- Air Quality
- Policies Involving Green Technologies and Environmental Engineering
- Renewable Energy

IGI Global is currently accepting manuscripts for publication within this series. To submit a proposal for a volume in this series, please contact our Acquisition Editors at Acquisitions@igi-global.com or visit: http://www.igi-global.com/publish/.

The Advances in Environmental Engineering and Green Technologies (AEEGT) Book Series (ISSN 2326-9162) is published by IGI Global, 701 E. Chocolate Avenue, Hershey, PA 17033-1240, USA, www.igi-global.com. This series is composed of titles available for purchase individually; each title is edited to be contextually exclusive from any other title within the series. For pricing and ordering information please visit http://www.igi-global.com/book-series/advances-environmental-engineering-green-technologies/73679. Postmaster: Send all address changes to above address.

Titles in this Series

For a list of additional titles in this series, please visit: www.igi-global.com/book-series

Green Finance for Sustainable Global Growth
Sang-Bing Tsai (University of Electronic Science and Technology of China Zhongshan Institute, China & Civil Aviation University of China, China & WUYI University, China) Chung-Hua Shen (Shih-Chien University, Taiwan) Hua Song (Renmin University of China, China) and Baozhuang Niu (South China University of Technology, China)
Business Science Reference • copyright 2019 • 397pp • H/C (ISBN: 9781522578086) • US $215.00 (our price)

Building Sustainability Through Environmental Education
Lynn A. Wilson (SeaTrust Institute, USA) and Carolyn N. Stevenson (Purdue University Global, USA)
Engineering Science Reference • copyright 2019 • 335pp • H/C (ISBN: 9781522577270) • US $195.00 (our price)

Unique Sequence Signatures in Plant Lipolytic Enzymes Emerging Research and Opportunities
Nihed Ben Halima (University of Sfax, Tunisia)
Engineering Science Reference • copyright 2019 • 197pp • H/C (ISBN: 9781522574828) • US $145.00 (our price)

Handbook of Research on the Adverse Effects of Pesticide Pollution in Aquatic Ecosystems
Khursheed Ahmad Wani (Government Degree College Bijbehara, India) and Mamta (Jiwaji University, India)
Engineering Science Reference • copyright 2019 • 538pp • H/C (ISBN: 9781522561118) • US $295.00 (our price)

Harnessing Marine Macroalgae for Industrial Purposes in an Australian Context Emerging Research and Opportunities
Göran Roos (Economic Development Board of South Australia, Australia) Anthony Cheshire (Balance Carbon Pty Ltd, Australia) Sasi Nayar (South Australian Aquatic Sciences Centre, Australia) Steven M. Clarke (South Australian Research and Development Institute, Australia) and Wei Zhang (Flinders University, Australia)
Engineering Science Reference • copyright 2019 • 291pp • H/C (ISBN: 9781522555773) • US $195.00 (our price)

Advanced Methodologies and Technologies in Engineering and Environmental Science
Mehdi Khosrow-Pour, D.B.A. (Information Resources Management Association, USA)
Engineering Science Reference • copyright 2019 • 506pp • H/C (ISBN: 9781522573593) • US $295.00 (our price)

Transitioning Island Nations Into Sustainable Energy Hubs Emerging Research and Opportunities
Catalina Spataru (UCL Energy Institute, UK)
Engineering Science Reference • copyright 2019 • 172pp • H/C (ISBN: 9781522560029) • US $135.00 (our price)

Recent Challenges and Advances in Geotechnical Earthquake Engineering
T.G. Sitharam (Indian Institute of Science, India) and J. S. Vinod (University of Wollongong, Australia)
Engineering Science Reference • copyright 2019 • 315pp • H/C (ISBN: 9781522569480) • US $195.00 (our price)

IGI Global
DISSEMINATOR OF KNOWLEDGE

701 East Chocolate Avenue, Hershey, PA 17033, USA
Tel: 717-533-8845 x100 • Fax: 717-533-8661
E-Mail: cust@igi-global.com • www.igi-global.com

Editorial Advisory Board

Table of Contents

Detailed Table of Contents

The method has been used to estimate the values of the optimal parameters of the functioning system. Thus, in the poultry house of 78m x 18m, four lines of LED lamps have been installed for illumination, 40 pieces per each line. With the computer calculations of illuminance, estimates of the optimal values have been obtained: the total number of LEDs is equal to 1920 and allocated over 120 lamps instead of 160 previously used, which allows predicting a fall in actual energy consumption by 25% with an irregular illumination coefficient being 1.13. Based on the developed and manufactured LED lamp, the applicability of the mathematical model has been experimentally justified.

Each modern tractor should be adapted to perform an increasing number of various traction, transportation, and other works. Therefore, their designs are becoming more complex. To improve tractor productivity, the power-to-weight ratio and the speed of its units are constantly increasing. This inevitably improves dynamic loading of the chassis and the transmission parts, the level of the generated oscillations, which increases the vibration load of the transmission. One of the promising ways to solve this problem is to introduce elastic links into the tractor transmission closer to an engine or a clutch. Installed closer to the engine, the elastic element in the transmission of the tractor absorbs the impact of the shocks, protecting the tractor engine from large dynamic loads. The material of the chapter contains studies confirming the efficiency of the use of an elastic mechanism (closer to the engine) in the tractor transmission of a small class of traction (14 kN).

The concept and principle solutions create breakthrough electro-robotized technology organic food ECO-1m used in plants for the production of ecologically clean meat and dairy food. Breakthrough technology overcomes competition and international sanctions, providing export expansion due to the novelty of domestic solutions, supported by patents for new robotic machines and equipment. ECO-1m for the first time used highly reliable, technically advanced serial technique of the sectors of the defense industrial complex, aerospace, and electrified railways. The stages and results of fundamental research on the improvement of technical solutions and a list of implemented automated farms, where the parameters of breakthrough technology ECO-1m. Technology is a modular plant consisting of 12 departments placed on 2500ga of arable land, and 3000 head of mixed aged cattle, which contains on cultural pastures. According to the indicators of the developed business plan, the plant has 3-3.5 times higher productivity, with a profitability of 30-40%.

In the chapter, both informational and system-based approaches have been applied to the analysis of grain drying process as an interaction of two systems, namely, "drying agent" and "grain layer." It made it possible to evaluate the information content of the process. Analytic expression has been obtained for grain layer informational entropy variations which enabled to make the conclusion that an adequate physical description of grain drying process has to include ambient air parameters control, as well as that of drying agent's parameters in its exit out of a grain layer, as well as parameters of both drying agent and grain in one point of grain layer. A mathematical model of heat-and-moisture exchange in dense grain layer has to be applied. As a result of modeling, it has been proved that the sensors of both drying agent and grain parameters have to be located at a distance of 10 to 11 cm from the central cylinder of the forced-aerated drying bin.

This chapter presents an analysis of factors influencing the heat and moisture exchange for their further use in experimental research to study the process of grain drying with active ventilation. The temperature and velocity of the drying agent and grain moisture content are determined. One centimeter (1 cm) thick layer is considered to be a thin layer. This size appears to be very manageable to transfer the regularities of the drying process to a thicker layer when modeling the drying process in a thick layer. The experiment was carried out for three drying modes: classical mode (i.e., drying of grains with natural and heated air), with a constant concentration of air ions in the drying agent, with cyclic (periodic) presence of air ions in the drying agent. Using the regression equations produced a so-called electro-activation criterion can be described. The dependencies obtained would help to optimize the drying process with respect to drying time criterion.

The reclamation of new areas and distant lands with limited energy and material resources demands using resource- and energy-saving technologies. One of such technologies can be magnetic treatment of irrigating water and artificial manure solutions, and presowing treatment of seeds of agricultural crops in magnetic field. The authors found that magnetic field accelerates the velocity of chemical reactions, enhances salts and fertilizers solvability, and increases oxygen concentration in a solution. Magnetic field improves membrane cells permeability which accelerates molecules and ions diffusion through it. This process improves the ability of a seed to absorb water and increases the concentration of oxygen in a cell. Presowing treatment of seeds increases their ability of germination by 26–50%, sprouting by 20–30%. The best regime of water solutions and seeds of agricultural crops treatment in magnetic field is when magnetic induction is 0.065 Tl and the velocity of a solution is 0.4 m/s.

The reviewed problem deals with the metrological support of sensory equipment using cameras with standard and infrared lenses. The elaborated method to use UAVs for remote shooting of agricultural crops and obtaining images is to be used to determine the level of nitrogen provision for grain crops has been developed. For radiofrequency correction of lighting changes while mineral nutrition state monitoring, it is recommended to use a method with the application of official data to be calculated on the basis of camera's regular camera exposure meter measurements. It is advisable to use an infrared lens for monitoring the state of individual crops using digital non-specialized cameras instead of optical one. Exampling GoPro Hero 4 digital camera, it has been found that there are three channels to use IR lens which have the characteristics of fixing radiation either in the infrared range or in the red area of the optical range.

Hexaqua magnesium chlorine, bishofite, almost never occurred in the fossil salt deposits until recently, while industry has used artificially produced magnesium chloride. The discovery of bishofite deposits in the Lower Volga region (both for raw materials purity of MgCl2 up to 95-98%, and for reserves about 500 billion tons) was a unique event. The use of bischofite consists in the salt compositions of sand to protect roads from icing yet. The chapter represents the results summary of the bishofite solutions electrochemical processing with the purpose of obtaining solutions with the antibacterial properties complex the use of which can be recommended for pre-plant seeds treatment, plant protection from fungal effects, which allowed, as shown by experiments on the crops cultivation, to increase yields up to two-three times, and also to use for disinfection of domestic and industrial wastewater in agricultural enterprises. The technological characteristics of the using electrochemical device are considered in the chapter.

An important factor of impact on the decrease in operational injuries is the high-quality and timely training, certification, preparation, and retraining of personnel in safety of production. Procedures of training and control of knowledge of various directions of safety of production are regulated by the existing normative documents developed by authors of the chapter . The systems of e-learning of personnel for safety of agro-industrial production have to be built in digital technologies of management of agro-industrial production and have the modular structure allowing the customization of systems on requirement of concrete production.

The object of the research in the chapter is green vegetable crops grown in a manmade environment. The purpose of the presented research is to study the process of growing plants in a man-made environment, the main influencing factor of which is the man-made sources (LEDs) emission, as well as to design and construct the installation (irradiation chamber) for implementation of the studied technology. The chapter presents the results of experiments conducted by a team of researchers, which allowed to determine the technological process indicators of growing green vegetable crops in a man-made environment, and also to develop designs for small-sized irradiation chambers with automatic control, both for laboratory research and for small business and private use. Studies were conducted in different regions of the country and combined the efforts of several scientific groups of researchers. The chapter presents the results of researching these areas conducted by the authors, which are completely original.

The object of research in the chapter is the plant tissue of weeds, fruits, vegetables, melons, etc. The purpose of the presented research is to study the process of electro-impulsive irreversible damage of plant tissue as the basis for energy saving, efficient electro-technologies with minimal negative effects of chemical components on the environment and food products. The chapter presents the results of experiments carried out by the team of researchers that allowed it to determine the technological indices of the irreversible electro-impulse damage of plant tissue (weeds, fruits, vegetables, melons, etc.), which is the basis for developing the technology of electro-impulse destruction of weeds, electro-impulse plasmolysis of fruits and melons cultures. This research was conducted in the southern region of the country and combined the efforts of several scientific groups of researchers.

Foreword

In order to meet these requirements of a steadily new involvement and economic circulation of new lands and territories – particularly for food supply, economic and managerial activities and their smart organization are continuously needed. Those territories and pieces of land are, by a rule, placed far away from the main populations and their activities whereas, at the same time, centers of energy supply are centralized. Agriculture is regarded as one of the most suitable activities within those territories. In order start a business there, it is easiest to get involved in the production of agricultural products through small-scale enterprises, with a small number of persons employed. Small-scale production permits to quickly deploy work, it ensures, facilitates and speeds up the sale of products. Such activities highly necessitate particular technical support measures for both the livelihood of people dwelling there and the organization a highly efficient and cost-effective production processes. Of course, in those territories it is most convenient and fast when deciding for and providing activities related to the production of agricultural goods. Such companies can more easily be provided with energy, especially, with renewable energy sources. However, existing technologies of energy supply, cropping and processing of crops are often hard to adapt to a use in small and medium-sized enterprises. Manufacturers and developers of technical facilities and technologies and scientific organizations engaged in the development of technologies and equipment for the power industry on the countryside, production processes of various types, and problems of housing and communal solutions, raise a great interest in new efficient techniques of conducting agricultural business, in particular, at small scale and for problems of the rural population.

The book describes a series of technological methods and various technical means allowing for an organization of production of different goods in landscapes and regions that have become newly allocated for economic activity, or cultivated again for those purposes. The needed and identified technologies and means are adapted to be used in production at small scale, and in companies of small and medium-sized business. Furthermore, the book also considers the design of technical solutions to guarantee a vital activity in new territories with a high degree of comfort and high standards of living conditions. In fact, this work pays a lot of attention to the organization of efficient production processes and to social comfort, even when facing scarce energy resources and the need to a rational use of them.

It is a great achievement of this book to having satisfied all the aforementioned needs, as not any similar work had been complied before for the sake of developmental issues of rural business designed in the intended mode as explained before. Herewith, this book should cause a high attraction and an attention by many specialists of rural agribusiness activities in remote areas, of a wide range of practical management projects, and by businessmen as well as researchers involved in the activities investigated and designed in this new work.

Modelling and solution of the global problem described above requires the creation of technological and organizational preconditions. Today's state-of-knowledge and -art in science and technology allows the offer of new techniques and equipment for perspectives of electro-technologies effective utilization. This should enable manufacturing of agricultural commodities and permitting a comfortable living of the population. This book has been open to cover all problem complexities in the decision making on the formation of a distributed economy and the development of new territories as a core element of this problem.

In this compendium, innovative technologies which provide an effective agroengineering technologies for a successful development of new territories for realization of the concept of a distributed economy, are described. A particular interest has been given by the fact that a considerable part of the material presented in the book was unpublished so far in Anglo-Saxon literature and, consequently, it was hardly available for a wide range of interested readers in many countries worldwide.

The readership of this work was targeted as a wide range of persons, such as students of agro-engineering and power specialists, experts and heads of municipal unions, managers of the ministries and other organizations responsible for development of new territories, the separate businessmen who are starting activities in new territories, and many others, as well as organizations such as large international consortia specialized in various businesses in rural areas, primarily related to -but not limited to- the production of agricultural products, including food. This very useful book will serve for small commodity producers on the countryside, various entrepreneurs of medium and small companies or projects as well as for inter-regional public organizations that assist in the enhancement of new technologies in the social sphere, regarding new environmentally friendly and sustainable development, self-determined and at the same time with responsibility for the whole and wider population. In general, this book intends to represent newest advanced achievements in agro-engineering related technical means which will be useful to a broad variety of readers and doubtlessly have a positive impact on a solution of sustainable development in new rural territories.

This work provides a reference of premium quality to graduate and postgraduate students, decision makers and investigators in private domains, universities, traditional and emerging industries - on agriculture, forestry, tourism and energy sector and beyond, for governmental and non-governmental organizations, in the areas of various Operational Research, Artificial Intelligence engineering, management, business and finance, where ever one has to represent and solve uncertainty-affected practical and real-world problems related to rural, remote, developing and emerging regions and territories. In the coming times, mathematicians, statisticians, computer scientists, economists, game theorists, physicists, chemists, geologists, representatives of environmental, civil, electrical and electrical engineering, but also biologists, neuroscientists, social scientists, scientists on natural resources, landscape architects, and representatives of the humanities, are cordially welcome to join this discourse and enter the collaboration for reaching improved and sustainable solutions. It has been well understood that predictability under uncertain environments and rapidly changing scenarios exercise a core request and a challenge in all fields of engineering, science and management. In this respect, this book is following a very modern perspective; finally, it promises to become very significant both in academia and in practice, and very fruitful and important for humankind.

The topics of this book's chapters were open, but not limited to:

- Advanced technologies in the production of grain,
- Energy-efficient thermal processes in agricultural production,

- Innovative technologies in the production of milk and meat,
- Advanced technologies for processing agricultural products,
- Intensification of technical processes by electric, magnetic and other influences,
- Diagnostic and monitoring methods for energy equipment status in rural conditions,
- Energy saving system for illumination of greenhouse plants,
- Energy efficient electric equipment for heat supply in Agriculture,
- Hydrocarbon local Recovery,
- Electric pulse cultivation,
- Pre-sowing Seed Treatment in a Magnetic Field,
- Use of Microwave Energy at Thermal Treatment of Grain Crops,
- Scientific and methodological support of rural electric system operation,
- Biological Objects Reactions Management,
- Solutions for diesel, wind and solar power plants joint use.

Let me please underline the high importance of this book for everyone related with it so far, and for all its forthcoming readers and appliers. Indeed, I trust that in the years to come, *Operational Research, Optimization* and *Optimal Control* will remain in the center of giving *Decision-Making Aid* and *Support* which this book targets at. In this context, I especially mention *Stochastic Optimal Control* under *Impulses* and *Regime Switching* of economic and cultural kinds. Here, my coauthors and I are working on scientifically, towards further inclusion of real-word situations and cases, with all their immense uncertainties and with the "human factors" even, into an *optimal decision-making*.

To all the *authors* of this valuable book, I convey my appreciation and thanks for having shared their expertise and idealism with the academic community and, eventually, with humankind. Moreover, I extend our gratitude to the two *editors* of this compendium, *Prof. Dr. Valeriy Kharchenko* and *Prof. Dr. Pandian Vasant*, for their dedicated hard-work and vision, of having gained an impressive variety of prosperous contributions. Finally, I am very thankful to the publishing house of IGI Global, for making possible and become reality a premium book of real-life significance and of potential high impact for the world of tomorrow, for our present generation and for the following generations.

Now, I wish us all a lot of pleasure and gain when browsing through and studying this interesting book, and I hope for a remarkable benefit going to be received from it scientifically, personally and societally.

Gerhard-Wilhelm Weber
Poznan University of Technology, Poland
November 2018

Gerhard-Wilhelm Weber *is a Professor at Poznan University of Technology, Poznan, Poland, at Faculty of Engineering Management, in the Chair of Marketing and Economic Engineering. His research is on mathematics, operational research), finance, economics, optimization and control, neuro-, bio- and earth-sciences, medicine, data mining, analytics, artificial intelligence, education and development; he is involved in the organization of scientific life internationally. He received Diploma and Doctorate in Mathematics, and Economics / Business Administration, at RWTH Aachen, and Habilitation at TU Darmstadt (Germany). He replaced Professorships at University of Cologne, and TU Chemnitz, Germany. At Institute of Applied Mathematics, Middle East Technical University, Ankara, Turkey, he was a Professor in both Financial Mathematics and Scientific Computing, and Assistant to the Director, and has been a member of five further graduate schools, institutes and* *departments of METU. G.-W. Weber has different affiliations at, e.g., Universities of Siegen (Germany), Federation University (Ballarat, Australia,, University of Aveiro (Portugal), University of North Sumatra (Medan, Indonesia), Malaysia University of Technology, Chinese University of Hong Kong, and KTO Karatay University (Lean & Green SCM Lab; Konya, Turkey), at EURO (Association of European OR Societies) where he is "Advisor to EURO Conferences" and IFORS (International Federation of OR Societies), where he is member in many national OR societies and working groups, at Pacific Optimization Research Activity Group, etc. G.-W. Weber has (co-)supervised many MSc. and PhD. students, (co-)authored numerous books, special issues and articles, and given many presentations from a diversity of subject areas, on theory, methods and practice. He has been a member of many international editorial and award boards; he participated at and was in charge of numerous research projects; G.-W. Weber received various recognitions and awards by students, universities, conferences and scientific organizations, nationally and internationally.*

Preface

INTRODUCTION

Nowadays circumstances year by year put forward new problems and challenges. Mankind food and other vital goods consumption steadily grows. To meet these requirements constant involving in economic circulation of new lands and territories, suitable for the economic activities organisation is required.

As a rule these territories and pieces of lands disposed far of main populations activity and in the same time from centralized centres of energy supply. The most suitable activity for these territories is agriculture

Start a business in such territories is easiest if you are engaged in the production of agricultural products on small-scale enterprises, with a small number of employed people. Small scale production makes it possible to quickly deploy work, easier and faster to ensure the sale of products.

Such activities require special technical support measures both for the livelihood of people living there and for organizing a highly efficient and cost-effective production process. Naturally, in such territories it is easiest and quickest to deploy activities related to the production of agricultural products. Such enterprises are easier to provide with energy resources, in particular, using renewable energy sources.

Existing technologies of energy supply, cropping and processing of crops can not always be easily adapted to use in small and medium-sized enterprises.

There was such a situation that information of this kind is needed not only directly by the worker in the village. The manufacturers and developers of such technical facilities and technologies and scientific organizations that are engaged in the development of technologies and equipment for the power industry of the village, production processes of various types, as well as the problems of housing and communal solutions, are showing great interest in new efficient means of conducting agricultural business, especially on a small scale, problems of the rural population.

An important issue is the organization of reliable power supply in an autonomous mode. The use of renewable energy sources can play an important role in this issue. And this means that information on the possible technologies of autonomous generation of electric and thermal energy, its rational consumption through the use of energy-saving technologies will inevitably be in demand. Of great importance in the organization of active activities in rural areas can also be the use of intensive technologies for the cultivation of products and its processing on site with the least expenditure of energy, the amount of which at remote sites may be critically small.

OBJECTIVE OF THE BOOK

The book includes a description of a series of technological methods and a number of technical means that ensure the organization of production of various products in the territories newly allocated for economic activity or recultivated for these purposes. The specified technologies and technical means are adapted to application in production of small scale, at the enterprises of small and medium business.

Much attention is paid to work aimed at organizing an efficient production process and social comfort in the face of scarcity of energy resources and the need to rationally use them.

It is an obvious fact that any similar published books complied altogether issues of rural business development concerned in the intended mode doesn't exist. No doubts that the book being published attract attention of a huge number specialists related to rural agribusiness activity on remote territories and looks to be very interesting for wide range of practical managements and businessman as well as scientists involved in activity connected with considered in the book technical field.

The particular interest to the book may be connected with the fact that the materials represented in the book, mainly was not published in English yet and consequently was not available for readers throughout the world.

TARGET AUDIENCE

The book will be useful to a wide range of persons, such as students of agroengineering and power specialities, experts and heads of municipal unions, managers of the ministries and other organisations responsible for development of new territories, the separate large businessmen beginning business at new territories and many others as for organisations like large international consortia specializing in various business in rural areas, primarily related to but not limited to the production of agricultural products, including food. A very useful book will be for small commodity producers in the countryside, various entrepreneurs of medium and small businesses as well as for inter-regional public organizations for assistance to advancement of new technologies in social sphere in view of environmentally friendly sustainable development. In general the book is intended for representation of the wide public advanced achievements in the field of agroengineering and related to the rural business energy technical means which will be useful to a wide range of readers and doubtless positive impact on a solution of a problem of a sustainable development of new rural territories.

Topics reflected in the book:

- Advanced technologies in the production of grain
- Innovative technologies in the agricultural production
- Energy-efficient thermal processes in agroengineering
- Innovative technologies in the production of milk and meat
- Advanced technologies for processing agricultural products
- Intensification of technical processes by electric, magnetic and other influences
- Solar thermal energy installations
- Energy saving system for illumination of greenhouse plants
- Energy efficient electric equipment for heat supply in Agriculture

- Electric pulse cultivation
- Pre-sowing Seed Treatment in a Magnetic Field
- Use of Microwave Energy at Thermal Treatment of Grain Crops
- Scientific and methodological support of rural electric system operation
- Biological Objects Reactions Management
- Solutions for diesel, wind and solar power plants joint use

A significant part of the book is devoted to the issues of application of electrical technologies in agroengineering processes that provide increased yields, increased productivity, improved working conditions, etc.

It is necessary to note the fact that a significant amount of the material presented in the book is devoted to works made in different countries, but not published in English. This will allow the English-speaking reader to get to know them. The creators of the book hope that these sections of the book will be useful to English-speaking readers and will be in high demand.

In Chapter 1, the Application of Electrophysical Effects in the Processing of Agricultural Materials is considered. Crop production is largely associated with a number of external factors that affect productivity, quality and cost of production. A significant part of the cost (20-40%) occurs on post-harvest processing, storage and preparation for feeding. These processes include drying, decontamination, steaming, micronisation, etc. The work describes the electrotechnology impacts and their use in processes of grain processing. The article deals with the classification of electrophysical methods of influence, correlated with the processes in which they can be applied. It is described the required levels of exposure and the resulting values of productivity, intensity and other parameters, qualitatively describing the specific processes. The factors which can positively affect grain production include: the use of ozone, ions, infrared and microwave fields, etc.

Chapter 2 is devoted to the consideration of theoretical Prerequisites for Subsurface Broadcast Seeding of Grain Crops in the Conditions of Pneumatic Seed Transportation to the Coulters. Only 30 – 40% of the plant productivity potential is realized in crop production technologies, which leads to a situation where the growth of field yields lags behind the growth of production costs. Increasing the realization of the productivity potential of cultivated plants is a way to improve the efficiency of the crop industry. The task of sowing machines is the uniform placement of seeds on the field area, while plants get their living space and, accordingly, form a crop. The practice of using such working bodies has shown that to achieve the necessary seeding regimes under the coulter elements, the use of additional sources of kinetic energy is required.

The Chapter 3 considers the UV-based indoor disinfecting system. It is shown that when contagious diseases occur, there is a tangible threat of rampant spread of infection, incurring huge economic losses in animal deaths and decreased animal productivity. Thus, preventing pathogenic flora concentration in rooms where birds and animals are raised from exceeding permissible levels ranks first among veterinarian and sanitary concerns. When birds are kept on the floor during feeding, germ and dust concentration increases nine to ten times against normal. Ample research shows that ultraviolet (UV) radiation possessing a bactericidal effect is the most promising and environment-friendly method of cleansing the air from harmful germs.

Chapter 4 considers energy-saving electrical installations for heat supply of agricultural objects. Electrical energy in the thermal processes of agricultural production has a number of technical and technological advantages in comparison with other energy carriers. At the same time electricity is the most high-quality, but expensive type of energy. The effective use of electrical energy in thermal processes is possible with the implementation of all its advantages over fuel energy. This requires a system of technical, economic and energy analysis and the selection justification of the most effective systems and technical means of heat supply of stationary agricultural technological processes. The paper presents functional and technological schemes, physical models, the experimentally obtained dependencies, parameters and operating regimes of the developed energy-saving electrical equipment and systems for the main thermal technological processes.

In Chapter 5, a method for solving the problem of the uniform led illumination for the floor maintenance of birds is proposed. On the basis of physical modeling, the dependences have been obtained between the geometric dimensions of the poultry house and the coefficient of uniformity of illumination with the purpose of optimization of the parameters on which the energy consumption and quality of illumination depend. The method has been used to estimate the values of the optimal parameters of the functioning system. Thus, in the poultry house of 78m x 18m four lines of led lamps have been installed for illumination, 40 pieces per each line. With the computer calculations of illuminance, estimates of the optimal values have been obtained: the total number of LEDs is equal to 1920 and allocated over 120 lamps instead of 160 previously used, that allows predicting a fall in actual energy consumption by 25% with an irregular illumination coefficient being 1.13. Based on the developed and manufactured led lamp the applicability of the mathematical model has been experimentally justified.

In Chapter 6, it is shown that each modern tractor should be adapted to perform an increasing number of various traction, transportation and other works. Therefore, their designs are becoming more complex. To improve tractor productivity, the power-to-weight ratio and the speed of its units are constantly increasing. This inevitably improves dynamic loading of the chassis and the transmission parts, the level of the generated oscillations, which increases the vibration load of the transmission. One of the promising ways to solve this problem is to introduce elastic links into the tractor transmission closer to an engine or a clutch. Installed closer to the engine, the elastic element in the transmission of the tractor absorbs the impact of the shocks, protecting the tractor engine from large dynamic loads. The material of the chapter contains studies confirming the efficiency of the use of an elastic mechanism (closer to the engine) in the tractor transmission of a small class of traction (14 kN).

Chapter 7 is devoted to consideration of Electric robotized organic technology for livestock production on a pasture field. The concept and principle solutions create breakthrough elektrorobotized technology organic food ECO-1m used in plants for the production of ecologically clean meat and dairy food. Breakthrough technology overcomes competition and international sanctions, providing export expansion due to the novelty of domestic solutions, supported by patents for new robotic machines and equipment. ECO-1m for the first time used highly reliable, technically advanced serial technique of the sectors of the defense industrial complex, aerospace, and electrified railways. The stages and results of fundamental research on the improvement of technical solutions and a list of implemented automated farms, where the parameters of breakthrough technology ECO-1m. Technology is a modular plant consisting of 12 departments placed on 2500ga of arable land, and 3000 head of mixed aged cattle, which contains on cultural pastures. According to the indicators of the developed business plan, the plant has 3-3.5 times higher productivity, with a profitability of 30-40%.

In presented at Chapter 8 material, both informational and system-based approaches have been applied to the analysis of grain drying process as an interaction of two systems, namely, 'drying agent' and 'grain layer'. It made it possible to evaluate the information content of the process. Analytic expression has been obtained for grain layer informational entropy variations which enabled to make the conclusion that an adequate physical description of grain drying process has to include ambient air parameters control, as well as that of drying agent's parameters in its exit out of a grain layer, as well as parameters of both drying agent and grain in one point of grain layer. Besides, a mathematical model of heat-and-moisture exchange in dense grain layer has to be applied. As a result of modeling, it has been proved that the sensors of both drying agent and grain parameters have to be located at a distance of 10 to 11 cm from the central cylinder of the forced-aerated drying bin.

Chapter 9 presents an analysis of factors influencing the heat and moisture exchange for their further use in experimental research to study the process of grain drying with active ventilation. The temperature and velocity of the drying agent and grain moisture content are determined. One centimeter (1 cm) thick layer is considered to be a thin layer. This size appears to be very manageable to transfer the regularities of the drying process to a thicker layer when modeling the drying process in a thick layer. The experiment was carried out for three drying modes: classical mode, i.e. drying of grains with natural and heated air; with a constant concentration of air ions in the drying agent; with cyclic (periodic) presence of air ions in the drying agent. Using the regression equations produced a so-called electro-activation criterion can be described. The dependencies obtained would help to optimize the drying process with respect to drying time criterion.

Chapter 10 is dedicated to consideration very important issues concerning innovative approaches for agricultural processes intensification. The reclamation of new areas and distant lands with limited energy and material resources demands using resource- and energy-saving technologies. One of such technologies can be magnetic treatment of irrigating water and artificial manure solutions, and presowing treatment of seeds of agricultural crops in magnetic field. We found out that magnetic field accelerates the velocity of chemical reactions, enhances salts and fertilizers solvability, and increases oxygen concentration in a solution. Magnetic field improves membrane cells permeability which accelerates molecules and ions diffusion through it. This process improves the ability of a seed to absorb water and increases the concentration of oxygen in a cell. Presowing treatment of seeds increases their ability of germination by 26–50%, sprouting by 20–30%. The best regime of water solutions and seeds of agricultural crops treatment in magnetic field is when magnetic induction is 0.065 Tl and the velocity of a solution is 0.4 m/s.

The reviewed problem in Chapter 11 deals with the metrological support of sensory equipment using cameras with standard and infra-red lenses. Own elaborated method to use UAVs for remote shooting of agricultural crops and obtaining images is to be used to determine the level of nitrogen provision for grain crops has been developed. For radiofrequency correction of lighting changes while mineral nutrition state monitoring, it is recommended to use a method with the application of official data to be calculated on the basis of camera's regular camera exposure meter measurements. It is advisable to use infra-red lens for monitoring the state of individual crops using digital non-specialized cameras instead of optical one. Exampling GoPro hero 4 digital camera it has been found that there are three channels to use IR lens which have the characteristics of fixing radiation either in the infrared range or in the red area of the optical range.

In Chapter 12, an application of solutions of the Electrochemical Processed Mineral "Bishofit" in Plant Production is considering. Hexaqua magnesium chlorine – bishofite, almost never occurs in the fossil salt deposits until recently, while industry has used artificially produced magnesium chloride. The discovery of bishofite deposits in the Lower Volga region (both for raw materials purity of MgCl2 up to

95-98%, and for reserves about 500 billion tons) was an unique event. The use of bischofite consists in the salt compositions of sand to protect roads from icing yet. The report represents the results summary of the bishofite solutions electrochemical processing with the purpose of obtaining solutions with the antibacterial properties complex the use of which can be recommended for pre-plant seeds treatment, plant protection from fungal effects, which allowed, as shown by experiments on the crops cultivation, to increase yields up to two - three times, and also to use for disinfection of domestic and industrial wastewater in agricultural enterprises. The technological characteristics of the using electrochemical device are considered in the chapter.

In Chapter 13, electric robotized organic technology for livestock production on a pasture field is considering. The concept and principle solutions create breakthrough elektrorobotized technology organic food ECO-1m used in plants for the production of ecologically clean meat and dairy food. Breakthrough technology overcomes competition and international sanctions, providing export expansion due to the novelty of domestic solutions, supported by patents for new robotic machines and equipment. ECO-1m for the first time used highly reliable, technically advanced serial technique of the sectors of the defense industrial complex, aerospace, and electrified railways. The stages and results of fundamental research on the improvement of technical solutions and a list of implemented automated farms, where the parameters of breakthrough technology ECO-1m. Technology is a modular plant consisting of 12 departments placed on 2500ga of arable land, and 3000 head of mixed aged cattle, which contains on cultural pastures. According to the indicators of the developed business plan, the plant has 3-3.5 times higher productivity, with a profitability of 30-40%.

In Chapter 14, important factor of impact on decrease in operational injuries is considered. It is also production the caused incidence of employees of the agro-industrial organizations and enterprises high-quality and timely training, certification, preparation and retraining of personnel in safety of production are. Procedures of training and control of knowledge of various directions of safety of production are regulated by the existing normative documents developed by authors of the present chapter now. The systems of e-learning of personnel for safety of agro-industrial production have to be built in digital technologies of management of agro-industrial production, have the modular structure allowing customizing systems on requirement of concrete production.

The object of research in the Chapter 15 is the plant tissue of weeds, fruits, vegetables, melons etc. The purpose of the presented research is to study the process of electroimpulsive irreversible damage of plant tissue as the basis for energy saving, efficient electrotechnologies with minimal negative effects of chemical components on the environment and food products. The chapter presents the results of experiments carried out by the team of researchers that allowed us to determine the technological indices of the irreversible electroimpulse damage of plant tissue (weeds, fruits, vegetables, melons, etc.), which is the basis for developing the technology of electroimpulse destruction of weeds, electroimpulse plasmolysis of fruits and melons cultures. This research was conducted in the southern region of the country and combined the efforts of several scientific groups of researchers.

The subject of Chapter 16 is the Control Of Advanced Fodder Disinfection In Terms Of Economic Criteria consideration. Upon long-term storage germinated grain with 50-60% humidity begins to mold and rot. Therefore, it should be fed off to livestock as soon as possible after its germination. To extend storage life of germinated grain it is necessary either to dry it to 14% relative humidity, or to carry out its disinfection. Grain forms the basis of combined fodder in agriculture. Some technological solutions are available to provide fodder disinfection optimized in technological or economic terms, as well as rational energy use, including disinfection by high-speed electron bunches, ultra-high frequency electric waves, combined IR and conductive (contact) electric heating, and toxic ozone gas. There also exists a similar solution for fodder premises disinfection through controlling excessive pressure of their internal air. In control by technological (production) criterion, the mode of minimal integrated cost of calculated loss of quality (expenditures) of products being disinfected (fodder) is determined, so the highest livestock productivity in the current context is obtained.

The editors express strong confidence that the results presented in this book will undoubtedly be in demand by a wide circle of readers and will help spread the information presented in the book to a wide circle of specialists all over the world, and the exchange of scientific knowledge through this book will contribute to the overall progress of the agro-engineering industry in many countries.

Valeriy Kharchenko
Federal Scientific Agroengineering Center VIM, Russia

Pandian Vasant
Universiti Teknologi PETRONAS, Malaysia

Acknowledgment

The editors would like to acknowledge the efforts of all the people which were involved in the activity on creation of this book. Particularly they would like to thank each one of the authors for their contributions. Our sincere gratitude goes to the chapters' authors who contributed the time and expertise to this book. Special gratitude should be given to the people which beside the preparing their chapters took part in the reviewing process. Without invaluable support of all of them this book would not have become a reality. Among all these persons special gratitude should be given to such persons as A.Vasiliev, V. Bolshev, T. Gallyamova, D. Budnikov, etc.

A lot of thanks to the editorial advisory board members which have made a lot for making the high quality of the book. Specialy needs to underline essential role of Editorial Board member, professor Gerhard-Wilhelm Weber for excellent reviews and invaluable assistance.

The editors are confident that this book will certainly be useful to readers, because, among other things, it gives them the opportunity to get acquainted with the results of research that has not been published in English-language literature so far. Therefore, it would be wrong not to mention the terrific role that IGI Global publishing house has played in order for this information to become published .

Finally, many thanks the editors would like to express to the entire staff of IGI Global publishing houses who helped the editors and authors in preparing the manuscript of this book professionally and extremely friendly, and have made a lot to make this book the edition of premium quality.

Valeriy Kharchenko
Federal Scientific Agroengineering Center VIM, Russia

Pandian Vasant
UniversitiTeknologi PETRONAS, Malaysia

Chapter 1
The Application of Electrophysical Effects in the Processing of Agricultural Materials

Dmitry Budnikov
https://orcid.org/0000-0002-8729-1300
Federal Scientific Agroengineering Center VIM, Russia

Alexey Nikolaevich Vasiliev
https://orcid.org/0000-0002-7988-2338
Federal Scientific Agroengineering Center VIM, Russia

Alexey Alekseevich Vasilyev
Federal Scientific Agroengineering Center VIM, Russia

Konstantin Sergeevich Morenko
Federal Scientific Agroengineering Center VIM, Russia

Ihab S. Mohamed
INRIA Sophia Antipolis - Méditerranée, France

Aleksandr Belov
Federal Scientific Agroengineering Center VIM, Russia

ABSTRACT

Crop production is largely associated with a number of external factors that affect productivity, quality, and cost of production. A significant part of the cost (20-40%) occurs on post-harvest processing, storage, and preparation for feeding. These processes include drying, decontamination, steaming, micronisation, etc. The chapter describes the electrotechnology impacts and their use in processes of grain processing. The chapter deals with the classification of electrophysical methods of influence, correlated with the processes in which they can be applied. It describes the required levels of exposure and the resulting values of productivity, intensity, and other parameters, qualitatively describing the specific processes. The factors which can positively affect grain production include the use of ozone, ions, infrared, and microwave fields.

DOI: 10.4018/978-1-5225-7573-3.ch001

INTRODUCTION

The loss of grain crop during cultivation, harvesting, post-harvest processing and storage is in the range of 5-40% according to various estimates. From our point of view, the greatest interesting are the losses during post-harvest processing and storage. They form about 80% in the structure of these losses. These losses are mainly related to irregularities in the processing technology of storage and post-harvesting. However, in cases of high average yields and the associated reduction in the cost of crop production, it is cheaper for producers to leave part of the crop that requires drying. This happens because of the high cost of post-harvest heat treatment of grain and drying in particular.

BACKGROUND

The average annual loss of grain is about 10% of total production according to the Food and Agriculture Organization (FAO). At the same time, for some less technologically advanced countries this indicator reaches 30-50%. In addition to the loss of the crop volume in case of irregularities in the technological processes of processing and storage, there is a decrease in the quality, sowing and nutrition indicators of grain. The highest level of losses is typical for farms with a low level of technical base and outdated equipment used for processing and storage of grain. Technically, there can be reached the level of losses of 1-2%, which is typical for countries with a high level of technological advance (Germany, Canada, USA).

Losses of grain mass could be divided into mechanical and biological. Some types of losses cannot be avoided, for example, natural loss of grain mass, loss of dry matter due to transpiration. Others are the result of irregularities in post-harvest processing and storage. It is necessary to fight against such losses both organizationally and technologically. The values of natural losses of grain and grain products during storage and transportation are determined by the standard method. Also, do not forget about the pests of grain stocks, whose life activity increases in the overall level of loss of grain mass.

In case of irregularities of storage technology there appears losses of grain due to increased breathing, self-heating, as well as the growth of microorganisms. In cases of increase in humidity or temperature of grain, and especially at joint action of these factors there is a danger of crop damage. This happens due to the sharp intensification of grain respiration at humidity above the standard, accompanied by a loss of dry matter. Air humidity levels for different crops are slightly different and humidity levels that characterize the intensification of breathing and reduce in the safe storage time are different. Thus, millet seeds with a moisture content of 14-15% breathe 2-4 times more intensively than with a moisture content of 13%. Raw wheat grain with a moisture content of more than 17% breathes 20-30 times more intense than dry (14%). In wheat, triticale, rye and barley, the maximum respiration rate corresponds to a temperature of 50...60 °C (almost regardless of humidity). However, with increasing grain moisture from 14 to 22%, the intensity of respiration increases by almost 20 times. If the moisture content of rye grain increases from 15 to 20%, the intensity of its breathing increases 35 times.

ELECTROPHYSICAL EFFECTS IN THE PROCESSING OF AGRICULTURAL MATERIALS

Post-harvest processing is necessary to reduce this type of loss, to increase the safety of the crop and to increase the period of its safe storage. First of all, it includes cleaning and drying of grain to standard moisture.

Classification of Drying Methods

Drying is a heat and mass transfer process of removing liquid from solids, liquids or their mixtures by evaporation. The most often liquid to be removed is moisture or volatile organic solvents.

The drying process in the most general case is as follows: the heated gas stream (drying agent), giving the heat to the material being processed, absorbs the liquid evaporated by it, removing it from the total mass of the substance. Often drying is the last step in the production process, immediately preceding the sale or packaging of products.

There are different variants of drying methods classification depending on the process to be used. The choice of the method depends on the physical condition of the raw material, its chemical composition, the required properties of the final product and the efficiency of the process (Keey R.B., 2011; Mohapatra D. & Mishra S., 2011; Ospanov A.B., Vasilev A.N. at al., 2016).

According to the method of exposure of the drying agent there is:

- Natural drying is drying outdoors in natural light and air, without human influence on the factors intensifying the process (product temperature and drying agent (air), pressure, speed of the drying agent, humidity, etc.). This method are using for drying fruits, berries, mushrooms and fish in regions with suitable climatic conditions.
- Artificial drying is processing in special machines (dryers), with a forced change in the factors affecting the intensity of the process (temperature, pressure, humidity, the geometric dimensions of the drying object, etc.).

According to the pressure in the working chamber there is:

- Atmospheric as a rule, the drying agent is atmospheric air with a pressure deviation in the drying chamber not higher than 49 MPa.
- Vacuum drying is processing in vacuum.
- Drying under overpressure.

According to the method of heat supply to the wet material, the dryers are classified into:

- Convective is thermal energy is transferred by convection;
- Conductive (contact) is thermal energy is transferred by thermal conductivity;
- Wave:
    - Thermal radiation is thermal energy is transferred by thermal radiation;

    - High frequency is heat energy is converted from electrical energy inside the dried material;
    - Acoustic
- Combined is heat transfer is carrying out by combinations of the above methods.

Depending on the mutual moving direction of the dried material and the drying agent the dryers are classified in:

- Straight-flow is the moving directions of the dried material and the drying agent are the same;
- Counter-current is the moving directions of the dried material and the drying agent are opposite;
- Cross-flow is the moving direction of the dried material are perpendicular to the direction of the drying agent.

By type of drying agent there are:

- Apparatus using heated air;
- Installations using flue or inert gases;
- installations using air mixed with flue gases;
- Installations used superheated or saturated steam;
- Installations using liquid heat transfer fluid;
- Installations using electric current.

According to the method of heating the drying agent there are:

- Apparatus using heated air;
- Installations with steam heaters;
- Installations with fire heaters;
- Installations with liquid fuel furnaces;
- Installations with gas fuel furnaces.

By the multiplicity of use of the drying agent there are:

- Installations with single use of heated air; drying agent;
- Installations with repeated use of heated air.

According to the form of the drying object there are:

- Installations for solid materials (large, small, dusty);
- Installations for flexible materials (yarns, films, fabrics, nonwovens);
- Installations for liquid materials;
- Installations for pasty products.

According to the operation mode, the installations are classifying in:

- Periodic;

- Continuously working.

According to the hydrodynamic mode:

- There are dryers with a fixed, suspended and mixed layer of material.

According to the design, the installations are classifying in:

- Chamber, screw, pneumatic pipe, plate, rotary drum, belt, shaft, eddy, fluidized bed, aeropostale, roller, vibrating, spraying, etc.
- Almost all of the presented drying methods are applying in the drying of agricultural products.

Selection of Dryers

In view of the enormous choices of dryer types one could possibly deploy for most products, selection of the best type is a challenging task that should not be taken lightly nor should it be left entirely to dryer vendors who typically specialize in only a few types of dryers. The user must take a proactive role and employ vendors' experience and bench scale or pilot-scale facilities to obtain data, which can be assessed for a comparative evaluation of several options. A wrong dryer for a given application is still a poor dryer, regardless of how well it is designed. Note that minor changes in composition or physical properties of a given product can influence its drying characteristics, handling properties, etc., leading to a different product and in some cases severe blockages in the dryer itself.

Tests should be carried out with the "real" feed material and not a "simulated" one where feasible.

A number of features must be taken into account including pre-treatment stages and subsequent processing stages (waste drying agent utilization and recovery, further product processing, product cooling, etc.) when selecting equipment for each specific process line. The optimal choice of the dryer largely depends on the correct description of the processes at these stages. Product quality requirements do not always allow choosing the least expensive option, based solely on considerations of heat and mass transfer. So drying with a mixture of flue gases has an impact on the quality of products (Budnikov D.A. & Vasiliev D.A., 2018; Fadhel M. I. et al., 2011; Jerzy R. et al., 2011; Nelson S. O., 2015). Often product quality requirements have an excessive impact on the selection process.

The simple decision trees for the selection of periodic and continuous dryers proposed by Van't Land (1984) are presented in figures 1 and 2.

Currently, the selection of equipment for heat treatment is basing not only on the provision of a given performance, but also on the achieving the optimal flow of the passing of the process for the specified factors. The factors may be the minimum energy consumption, maximum drying rate, maximum profit and others, depending on the specific farm. Modern equipment allows the use of solar energy, heat pumps, electrophysical factors, etc.

Intensifying Factors

The intensification of the processes of heat treatment of agricultural materials (drying, disinfection, frying) should be based on the ways that take into account the physical laws of the phenomena of internal and external heat and moisture exchange. The greatest effect on the intensification of these processes

Figure 1. Decision tree for the selection of a batch dryer (van't Lan, 1984)

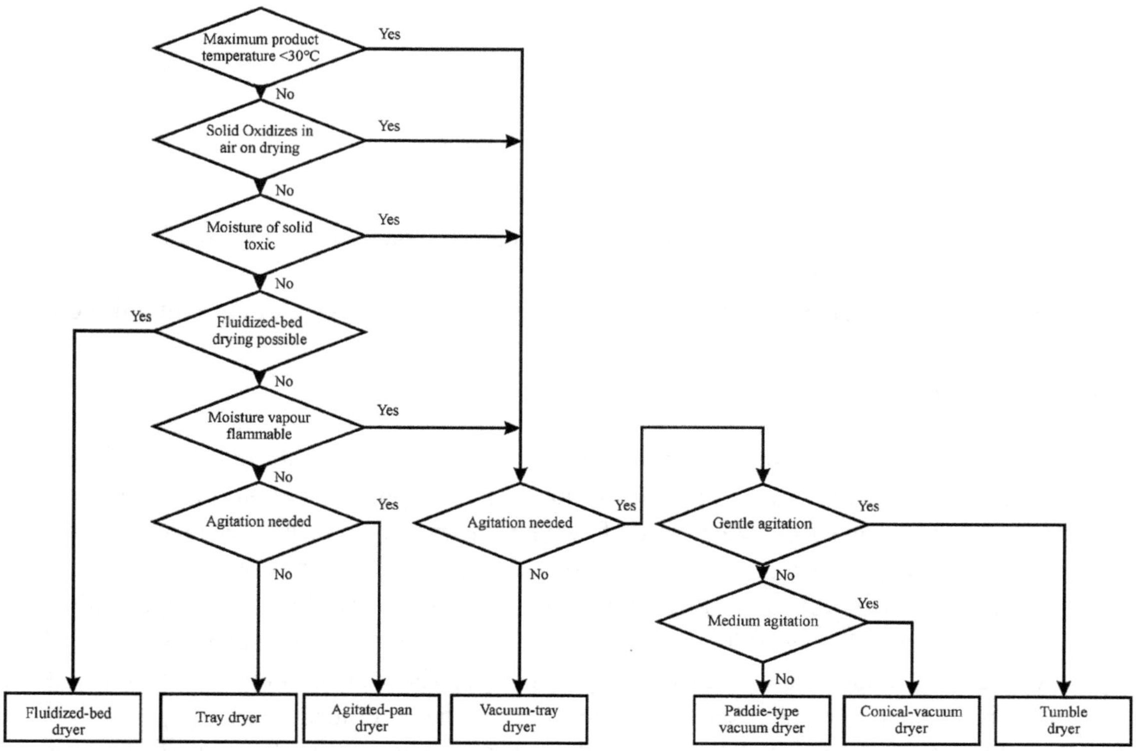

Figure 2. Decision tree for the selection of a continuous dryer (van't Lan, 1984)

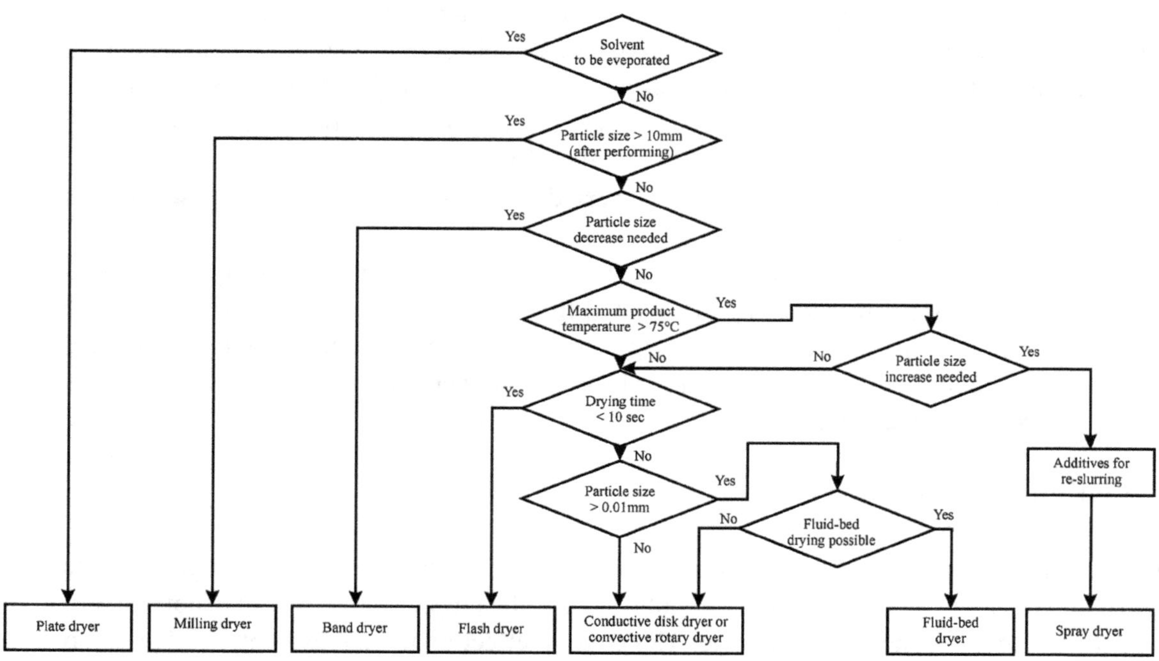

can be obtained through a combination of several factors (technological methods) used for heat treatment of grain in the operating and designing installations (De Freitas et al., 2017; Vasant P. et al., 2017).

Generalization of the literature data allows reducing the ways of heat treatment intensification to the following form (figure 3):

In the field of internal moisture transfer, the intensification of processes can be achieved by increasing the temperature of the grain and eliminating the inhibitory effect of thermal conductivity. There could be used electrotechnological effects such as microwave, IR. The combination of these and other factors, such as air ions, ozone, ultrasonic, etc., along with the use of impulse actions can provide additional benefits in the energy costs and in the productivity of the process. Therefore, the use of microwave treatment leads to a number of technical advantages of the corresponding technological processes (De Freitas et al., 2014; Granella Suian Jose et al., 2018; Fernandes F.A.N. et al., 2016; Mohapatra, D., & Mishra, S., 2014; Tireki, 2006; Vasant P., 2016):

- the humidity uniformity in the volume of processed products significantly increases, the safety of food during drying and heat treatment increases;
- the final moisture level of the dried product can be reduced at a lower cost, compared with convective processing during microwave drying if it is necessary;
- it is possible to significantly reduce the drying time and reduce the energy intensity of the process (especially for processes in which it is necessary to obtain low final moisture).

In the field of external moisture exchange, the process could be intensified by increasing the temperature and speed of the drying agent, as well as increasing the active surface of the grains involved in the process of heat and moisture exchange with the drying agent. For this purpose, drying machines with a suspended layer of material with different hydrodynamics are used, which determines the stay time of

Figure 3. Ways to intensify the heat treatment of agricultural products

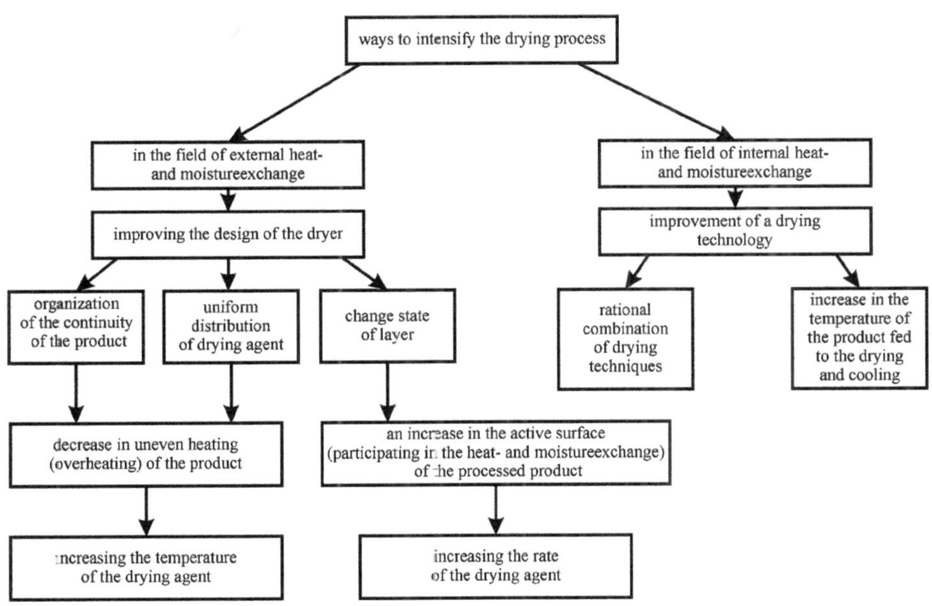

the dried material in the apparatus, the acceptable temperature of the drying agent, the intensity of heat and mass transfer processes and the energy costs for drying. The use of active hydrodynamic modes can significantly intensify the drying process without reducing the economic efficiency, ensure high quality of the finished product and complete safety and processability (Keey R.B., 2011).

Electrophysical Factors of Intensification

As electrophysical influence factors used to achieve the required parameters of the drying process the ozone-air mixtures, air ionization of the drying agent, ultrasound, infrared radiation (IR), high-frequency (HF) and ultra-high-frequency (microwave) radiation are used (Vasant P. at al., 2011).

Ozone

Recently, the development of methods and devices that use ozone in agriculture has received a great deal of attention worldwide, especially in the United States, France, Japan and other countries.

Ozonation in the food industry is widespread in Japan, Australia, France. In the United States, ozone was allowed as a safe tool in 1997 by a decision of the government commission to use in existing and potential technologies related to food storage and processing (Bourgin M. at al., 2011; Raila Algirdas at al., 2006).

Ozone is becoming popular due to the growing demand for products grown without the use of pesticides, stimulants, antibiotics, and other toxic substances for humans and public concern about the environmental hazards of food production technologies.

Currently, ozone technology worldwide interest in the following areas: cleaning and disinfection of drinking water as well as sewerage and industrial effluents; deodorization and air cleaning of various production facilities and gas emissions in various industries; neutralization of toxic substances; increasing shelf life and improving the biological value of feed for animals and food for man; sterilization and packaging of surgical dressings in pharmaceutical industry; stimulation of embryonic development and increasing of hatchability of young animals in poultry; improving the quality of eggs and its fertilization in fisheries; prevention and treatment of various diseases in medicine and veterinary medicine.

The results of recent studies in the field of medicine and veterinary medicine clearly indicate a positive impact of ozoned materials on the natural resistance of living organisms both in oral and parenteral methods of ozone injection. It is noted that strict control over the ozone content in the injecting solution is mandatory. It was found that the optimum ozone dosage for calves when intravenously injecting is 0.015 mg/kg, muscle injecting – 0.04-0.06 mg/kg, oral injecting – 0.09 mg/kg. The ozone acts as a biostimulant there (M. Bourgin at al., 2011; Raila Algirdas at al., 2006).

Table 1 shows the effects of different ozone concentrations on some biological objects.

Ozone is an active oxidizing agent with pronounced antibacterial action, it have been proposed for disinfection of air and water by professor Dobroslavin A. P. in 1874. In 1909 it was used for prolonged storage of meat in the refrigerator in Cologne. Almost simultaneously, in 1911 in St. Petersburg the first European ozone-water station was put into operation. For this purpose, ozonation stations with a capacity of thousands of cubic meters of water per day were built. Ozone destroys microflora, deodorizes and brightens the treated water, gives it the properties of natural pure water in taste and color. It turns back into oxygen (during 40-60 minutes), and saturated with oxygen water is a favorable environment for the growth of microorganisms. For this reason, ozoned water has not used for disinfection.

Ozone not only kills bacteria, viruses, mold, but also decomposes toxins and effectively destroys odors. The efficiency of cleaning from microbes and micro-fungi, pollutants and unpleasant odors (ammonia, hydrogen sulfide) reaches 90%.

Periodic ozonation of air in the production facilities can reduce the content of harmful gases (ammonia, hydrogen sulfide) by 80-85%, reduce the contamination with bacteria and mold by 80-90%.

Special experiments have shown that the use of ozonation to prevent bronchopulmonary diseases (bronchitis, pneumonia) in animals leads to no practically occurrences of them.

Ozone concentrations at the MPC level can stimulate the vital activity of living organisms, for example, this is used in the sanitation of rooms with animals and plants to improve the comfort of their stay. Higher than MPC ozone concentrations inhibit the activity of harmful organisms and eliminate harmful pollution.

When keeping animals and poultry with a high concentration of ones per unit area, the air in the room is significantly contaminated with inorganic and organic dust, organic compounds: ammonia, hydrogen sulfide, carbon dioxide and other harmful substances.

Currently, the usual way to remove harmful substances from livestock rooms is forced ventilation, which entails high energy costs.

Ozone leaves no harmful products and is environmentally friendly. The unused ozone dissolves back to a more stable form of oxygen within a few tens of minutes.

It should be noted that the studies conducted by the Electric Power Research Institute (EPRI, USA) on the order of the U.S. Food and Drug Administration (FDA) found that food processed with ozone do not form any substances that have mutagenic or carcinogenic properties. Therefore, the FDA has certified ozone as a disinfectant and sanitizer for use without any restrictions in the food industry of USA.

Consequently, ozone has received the status of "Generally Recognized as Safe" (GRAS), which opens wide horizons for the use of a gas mixture containing ozone in agricultural production.

Table 1. Ozone effects on various biological objects

Ozone Concentration, mg/m^3	Impact on Biological Objects
4.0-6.0	Inhibits the mold growth
6.0	80% of microflora dies during 3.5 hours
6.0-10.0	Bronchitis, focal catarrhal pneumonia, degenerative changes in the liver grow in rats.
200-400	For disinfection of rooms and carton boxes under processing 15-30 minutes kills adult parasites (mites, cockroaches, bedbugs, fleas)
400	It needs to process for 4 minutes to inactivate the virus of poliomyelitis. Colon bacillus dies within 3 hours of treatment.
1600	Almost all types of Colon bacillus in the water die in 10 minutes.
2000	Disinfection of suspensions of almost all types of bacteria in water for one minute. Harmful effects on amoebas and liquid crystal forms of multicellular organisms (kristosoma, mansoni, bilharriose). For complete disinfection of premises requires exposure 1-2 hours.

Air Ions

In addition to the classical high-temperature drying methods in mine dryers, as well as other less energy-intensive but also less productive methods, such as drying by active ventilation, there are methods that were developed with the use of alternative energy sources and the use of electrical technologies in the intensification of low-efficiency methods and reducing the cost of high-performance methods (Vasiliev A.N., 2018).

The saturation of the drying agent with air ions when active ventilation of grain are using means the saturation of the drying agent with air ions of positive or negative polarization. In this case, the supply of ionized drying agent can occur continuously or by pulses. Some drying curves which was the results of the experiment are shown below in figures 4–7 (Vasiliev A.N. at al., 2018).

The comparison of the drying curves (figure 4a and figure 4b) leads to note that the change in the drying agent speed affects the drying speed of the grain and the ratio of the moisture removind for different modes. Thus, for the drying agent speed of 0.6 m/s, the drying time of grain to 14% humidity, under the cyclic mode of exposure to air ions, is 1.2 hours, at constant exposure of air ions - 2 hours, and with classical drying - 1.7 hours.

The rate of grain drying in the cyclical influence of aeroions on a range of grain moisture content from 16% to 15% reduces at air speeds of 1 m/s. The speed of drying under constant influence of aeroions remains constant, and under the classical drying it increases. However, at an air velocity of 1 m/s, the drying time to 14% humidity both in the classical mode and with a constant supply of air ions did not become less than the drying time than in the cyclic supply of air ions and was 1.2 hours.

These results allow us to say that when drying the grain layer with atmospheric air, as it is in the case with drying by active ventilation, the increase in the speed of the drying agent from 0.6 m/s to 1.0 m/s is equivalent to the speed of grain drying as with the use of a cyclic mode of action of air ions. Given the fact that increasing the speed of drying agent the cost of electricity will be higher than under the mode of cyclic exposure to air ions, the mode with cyclic exposure can be recommended to use with the active ventilation of grain with humidity up to 16%.

When the temperature of the drying agent increases to 34% (figure 5), the air filtration rate in the grain layer is affecting at the moisture removal rate again.

Figure 4. Drying curves for raw data with three drying modes: a) Winitial=16%, T=20 °C, V=0.6 m/s; b) Winitial=16%, T=20 °C, V=1 m/s

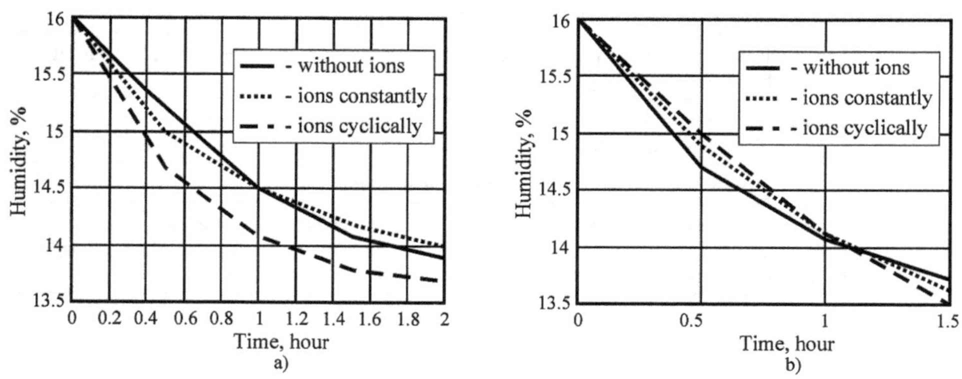

Figure 5. Drying curves for raw data with three drying modes: a) Winitial =16%, T=34 °C, V=0.6 m/s; b) Winitial =16%, T=34 °C, V=1 m/s

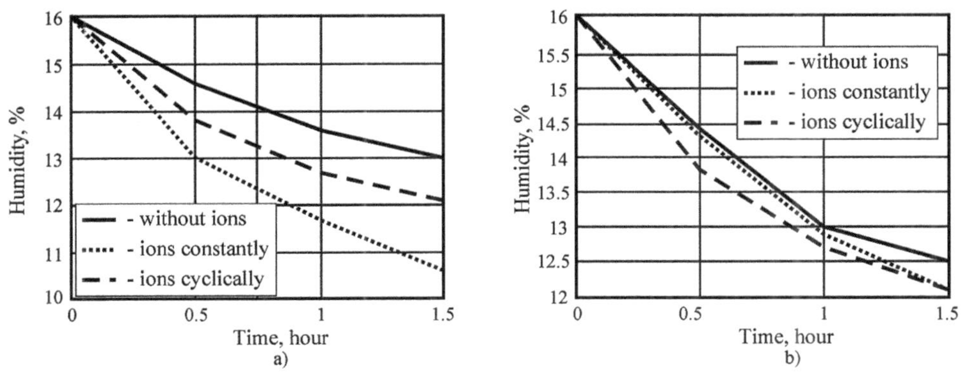

Thus, at an air velocity of 0.6 m/s, the highest drying rate is observed under the constant influence of air ions. In this mode, the grain was dried to 14% for 0.32 hours, with cyclic exposure – for 0.45 hours, and in the classical mode – for 0.75 hours. The increase in air speed to 1.0 m/s reduced the drying time in the classical mode to 0.62 hours, and with constant exposure to air ions – increased to 0.6 hours. Under the mode of cyclic impact of air ions, the grain drying time to 14% remained equal to 0.45 hours.

These results allow us to say that at low-temperature drying of grain with moisture content up to 16% the most effective are the modes with the use of electroactivated air, but which of the modes to prefer depends on the speed of the drying agent.

Grain drying curves by active ventilation by atmospheric air were obtained with the initial grain moisture content of 24% in accordance with the plan of experimental studies. The curves are shown on figure 6.

The graphs show that the shortest drying time of the grain is observed under the cyclic action of air ions. However, at an air speed of 0.6 m/s, when drying grain to 17%, the rate of moisture removing was greater than in the classical drying mode. After it was decreased to the drying rate as with the constant concentration of air ions.

Figure 6. Drying curves for raw data with three drying modes: a) Winitial =24%, T=20 °C, V=0.6 m/s; b) Winitial =24%, T=20 °C, V=1 m/s

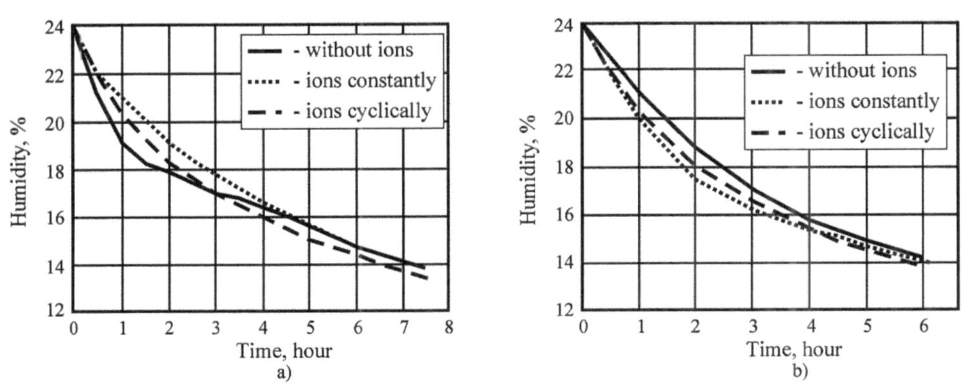

When speed of the drying agent increasing to 1 m/s, it was observed a higher rate of moisture removal to 16% humidity under the mode of constant exposure to air ions. Further, the drying rate became greater with the cyclic effect of air ions.

These results allow to say that during drying by active ventilation of grain with moisture content of 14-24%, the drying mode must be chosen depending on the current grain moisture and air speed.

The drying curves showed more unambiguous results (figure 7) at low-temperature grain drying, when the atmospheric air was heated to 34 °C. Thus, at an air velocity of 0.6 m/s and at 1 m/s, the highest rate of moisture removal is observed in the mode of cyclic exposure to air ions.

Therefore, we conclude that when drying grain by heated air with moisture content of grain of 16-24%, it is advisable to apply a cyclic mode of exposure to increase the rate of moisture collection.

The analysis of drying curves allowed to determine which drying mode is advisable to use to different initial data of the process. Nevertheless, the driving force of the process of the drying intensification when using electro activated air cannot be seen from the curves of drying. This suggests the feasibility of studying the use of electrical technologies in the processes of post-harvest processing of grain and feed.

Ultrasonic

A promising area of intensification of technological processes is the use of energy of mechanical vibrations of high-intensity ultrasonic frequency.

Currently, a new area of knowledge, named bioacoustics, and studying the forms, methods and systems of sound communication, is actively developing. It turns out that many living beings in the process of evolution have learned to use ultrasonic vibrations for orientation in space (Fernandes, F. A. N. at al., 2016; He, Z., Zhang, Y. et al., 2016).

Scientists have identified many different examples of the use of ultrasound by animals. Most often, these are warning signals about the danger, the expression of threat, satisfaction, victory, etc.

The most widely used ultrasound is the inhabitants of the seas. It is determined that in water the ultrasonic spreads with the speed of 5300 km/h. Nothing can move in water faster than ultrasonic vibrations. The source with a power of 100 kW audible at a distance of 15 km in the air, the source with 1 kW of power extends to 100 km in water. Water is transparent to ultrasound, as such as air to light. Vibrations emitted by fish, shrimp and other marine animals allow fishers to detect them. The same radiation al-

Figure 7. Drying curves for raw data with three drying modes: a) $W_{initial}$ =24%, T=34 °C, V=0.6 m/s; b) $W_{initial}$ =24%, T=34 °C, V=1 m/s

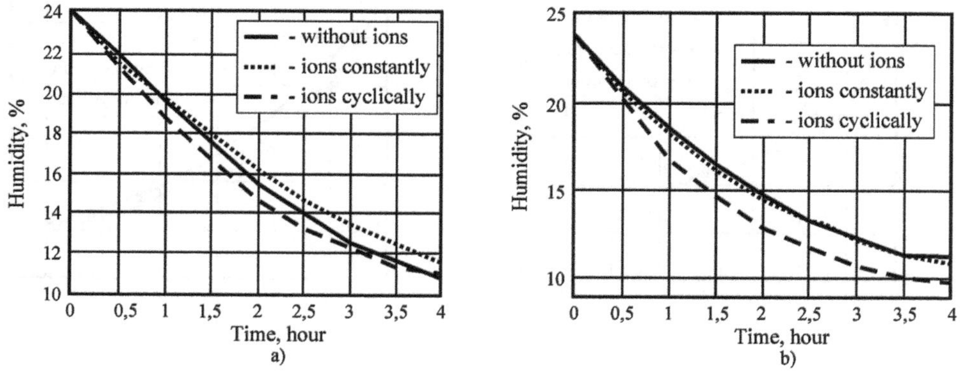

lows determining the location of the shoal of fish and its size (Fernandes, F. A. N. at al., 2016; He, Z., Zhang, Y. et al., 2016).

The practical application of ultrasound is developing in two directions:

1. The use of low-intensity waves (low-energy oscillations that do not lead to irreversible changes in the materials and bodies through which they are distributed) for monitoring, measurement, research of the internal structure of materials and products (level gauges, flow meters, analyzers of gas, liquid and solid substances composition, flaw detectors).
2. Application of high-energy oscillations - high-intensity waves - for active influence on substances and changing their structure and properties.

The biological effect of ultrasound may also be based on secondary physical and chemical effects. So, it can occur mixing of the intracellular structures in the formation of acoustic flows. Cavitation leads to the rupture of molecular bonds in biopolymers and other vital compounds and to the increasing of redox reactions. Ultrasound increases the permeability of biological membranes, resulting in the acceleration of metabolic processes by diffusion. All of these factors in real conditions affect on biological objects in a particular combination together, and therefore it is difficult, and sometimes impossible, to study separately the processes of different physical nature.

Currently, ultrasound is one of the methods of intensification of catalysis. Ultrasound radiation is associated with cell damage, but the beneficial effects of sound on the enhancement of catalysis and the life of living cells also have been proven.

Cavitation in cell suspensions. When the intensity of ultrasound increases to values, when the mechanical forces, comparable with the strength of cell membranes, appears in medium, the process of cell destruction begins. The effect occurs if the amplitude of the bubble pulsation increases to a certain value. This value is different for different cells, depending on their shape and size, as well as the strength of the cytoplasmic membrane and the presence of the cytoskeleton.

Ultrasonic cell disintegration is widely used in biotechnology, in biochemical and virological studies for the isolation of individual substances and cell fragments, as well as in laboratory diagnostics to determine the mechanical resistance of cell membranes.

The ultrasound effect on proteins. It is known that there is a change in the structure, shape and function of the protein molecule in the ultrasonic field. The direction of these changes depends on the structure of protein and protein end groups and the properties of the gas contained in the aqueous solution.

It was found that the effect of ultrasound on biomacromolecules (nucleoproteins, nucleic acids, lipoproteins, etc.) depends on the nature of the gas present in the solution of the studied substances. There is a process of degradation of biomacromolecules, causing inhibition of their biocatalytic activity in the presence of oxygen. These processes are accompanied by a decrease in the viscosity of the solutions of these substances. An increase in the molecular weight of biomacromolecules is observed in the presence of hydrogen while maintaining their native properties, in particular proteins such as serum albumin, trypsin and pepsin.

The ultrasound effect on enzymes. It was found that changes in enzymes during ultrasonic irradiation are depend on not only the structure of these proteins, but also the nature of the presented gas. In this regard, ultrasound does not always have an inactivating effect on them. Inactivation of enzymes during voicing occurs if this process proceeds in the presence of oxygen; in an environment saturated with hydrogen, inactivation does not occur. It was established in ultrasonic processing of such enzymes as trypsin, pepsin, tyrosinase and others. Some enzymes, such as catalase, are not inactivated by ultrasound at all. Oxidase is more sensitive to ultrasound, while reductase, catalase and amylase have sufficient stability.

The sterilizing effect of ultrasound is associated with the destructive effect of ultrasonic vibrations on microorganisms, which allow using the ultrasound for sterilization and disinfection. For example, the sterilization of milk while voicing for 15-60 seconds delay its souring for 5 days, the vitamins in the milk are preserved. In 1 cm3 of milk sterilized by ultrasound contains an average of 18 CFU, while after conventional pasteurization for 1 hour it remains about 3000 CFU in 1 cm^3.

The mechanism of the sterilizing action of ultrasound is very complex and not fully opened. Obviously, cavitation is the leading factor. The phenomenon of cavitation occurs primarily where the strength of the liquid is the lowest, i.e. at the interface of the cell-liquid media. When a cavitation bubble is formed on the cell surface, the structure of the wall is drawn into the cavity at the moment of its reduction. In the subsequent phase, when the cavity is slammed, a powerful hydrodynamic shock occurs, reaching tens of MPa.

Positive results of water disinfection by means of ultrasonic vibrations are revealed; it is possible to reach full sterilization of water without application of any chemical reagents within 5 minutes. There is also a positive effect of ultrasonic vibrations for sterilization of canned food and juice.

It was found that the ultrasound effect increases the sensitivity of microorganisms to disinfectants. The sterilizing effect in the treatment of aqueous suspensions of Escherichia coli bacteria, which were previously exposed to ultrasound, was achieved at significantly lower concentrations of chlorine, formalin and other disinfectants. This is associated with the change ("loosening") of macromolecular structures that are part of the shell of the voiced microorganisms, which leads to a violation of the permeability of the membranes of living cells.

Almost all microscopic plants and organisms die if exposed to high-intensity ultrasound. This fact is now considering as an alternative safe way to purify water and food.

Escherichia coli, Salmonella typhi, diphtheria (Corynebacterium diphtheriae), hay (Bacillus subtilis) sticks, tetanus (Tetanus bacillus), Salmonella, Staphylococcus, trypanosomes (Trypanosoma bacillus), Trichomonas (Trichomonas), the causative agent of typhus (Typhus recurrens) and other are destroying by ultrasound. High-intensity ultrasound has a destructive effect on Tobaco mosaic virus, encephalitis (Encephalitis virus), typhoid, influenza (Influentia). Bacteriophages of large sizes are also sensitive to the action of ultrasound. The most resistant to ultrasound from the pathogenic microorganisms is the various strains of Tuberculosis bacillus.

The sterilizing effect of ultrasound on microorganisms appears at frequencies of 20 kHz and above, at an intensity of more than 0.5 W/cm^2 in the cavitation mode of exposure.

The ultrasound could be used for sterilization of drugs prepared in the factory and in the pharmacy. The use of ultrasound with frequency to 490 kHz, and intensity up to 20 W/cm^2 for 5 min achieved complete sterilization of eye drops: 0.25% solution of zinc sulfate, 1% aqueous solutions dwelt, platifillina tartrate, pilocarpine hydrochloride in pharmacy. Analysis of the prepared solutions showed the preservation of the authenticity and quantitative content of drugs in solutions before and after ultrasound exposure.

During ultrasonic treatment of aqueous suspensions of mycobacteria with a frequency of 20 kHz, there are destroyed about of 93% of mycobacteria, and at high frequency ultrasonic action (612 kHz) this value reaches 35.5%.

The ultrasonic capillary effect is a phenomenon of increasing the depth and speed of liquid penetration into the capillary channels under the influence of ultrasound (in comparison with the depth and speed caused only by capillary forces). The discovery of the ultrasonic capillary effect belongs to the Belarusian scientist academician E. G. Konovalov (1961). Glass tubes with a diameter of 2 mm and an input horn were used in his experiment. The rise of the liquid level (above the height of capillary level), proportional to the force of the sound was observed at normal incidence of the waves on the hole of the mouthpiece in the tube. It was found in a series of works by E. G. Konovalov that the additional rise under the influence of ultrasound linearly increases with increasing temperature and diameter. The maximum height and rate of ascent were observed when the capillaries came into contact with the surface of the emitter. Studies have suggested that the ultrasonic capillary effect is determined by the impact of cumulative jets or in a simplified version of the pressure that occurs when the cavitation bubbles close to the mouth of the capillary. Like the ultrasonic capillary effect, the phenomenon of "reverse" ultrasonic capillary effect, which has found wide application, including in medicine, is known.

Pre-Sowing Seed Treatment

The high yield requires high quality planting material with a high percentage of germination. To do this, the seeds are pre-treated before planting (soaking, etching, etc.). Pre-sowing seed treatment allows to intensify the germination process, destroys harmful microorganisms (Kaasová J. at al., 2002; Nelson, S. O., 2015; Vasiliev, A.N. at al., 2018).

The methods used for pre-sowing seed treatment have a number of disadvantages: the duration of the soaking process, the inability to separate the germinating seeds from the non-germinating ones before they germinate. In this regard, the improvement and further development of existing methods of pre-sowing seed treatment has a great interest.

In the 30-40s of the last century in the Soviet Union, the influence of ultrasonic vibrations on the process of plant development was studied. Studies have shown that the use of ultrasonic vibrations has a beneficial effect on the process of seed germination and subsequent development of plants.

Thus, ultrasound-treated barley grains give germination 2-3 days earlier than the control planting, the ear length and the number of grains in it increase by 30%.

The process of ultrasonic seed treatment is as follows:

1. the batch of seeds is placed in a technological volume with water, the volume of seeds should not exceed 30% of the volume of water;
2. the necessary microelements are introduced;
3. ultrasound treatment is performed within 5-10 minutes.

Thus, ultrasonic pre-sowing treatment of seeds is characterized by high productivity.

In addition, the ultrasonic pre-sowing treatment of seeds has one more conclusive advantage which enhances its value. After pre-sowing seed treatment the empty, non-viable seeds remain floating on the surface of the liquid, and germinating seeds settle to the bottom. Thus, it can be quite easy to separate non-viable seeds from germinating ones.

Unfortunately, the mechanism of ultrasonic action on grains and seeds has not been fully investigated. What is clear is that ultrasound is able to stimulate the vital forces inherent in nature in every crop.

An example of the results of existing studies of ultrasonic treatment of melon seeds, tomatoes, corn, wheat, buckwheat are shown in table 2.

Table 2. Results of experiments on pre-sowing seed treatment

Plant Species	Yield Increase, %
Melon	45
Tomatoes	15
Corn	35
Wheat	25
Buckwheat	30

SHF Treatment

It is known that microwave heating leads to an increase in the pressure of water vapor inside the grain, but it is not investigated the coefficient of convective heat exchange, which depends on the intensifying effect, and also determines the rate of moisture transfer in the material (Magdalena Zielinska & Marek Markowski, 2017; Nelson S.O., 2008; Ranjbaran M. et al., 22).

There is little mention of the use of microwave intensification for drying in active ventilation bins. There are not given models of drying by active ventilation during microwave activation in cases that mention this combination.

Analysis of works on drying using microwave fields showed that many existing installations use powerful generators of microwave energy used in the short-term mode (2-3 seconds), such solutions lead to very rapid heating of moisture in the dried material, which can lead to damage of grain. The use of resonator chambers in which the grain occupies a part of the microwave activation zone was also studied. Another solution was the use of consistently located low-power magnetrons, for example, in the case of a conveyor dryer, but due to the high speed of the belt and the uneven distribution of the microwave field in the material, the grain layer must be thin (2.5-3 cm). This leads to an increase of the conveyor length to ensure the heating of the material to the required temperature (He Z. et al., 2016; Warchalewski J. R. et al., 2010; Zhao Y. et al., 2017).

Existing dryers using combined action together with the treatment by microwave field cover almost the entire range of dryer types. However, the intensifying effect of the microwave field for the intensification of drying in the active ventilation bins is still not used. However, this direction seems to us promising, since active ventilation is economically advantageous and the use of microwave energy can accelerate the drying process, making it even more economical and convenient for use in farms. In addition, active ventilation is multifunctional and could be used not only for drying, but also for cooling, degassing, elimination of self-heating centers, etc. (Trygve Magne Eikevik et al., 2012; Wang Y., 2011; Zielinska,M. et al., 2012; Zielinska,M. et al., 2015).

The aim of using the microwave with the active vent is creating the weevil gradients of temperature and humidity aimed in one direction; creating gradients of temperature and moisture pressure inside the grains leads to enhanced dehumidification.

Simplified mechanism of moisture removal from the grain at high-frequency energy supply can be presented as follows (figure 8). High intensity dielectric heating of the grain leads to the fact that inside the grain there is excessive (compared to atmospheric) pressure of steam P_{sur}. The created pressure gradient is the driving force of moisture transfer from the inner layers to the surface. The value of the pressure

Figure 8. Mechanism of moisture removal from grain during high-frequency drying: 1 – grain; 2 – moisture evaporation zone

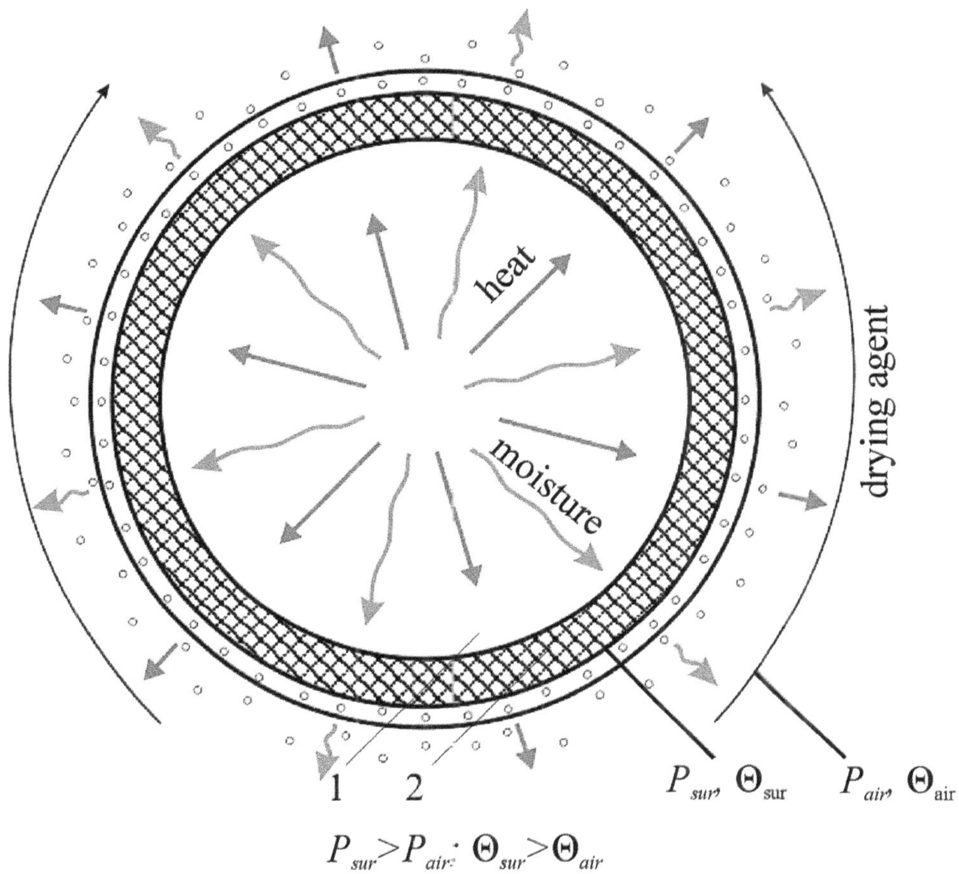

gradient and the rate of moisture transfer depends on the parameters of the microwave electromagnetic field. At their high values the excess pressure can increase to such an extent that it will meets the high resistance of the grain structure and leads to its destruction. Thus, when drying grain with SHF (Super High Frequency) EMF (electromagnetic field), the internal moisture transfer can be significantly (compared to the conditions of convective drying) intensified, but only up to some critical value, limited by the condition of preservation of the grain structure and its quality as a whole.

Moisture moved from the inside to the surface of the grain evaporates into the environment. The speed of this evaporation or external moisture transfer will be determined not so much by the parameters of SHF EMF, but by the parameters of the environment. A mandatory condition for the moisture removal from the surface of the grain, as in convective drying, is the presence of a difference between the partial pressure of steam at its surface P_{sur} and in the environment P_{air} ($P_{sur}>P_{air}$). However, if during the convective drying the temperature of the environment (drying agent) Θ_{air} excesses the temperature of the surface of the grain Θ_{sur} ($\Theta_{air}>\Theta_{sur}$), these condition of external moisture transfer is not significant when high-frequency drying and Θ air may be less than Θ_{sur} (i.e. $\Theta_{air} < \Theta_{sur}$).

Therefore, grain drying by the energy of the SHF EMF, as well as convective, is characterized by a combination of two phenomena – internal and external moisture transfer. However, in comparison with the convective, microwave grain drying provides better conditions for internal moisture transfer than the external, and the drying process limits the external moisture transfer, the speed of which is determined by the parameters of the environment (air). P_{sur}, Θ_{sur}

The complexity of the calculation not only the pressure gradient, but also the pressure of the liquid and steam in the grain are described above. Therefore, it is advisable to consider the relationship of changes in pressure in the grain with the change in temperature difference (Θ_{centre} - Θ_{sur}).

The driving forces of the drying process of microwave processing grain are the temperature difference between the center and the surface of the grain, and the vapor pressure of the liquid in the grain P. A linear dependence of and P at microwave exposure are theoretically proved, the dependences are given to determine the temperature Θ_{sur} and Θ_{centre} at different moments of time of microwave exposure. Therefore, experimental confirmation of the temperature distribution inside the grain is required. The problem is that the temperature of the material must be measured under the microwave field. Previously, for these purposes, kerosene thermometers were used, which could operate in a microwave field with sufficient accuracy. However, they cannot be used to measure the temperature inside the grain. During various studies on microwave processing and drying of grain the temperature measurement of the material was carried out by using thermocouples, but immersing them in the product after the switching off the microwave field. The earlier analytical calculations show that the temperature rise time inside the grain is not more than 20...60 seconds. Time of temperature equalization inside the grain with the temperature of the drying agent after the termination of the microwave, depending on the grain moisture can vary from 5 to tens of minutes. With this dynamics of heating and cooling, the initial moments of the process are especially important. The procedure of entering a thermocouple in the test layer after the termination of the microwave radiation does not allow to capture the dynamics of the heating of the grain layer and the single grains and provides a large error in the measurement of cooling of the grain layer. In addition, it is almost impossible to measure the temperature difference Θ_{centre}-Θ_{sur}.

Continuous and Periodic Action of Microwave

When drying and disinfecting grain using the microwave field, it is very important to distribute moisture over the grain volume. The closer moisture to the grain surface, the lower energy consumption for drying. If the outer layers of grain are more moistened than internal layers, then intense heating takes place in the outer layers, which leads to disinfection of the grain surface.

Laboratory experiment was carried out for the comparative assessment of influence of microwave field on the rate of wheat drying. The factors were: initial humidity – 4 levels (16, 18, 22, 26%); drying agent speed – 3 levels (0.2, 0.6, 1 m/s); drying agent temperature – 2 levels (15, 30°C). Operation modes of the microwave fields source: switched off, permanent microwave effects with the limitation in 55°C, pulse switching with the limitation in 55°C. The response function is the current moisture content of the material.

Experimental studies have allowed to obtain drying curves, an example of which is presented on figure 9.

The required coefficient of drying characterizes the amount of moisture removed from a material unit per unit time and characterizes the drying process intensity. The processing of the experimental studies results was carried out using a package of applied Matlab software. As a result, a set of dependencies

Figure 9. Experimental drying curves

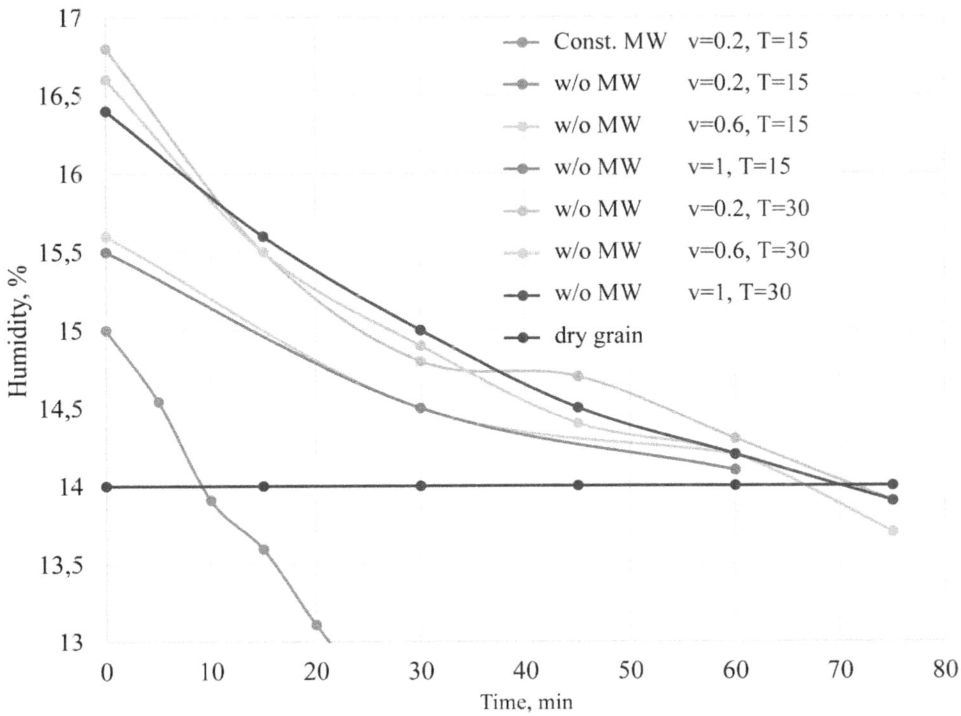

of the drying coefficient, *K*, on the process parameters ($K = f\,(T,\ v,\ W)$) was obtained in the form of regression models of the form:

$$K = A_1 + A_2 \cdot T + A_3 \cdot v + A_4 \cdot W + A_5 \cdot T \cdot v$$
$$+ A_6 \cdot T \cdot W + A_7 \cdot W \cdot v + A_8 \cdot T^2 + A_9 \cdot v^2 + A_{10} \cdot W^2 \quad (1)$$

The analysis of the experimental data makes it possible to judge that an increase in the unit capacity is compensated by the acceleration process, which leads to a decrease in energy intensity of the drying process.

Periodic Exposure to Microwave

There are important production tasks, such as increasing the speed of drying, reducing the uneven drying of grain on the layer and the energy intensity of the process of active grain ventilation in the bunker, the solution of which is carried out by various methods. The use of available experience has allowed to investigate the application of the principle of recirculating drying for these purposes. There is a redistribution of moisture between the grains of different moisture content in the recirculation zones of the dryers. First, there is a period of binning, during which the specified redistribution occurs, then the grain is dried with high temperature air. Redistributed moisture between grains is not bonded, so the moisture is carried out with less energy than in conventional, high-temperature drying (Budnikov D. A., 2015).

The grain from the layers located closer to the central air duct is mixed with the grain located closer to the outer cylinder of the hopper to implement this principle in the bunkers of active ventilation.

Since the active ventilation uses atmospheric air, the grain temperatures are not high, so the recirculation will not give a significant redistribution of temperatures between the grains. Intensification of the recycling process is possible due to the selective action of the microwave field. When processing the grain mixture, wet grain should be heated more quickly. The vapor pressure created in the grains should stimulate the intensive removal of moisture from the grain surface into the intergranular space. More dry grain should absorb moisture from the intergranular space.

Microwave recirculation processes do not have analytical description, in this regard, it is desirable to obtain a mathematical model of heat and moisture exchange in the grains in the microwave recirculation zone, which will allow to perform their preliminary calculations.

The system of differential equations of heat and moisture transfer under microwave heating has the form:

$$\left.\begin{aligned} \frac{\partial \Theta}{\partial \tau} &= a \cdot \nabla^2 \Theta + \frac{\varepsilon}{c \cdot r'} \cdot \frac{\partial U}{\partial \tau} + \frac{Q_v}{c \cdot \rho_0}; && (2.1) \\ \frac{\partial U}{\partial \tau} &= a_{m_2} \cdot \nabla^2 U + a_{m2} \cdot \delta_2 \cdot \nabla^2 \Theta + \varepsilon \frac{\partial U}{\partial \tau}; && (2.2) \\ \frac{\partial P}{\partial \tau} &= a_p \cdot \nabla^2 P + \frac{\varepsilon}{c_v} \frac{\partial U}{\partial \tau}. && (2.3) \end{aligned}\right\} \quad (2)$$

where Θ – temperature of grain, °C; τ – microwave exposure time, s; a – coefficient of thermal diffusivity, m^2/s; ∇ - Hamiltonian operator; ε – coefficient of liquid-vapor phase transformation;

c – specific heat of grain, kJ/kg·°C; r' – specific heat of vaporization, kJ/kg; Q_v – specific power dissipated in the dielectric under the influence of the microwave field, W/m^3; ρ_0 – density of dry matter of grain, kg/m^3; a_{m2} – coefficient of fluid diffusion, m^2/s; δ_2 – relative coefficient of thermal diffusion; P – excess pressure in the sample, Pa; P_H – the vapor pressure of the material at a given moisture content, Pa; c_v – body capacity relative to moist air, Pa^{-1}; a_p – coefficient of convective diffusion of vapor, m^2/s.

The impact of the microwave field in the zone of microwave recirculation on the grain begins immediately after loading it with processing grain. Heat and moisture exchange in the layer are describing by the system (2). The grains in the layers are heating to different temperatures depending on their humidity. There will be changes in equations 2.1 and 2.3 after the termination of the microwave field, when the grain temperature reaches the desired value. Since the effects of the microwave field is vanished, then the third term from equation (2.1) eliminates. After removing the microwave impact in the grain there occurs the relaxation of the pressure, so the coefficient of convective diffusion of vapor a_p in the equation (2.3) must be replaced by a_{pr} - coefficient of convective diffusion when the pressure relaxation has occurred.

In this case, the system of equations takes the form:

$$\left.\begin{array}{ll}\dfrac{\partial\Theta}{\partial\tau}=a\cdot\nabla^2\Theta+\dfrac{\varepsilon}{c\cdot r'}\cdot\dfrac{\partial U}{\partial\tau}; & (3.1)\\ \dfrac{\partial U}{\partial\tau}=a_{m_2}\cdot\nabla^2 U+a_{m2}\cdot\delta_2\cdot\nabla^2\Theta+\varepsilon\dfrac{\partial U}{\partial\tau}; & (3.2)\\ \dfrac{\partial p}{\partial\tau}=a_{pr}\cdot\nabla^2 P+\dfrac{\varepsilon}{c_v}\dfrac{\partial U}{\partial\tau}. & (3.3)\end{array}\right\} \quad (3)$$

In this period there is a redistribution of temperature between grains and air in the intergranular space, between the grains, inside the grain. As previously noted, the grains in the layers are heated to a different temperature depending on the humidity, but their temperature is higher than the temperature of the intergranular space. Therefore, after the termination of the microwave field, the intergranular space receives heat from all grains.

The wettest grains give heat and moisture to the intergranular space. If the heating temperature of less wet grains is also higher than the temperature of the intergranular space, they also give off heat. As a result of the use of microwave heating, the temperature and humidity of all the grains of the layer should be equalized, the hygrothermal equilibrium should be established, when the temperature of all grains and the temperature of the intergranular space are the same. The moisture content of all grains is equal and the amount of moisture absorbed from the intergranular space is equal to the amount of moisture given by the grains. When the state of hygrothermal equilibrium is reached during microwave heating, the processes of heat and moisture exchange will be described by the following system of equations:

$$\left.\begin{array}{ll}\dfrac{\partial\Theta}{\partial\tau}=a\cdot\nabla^2\Theta+\dfrac{Q_v}{c\cdot\rho_0}; & (4.1)\\ \nabla^2 U=\delta_2\cdot\nabla^2\Theta; & (4.2)\\ \dfrac{\partial p}{\partial\tau}=a_p\cdot\nabla^2 P. & (4.3)\end{array}\right\} \quad (4)$$

Since before reaching the state of hygrothermal equilibrium in the microwave recirculation zone, moisture is not carried out, then in the system of equations (4) there are not presented all the terms containing $\partial U/\partial\tau$. Under the influence of microwave field temperature (4.1) is growing inside each grain, increasing the humidity drop across the cross section (4.2) and increasing vapor pressure (4.3). After removing the microwave impact, the system of equations will take the following form at the hygrothermal equilibrium:

$$\left.\begin{array}{ll}\dfrac{\partial\Theta}{\partial\tau}=a\cdot\nabla^2\Theta; & (5.1)\\ \nabla^2 U=\delta_2\cdot\nabla^2\Theta; & (5.2)\\ \dfrac{\partial p}{\partial\tau}=a_p\cdot\nabla^2 P. & (5.3)\end{array}\right\} \quad (5)$$

Thus, the process of microwave recirculation is described by four systems of equations (2) ... (5).

In order to take full advantage of the achieved effect of "water back" from the center of the grain to the surface in the zone of microwave recirculation, it is need to start the ventilation process immediately. The process of heat and moisture exchange during convective drying is usually described by the following system of equations:

$$\left.\begin{array}{lr} \frac{\partial T}{\partial \tau} - 3600 \cdot V \frac{\partial T}{\partial \tau} = -\frac{\gamma_{grain} \cdot A_{grain}}{\gamma_{air} \cdot A_{air} \cdot \varepsilon_{grain}} \cdot \frac{\partial \Theta}{\partial \tau} - \frac{\gamma_{air} \cdot r'}{\gamma_{7} \cdot A_{air} \cdot \varepsilon_{grain}} \cdot \frac{\partial W}{\partial \tau}; & (6.1) \\ \frac{\partial W}{\partial \tau} = -\frac{\gamma_{air} \cdot \varepsilon_{grain}}{10\gamma_{grain}} \cdot \left(\frac{\partial F}{\partial \tau} + 3600 \cdot V \cdot \frac{\partial F}{\partial \tau}\right); & (6.2) \\ \frac{\partial T}{\partial \tau} + 3600 \cdot V \cdot \frac{\partial T}{\partial \tau} = -\frac{\alpha S_{v}}{\gamma_{air} A_{air} \varepsilon_{grain}} \left('' - \Theta\right); & (6.3) \\ \frac{\partial W}{\partial \tau} = \frac{\beta \cdot S_{v}}{\gamma_{grain}} \left(P_{grain} - P_{air}\right), & (6.4) \end{array}\right\} \quad (6)$$

where T – drying agent temperature, °C; τ – microwave heating time, s; V – drying agent speed, m/s; γ_{grain} – bulk weight of grain, kg/m3; γ_{air} – specific weight of air, kg/m3; c_{air}, c_{grain} – heat capacity of air and grain, J/kg•°C; ε_{grain} – porosity of the grain layer; Θ - temperature of grain, °C; r' – specific heat of vaporization, kJ/kg; W – moisture content of grain, %; F – moisture content of drying agent, g/kg; – heat transfer coefficient, J/kg•s•°C; S_v – specific surface of seeds, 1/m; P_{grain}, P_{air} – water vapor pressure in the grains and in air, Pa.

Equation (6.4) describes the change in grain moisture due to the vapor pressure in the grain P_{grain} and in the drying agent P_{air}. Since after the microwave effect the change of the vapor pressure in the grain is described by the equation (5.2), for this case the differential equation of humidity change will be changed, transforming the equations (6.3), taking into account (5.1) we obtain:

$$\left.\begin{array}{lr} \frac{\partial T}{\partial \tau} - 3600V \frac{\partial T}{\partial \tau} = -\frac{\gamma_{grain} \cdot A_{grain}}{\gamma_{air} \cdot A_{air} \cdot \varepsilon_{grain}} \cdot \frac{\partial \Theta}{\partial \tau} - \frac{\gamma_{air} \cdot r'}{\gamma_{grain} \cdot A_{air} \cdot \varepsilon_{grain}} \cdot \frac{\partial W}{\partial \tau}; & (7.1) \\ \frac{\partial W}{\partial \tau} = -\frac{\gamma_{air} \cdot \varepsilon_{grain}}{10\gamma_{grain}} \cdot \left(\frac{\partial F}{\partial \tau} + 3600 \cdot V \cdot \frac{\partial F}{\partial \tau}\right); & (7.2) \\ \frac{\partial T}{\partial \tau} + 3600 \cdot V \cdot \frac{\partial T}{\partial \tau} = -\frac{\alpha S_{v}}{\gamma_{air} A_{air} \varepsilon_{grain}} \left(T - \int_{0}^{\tau_1} a \nabla^2 \Theta \, d\tau\right); & (7.3) \\ \frac{\partial W}{\partial \tau} = \frac{\beta \cdot S_{v}}{\gamma_{grain}} \left(\int_{0}^{\tau_1} a_{pr} \cdot \nabla^2 P_{grain} \, d\tau - P_{air}\right). & (7.4) \end{array}\right\} \quad (7)$$

Using systems (2) ... (6), and also knowing that the temperature of the center, Θ_{center}, °C and surfaces, Θ_{sur}, °C are determined by the following dependences:

$$\Theta_{center} = \Theta_0 + \frac{2Q_v \cdot R^2}{\lambda \cdot \pi^2}\left(e^{K\tau} - 1\right), \tag{8}$$

and

$$\Theta_{sur} = \Theta_0 + 0{,}687 \cdot \frac{Q_v \cdot R^2\left(e^{K\tau} - 1\right)}{\lambda \cdot \pi^3}, \tag{9}$$

we obtain a dependence for determining the temperature of the intergranular space:

$$T = \Theta_0 + 0{,}1124 \cdot R^2 \frac{\frac{Q_{v_{dry}}}{\lambda_{dry}} \cdot K_r + \frac{Q_{v_{wet}}}{\lambda_{wet}}}{1 + K_r} \cdot \left(\left(\frac{\gamma_{air} \cdot A_{air} \cdot K}{\alpha \cdot S_V} + 1\right) \cdot e^{K \cdot \tau} - 1\right) \tag{10}$$

where Θ_0 – initial grain temperature, °C; R – grain radius, m; Q_v, Q_{vdry}, Q_{vwet} – specific power dissipated in the dielectric under the influence of the microwave field, dry- in dry (recycled) grain, wet - in wet grain, W/m³; λ, λ_{dry}, λ_{wet} – coefficient of thermal conductivity, dry – dry (recycled) grain, wet – wet grain, W/m•°C; K_r – coefficient of recirculation, γ_{air} – specific gravity of air, kg/m³; c_{air} – heat capacity of air, J/kg•°C; K – separation factor, s^{-1}; α – coefficient of heat transfer, J/kg•s•°C; S_v – specific surface of seeds, 1/m.

The obtained mathematical model of processes occurring in the grain at microwave recirculation allows performing preliminary calculations to determine the grain processing modes for its effective drying.

Using data for the initial grain moisture content of wheat equal to 22%, moisture content of dried grain equal to 14%, the coefficient of recirculation equal to 3, the initial grain temperature equal to 20 °C, we have calculated the temperature change in the center of the "wet" grain, on its surface and in the intergranular space (figure 10).

It should be noted that these graphs are more qualitative, but allow us to predict the dynamics of temperatures in the grain layer. So, presented graphs allow us to say that the first exposure to the microwave field on the "wet" grain temperature in its center and on the surface changes according with the previously described laws. The center of the grain is heating more, because it has a higher humidity than the surface. The temperature in the intergranular space increases not so much as in the center and even the surface of the grain. This confirms the previously stated assumption that it is not possible to achieve a uniform distribution of the temperature of the intergranular space and temperature equalization between the grains in one-step microwave exposure.

CONCLUSION AND FURTHER RESEARCH DIRECTIONS

We could conclude the following based on the material presented:

- The existing set of drying equipment allows us to choose the installation for almost all existing processes, but often it has increased energy costs.

Figure 10. Calculated temperature variation curves in the center, Θ_{center}, on a surface, Θ_{sur}, "wet" grains and in intergranular space, T

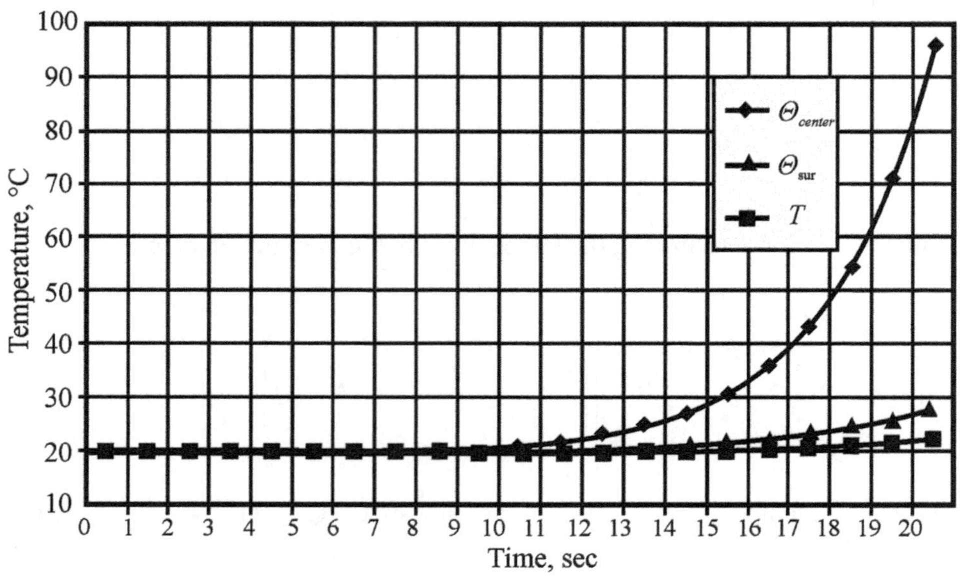

- A large number of factors, including electrophysical ones, can be applied for intensification of processes taking place during processing of agricultural products in the specified equipment. At the same time, the choice of a particular factor depends on the requirements of the process and the characteristics of the equipment.
- Ozone, air ions, IR and microwave effects are effectively applied to intensify heat and mass transfer processes.
- Ozone-air mixtures and microwave fields are applied for disinfection of agricultural products.

The choice of a technological impact depends on the possibility of its use.

Further research directions could be:

- Calculation of the energy balance of the intensification technologies applied to specific equipment and technological lines.
- Study of the processed materials properties (thermal, electrical, etc.) for existing factors and determine the possibility of effective application and control of technological processes during the processing.

REFERENCES

Algirdas, R. (2006). Application of ozone for reduction of mycological infection in wheat grain. *Annals of Agricultural and Environmental Medicine, 13*(2), 287–294. PMID:17196003

Bourgin, M., Violleau, F., Debrauwer, L., & Albet, J. (2011). Ozonation of imidacloprid in aqueous solutions: Reaction monitoring and identification of degradation products. *Journal of Hazardous Materials, 190*(1-3), 60–68. doi:10.1016/j.jhazmat.2011.02.065 PMID:21524846

Budnikov, D. A. (2015). Issledovanie raspredelenija naprjazhennosti SVCh polja v zernovom sloe [Investigation of the distribution of the microwave field intensity in the grain layer]. *Inzhenernyj vestnik Dona*, 3. Retrieved from http://www.ivdon.ru/ru/magazine/archive/n3y2015/3234

Budnikov, D. A. (2015). Issledovanie raspredelenija naprjazhennosti SVCh polja v zernovom sloe [Investigation of the distribution of the microwave field intensity in the grain layer]. *Inzhenernyj vestnik Dona*, 3. Retrieved from http://www.ivdon.ru/ru/magazine/archive/n3y2015/3234

Budnikov, D. A., & Vasiliev, D. A. (2018). The Use of Microwave Energy at Thermal Treatment of Grain Crops. In Handbook of Research on Renewable Energy and Electric Resources for Sustainable Rural Development. Hershey, PA: IGI Global. Doi:10.4018/978-1-5225-3867-7.ch020

De Freitas. (2014). Development of a solid-liquid extraction method with low-temperature partitioning for the determination of insecticides in ozonized maize grain. *Quimica Nova, 37*(2), 238–243.

De Freitas. (2017). Degradation kinetics of pirimiphos-methyl residues in maize grains exposed to ozone gas. Journal of Stored Products Research, 1-5.

Eikevik, T. M. (2012). Microwave-Assisted Atmospheric Freeze Drying of Green Peas: A Case Study, Drying Technology. *International Journal (Toronto, Ont.), 30*(14), 1592–1599.

Fernandes, F. A. N., Rodrigues, S., García-Pérez, J. V., & Cárcel, J. A. (2016). Effects of UltrasoundAssisted Air-Drying on Vitamins and Carotenoids of Cherry Tomatoes. *Drying Technology, 34*(8), 986–996. doi:10.1080/07373937.2015.1090445

Granella, S. J., Christ, D., Werncke, I., Bechlin, T. R., & Machado Coelho, S. R. (2018). Effect of drying and ozonation process on naturally contaminated wheat seeds. *Journal of Cereal Science, 80*, 205–211. doi:10.1016/j.jcs.2018.03.003

He, Z., Zhang, Y., Wang, Z., Zhao, Z., & Yi, S. (2016). Reducing Wood Drying time by Application of Ultrasound Pretreatment. *Drying Technology, 34*(10), 1141–1146. doi:10.1080/07373937.2015.1099107

Jerzy, R. (2011). Changes in Microwave-treated Wheat Grain Properties. *Advances in Induction and Microwave Heating of Mineral and Organic Materials.*

Kaasová, J., Hubáčková, B., Kadlec, P., Příhoda, J., & Bubník, Z. (2002). Chemical and biochemical changes during microwave treatment of wheat. *Czech Journal of Food Sciences, 20*(2), 74–78. doi:10.17221/3513-CJFS

Keey, R. B. (2011). *Dryers*. Thermopedia. doi:10.1615/AtoZ.d.dryers

Mohapatra, D., & Mishra, S. (2011). Current trends in Drying and Dehydration of Foods. In *Food Engineering*. NOVA Science Publishers Inc.

Mohapatra, Giri, & Kar. (2014). Effect of Microwave Aided Disinfestation of Callosobruchus Maculatus on Green Gram Quality. *International Journal of Agriculture and Food Science Technology*, *5*(2), 55–62.

Nelson, S.O. (2004). RF and microwave permittivities of insects and some applications. *URSI EMTS*, 1224-1226.

Nelson, S. O. (2008). Dielectric properties of agricultural products and some applications. *Research in Agricultural Engineering*, *54*(2), 104–112. doi:10.17221/1000-RAE

Nelson, S. O. (2015). *Dielectric Properties of Agricultural Materials and Their Applications*. Academic Press.

Ospanov, A. B., & Vasilev, A. N. (2016). Controlling reactions of biological objects of agricultural production with the use of electrotechnology. *International Journal of Pharmacy and Technology, 8*(4), 26855-26869.

Ospanov, A. B., & Vasilev, A. N. (2016). Changing parameters of the microwave field in the grain layer. *J. Eng. Applied Sci.*, *11*(S1), 2915-2919.

Ranjbaran, M., & Zare, D. (2013). Simulation of energetic- and exergetic performance of microwaveassisted fluidized bed drying of soybeans. *Energy*, *59*, 484–493. doi:10.1016/j.energy.2013.06.057

Tireki, Sumnu, & Esin. (2006). Effect of Microwave, Infrared and Infrared-assisted Microwave Heating on the Drying Rate of Bread Dough. *American Journal of Food Technology*. Doi:10.3923/ajft.2006.82.93

Vasant, P., Alparslan-Gok, S. Z., & Weber, G.-W. (2017). *Handbook of Research on Emergent Applications of Optimization Algorithms*. IGI Global.

Vasant, P., Barsoum, N., & Webb, J. (Eds.). (2011). *Innovation in Power, Control, and Optimization: Emerging Energy Technologies*. IGI Global.

Vasant, P., & Voropai, N. (2016). *Sustaining Power Resources through Energy Optimization and Engineering*. IGI Global. doi:10.4018/978-1-4666-9755-3

Vasiliev, A. N. (2018). Increasing Efficiency of Grain Drying With the Use of Electroactivated Air and Heater Control. In Handbook of Research on Renewable Energy and Electric Resources for Sustainable Rural Development. Hershey, PA: IGI Global. Doi:10.4018/978-1-5225-3867-7.ch011

Wang, Y. (2011). Review of dielectric drying of foods and agricultural products. *International Journal of Agricultural and Biological Engineering*, *4*(1). Retrieved from http://www.ijabe.org

Warchalewski, J. R., & Gralik, J. (2010). Influence of microwave heating on biological activities and electrophoretic pattern of albumin fraction of wheat grain. *Cereal Chemistry*, *87*(1), 35–41. doi:10.1094/CCHEM-87-1-0035

Zhao, Y., Wang, W., Xie, J., Zheng, B., Miao, S., Lo, Y. M., ... Tian, Y. (2017). Microwave Vacuum Drying of Lotus Seeds: Effect of a Single-Stage Tempering Treatment on Drying Characteristics, Moisture Distribution and Product Quality. *Drying Technology*, *35*(13), 1561–1570. doi:10.1080/07373937.2016.1261890

Zielinska, M. (2012). A multi-stage combined heat pump and microwave vacuum drying of green peas. *Journal of Food Engineering*. doi:10.1016/j.jfoodeng.2012.10.047

Zielinska, M., & Markowski, M. (2017). The effect of microwavevacuum, ultrasonication and freezing on mass transfer kinetics and diffusivity during osmotic dehydration of cranberries. *Drying Technology*. doi:10.1080/07373937.2017.1390476

Zielinska, M., Sadowski, P., & Błaszczak, W. (2015). Thawing and Microwave-Assisted Drying of Blueberries (Vaccinium Corymbosum L.). *Lebensmittel-Wissenschaft + Technologie*, *62*(1–2), 555–563. doi:10.1016/j.lwt.2014.08.002

Chapter 2
Theoretical Prerequisites for Subsurface Broadcast Seeding of Grain Crops in the Conditions of Pneumatic Seed Transportation to the Coulters

Sergey Senkevich
https://orcid.org/0000-0001-6354-7220
Federal Scientific Agroengineering Center VIM, Russia

Pavel Aleksandrovich Ivanov
Don State Agrarian University, Russia

Pavel Vladimirovich Lavrukhin
Don State Agrarian University, Russia

Zarifdzhan Yuldashev
Tajik Agrarian University, Tajikistan

ABSTRACT

Only 30-40% of the plant productivity potential is realized in crop production technologies, which leads to a situation where the growth of field yields lags behind the growth of production costs. Increasing the realization of the productivity potential of cultivated plants is a way to improve the efficiency of the crop industry. The task of sowing machines is the uniform placement of seeds on the field area, while plants get their living space and, accordingly, form a crop. The practice of using such working bodies has shown that to achieve the necessary seeding regimes under the coulter elements, the use of additional sources of kinetic energy is required.

DOI: 10.4018/978-1-5225-7573-3.ch002

INTRODUCTION

Modern agricultural production requires to increase yields of major crops while reducing the cost of their production. This problem can be solved in the creation and implementation of agricultural technologies in a new, more highly productive level. This method allows to realize 60-70% of the biological potential of cultivated plants to obtain high-quality products, and improvement of agricultural machinery (Krasnoshchekov, 2002; O'Leary& Garry J., 2018).

BACKGROUND

When solving the most important task to increase the grain crops yield, one of the main ways is to improve the sowing operation (De Vita, Colecchia, Pecorella, &Saia, 2017; Ewbank&Buckby2002). The most common methods of sowing – ordinary, have a number of significant drawbacks, one of which is the uneven distribution of seeds on the nutrition area.

The works of researchers and farmers have proved that the most of them take into account and meet the biological needs of grain in the size and shape of areas nutrition plant (and, consequently, the need for nutrients and solar energy) allows intrasoilbroadcast seeding. The transition to this method of seeding is also necessary to reduce the energy intensity of soil cultivation and the seeding time, which have been confirmed by numerous studies.

The most common working body when performing intrasoilbroadcast seeding is a hoe, that creates the conditions for the distribution of seeds inside space of coulter across the width of the hoe.

All working bodies performing this type of seeding are divided according to the way of kinetic energy transfer to the seeds during distribution. They are working bodies with passive distribution, which use kinetic energy of the seeds fall, and working bodies with active distribution, which use additional sources of energy for sowing seeds.

The main disadvantage of these working bodies is the low range drop distance seeds into inside space of coulter and the complexity of the proposed designs (Lavrukhin, 2000).

Recently, sowing units using kinetic energy of air flow for transporting seeds to coulters have become widespread. The seeds acquire a large speed, which allows to increase the range of garbage seeds into inside space.

The task of obtaining uniform placement of seeds into inside space of coulter when using the kinetic energy of the flow. To solve this problem, it is necessary to determine the main dependences of the kinetic energy transfer from the air flow to the seeds and to determine the necessary directions of the seed movement during the energy transfer.

Overview of Research Made in Broadcast Seeding

There have been considered two main technologies of seeding: surface and subsurface broadcast seeding. Sifting of seeds on the surface with their subsequent or simultaneous imposition into the soil can be done by aviation, by various tillage tools (spreaders of mineral granulated fertilizers, special seeders) or by hand.

Such seeding technologies are utilized in rice-growing regions of Russia (the Premorsky Area), in the republics of the Middle Asia, in Kazakhstan, and some other countries (Birma, Indonesia, China).

When using this technological principle (when seeds are distributed on the surface of the soil) the seed covering is done by plowing the upper layer of soil with seeds, or isn't done at all. As a result, the quality of seeding is poor and it does not meet the main requirements of agricultural technology. It should be added that the mechanical and technological foundations of the sowing technologies on the surface are practically not developed, excepting the spreaders of mineral fertilizers. Their working process is widely developed for various types of mineral fertilizers both theoretically and practically.

In particular, the working process is divided into a number of consecutive elements. They are given mathematical dependencies to describe the corresponding processes and to determine the main parameters. The velocity and direction of the flight of a particle after leaving the disk blade is determined from the differential equation:

$$\xi = r\omega^2 \cos\psi - fg \pm fr\omega^2 \sin\psi - 2f\omega\xi', \tag{1}$$

where f is coefficient of friction of particle on disk; ξ is axis of coordinates along the blade; r is radius vector of a particle; ψ is blade angle.

Simplified solutions of equation have been offered:

$$\omega t_c = \frac{\cos\phi}{-1-\sin\phi} \cdot \ln\frac{2\left[r_0 \cos\left(\phi \pm \psi_0\right) + l\cos\phi\right]}{r_0\left(1+\sin\phi\right)\cos\left(\phi \pm \psi_0\right)}, \tag{2}$$

$$v_e = \frac{\omega}{1-\sin\phi}\left[r_0 \cos\left(\phi \pm \phi_0\right) + l\cos\phi\right], \tag{3}$$

where l is blade length.

V.A. Chernovolov's investigations (Skolzayev & Chernovolov, 1969; Skolzayev & Chernovolov, V.A. 1969) are of a special interest. They consider the question of the uniformity of the fertilizers distribution over the sifter width when they were distributed along the radii, determined by the density function $p(\rho)$ and in the throwing sector, determined by the density function $p(\alpha)$.

The distribution of fertilizers in the throwing sector is characterized by a function $q_\alpha = f(\alpha)$, and the distribution of fertilizers along the sifter width can be characterized by a function $q_F = f(X)$

The fertilizer rate can be determined as follows

$$q_F = \int_0^{t_1} \frac{1}{\rho} q_\alpha p(\rho) dt\,. \tag{4}$$

With the central loading of the apparatus, the distribution of fertilizers in the throwing sector is uniform, i.e. $q_\alpha = \frac{Q_\alpha}{2\alpha_m} = const$. In this case, the approximate solutions of equation (4) at $0 \angle \left|X\right| \angle \left|\rho_2 \sin\alpha_m\right|$ have the form

$$q_F \approx \frac{q_a(\alpha_1 - \alpha_2)}{V_m \rho_{cp}(\sin\alpha_1 - \sin\alpha_2)}, \tag{5}$$

and at $\left|\rho_2 \sin\alpha_m\right| \angle \left|X\right| \angle \left|\rho_1 \sin\alpha_m\right|$ have the form

$$q_F \approx \frac{2q_c(\alpha_1 - \alpha_2)}{V_m(\rho_1 + a)(\sin\alpha_1 - \sin\alpha_2)} \int_a^{\rho_1} p(\rho)\,d\rho\,. \tag{6}$$

For a machine with non-central loading, the fertilizers distribution in the throwing sector is close to normal and therefore

$$q_\alpha = Q_\alpha f(z_\alpha), \tag{7}$$

where $f(z_\alpha)$ is a Laplace-Gaussian function.

In this case, the equation (4) can be solved only by approximate methods.

To determine the unevenness of the sifting for different width, that is, for different overlaps, use the following equation:

$$H_l = \left(\sum_1^k \frac{q_{Fi}}{\sum q_{Fi}} - \frac{k}{n}\right) \cdot 200\,, \tag{8}$$

where H_l is unevenness; k is a number of relations $\frac{q_{Fi}}{\sum q_{Fi}}$, large $\frac{1}{n}$; $n = \frac{B_p}{b_y}$ is number of catchers, on working width of sifting B_p; b_y is width of one catcher.

The following two groups of technical tools for subsoil spreading can be divided according to the method of giving kinetic energy to the seeds for their scattering. They are the working organs with passive distribution which use kinetic energy of seed falling and the working bodies with active distribution which use additional energy sources for seed sifting.

When combined machines and working tools appeared, there were designed band drills. They scattered seeds by a band and distributed them along the furrow bottom in the sub-coulter space. Many authors consider this method of seeding efficient, as it improves the yield of grain crops on 30% (Bakhmutov, 1980;Machnev, 2012; Siracusa, Laura, Gresta, Fabio,&Sperlinga, Elisa. 2017).

In coulters with passive distribution of seeds in width, the effect is achieved by reflecting from the blades, made in the form of various swiveling surfaces, or volumetric geometric bodies (Malev, 1971; Nogtikov, Glotov & Sazonov, 1996; Nogtikov & Sazonov, 1995).

This direction is mostly developed in theory. In the works of different authors, the broadcast seeding is considered according to the effect of such factors as the recovery coefficient (when a seed strikes the reflector), the coefficient of friction (when a seed slides over the blade surface), air resistance, etc. The greatest attention has been paid to the problem by I.T. Kovrikov, A.N. Semenov, F.V. Grischenko, M.K. Malev and other researchers.

To determine the range of throwing back A.N. Semenov (Semenov, 1959) considered the process as the impact of a material point on a solid surface, set at an angle to the horizon. Initially, a direct central blow was taken into account. The noncentral impact was taken into account by the angle between the reflector plane and the tangent to the trajectory of the seed fall γ0. He obtained an analytical dependence to determine the range of throwing back after the impact, the direction and the speed of grain flight after the impact. A.N. Semenov took into account the recovery coefficient and the coefficient of grain friction on the material of the throwing back surface:

$$X_3 = \frac{U\cos\alpha_3}{g}\left(\sqrt{U^2\sin\alpha_3 + 2gY_1} - U\sin\alpha_3\right). \tag{9}$$

The speed is determined from the equations:

$$\cos\alpha_3 = \frac{v_y\left[\cos\alpha_0\left(1+\varepsilon'\right)\left(\cos\alpha_0 + f\sin\alpha_0\right)\right]}{U}, \tag{10}$$

$$\sin\alpha_3 = \frac{v_y\left[1-\cos\alpha_0\left(1+\varepsilon'\right)\left(\cos\alpha_0 + f\sin\alpha_0\right)\right]}{U}. \tag{11}$$

F.V. Grishchenko (Grishchenko, 1963; Grishchenko, 1970) determined the seeds flight after striking a plate as the motion of a material point at an angle to the horizon, not taking into account air resistance. He gave an analysis of the seed distribution in the soil. Based on the study, a conclusion about the necessity of seedbed equalization has been made.

To justify the form of a distributing tool Z.I. Tuvalev (Tuvalev, 1987) investigated the nature of seed movement in sub-coulter space of the no-row spade coulters. Taking into account the initial conditions, he proposed the equation of a seeder part:

$$y = K_1x + K_2\left(1 - l^{K_3x}\right). \tag{12}$$

With the necessity to place the trajectory of the falling seeds in a confined space, the curvature of the part in the lower part is reduced, for which a curve with a constant radius of curvature has been used:

$$R^2 = \left(\eta + R^2\right) + \xi. \tag{13}$$

The question of broadcast seeding with passive seed distribution has been considered more profound and deep by I.T. Kovrikov (Kovrikov, Nogtikov & Sazonov, 1995; Kovrikov, 1983). According to him the spatial seed distribution during sowing is influenced by a complex of factors. The main of them are coulter system operation, the construction of the furrow former and the seed distributor, the type of the coulter attachment, and the quality of the seeding tools.

The main direction of stubble seeders development was obtained from the conditions that equal areas of seed supply should correspond to equal areas of nutrition. This condition will be fulfilled if every distributor sector supplies seeds along the directions of the tracks dividing the coulter width into equal parts.

For these conditions, the author gives a model of the distributor, which provides for a uniform seed distribution over the area of nutrition.

$$\begin{gathered} Y=\left(B/2\pi R-1\right)\Big[\left(R+a'\right)\ln\left(R+a'-\sqrt{R^2-x^2}\right)+\sqrt{R^2+x^2}+ \\ +B\big/2\pi\left(-1\big/2\arcsin^2 x\big/R+3(R+a')\right)/\sqrt{9R^2+6a'}\ln(\arcsin^2 x\big/R+, \\ +2\sqrt{(9R+6a')/(R-6)}/\arcsin x/2-\sqrt{(9R+6a')/R-6)}\Big]+C \end{gathered} \tag{14}$$

where R is radius of the feed section (seed drill tube); a`is coordinate of rear point of a supplier; B is width coulter.

Approximate function has been determined as

$$\phi\left(x\right)=1+0{,}15x+0{,}3x^2, \tag{15}$$

and has been determined as a parabola equation

$$x^2=2\times1{,}6y. \tag{16}$$

The seed bouncing range in the sub-coulter space is described by the system

$$\begin{cases}\left(x_0+v_0^2\sin2\alpha/2g+1/g\sqrt{v_0^2\sin\alpha_0+2gz_0}\right)\geq B/\cos\Theta \\ tg\alpha_0\left(x-x_0\right)-g\left(x-x_0\right)/2v_0^2\cos^2\alpha_0+z_0\leq H-xH\cos\Theta/B\end{cases}, \tag{17}$$

where $z=H-xH\cos\Theta/B$ is the equation of the arch of the sub-coulter space with the worst conditions for the seed flight; H is arch height at the level of the axis of coulter symmetry ;$\alpha_0$is angle determining the direction of the initial velocity ;θ is angle determining the plane of extreme seeds flight.

According to the research results A.A. Nogtikov and V.A. Golivets (Nogtikov & Golivets, 1997) came to the conclusion that in order to realize the subsoil broadcast seeding, there is a need in a combined coulter. It must consist of a universal a-hoe blade (goose foot share) with a passive seed distributor throughout the width of seeding located in its sub-coulter space.

According to the researchers, the distributor must have a divider and a throwing back element. The divider can be represented by a three-edged wedge with a flat back edge and cylindrical sides.

The authors make the assumption that on the surface of the sides during sliding the seed is held by friction. Having compiled the differential equations of motion in the projections on the axes of the natural trihedral, and having solved them taking into account the initial conditions at t = 0, $\alpha = 0$, $\alpha' = V_0/r$, the authors obtained:

$$v = \left(2gr\cos\beta\left[\left(1-2f^2\right)\sin\alpha + 3f\cos\alpha\right] / \left(1+4f^2\right) + v_0^2 - 6gfr\cos\beta / \left(1+4f^2\right)e^{-2f}\right)^{1/2}, \tag{18}$$

where f is dynamic coefficient of seed friction of seeds through metal; v_0 is initial speed of seed movement.

In general, for all considered coulters where the kinetic energy of seeds is used for sifting, the range of sifting can be determined in accordance with the dependences:

$$L = hctg\gamma, \tag{19}$$

where L is range of seed sifting; h is height of seed sifting; γ is angle of seed sifting, as well as

$$h = v_0^2 / 2g\cos ec^2\gamma, \tag{20}$$

where v_0 is speed of seed flight.

Analysis of the proposed dependencies shows a limited range of seed sifting without changing a coulter's form. In addition, the width of the band depends on the kinematic ratio of seeder's speed and seed flight after throwing back.

Therefore, in practice, it is not possible to obtain a wide band of sown seeds. The conditions of the coulters require their arrangement with a row spacing of 22.8 cm, and the width of the band is 12 ... 14 cm.

As a rule, when seeding with such coulters, the seeds are located in the furrow a little bit thicker in the middle part (Lyubushko & Zvolinskiy, 2001), and thus one can't achieve necessary uniformity in seed distribution.

Active coulters are equipped with devices that, acting on the seeds, give them of additional speed and acceleration. At the same time, the required uniformity of seed distribution can be achieved. In addition, the use of active spreaders allows increasing the width of the grab.

The question of the active seed distribution in sub-coulter place was thoroughly studied by P.V. Lavrukhin (2000). As an active element of the working organ, there was a metal plate of an electromagnetic vibrator receiving power from the on-board power supply of the tractor.

The required amplitude-frequency characteristic of the vibrations of the vibrator plate is achieved by feeding an alternating current signal to the solenoid of the vibrator with a given frequency, formed by a low-frequency synthesizer.

Thus, a controlled variable frequency of vibrations of the vibrator plate is achieved. It transmits impulses of different strength to the sifting seeds and ensures a different range of their flight in the sub-coulter space.

Vibration pattern of the vibrator plate from the effect of the out-of-use torque of the electromagnetic field occurring in the coils of the vibrator solenoid was analyzed on the basis of the differential equation:

$$I_{nb}\phi'' + 2b\phi' + c\phi = BI_m S\cos\omega_x t, \tag{21}$$

where I_{nb} is moment of inertia of the vibrator plate; $\varphi$is range (amplitude); B is magnetic induction; $C$is coefficient of the system's rigidity; b is damping factor; I_m is amperage; S is area of the frame; ω_x is driven frequency; t is time.

The work of the "vibrator plate-seed" system at the moment of collision is described by differential equations:

$$\begin{cases}\left(m_1+m_2\right)\cdot x''+B_x\cdot x'+C_x\cdot x=m_0 l\cdot\omega_0^2\cdot\cos\alpha_1\cdot\cos\omega_0 t\\ \left(m_1+m_2\right)\cdot y''+B_x\cdot y'+C_x\cdot y=m_0 l\cdot\omega_0^2\cdot\sin\alpha_1\cdot\cos\omega_0 t\end{cases}, \tag{22}$$

where m_1 is vibrator plate weight; m_0 is seed weight; r is radius of the funnel; ω_0 is frequency of vibrations of the vibrator plate.

The direction of the seed affected by the instant force impulse transmitted to the seed by the side of the vibrator plate at the moment of impact and exciting the flight of the seed along the free trajectory is formed in the following way. When the seed strikes the plate, vibrations of the 'plate-seed' system occurs along an elliptical trajectory, one of whose axes is rotated about the OX axis by an angle α_1 determined by the equation:

$$\alpha_1=\frac{1}{2}arctg\frac{2X_aY_a\left(\phi_y-\phi_x\right)}{X_a^2+Y_a^2}. \tag{23}$$

The angle α_1 is taken as a direction of further seed movement. This angle can obtain both positive and negative values.

Mechanic-technological foundations of the process of active seed distribution are considered by N.M. Bespamyatnova. She found out that the control action in the executive subsystem of seed placement by the power area is formed by introducing the integrating unit. The unit provides a negative phase shift (-90), with a frequency characteristic of 25-45 c^{-1}, and makes it possible to obtain a required seed sifting quality with the width of coulter grab up to 0.7 m.

In her works she comes to the conclusion that the uniform seed distribution is possible with a controlled active energy supply from the outside source to the flow of sifting seeds, giving the latter the additional modulo and flight speed vector variables (Bespamyatnova, 2010; Bespamyatnova, 2002).

General Model of Seeder Machine Operation for Intrasoil Broadcast Seeding

If we consider the seeder's work with the working organs of the broadcast seeding as a reaction of the dynamic system to the input and control actions, the design scheme can be represented by the 'input-output' principle proposed by A.B. Lurye (Lurye, 1979).

In this case, the general plan of the seeder for this type of seeding will look like a system that converts the input vectors of operating and control functions into output variables (Figure 1).

The input vector-functions F are the resistance to movement *R (t),* the characteristic of the seeding material *k*, etc. (the number of components depends on the degree of their recording).

Figure 1. General scheme of a seeder machine for intrasoil broadcast seeding

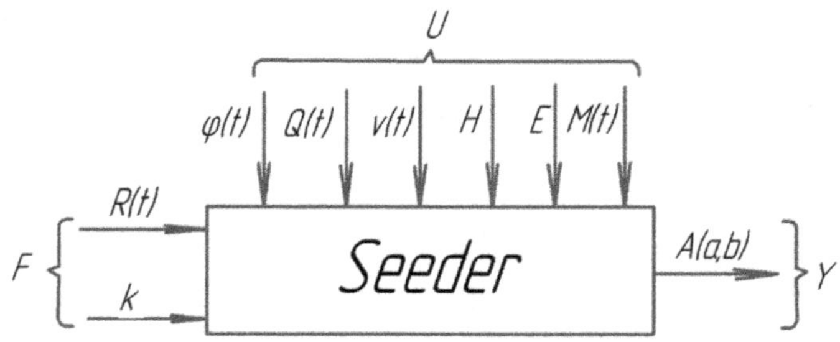

The control vector *U* includes the control of the movement (tractor turns) ϕ (t), the change of seeding material weight *M (t),* speed change *v (t),* the state of adjustment of the working mechanisms *H,* flow energy *E,* etc. The output vector is a random field of seeds *A (a, b)* distributed in the sub-coulter space *A (a, b).*

In the element-by-element examination of the machine, the output vectors of the *Y* functions of some elements of the seeder are input *F* for others and they form a certain quality indicator if their cause-effect relationships are combined.

Thus, for a pneumatic seeder of an intrasoil broadcast seeding equipped with coulters with seed spreaders, we can proposea functional scheme. The scheme takes into account the following factors: the change in the amount of seeding material, its dosing, transportation to the coulters by an air stream, separation of the seed stream through the seed-drill tubes, the formation of flow density with the help of the guide, the interaction of the delivered seeds with the distributor at the outlet from the seed-drill tube and with the coulter elements, the interaction of the coulters with the soil, seed plowing (Figure 2).

The process of forming the seed distribution quality by the coulter is represented by the diagram in the figure 3.

The separated part of the seed material *q (t),* transported in the seed-drill tube by the air stream, falls on the distributor *1*, and interacting with its surface, receives a direction and a trajectory of motion. After throwing back from the inner elements of the coulter *2*, the soil falling from the outer elements of the coulter *3* and forms a random field of seed distribution along the grab width and their covering depth.

Figure 2. Functional scheme of a seeder: 1 – frame with wheels and seed boxes; 2 – dosing devices; 3 – devices transporting seeds to coulters; 4 – coulters; 5 – soil; B – source of energy of a moving air stream (ventilator)

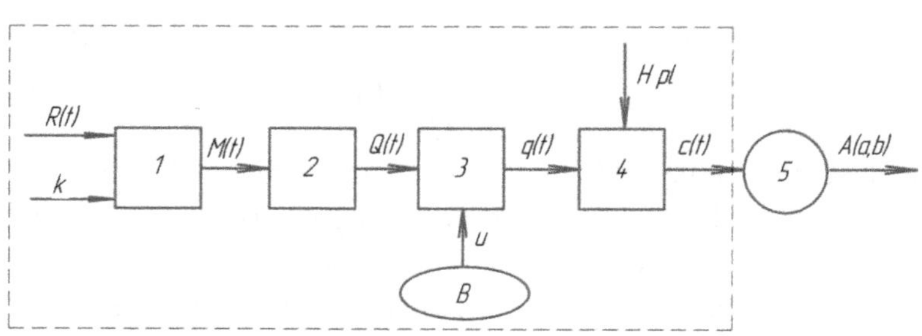

Figure 3. Calculation scheme of a coulter: 1 – seed distributor; 2 – inner elements of the coulter; 3 – outer elements of the coulter

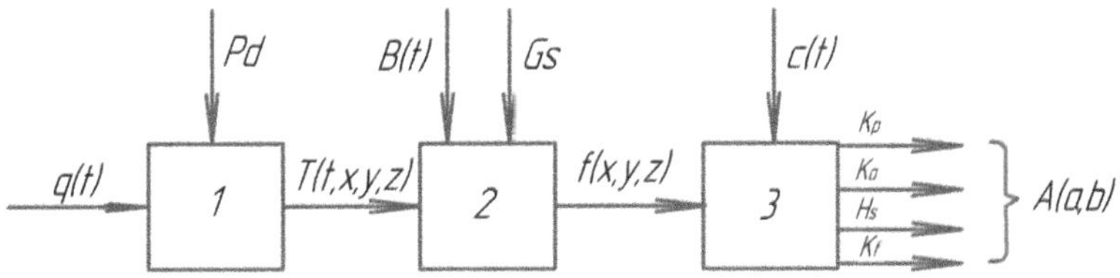

This scheme is close to the real conditions of the functioning of the working body, but due to its multidimensionality there are certain difficulties for the study. In order to simplify the scheme, it is necessary to isolate the most significant effects and internal relationships that form the output function indicators and reduce their number by consistent assumption.

For this study, we can assume a constant grain flow Q (t), as a violation in dosing and feeding to the coulters leads to a process disruption, which is unacceptable.

According to the requirements for the machines, the adjustment of their mechanisms must not be changed during the technological process.

Thus, we will think that the initial adjustments to the seeder mechanisms *H* are not changed, assuming that the seeding is done through the prepared background.

It can be assumed that the vibrations transmitted to the coulter from the side of the carrying system do not have any significant influence on seed distribution. Also, in the first assumption, we think that the soil falls from the elements of the coulter during the contact and it does not acquire velocities that can essentially move the seeds along the bottom of the furrow, or allow them to deviate from the site of their fall.

After the accepted assumptions, the process of seed sifting can be represented by the scheme in *Fig. 4*, where the vector of external effect *q (t)*, characterizing part of the seeding rate per one coulter, enters the system.

In the process of operation, the parameter *q (t)* under the influence of the distributor parameters P_d, as well as the static vector G_s. The vector determines the geometry of the sub-coulter space and undergoes a distribution along the width of the coulter's sifting chamber.

Figure 4. Scheme of seed sifting process

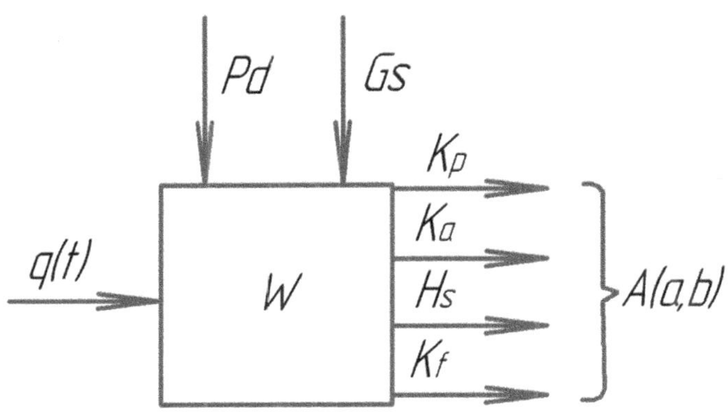

Then the set of sawn seeds can be represented by a random field *A (a, b),* i.e. by a seed distribution function taking into account the spatial orientation in the horizontal plane *(a, b).* The quality of seed distribution in the sub-coulter space can be estimated with the coefficients K_p, K_a, H_s, K_f according to a known technique (Lavrukhin, 2003).

Since the vector G_s determines the configuration of the sub-coulter space and its main parameters, its control effect appears when the seed particles are thrown back from the internal elements of the coulter.

Without affecting the seed distributor work, the vector G_s is a limiting factor. Its effect on seed distribution in the sub-coulter space should be analyzed after considering the movement and seed falling from the distributor. The influence of the geometry of the sub-coulter space on the seed distribution was thoroughly considered in the work of Lavrukhin (2000).

Finally, the seed sifting process can be represented by the scheme given in Figure 5.

The vector *q (t)* enters the system, which, under the effect of the vectors P_d, which give the direction and speed to seeds leaving the distributor, is subjected to the action of the static vector G_s.

Then, the seeds with direction and speed move along the trajectories of the forced motion. A part of them is thrown back from the inner elements of the coulter and after falling to the furrow bottom and final stop, it forms a random field *A (a, b).*

In this model, it is not possible to determine the effect of the seed flow direction in the sub-coulter space on the seed distribution. This issue should be considered separately.

Thus, in order to determine the qualitative parameters of seed sifting in the sub-coulter space, it is necessary to consider the transfer of kinetic energy from the airflow to moving seeds; to determine the effect of the distributor parameters, pneumatic transport regimes on the speed, range of the seed flight and their distribution along the width of the grab in the sub-coulter space.

Principle of Coulter Work for Intrasoil Broadcast Seeding of Grain Crops in Conditions of Pneumatic Transportation of Seed Material

To choose the criterion for evaluating both the seed-drill tube and the spreader of the seeding device for the intrasoil broadcast seeding of grain crops, the seeds should smoothly go through the seed-drill tube to the top of the distributor, receive the given direction of movement, and when leaving it, receive proper flight speed for their distribution over the entire width of the working element.

To ensure smooth entrance of seeds on the distributor surface, the angle of falling should not exceed the angle of seed friction against the distributor surface. Otherwise, the seeds from the distributor surface

Figure 5. Calculation scheme of a working body of broadcast seeding: 1 – pneumatic transporting system; 2 – seed distributor; 3 – coulter housing

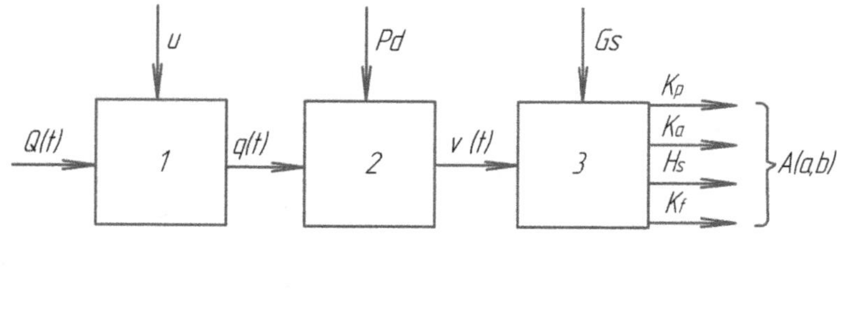

will bounce, their trajectories will be unpredictable, that will lead to a disruption of the uniformity of distribution in the sub-coulter space.

The technological process of intrasoil broadcast seeding with the use of residual kinetic energy during their pneumatic transportation was considered using a coulter for that method of seeding (Figure 6) according to AS No. 89324(Lavrukhin & Ivanov, 2009). The coulter contains a goose foot share *1* which is fastened to a pipe/tube/seed-drill tube *2*. The tube consists of vertical and inclined rectilinear sections, where the angle α of the inclination of the seed-drill tube *2* with the vertical axis does not exceed the angle of particles friction with respect to the material of the seed-drill tube. In its lower part a seed distributor *3* of a symmetrical configuration is installed, which is a polyhedral pyramid with curved sections in the lower part.

The operating principle of this coulter is as follows. When the seeder moves along the field, the goose foot share *1*, fixed on the seed-drill tube *2*, cuts the soil layer at a given depth and raises it. The sown material, transported by the air flow, enters the vertical, and then the inclined section of the seed-drill tube *2*, and flows smoothly to the seed distributor *3*.

The curvilinear part of the seed distributor *3* has a horizontal section at the outlet. On the opposite side of the seed-drill tube *3*, in the lower part, a seed-guide *4* is installed, which forms seed flow.

Along the way, the sowing material falls on the seed-guide *4*, which forms a flow and directs it to the seed distributor *3* and the material enters the surface of the seed distributor *3* at an angle less than the friction angle.

The seed distributor *3* is installed in such a way that the projections of the seed velocity vectors on the direction of movement coincide with this direction. Sliding on the seed distributor *3*, the sowing material changes the direction of movement and is evenly distributed over the sifting zones in the horizontal plane over the grip width of the goose foot share.

Figure 6. A coulter for broadcast seeding: a – general appearance of a coulter; b – distributor position in a seed-drill tube; c – seed distributor

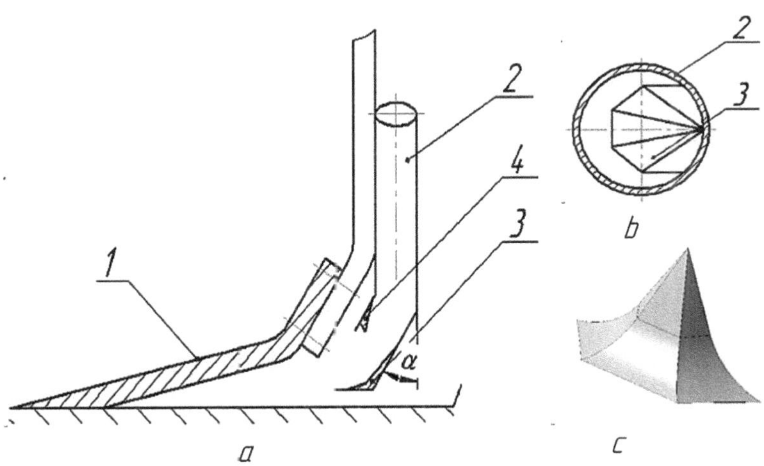

Seed Movement in the Air Flow in the Vertical And Inclined Rectilinear Segments of the Seed-Drill Tube

Pneumatic transportation can be realized under different regimes, determined by the flow parameters. The main of them are the speed of the carrier, the mass or volume ratio of the solid and gaseous components in the flow, called concentration, and the amount of the excess pressure.

Depending on the value of these parameters, there are two regimes of pneumatic transportation of bulk cargos: pneumatic transport in the discharged phase (or pneumatic transport with low concentrations) and pneumatic transport in the dense phase (or pneumatic transport with high concentrations).

The first regime occurs as a result of the aerodynamic effect of the air flow on solid particles. The seed transportation takes place under an air flow effect. Its magnitude is characterized by the force of the air flow or by aerodynamic force. In the general, the aerodynamic force can be expressed by the dependence (Zuyev, Levachev & Lotkov, 1987)

$$F_a = f \cdot F_1 \rho (u - v_c)^2 + \frac{kA\rho(u - v_c)^2}{2}, \tag{24}$$

where f is coefficient of air flow friction against a seed; F_1 is value of friction surface, m^2; u is speed of air flow, m/s; v_c is speed of a seed placed in the flow, m/s; k is coefficient of resistance when a seed is affected by an air jet (coefficient of aerodynamic resistance); A is front profile of a seed, m^2; ρ is air density, kg/m^3.

In equation (24), the first summand expresses the force acting on the seed from the direction of the air flow which is equal to the friction force of the air against the seed. The second one corresponds to the loss of the kinetic energy of the carrier, which is consumed to overcome the seed inertia.

The first component of the equation has a predominant value in comparison with the second one at low airflow speed (under the laminar regime). Pneumatic transportation takes place under turbulent airflow, when the Reynolds number is $Re=10^4 - 10^6$(Zuyev, Levachev & Lotkov, 1987). In this case, the second component of the equation is predominant. If the value of the first component of the equation is neglected, we'll obtain:

$$F_a = \frac{kA\rho(u - v_c)^2}{2}, \tag{25}$$

where $k = \frac{k_n 2m}{A\rho}$,where $k_n = \frac{g}{c^2}$ is soaring ratio, 1/m; c is speed of seed winding or critical velocity, m/s; $A = \frac{\pi \cdot d^2}{4}$; $\rho = 0,464 \frac{B}{273 + \mathrm{t}}$, where B is barometer pressure, mm Hg; t is air temperature, C.

If the expressions are put into equation (25), we'll obtain

$$F_a = k_n m (u - v_c)^2. \tag{26}$$

The dependence (26) is valid when air flow velocities are of 20 m/s and more. In this case, the process of pneumatic transportation occurs at lower speeds, so the aerodynamic force is proportional to the first degree of airflow velocity (Lachuga & Ksendzov, 2000), i.e.

$$F_a = k_a m(u - v_c),$$

where k_a is air dynamic coefficient, s^{-1}; u is air flow speed (velocity), m/s; v_c is speed of seed movement in the air flow, m/s.

During pneumatic transportation, seeds pass through the sections of the seed-drill tube and the distributor, on each of which there is a change of their speed and direction of movement.

To obtain reliable information about the kinematic parameters of the seeds, it is necessary to take into account their movement in the vertical and inclined sections of the seed-drill tube, and along the inclined rectilinear and curvilinear sections of the distributor; and to determine the flight range and the trajectory of its movement in the sub-coulter space when they leave distributor surface.

In the process of approximate analysis of the particle motion in the air flow, we accept the following certain assumptions:

- the air flow is stationary in time and along the section of the seed-drill tube, i.e. its average velocity is constant;
- soaring ratio kn = const, as for the orientation of particles in the vertical air flow, their largest cross sections are settled perpendicularly to the axis of the flow;
- only the motion of an isolated particle is considered;
- particle rotation is not taken into account.

Let us consider the problem of choosing a profile curve that satisfies the formulated criterion. Any mobile coordinate system rigidly connected with the seed-drill tube can be considered inert with a high degree of accuracy. The seed movement with respect to it is described in the same way as its motion with respect to the fixed frame of reference (Nogtikov & Sazonov 1995).

The linear seed sizes are negligibly small in comparison with the distances gone by them in the seed-drill tube and the distributor. Thus, the seed movement can be described with accuracy sufficient for practice as the motion of a material point along the intersection line of the internal surface of the seed-drill tube (distributor) with its plane of symmetry.

Consider the movement of the seed in the air flow in the vertical section.

Gravity and aerodynamic forces are applied to the seed. Thus, the speed of seed movement in the seed pipe can be determined from the equation compiled on the basis of these forces acting on the seed (Figure 7)

$$m\frac{d^2x}{dt^2} = F_a + P,$$

that is

Figure 7. Scheme of forces acting on the seed while moving in a straight vertical section of theseed pipe

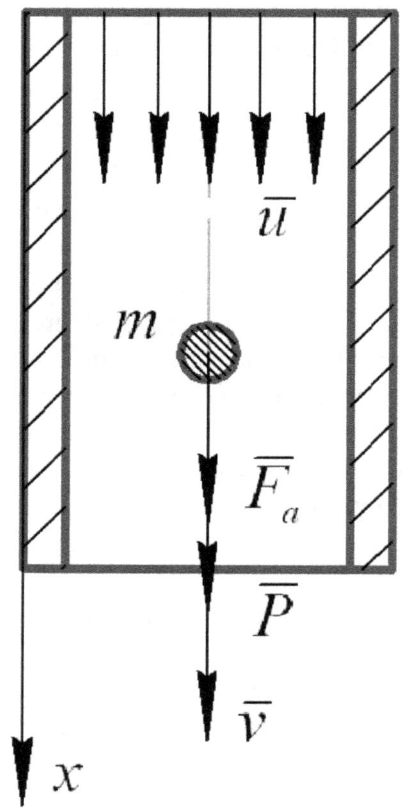

$$\frac{d^2x}{dt^2} = k_n(u - v_c) + g\,.$$

Because $\frac{d^2x}{dt^2} = \frac{dv_c}{dt}$, in that case

$$\frac{dv_c}{dt} = k_a(u - v_c) + g\,.$$

Divide the variables and integrate the resulting differential equation

$$\int \frac{dv_c}{k_a u + g - k_a v_c} = \int dt\,.$$

Then

$$\ln\left|u+\frac{g}{k_a}-v_c\right|=-k_a t+c_1.$$

At $t=0$, $v_c=v_0$ $c_1=\ln\left|u+\frac{g}{k_a}-v_0\right|$.

Accordingly,

$$\ln\left|u+\frac{g}{k_a}-v_c\right|=-k_a t+\ln\left|u+\frac{g}{k_a}-v_0\right|.$$

The speed flight of the seed

$$v_c=u+\frac{g}{k_a}-5^{-k_a t+\ln\left|u+\frac{g}{k_a}-v_0\right|}. \tag{27}$$

Gravity, friction and aerodynamic forces are applied to the seed. The speed of the seed in the seed pipe can be determined from the equations compiled on the basis of the forces acting on the seed (Figure 8).

$$m\frac{d^2x}{dt^2}=F_a-F_{mp}+P\cdot\cos\alpha$$
$$m\frac{d^2y}{dt^2}=N-P\cdot\sin\alpha$$

Because $a_y=\frac{d^2y}{dt^2}=0$, in that case $N-P\cdot\sin\alpha=0$, that is $N=P\cdot\sin\alpha$.

Then

$$m\frac{d^2x}{dt^2}=F_a+P\cdot\cos\alpha-f\cdot P\cdot\sin\alpha,$$

$$\frac{dv_c}{dt}=k_a(u+\frac{g}{k_a}(\cos\alpha-f\cdot\sin\alpha)-v_c).$$

Further, similar to the previous calculations.
Consequently, the speed of seed flight

$$v_c=u+\frac{g}{k_a}(\cos\alpha-f\sin\alpha)-e^{-k_a t+\ln\left|u+\frac{g}{k_a}(\cos\alpha-f\sin\alpha)-v_0\right|}. \tag{28}$$

Figure 8. Diagram of the forces acting on the seed when moving in the air flow in an inclined straight section of the seed pipe

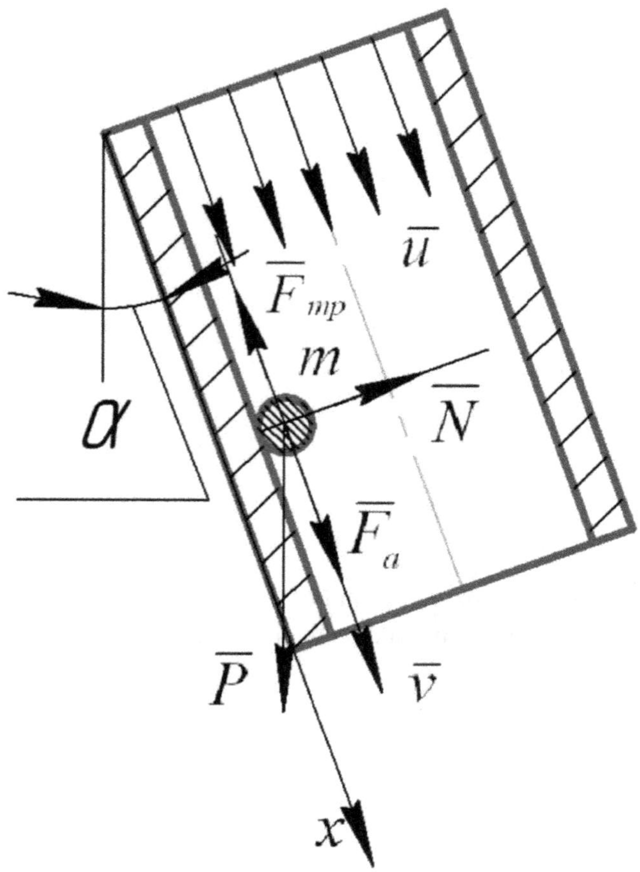

From the dependence (28), it can be seen that the seed movement proportionally increases its speed over time with an increase in the air flow speed and the k_a coefficient, and with a decrease in the friction coefficient.

The Movement of Seeds in the Air Flow Along the Inclined Straight and Curved Sections of the Distributor

The seed mass *m* moves along a straight section of the distributor, which is located at an angle δ=90-(α+β) to the horizontal with an initial velocity v_0, then along the arc BC circle radius R with a Central angle φ, (Figure 9).

Plot AB touches the arc of the circle at point B. It is necessary to determine the speed of the seed at point C when it descends from the distributor.

To study the motion of the seed in the arc BC of circle it is necessary to know its initial velocity v_B at point B. This speed is finite when moving along the rectilinear section of AB, so we first consider the movement of the seed in this section and determine the speed v_B at point *B*.

Figure 9. Scheme of inclined rectilinear and curvilinear distributor sections

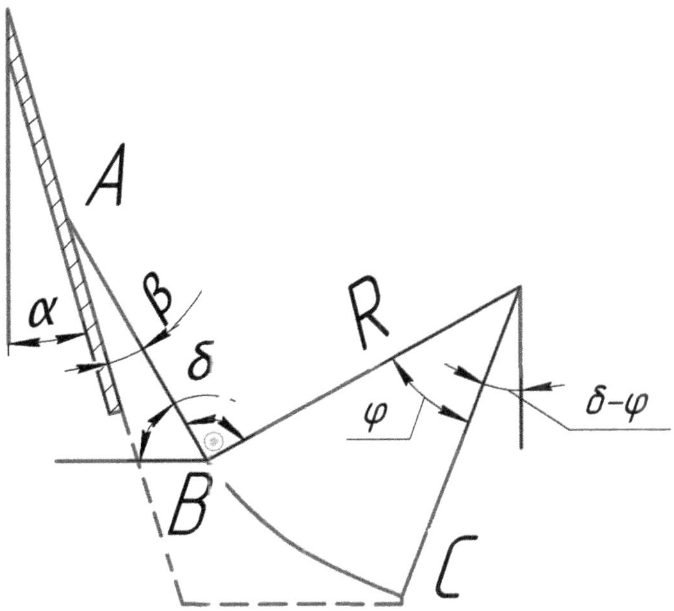

To the seed forces of gravity $P = mg$, friction $F_{mp} = fN$ and aerodynamic $F_a = k_0 m(u - v_c)$ are applied. Let us direct the *l* axis along the AB line and the *s* axis perpendicular to it (Figure 10).

The differential equations of motion in the projections on the l and s axes have the form:

$$m\frac{d^2l}{dt^2} = F_a - F_{mp} + P \cdot \sin\delta$$

$$m\frac{d^2s}{dt^2} = N - P \cdot \cos\delta .$$

Because $\frac{d^2s}{dt^2} = 0$, in that case $N - P \cdot \cos\delta = 0$, therefore $N = P \cdot \cos\delta$.

Then

$$m\frac{d^2l}{dt^2} = mk_0(u - v_c) + mg \cdot \sin\delta - fmg \cdot \cos\delta .$$

Divide by *m*, get

$$\frac{d^2l}{dt^2} = k_0(u - v_c) + g \cdot \sin\delta - f \cdot g \cdot \cos\delta .$$

Figure 10. Diagram of the forces acting on the seed when moving in the air flow along an inclined straight section of the distributor

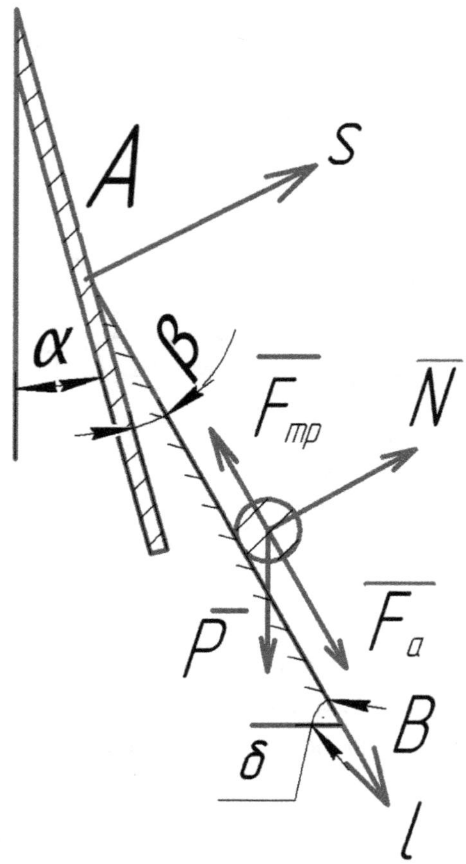

Since $\frac{d^2 l}{dt^2} = \frac{dv_A}{dt}$, in that case

$$\frac{dv_A}{dt} = k_0 (u - v_c + \frac{g}{k_0} (\sin\delta - f\cos\delta)).$$

Divide the variables and integrate the resulting differential equation

$$\int \frac{dv_A}{u - v_c + \frac{g}{k_0}(\sin\delta - f\cos\delta)} = k_0 \int dt$$

$$\ln\left| u + \frac{g}{k_0}(\sin\delta - f\cdot\cos\delta) - v_A \right| = -k_0 t + c_3 .$$

At $t=0$, $v_c=v_0 c_3 = \ln\left|u + \frac{g}{k_0}(\sin\delta - f\cdot\cos\delta) - v_0\right|$.

Then

$$\ln\left|u + \frac{g}{k_0}(\sin\delta - f\cdot\cos\delta) - v_A\right| = -k_0 t + \ln\left|u + \frac{g}{k_0}(\sin\delta - f\cdot\cos\delta) - v_0\right|$$

The speed of the seed flight

$$v_c = u + \frac{g}{k_a}(\sin\delta - f\cos\delta) - e^{-k_a t + \ln\left|u + \frac{g}{k_a}(\sin\delta - f\cos\delta) - v_0\right|} \qquad (29)$$

To study the motion of the seed in the arc BC (Figure 11) use the differential equations of motion in the projections on the axis of the natural coordinate system.

According to Newton's second law:

$$m\vec{a} = \sum \vec{F}_k .$$

Project on the axis τ and n:

$$ma_\tau = \sum F_k^\tau$$

Figure 11. Diagram of the forces acting on the seed in motion in air flow through the curved section of the distributor

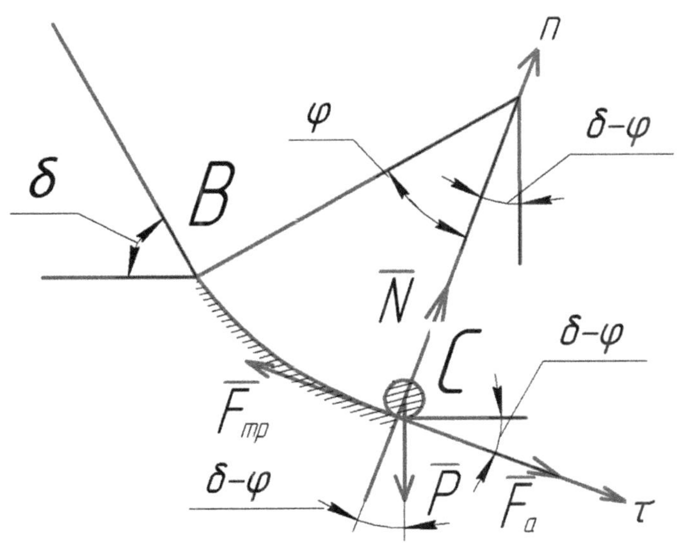

$$ma_n = \sum F_k^n,$$

where $a_\tau = \dfrac{dv_A}{dt}$;

$$a_n = \frac{v_c^2}{R}.$$

Substitute the acceleration values in the equations and project the forces on the coordinate axis

$$m\frac{dv_A}{dt} = F_a - F_{mp} + P \cdot \sin(\delta - \phi) \tag{30}$$

$$m\frac{v_A^2}{R} = N - P \cdot \cos(\delta - \phi). \tag{31}$$

From (31) $N = m(\dfrac{v_A^2}{R} + g \cdot \cos(\delta - \phi))$.
Substitute in (30), we get:

$$m\frac{dv_A}{dt} = m \cdot k_0(u - v_A) + mg \cdot \sin(\delta - \phi) - fm(\frac{v_A^2}{R} + g \cdot \cos(\delta - \phi)). \tag{32}$$

Divide by *m*, get

$$\frac{dv_A}{dt} = k_0(u - v_A) + g \cdot \sin(\delta - \phi) - f\frac{v_A^2}{R} - fg\cos(\delta - \phi)$$

$$\frac{dv_A}{dt} = -k_0 v_A - f\frac{v_A^2}{R} + k_0 u + g(\sin(\delta - \phi) - f \cdot \cos(\delta - \phi)).$$

Denote $k_0 u + g(\sin(\delta - \phi) - f\cos(\delta - \phi)) = a$, $-k_0 = b$, $-\dfrac{f}{R} = A$,

then:

$$\frac{dv_A}{dt} = a + bv_A + cv_A^2.$$

Divide the variables and integrate

$$\int \frac{dv_A}{a + bv_A + cv_A^2} = \int dt$$

$$\frac{1}{\sqrt{b^2 - 4ac}} \ln \frac{2cv_A + b - \sqrt{b^2 - 4ac}}{2cv_A + b + \sqrt{b^2 - 4ac}} = t + c_4, (b^2 \succ 4ac).$$

At $t=0$, $v_c = v_0$

$$c_4 = \frac{1}{\sqrt{b^2 - 4ac}} \ln \frac{2cv_0 + b - \sqrt{b^2 - 4ac}}{2cv_0 + b + \sqrt{b^2 - 4ac}}$$

Then

$$\frac{1}{\sqrt{b^2 - 4ac}} \ln \frac{2cv_A + b - \sqrt{b^2 - 4ac}}{2cv_A + b + \sqrt{b^2 - 4ac}} = t + \frac{1}{\sqrt{b^2 - 4ac}} \ln \frac{2cv_0 + b - \sqrt{b^2 - 4ac}}{2cv_0 + b + \sqrt{b^2 - 4ac}}$$

$$\ln \frac{2cv_A + b - \sqrt{b^2 - 4ac}}{2cv_A + b + \sqrt{b^2 - 4ac}} = \sqrt{b^2 - 4ac} \cdot t + \ln \frac{2cv_0 + b - \sqrt{b^2 - 4ac}}{2cv_0 + b + \sqrt{b^2 - 4ac}}$$

$$\frac{2cv_A + b - \sqrt{b^2 - 4ac}}{2cv_A + b + \sqrt{b^2 - 4ac}} = e^{\sqrt{b^2-4ac}\cdot t + \ln\frac{2cv_0 + t - \sqrt{b^2-4ac}}{2cv_0 + t + \sqrt{b^2-4ac}}}$$

$$2cv_A + b - \sqrt{b^2 - 4ac} = 2cv_A \cdot e^{\sqrt{b^2-4ac}\cdot t + \ln\frac{2cv_0+b-\sqrt{t^2-4ac}}{2cv_0+b+\sqrt{k^2-4ac}}} + b \cdot e^{\sqrt{b^2-4ac}\cdot t + \ln\frac{2cv_0+b-\sqrt{b^2-4ac}}{2cv_0+b+\sqrt{b^2-4ac}}}$$
$$+ \sqrt{b^2 - 4ac} \cdot e^{\sqrt{b^2-4ac}\cdot t + \ln\frac{2cv_0+b-\sqrt{b^2-4ac}}{2cv_0+b+\sqrt{b^2-4ac}}}$$

$$2cv_A - 2cv_A \cdot e^{\sqrt{b^2-4ac}\cdot t + \ln\frac{2cv_0+b-\sqrt{b^2-4ac}}{2cv_0+b+\sqrt{b^2-4ac}}} = b \cdot e^{\sqrt{b^2-4ac}\, t + \ln\frac{2cv_0+b-\sqrt{b^2-4ac}}{2cv_0+b+\sqrt{b^2-4ac}}} - b$$
$$+ \sqrt{b^2 - 4ac} \cdot e^{\sqrt{b^2-4ac}\cdot t + \ln\frac{2cv_0+b-\sqrt{b^2-4ac}}{2cv_0+b+\sqrt{b^2-4ac}}} + \sqrt{b^2 - 4ac}$$

$$v_A(2c - 2c \cdot e^{\sqrt{b^2-4ac}\cdot t+\ln\frac{2cv_0+b-\sqrt{b^2-4ac}}{2cv_0+b+\sqrt{b^2-4ac}}}) = b \cdot e^{\sqrt{b^2-4ac}\cdot t+\ln\frac{2cv_0+b-\sqrt{b^2-4ac}}{2cv_0+b+\sqrt{b^2-4ac}}} - b$$

$$+\sqrt{b^2-4ac} \cdot e^{\sqrt{b^2-4ac}\cdot t+\ln\frac{2cv_0+b-\sqrt{b^2-4ac}}{2cv_0+b+\sqrt{b^2-4ac}}} + \sqrt{b^2-4ac}$$

The seed flight speed

$$v = \frac{b \cdot e^{\sqrt{b^2-4ac}\cdot t+\ln\frac{2cv_0+b-\sqrt{b^2-4ac}}{2cv_0+b+\sqrt{b^2-4ac}}} - b + \sqrt{b^2-4ac} \cdot e^{\sqrt{b^2-4ac}\cdot t+\ln\frac{2cv_0+b-\sqrt{b^2-4ac}}{2cv_0+b+\sqrt{b^2-4ac}}} + \sqrt{b^2-4ac}}{2c - 2c \cdot e^{\sqrt{b^2-4ac}\cdot t+\ln\frac{2cv_0+b-\sqrt{b^2-4ac}}{2cv_0+b+\sqrt{b^2-4ac}}}}. \quad (33)$$

Taking into account that the seed descends from the distributor at an angle of0 degrees, that is, horizontally, that $\phi = \delta$,

Then

$$m\frac{dv_A}{dt} = F_a - F_{mp} \quad (34)$$

$$m\frac{v_A^2}{R} = N - P \quad (35)$$

From (35) $N = m\left(\frac{v_A^2}{R} + g\right)$

Substitute in (34), we get:

$$m\frac{dv_A}{dt} = m \cdot k_0(u - v_A) - fm\left(\frac{v_A^2}{R} + g\right) \quad (36)$$

Divide by *m*, get

$$\frac{dv_A}{dt} = k_0(u - v_A) - f\frac{v_A^2}{R} - fg$$

$$\frac{dv_A}{dt} = -k_0v_A - f\frac{v_A^2}{R} + k_0u - fg.$$

Denote $k_0 u - fg = a$, $-k_0 = b$, $-\dfrac{f}{R} = A$

Then

$$\frac{dv_A}{dt} = a + bv_A + cv_A^2$$

Divide the variables and integrate

$$\int \frac{dv_A}{a + bv_A + cv_A^2} = \int dt$$

$$\frac{2}{\sqrt{4ac - b^2}} arctg \frac{2cv_A + b}{\sqrt{4ac - b^2}} = t + c_5, (b^2 \prec 4ac)$$

At $t=0$, $v_c=v_0$

$$c_5 = \frac{2}{\sqrt{4ac - b^2}} arctg \frac{2cv_0 + b}{\sqrt{4ac - b^2}}$$

Then

$$\frac{2}{\sqrt{4ac - b^2}} arctg \frac{2cv_A + b}{\sqrt{4ac - b^2}} = t + \frac{2}{\sqrt{4ac - b^2}} arctg \frac{2cv_0 + b}{\sqrt{4ac - b^2}}$$

$$arctg \frac{2cv_A + b}{\sqrt{4ac - b^2}} = \frac{\sqrt{4ac - b^2}}{2} \left(t + \frac{2}{\sqrt{4ac - b^2}} arctg \frac{2cv_0 + b}{\sqrt{4ac - b^2}} \right)$$

$$arctg \frac{2cv_A + b}{\sqrt{4ac - b^2}} = \frac{\sqrt{4ac - b^2}}{2} \cdot t + arctg \frac{2cv_0 + b}{\sqrt{4ac - b^2}}$$

$$2cv_A + b = \sqrt{4ac - b^2} \cdot tg \left(\frac{\sqrt{4ac - b^2}}{2} \cdot t + arctg \frac{2cv_0 + b}{\sqrt{4ac - b^2}} \right)$$

The seed flight speed

$$v_A = \frac{-b + \sqrt{4ac - b^2} \cdot tg\left(\frac{\sqrt{4ac-b^2}}{2} \cdot t + arctg\frac{2cv_0 + b}{\sqrt{4ac - b^2}}\right)}{2c} \tag{37}$$

If we substitute the values of the given parameters in the equation (37), we'll show the effect of the air flow velocity, the friction coefficient and the k_a coefficient on the seed speed along the curved section of the distributor.

Figure 12shows, that the speed of the seed moving along the surface of the distributor decreases over time. With the increase in the air flow rate, the seed speed at the exit from the distributor increases. We show the dependence of the seed velocity at the exit from the distributor on the air flow velocity, taking into account the dependence graph shown in figure 12 and the dependence (37).

Trajectory and Flight Range of Seeds

To determine the trajectory and the dependence of the flight range of the seed on the air flow rate, it is necessary to consider the seed speed when throwing back from the distributor. As noted above, the proposed distributor is a multi-faceted pyramid in which the lower part of each face is an arch of a circle with a given radius (figure 14).

The tangent line to the arch at the point of seed exit from the distributor and the horizontal line form the angle *α0* of seed exit from the distributor. By varying the angle α_0, it is possible to adjust the trajectory and range of the seed after separation from the dispenser. Thus, if the lower extreme point of the

Figure 12. Graph of the seed velocity dependence on the curvilinear section of the distributor from time to time at k_a=0.1 c^{-1}, f=0.3 and u=2.5 m/s; 3.5 m/s; 4.5 m/s; 5.5 m/c; 6.5 m/s

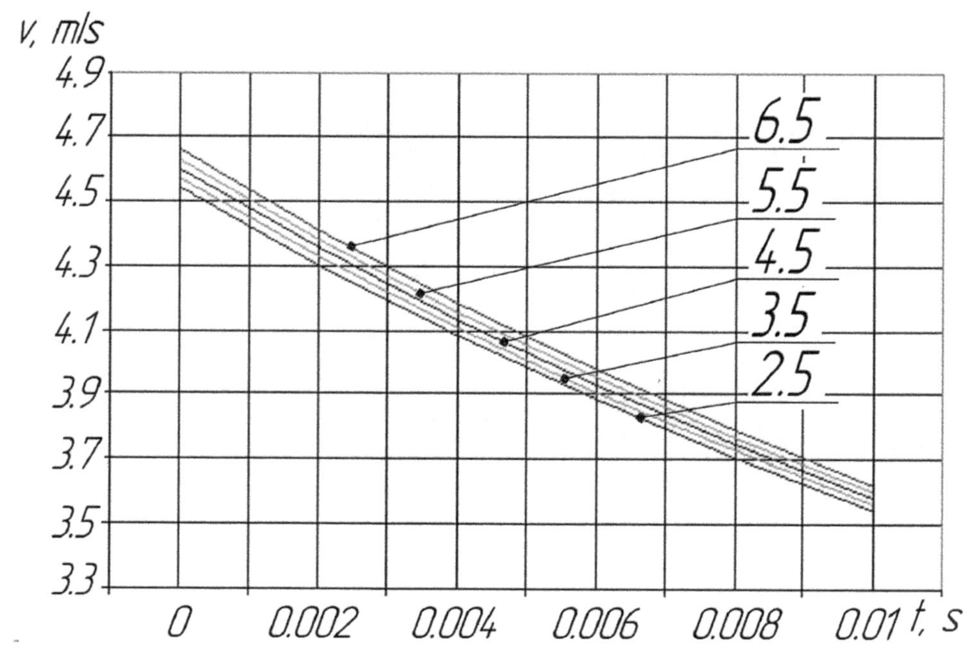

distributor arch is the point of contact with the horizontal line, that is, $\alpha_0=0$, the flight range of the seed is determined from the expression (Kovrikov & Skvortsov, 1977).

$$l = x_0 + \nu_c \sqrt{\frac{2h}{g}}, \tag{38}$$

where x_0 is the radius of the base of the distributor, m; h is height of the distributor installation, m.

From the formula (38) it can be seen that it's possible to increase the flight range in the conditions of the sub-coulter space by increasing the initial speed of the seed flight and the height h of the distributor lifting over the bottom of the furrow.However, the range of variation of these quantities is limited by the seeders' design and their operating characteristics.Thus, the height h is in the range of 10-20 mm and it is limited by the height of the inside space of coulter.The increase of the inside space of coulter height leads to an increase in the angle of setting the lateral plane of the coulter to the bottom of the furrow, as well as its width$B$, which deprives the coulter of functionality on medium and heavy soils.

When using the energy of free fall of seeds, the initial speed is practically not regulated, so it is necessary to choose the parameters of the coulter and the distributor, achieving the necessary values of the anglea_0 and the width of the seed distribution (at a given speed v_c).

Taking into account the influence of the Coulter on sieving (taking into account that $\alpha 0$ is not zero), the flight range of seeds can be determined from the expression (Kovrikov, 1976;Kovrikov&Skvortsov, 1977).

$$l = \frac{v_A^2 \sin 2\alpha_0}{2g} + \frac{v_A}{g}\sqrt{v_A^2 \sin^2 \alpha_0 + 2gh} + x_0 . \tag{39}$$

Taking into account that in the considered variant the seed descends from the distributor at an angle $\alpha_0=0$, we take the dependence for further studies (38). For the analysis of the formula (38), a graphical dependence of the seed flight range on the seed speed at the exit from the surface of the distributor was built.

Figure 13. Graph of the dependence of the seed velocity at the exit from the curved section of the distributor from the air flow velocity at $k_a=0.1\ s^{-1}$ and $f=0.3$

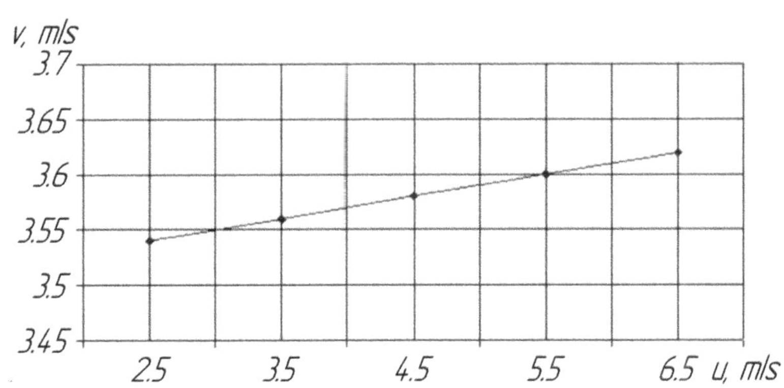

Figure 14. Flight range of the seed

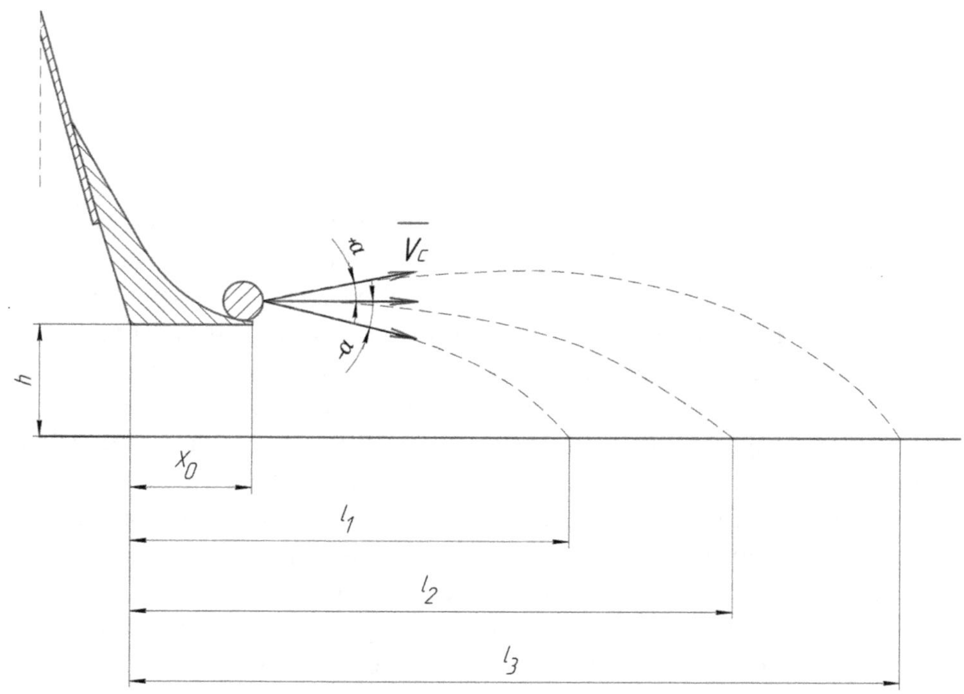

Figure 15 shows that the dependence has lineal character, and when the seed speed increases descending from the distributor the flight range of the seed also increases.

Taking into account the graphical dependence of the seed velocity on the air flow rate (figure 13) and the dependence of the flight range of the seed on its velocity at the exit of the distributor (formula 38), we obtain a dependence of the flight range of the seed on the air flow rate.

Figure 16 presents the graph of the dependence obtained by calculation on indirect indicators. Since it was not possible to obtain this dependence analytically, it was obtained by a graphic and analytical method.

$$l_m = 0,209 + 9 \cdot 10^{-4} u \,, \tag{40}$$

where l_m is the flight range of the seed, determined theoretically, *m*; *u* is air velocity in seed pipe, *m/s*.

This dependence shows that the increase in the air flow rate results in the increase of the flight range of the seed. The equations of seed motion after the descent from the distributor will have the following form, taking into account the initial conditions:

At $t=0,\ v_x=v_c;\ x=x_0$

$$x = v_c t + x_0 \,. \tag{41}$$

At $t=0,\ v_y=0;\ y=h$

Figure 15. Graph of the dependence of the flight range of the seed from the seed speed at the exit from the distributor h=0.005; 0.01; 0.015 m

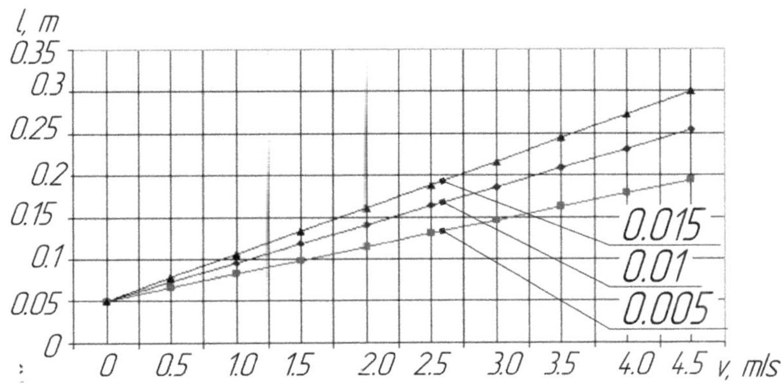

Figure 16. Graph of the dependence of the flight range of the seed from the air flow velocity at h=0.01 m

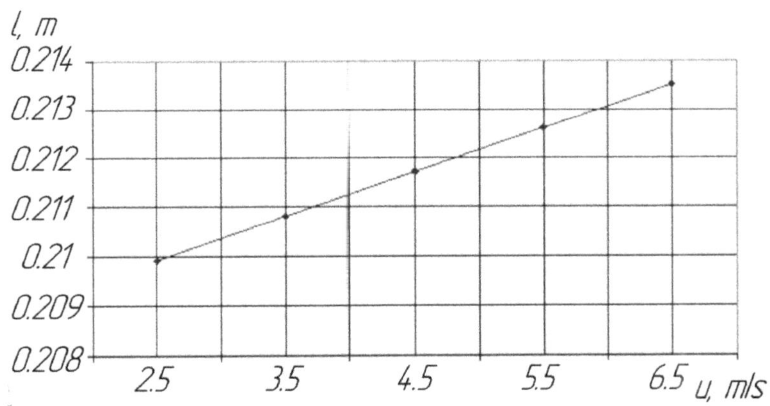

$$y = -g\frac{t^2}{2} + h. \tag{42}$$

Then, the flight trajectory of the seed

$$y = h - \frac{g}{2v_c^2}\left(x_0 - x\right)^2, \tag{43}$$

where x_0 is radius of the base of the dispenser seeds, m; v_c is initial flight speed of the seed at its descent from the distributor, m/s; h is height of the distributor installation, m.

If we substitute the values of all the parameters in the formula 43, we'll obtain the trajectory of the seed in the sub-coulter space at different speeds of the seed at the exit from the distributor and different values of the height of the distributor.

Figure 17. The trajectory and flight distance of a seed in sub-coulter space the seed at a speed v_c=3.6 m/s and the installation height distributor h=0.005 m, 0.01 m and 0.015 m

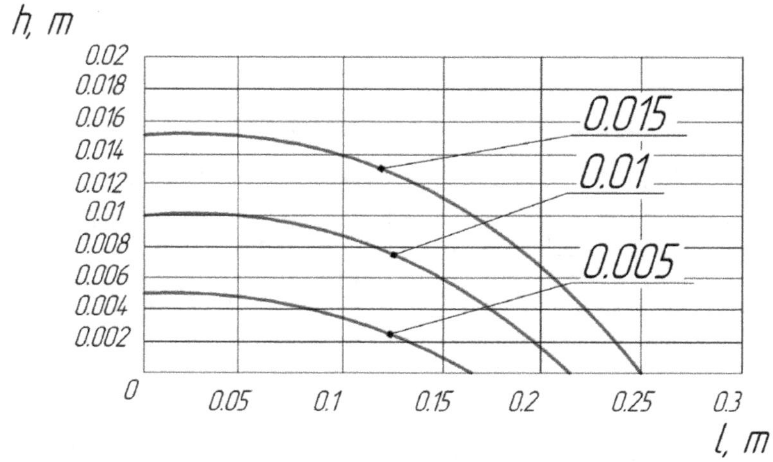

The Distribution of Seeds Along the Width of Coulter

The main indicator of the job quality seeder for intra-soil seed distribution sowing is the uniformity of the distribution of seeds over the entire width of the working body. To describe the process of particle distribution in the inside space of coulter we will use the theoretical positions developed by the professor V. A. Chernovolov (Skolzayev & Chernovolov, 1969).

In the process of seeder movement on the field, the seeds, moving along the seed pipe and the distributor, descend from the surface of the distributor at a speed and sow a strip of width (Figure 18). The width of the band seeding and the distribution of seeds depend on the speed of seed descending from the distributor and the seeding angle.

During the job coulter when the seeder is stationary, the seed distribution has a form of a half-ring. The dose of application q_F on a fixed site dF is determined by the integration of the formula (44)

$$q_F = \int_0^{t_1} I dt, \tag{44}$$

where t is the time of seeds application on the site.

The intensity of seed sowing on the site dF is determined by the ratio

$$I = \frac{dQ_F}{dF} = \frac{1}{\rho} q_\alpha p(\rho),$$

where $dQ_F = q_\alpha d\alpha\, p(\rho) d\rho$ is number of seeds dropped on the site dF per unit time; $dF = \rho d\alpha d\rho$ is elementary site, selected from sown area.

Figure 18. Scheme of seed distribution in the width of the coulter

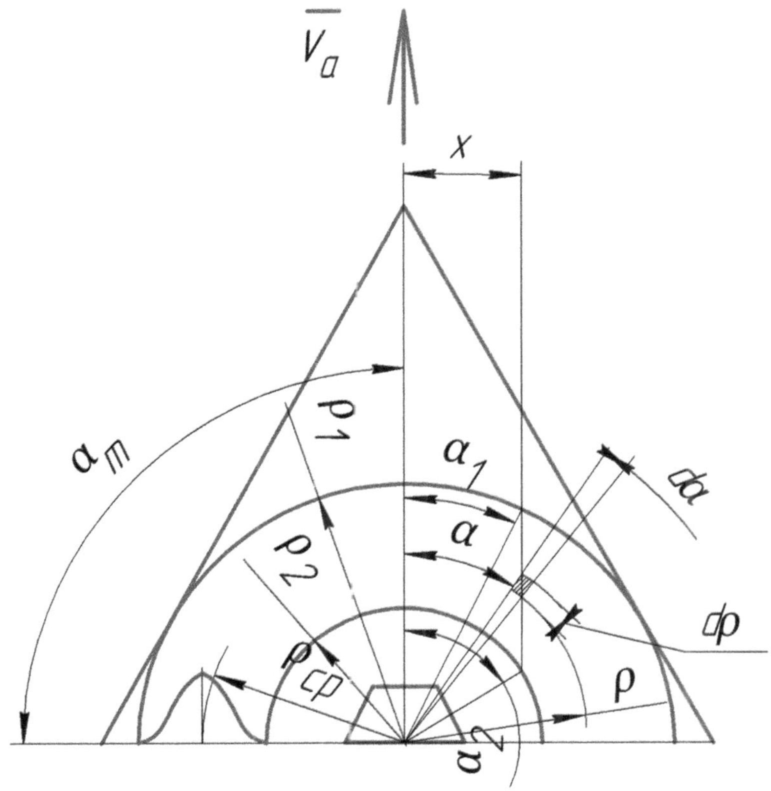

The dose of the fallen seeds at a distance *X* from the axis of symmetry of the distribution zone in the interval $0\angle\left|X\right|\angle\left|\rho_2 \sin\alpha_m\right|$ equal

$$q_F \approx \frac{q_a(\alpha_1 - \alpha_2)}{V_m\, \rho_{cp}(\sin\alpha_1 - \sin\alpha_2)},$$

and at $\left|\rho_2 \sin\alpha_m\right|\angle\left|X\right|\angle\left|\rho_1 \sin\alpha_m\right|$,

$$q_F \approx \frac{2q_a(\alpha_1 - \alpha_2)}{V_m(\rho_1 + a)(\sin\alpha_1 - \sin\alpha_2)}\int_a^{\rho_1} p(\rho)\,d\rho$$

where $q_\alpha = \dfrac{Q_\alpha}{2\alpha_m} = const$ is performance of the sowing unit, kg/sec×rad.

To reduce the error, the sector of throwing should be divided into several parts. The calculations of the dose q_F is done separately for each part of the sector, with equal intervals *X*. To obtain a total dose

q_F, all its particular values corresponding to a certain *X* are summed up. Thus, the accuracy of the result is quite high.

To build a graph $q_F = f(X)$, the method of numerical integration with the parabola formula can be used, but this process is quite time-consuming.

Having made the transition to the rectangular coordinates *X, Y* in the formula 4 and the replacement of the variable *t* by *Y*, we'll obtain

$$q_F = \frac{Q}{v_m} \int_0^{Y_k} \frac{1}{\sqrt{X^2 + Y^2}} f\left(arctg \frac{X}{Y}\right) f\left(\sqrt{X^2 + Y^2}\right) dY \, , \tag{45}$$

where Y_k is length of the longitudinal section line of the sowing zone at a given value of X.

The integral in the equation (45) for the distribution functions f(α(x, Y)) and f(ρ(x, Y)), which have practical distribution, is not expressed through elementary functions, but it can be calculated using a computer software in the MathCAD system. The size and position of the sieving zone at the known rules of distribution of the throwing angles and throwing ranges can be determined by the numerical characteristics of these random values. In this case, the seed distribution in the radius of the seeding zones occurs according to ordinary rules.

The rule of ordinary distribution is determined by two parameters: the mathematical expectation and the average square deviation of a random value. The mathematical expectation of throwing distances previously is previously selected by the obtaining a given width of the sieving. It is known (Skolzayev & Chernovolov, 1969) that the average square deviation of the throwing distances depends on the throwing distance and can be determined by a simple ratio:

$$\sigma_\rho = (0,25...0,3)\, M_\rho \, . \tag{46}$$

Numerical characteristics of the throwing angle is recommended to determine by the mathematical modeling. During the movement of the working body on the field, there is the symmetrical sowing area from the motion line, and $M_\alpha=0$ should be taken. The average square deviation of the sifting angle is recommended in the range of 0.8 ... 0.9 radians. Also, the method of mathematical modeling should be allowed to check the efficiency of these recommendations and the fulfillment of the requirements to the width and uniformity of seed sieving.

It is recommended to build graphs of the seed distribution on the width of the working body with the help of the program 'dose' (Chernovolov, 2007) performed in the MathCAD system, with different numerical characteristics of the throwing angle and seed flight range.

Analysis of the seed distribution graphs allows us to conclude that if we increase σ_a from 0.4 to 0.8 rad, the uniformity of the seed distribution across the width of the seeding band also increases. Therefore, the most preferable is the distribution at $\sigma_a=0.8$ radian.

The figures show that a uniform distribution over the entire width of the coulter is very difficult to achieve and therefore to obtain a uniform seed distribution over the area of the field. Thus it is necessary to make an overlap between the passages of the neighboring coulters. To estimate the irregularity of the sowing at different values of the working width, and, at different overlaps, you can use the dependence (47)

Figure 19. Distribution of seed norms in the sub-coulter space at ρ=0.25 m and values σ_a equal to 0.8; 0.67; 0.53 and 0.4 rad

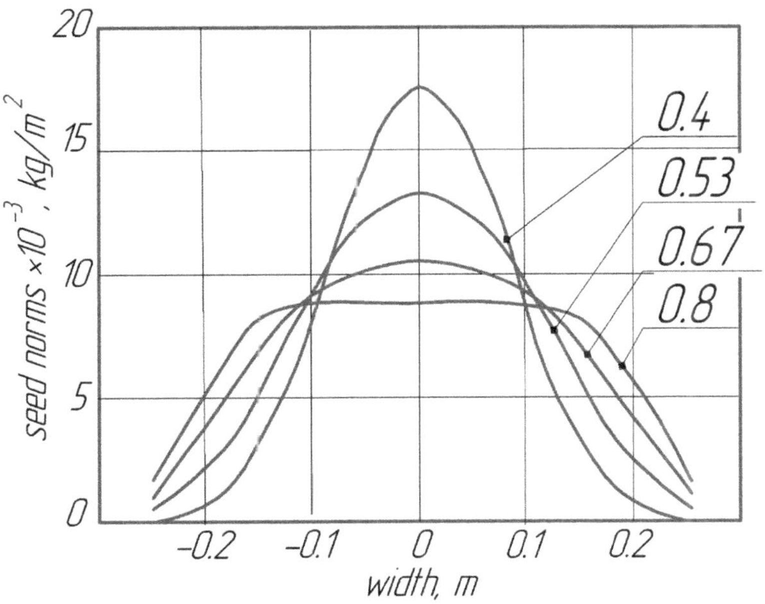

$$H_l = \left(\sum_1^k \frac{q_{Fi}}{\sum q_{Fi}} - \frac{k}{n} \right) \cdot 200 \qquad (47)$$

however, it is recommended to perform the choice of the optimal overlap of adjacent passages and to check the conditions of obtaining the desired width and uniformity of the sowing according to the program made in the system MathCAD as represented in the book Chernovolov (2007).

If the overlap passages are increased, the irregularity estimated by the coefficient of dose variation reduces (Figure 20) to the minimum value, and then again increases.

The width of the sowing with minimal irregularity will be called optimal. The sowing width, corresponding to the point of the first intersection of the lines of irregularity and agrotechnical tolerance, will be called effective.

The analysis of the graphs shows that it is impossible to obtain the required uniformity of sifting (with irregularity of 20%) for the scattering angle$2a_m$*=90*, but for the angles of 150 and 180 degrees, a guaranteed distribution is achieved at any width of the working body. For the dispersion angle of 120 degrees, the required uniformity of sifting can be achieved only if the width of the coulters is of 0.4 and 0.5 meters, and it is impossible with its width of 0.3 meters.

Figure 20. Graphs of uneven sowing when changing the overlap of adjacent passages and the permissible irregularity V_1=20% atρ=0.25 m

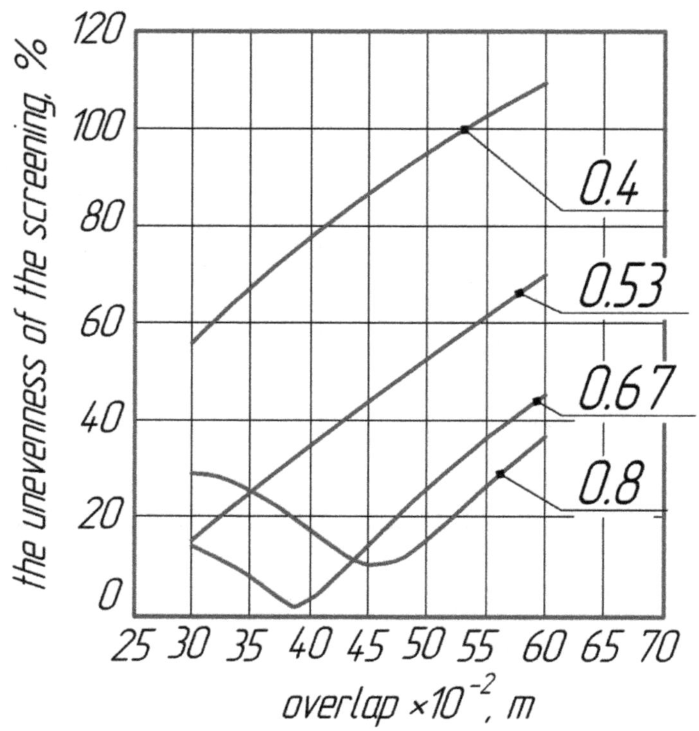

FURTHER RESEARCH DIRECTIONS

To obtain the required uniformity of the seed distribution, it is necessary to produce an overlap between the coulters. Satisfactory uniformity values for coulters with different widths are also obtained at angles of 150 and 180 degrees.

As a continuation of research for the subsequent practical application of this type of sowing, it is necessary to conduct an experiment to determine the effect of pneumatic transportation modes and distributor parameters on the uniformity of seed distribution over the field area.

We also consider it necessary to study the reaction of cultivated plants to changes in living conditions, as a result of their distribution over the field area, in order to optimize the parameters and seeding regimes.

CONCLUSION

1. The reserve to increase grain crops productivity is the use of intra-soil broadcast (non-row) method of seeding. This method allows to meet the agrotechnical requirements for the placement of seeds in the soil to the greatest extent.
2. The working process of seeding can be represented by a dynamical interaction of seeds moving in the air flow, and the elements of the passive distributor.

3. To select the criterion for assessing the profile line of the seed pipe and the distributor, it is necessary to take into account the condition of the unstressed entry of the seeds to the surface of the distributor, where the angle of incidence should not exceed the angle of friction of the seeds through the surface of the distributor.
4. When the seeds move in the air flow along the sections of the seed pipe and the distributor, there is a change in the seed speed and the direction of their movement. The value of the seed velocity with the throwing back from the surface of the dispenser, depending on the speed of the air flow may be determined according to the formula (37).
5. The required range of the seed flight in the sub-coulter space depends on the width of the working body and is achieved due to the required seed speed at the exit from the surface of the distributor and the height of the distributor.
6. The angle of seed sifting, which depends on the open angle of the distributor sides has a great effect on the uniformity of the seed distribution on the grab width of the working organ. The results that satisfy the requirements are obtained for the sifting angle of 150 and 180 degrees.

REFERENCES

Bakhmutov, V.A. (1980). Razmeshcheniye semyan po ploshchadi pri ryadovykh posevakh [seeds placement on the area at ordinary crops]. *Mekhanizatsiya i elektrifikatsiya sel khozyaistva*, *5*, 24-27.

Bespamyatnova, N. M. (2002). Nauchno-metodicheskiye osnovy adaptatsii pochvoobrabatyvayushchikh i posevnykh mashin [Scientific and methodical bases of adaptation of tillage and sowing machines]. Rostov-on-Don, Russia; NPK.

Bespamyatnova, N. M. (2010). Vibratsii v tekhnologicheskikh protsessakh [Vibrations in technological processes]. Zernograd: Russia, VNIPTIMESKh.

Chernovolov, V. A. (2007). Proyektirovaniye rabochego protsessa i modernizatsiya sborochnoy edinitsy mashiny dlya vneseniya mineralnykh udobreniy [Design of the working process and modernization of the Assembly unit of the machine for the application of mineral fertilizers]. In Uchebnoye posobiye po diplomnomu i kursovomu proyektirovaniyu. Zernograd, Russia: Academic Press.

De Vita, P., Colecchia, S. A., Pecorella, I., & Saia, S. (2017). Reduced inter-row distance improves yield and competition against weeds in a semi-dwarf durum wheat variety. *European Journal of Agronomy*, *85*, 69–77. doi:10.1016/j.eja.2017.02.003

Ewbank, N., & Buckby, L. (2002). Lynx engineering enters seed drill market with the seedline system. *Landwards, 82*(1), 153-155.

Grishchenko, F. V. (1963). Osnovy teorii dvizheniya semyan i raspredeleniye ikh v pochve pri bezryadkovom poseve [Fundamentals of the theory of seeds movement and their distribution in the soil at sowing without rows]. *Ryazanskiy SKhI, 10*, 31–33.

Grishchenko, F. V. (1970). Teoreticheskoye i eksperimentalnoye obosnovaniye seyalok-kultivatorov [Theoretical and experimental rationale of grain drills-cultivators]. In Materialy Nauchno-tekhnicheskogo soveta. Vypusk 28. Sostoyaniye i perspektivy razvitiya seyalok-kultivatorov. Moscow, Russia: Academic Press.

Kovrikov, I.T. (1976). Obosnovaniye nekotorykh parametrov raspredelitelya semyan i soshnika bezryadkovoy zernovoy seyalki [The rationale for some of the parameters of the distributor of seeds of scattering sowing coulters]. *Traktory i selskokhozyaystvennyye mashiny*, *4*, 26 - 28.

Kovrikov, I.T. (1983). Osnovnyye printsipy razrabotki raspredelitelnykh ustroystv podpochvenno-razbrosnykh soshnikov zernovykh seyalok [Basic principles of development of distribution devices of coulters of grain seeders subsurface-scattering]. *Traktory i selkhozmashiny*, *5*, 13-14.

Kovrikov, I.T., Nogtikov, A.A., & Sazonov, S.N. (1995). Povysheniye Effektivnosti ispolzovaniya kombinirovannykh pochvoobrabatyvayushche-posevnykh mashin [Increasing the efficiency of combined tillage and sowing machines using]. *Sovershenstvovaniye tekhnologiy i tekhn. sredstv dlya mekhanizatsii s.-kh. protsessov.-Vseros.N.-I. i proyekt.-tekhnol. In-t po ispolz. Tekhniki i nefteproduktov v sel. khoz.-ve*, *82*, 80 – 81.

Kovrikov, I.T., & Skvortsov, G.V. (1977). Sovershenstvovaniye soshnikov dlya bezryadkovogo poseva po sternevym fonam [Improvement of coulters for non-row sowing on stubble backgrounds]. *Traktory i selskokhozyaystvennyye mashiny*, *5*, 16 – 18.

Krasnoshchekov, N. V. (2002). O proyektirovanii tekhnologiy proizvodstva selsko-khozyaystvennoy produktsii [About design of production of agricultural products technologies]. *Prioritety mekhanizatsii rasteniyevodstva i zhivotnovodstva*, *138*, 11–20.

Lachuga, Yu. F., & Ksendzov, V. A. (2000). *Teoreticheskaya mekhanika* [Theoretical mechanics]. Moscow, Russia: Kolos.

Lavrukhin, P. V. (2000). *Sovershenstvovaniye protsessa podpochvennogo razbrosnogo raspredeleniya semyan* [Improvement of the process subsurface distribution of seeds] (PhD dissertation). Zernograd, Russia.

Lavrukhin, P. V. (2003). Issledovaniye sovremennykh sposobov i skhem razmeshcheniya semyan po ploshchadi polya [Study of modern methods and schemes of seed placement on the area of the field]. Otchet o NIR. / FGOU VPO AChGAA; NoGR01.20.0303698.

Lavrukhin, P. V., & Ivanov, P. A. (2009). *The coulter for scattering sowing*. RU patent No 89324 U1 A 01C 7/20. Moscow, Russia: Federal Institute of Industrial Property.

Lurye, A. B. (1979). *Modelirovaniye selskokhozyaystvennykh agregatov i ikh sistem upravleniya* [Modeling of agricultural units and their control systems]. Leningrad, Russia: Kolos.

Lyubushko, N.I., & Zvolinskiy, V.N. (2001). Zernovyye seyalki na rubezhe XXI veka [Grain drills at the turn of the XXI century]. *Traktory i selskokhozyaystvennyye mashiny*, *2*, 4 -7.

Machnev, A.B. (2012). Soshnik s napravitelem-raspredelitelem semyan dlya poseva zernovykh kultur [Coulter with guide-distributor of seeds for sowing of crops]. *Traktory i selkhozmashiny*, *7*, 42-43.

Malev, M.K. (1971). *Obosnovaniye parametrov rabochikh organov seyalok- kultivatorov dlya poseva na pochvakh* [Substantiation of parameters of working bodies of seeders-cultivators for sowing on soils exposed to wind erosion]. Kazan, Russia: Tr. Kaz. NII mekhanizatsii sel. khoz-va.

Nogtikov, A.A., Glotov, A.L., & Sazonov, D.S. (1996). Soshnik dlya vnutripochvennogo razbrosnogo poseva [The opener for subsurface broadcast seeding]. *Mekhanizatsiya i elektrifikatsiya sel. khoz.- va*, *2*, 29-30.

Nogtikov, A.A., & Golivets, V.A. (1997). Obosnovaniye parametrov bokovykh po-verkhnostey rassekate-lya [Justification of parameters of the side surfaces of the divider]. *Traktory i selskokhozyaystvennyye mashiny*, *11*, 31 - 32.

Nogtikov, A.A., & Sazonov, S. N. (1995). Obosnovaniye parametrov semyaprovoda dlya vnutripoch-venno - razbrosnykh soshnikov [parameters Justification of the seed pipe for inside scattering soil sowing coulters]. *Tekhnika v selskom khozyaystve*, *3*, 29-31.

Nogtikov, A.A., &Sazonov, S.N. (1995). Povysheniye effektivnosti raboty ras-predeliteley vnutripochven-no-razbrosnykh soshnikov [The efficiency of the distributors of subsoil-broadcast opener]. *Inzh.-tekhn. obespecheniye APK, 4*, 13-14.

O'Leary, G. J. (2018). Challenges and Responses to Ongoing and Projected Climate Change for Dryland Cereal Production Systems throughout the World. *Agronomy (Basel)*, *8*(4).

Semenov, A. N. (1959). *Zernovyye seyalki* [Grain drills.]. Moscow, Russia: Mashgiz.

Siracusa, L., Gresta, F., & Sperlinga, E. (2017). Strip tillage and sowing: Is precision planting indispens-able in silage maize. *Journal of Food Composition and Analysis*, *62*, 1–7. doi:10.1016/j.jfca.2017.04.005

Skolzayev, V.A., &Chernovolov, V.A. (1969). Elementy teorii raspredeleniya udobreniy diskovym tsentrobezhnym apparatom [Theory elements of the fertilizer distribution disc by centrifugal apparatus]. *Traktory i selkhozmashiny*, *2*, 15-17.

Skolzayev, V. A., & Chernovolov, V. A. (1969). Metodika tekhnologicheskogo rascheta tsentrobezhnykh diskovykh apparatov [Method of technological calculation of centrifugal disk devices]. Mekhanizatsiya i elektrifikatsiya selskokhozyaystvennogo proizvodstv, 12, 135-140.

Tuvalev, Z. I. (1987). *Issledovaniye parametrov zernovykh sternevykh seyalok v usloviyakh Zauralia Bashkirii* [Study of parameters of grain stubble seeders in the conditions of the trans-urals Bashkiria] (PhD dissertation). Chelyabinsk, Russia.

Zuyev, F. G., Levachev, N. A., & Lotkov, N. A. (1987). Mekhanizatsiya pogruzochno-razgruzochnykh. transportnykh i skladskikh rabot [Mechanization of loading and unloading, transport and storage works]. Moscow. Russia: VO Agropromizdat.

KEY TERMS AND DEFINITIONS

Coulter (Opener) for Undersoil: Scattering sowing is cultivator paw or hoe, designed to lift the top layer of soil and pruning the roots of weeds.

Passive Distribution of Seeds: Distribution of seeds in the undercoulter space using only the energy of their fall into the seed pipe.

Seeds Discard Distance: The distance the seed is discarded after hitting the coulter reflector.

Specific Nutrition Area: The area of the field provided by seeder to one plant.

Subsurface-Scattering Seeding: The location of the sown seed with their spreading under the soil layer.

Undercoulter (Opener) Space: The space under the soil layer formed by the horizontal elements of the coulter.

Uniform Distribution of Plants over the Field Area: The placement in which the distances between tend to be the same value.

Chapter 3
UV-Based Indoor Disinfecting System

Leonid Yuferev
Federal Scientific Agroengineering Center VIM, Russia

Alexander Sokolov
Federal Scientific Agroengineering Center VIM, Russia

Sergey Stepanovich Mironyuk
Uman National University of Horticulture, Ukraine

ABSTRACT

When contagious diseases occur, there is a tangible threat of rampant spread of infection, incurring huge economic losses in animal deaths and decreased animal productivity. Thus, preventing pathogenic flora concentration in rooms where birds and animals are raised from exceeding permissible levels ranks first among veterinarian and sanitary concerns. When birds are kept on the floor during feeding, germ and dust concentration increases nine to ten times against normal. Ample research shows that ultraviolet (UV) radiation possessing a bactericidal effect is the most promising and environmentally friendly method of cleansing the air from harmful germs.

INTRODUCTION

Increased raising of poultry, pigs, and other animals at large-scale, industrialised establishments as well as at smaller farms has led to a range of challenges, prominently including the prevention and elimination of contagious diseases in animals, especially respiratory diseases affecting young specimens.

When contagious diseases occur, there is a tangible threat of rampant spread of infection, incurring huge economic losses in animal deaths and decreased animal productivity. The problem is made worse by the fact that such diseases are chiefly caused by the association of a range of proven and semi-proven pathogens, with specific means of prevention producing little effect and chemotherapy, even in the form of sprays, useless due to their high toxicity when sprayed on a permanent basis. Respiratory diseases spread predominantly through air.

DOI: 10.4018/978-1-5225-7573-3.ch003

Figure 1. Germ contamination growth in an aviary (1 – overall bacterial contamination;2 – staphylococci;3 – colibacilli)

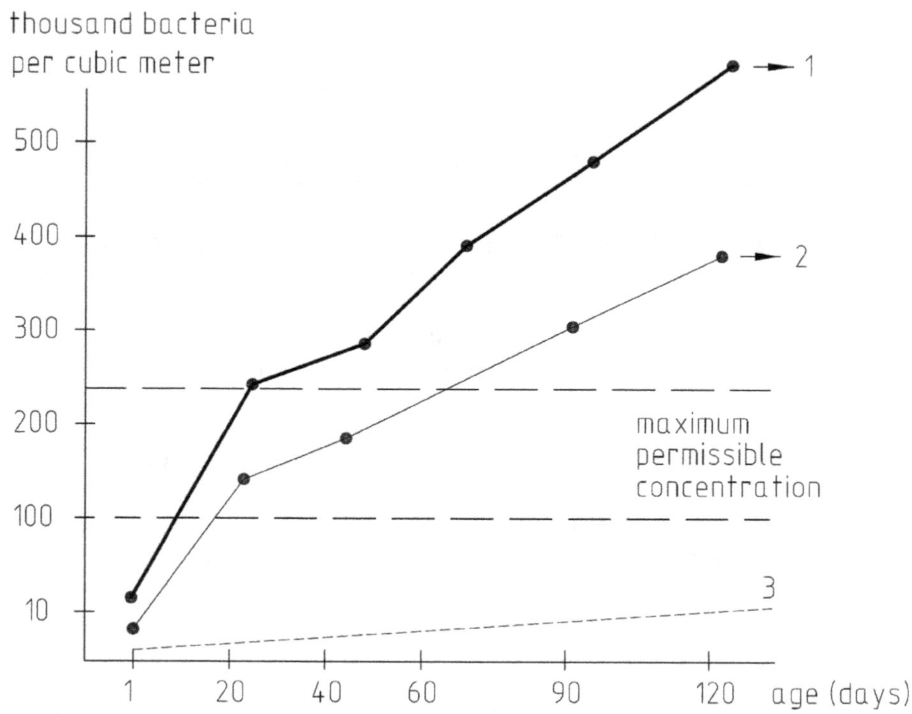

Thus, preventing pathogenic flora concentration in rooms where birds and animals are raised from exceeding permissible levels ranks first among veterinarian and sanitary concerns.

Large numbers of birds kept in enclosed spaces at large-scale, industrialised poultry farms lead to huge amounts of harmful substances being concentrated in the air in and out of the rooms.

Indoor air can become heavily contaminated with germs, including disease-causing ones, when infected birds are kept in the room in question or in the neighbouring room while due ventilation is not ensured.

Indoor germ concentration should never rise above sanitary norms, such as 100,000 bacteria per cubic metre for raising the young or 240,000 for keeping adult birds. Over 280,000 per cubic metre leads to sharply increased morbidity and mortality rates; at 910,000, morbidity reaches 25% and mortality is up to 10%.

When birds are kept on the floor during feeding, germ and dust concentration increases nine to ten times against normal.

As birds age, concentration of harmful substances and germs also increases. Thus, where week-old chicks were kept, between 45,000 and 65,000 germs were present in each cubic metre of air, but when the birds aged to between 120 and 150 days, the same value could reach one million (ref. Fig. 1).

Ample research shows that ultraviolet (UV) radiation possessing a bactericidal effect is the most promising and environment-friendly method of cleansing the air from harmful germs.

LITERATURE REVIEW

UV rays and their role in human and animal hygiene became the subject of many works written in Russia and abroad.

These works have demonstrated the effects that UV rays of various wavelengths have on germs. Experiments carried out at molecular, cellular, and bodily levels scientifically proved the use of UV rays for disinfecting outdoor environments and preventing infections from arising indoors.

However, in spite of significant achievements in the realm of hygienic use of UV rays, many veterinarian and sanitary practical aspects of it remain unaddressed, such as the increasing of disinfection efficiency species- and age-specifically, cutting costs in energy, and the identification of synergic effects of UV rays and ozone.

The bactericidal effect of sun rays was discovered over a century ago. It was also established then that said effect was caused by shortwave UV rays present in the solar spectrum. S. Band, 1901, F.L. Gates in 1928 were the first to discover that rays with the wavelength between 250 and 280 nm, selectively absorbed by nucleic acids, have the strongest destructive effect on cells).

Profound study of molecular mechanics and UV radiation effects on germs and viruses began in the 1940s, following the advent of artificial optical radiation sources.

The reaction of germs to radiation is complex, defined by a number of factors. These include the source of radiation (wavelength, intensity, etc.), duration of exposure, biological properties of germs (physiological condition, age, etc.), and the state of the surrounding environment. When UV rays' effect on bacteria was studied in 1935, subbacterial (or stimulating), minimal, and fully bactericidal doses were identified.

The creation of new and perfection of existing artificial sources of optical radiation as well as measuring methods and devices paved the way for calculating the quantitative correlation between UV radiation effects and its spectral parameters and exposure duration, as well as for studying the effect that radiation has on bacterial cells.

Ultraviolet can be used in the disinfection of glass containers for canning vegetables and fruits (Mironyuk 2015).

Different germs have different resistance to UV rays, largely dependent on their current condition. In the course of a study that lasted 10 to 30 minutes, unradiated cabinets yielded an average of 1,000 bacteria in the Petri dish, while in radiated cabinets the same figure amounted to 2 to 17.

Vashkov, 1973, established the UV ray energies as necessary to eliminate 90% of germs (Table 1).

Thus, 90% of relevant bacteria can be eliminated at 1.5–5 J/m^2 of UV irradiation.

In 1973, 30W UV lamps were used at a poultry farm to disinfect air. 14 lamps were hung at a height of 2m above floor, a total of one lamp per 30 m^2. Overall germ contamination measured after the lamps were activated is summed up in Table 2

It was also established that complete germ elimination is unattainable as the chickens in the room raise large amounts of dust, leading to more germ contamination.

More experiments on air disinfection at poultry farms were carried out in 1969. Screened lamps were used to irradiate air, and bacterial contamination decreased by 20.1% after one hour, by 37.2% after two hours, and by 60.7% after three hours of work of 30W UV lamps. It was noted that the animals and people present in the room constantly increased the number of germs, and more were brought in along with fodder, litter, or just from neighbouring rooms. After the lamps were switched off, bacterial

Table 1. Vashkov irradiation doses

Bacteria Type	Energy, J/m²
Bac. anthracis	4.52
Bac. megatherium	1.13
dysentery-causing bacteria	1.68
Colibacilli	3.00
Proteus	2.64
Staphyloccus albus	1.84
Staphylococcus aureus	2.28–4.95
Streptococcus viridans	2.00
Sarcina	19.70
Corynebacterium diphtheriae	3.37

Table 2. Germ contamination measurements

Exposure Time, min.	Air Germ Contamination, Thousands/m³	Remaining Germs, %
0	365	100
30	264	72.3
60	210	57.8
90	185	50.7
120	180	49.3
150	148	40.5
180	147	40.0

contamination increased by 7.1% during the next hour, and was up 21.6% and 37.4% respectively during the following two hours.

In 1978, Prokopenko used unscreened 30W UV lamps to disinfect air at a poultry farm where birds were kept in cages. He found out that, after one hour of lamp use, germ contamination in the air decreased from 217,000 to 105,000 per cubic metre, and half an hour later it was 83,000 per cubic metre.

There has been extensive research of air disinfection with shortwave UV rays in air ducts of poultry farms. Intake air processing devices consisted of two diffusers mounted into the pipe beneath the ceiling. Each diffuser contained thirty 30W lamps. It was found that irradiated air, at 15°C to 30°C and relative humidity of 40% to 60%, displays germ contamination rates reduced by 80% to 95%. Other climatic parameters of the farm were found to improve.

Indoor bacterial contamination was down 38% to 45%, ammonia contamination almost 70%, and carbon dioxide content was 30% to 40% lower. During the initial twenty days, little increase in fast negative ion content was observed. When chickens aged 1 to 60 days were exposed to better air, actual accretion increased by 7% to 10% while mortality fell by 2.5% to 8% and culling by 6%. The average bird weighed in at 16g more than in the comparison set.

The effectiveness of combined disinfection of colibacilli-contaminated air with sprays and UV radiation was also studied, revealing that 0.3g of iodine-containing chemical per cubic metre of air augmented by 60W UV lamps turned on at 600 to 1,200 mcW*sec/mm2 the colibacilli were completely destroyed within three minutes.

Table 3 summarises the data on various disinfection methods used by researchers at indoor poultry farms (Yuferev 2006).

It can be seen from the table that electric capacity of bactericidal UV lamps and their operational modes can vary greatly, resulting in widely divergent results. Not all authors quote such parameters as head count, ventilation profile, room size, or lamp mounting profile, making it impossible to summarise the results of their experiments and select the optimal mounting and operational mode for a given bactericidal UV lamp.

Research output and literature analysis showed that shortwave UV rays have a strong bactericidal effect and can be used broadly in medical and veterinary practices for negating the effects of airborne harmful substances. The use of said rays goes beyond improving the general environment and increases the productivity of fattening. However, data on how UV radiation affects germs residing in airborne drips or dust is insufficient (Prokopenko 2004).

Aside from the intensity of irradiation, the efficiency of air disinfecting can vary according to a large number of factors.

The efficiency of air disinfecting depends on the type of germs and the pattern of their presence in the air as well as on the room size; on the type, capacity, and location of the lamp; and on the uniformity of irradiation and air circulation speed.

In addition to the number of bactericidal lamps and the intensity of irradiation, efficiency depends on how the lamps are mounted and on the temperature and humidity in the room.

In every particular case, UV radiation efficiency should be calculated separately, factoring in all parameters affecting the process of exposing germs to radiation.

UV rays do not penetrate into solid objects deeper than 0.1 to 11 mm and only travel direct trajectories, meaning that any working space will have 'shadowed' areas which will receive no bactericidal irradiation. Besides, biocidal effects of UV rays decrease sharply as the distance from the source increases, and the area of effect is restricted to the surface of the object being irradiated, with its cleanliness being an important factor (Philips N.V. 1979).

Table 3. Radiation dose and mode study

Researcher	Room Type	Capacity, W/m³	Operation Mode	Air Disinfecting Method
Imiyev Y.I., Mugutdinov B.M.	Floor-pen	0.6	3 days, 1 hour daily	Unscreened bactericidal lamp irradiating the air in absence of people
Prokopenko A.A	Floor-pen	0.3	3 days, 1.5 hour daily	Unscreened bactericidal lamp irradiating the air in absence of people
Kozhurin V.M.	Floor-pen	0.3	3 hours daily	Unscreened bactericidal lamp irradiating the air in absence of people
Melyukov A.N.	Floor-pen	3–4	10–14 daily with a break of 1–2 hours every 3–4 hours	Screened bactericidal lamp irradiating the air in absence of people
Zakomyrdin A.A.	chickery	-	constantly	Intake air in ventilation ducts irradiated

When irradiation doses are high enough, cells die either immediately upon irradiation or soon after. In this case, broad and generalised damage to the cell structures can be observed, including coagulation and cytoplasm disintegration.

Biologically, UV rays have been proven to hinder RNA synthesis and disrupt DNA synthesis, resulting in inhibited synthesis in irradiated cells.

The spectrum of UV rays that suppress DNA, RNA, and protein synthesis coincides with the absorption spectrum of nucleic acids, possibly meaning that DNA and RNA can be damaged simultaneously.

The absorption spectrums of RNA and DNA add up to form the generalised spectrum of bactericidal effect (see Fig. 2) peaking at 260nm, which coincides with the spectrum of UV rays (254nm), possibly meaning that DNA and RNA can be damaged simultaneously (Yuferev, Alferova 2016).

UV rays can strongly effect birds. Their bactericidal and viricidal properties make them effective against harmful airborne substances at poultry and animal farms, improving the environment and increasing the productivity and safety of animals and birds.

When the air is ionised by UV rays, its physical profile is improved as dust particles and germs fall down, and atmosphere pollution is lowered.

Ultraviolet rays are a kind of electromagnetic waves and all their characteristics pertain to them.

A, B, and C are three subspectrums of UV radiation, having the wavelength between 320 and 400 nm, 260 through 320 nm, and 200 to 280 nm respectively. The latter carry a strong bactericidal effect.

The boundaries between the areas are tentative as attributes characteristic of each area also appear, but less prominently, in the other ones. The energy of the ultraviolet component of sunlight harms germs at both cellular and genetic levels; the same damage is done to humans, but it is restricted to their skin and eyes. Artificial sources of UV radiation generate much more concentrated rays than those present in normal sunlight.

Sources of UV rays are discharge lamps that, during an electric discharge, emit rays that include wavelengths between 205nm and 315nm (the rest of the spectrum is of secondary importance).

Figure 2. RNA and DNA absorption spectrums

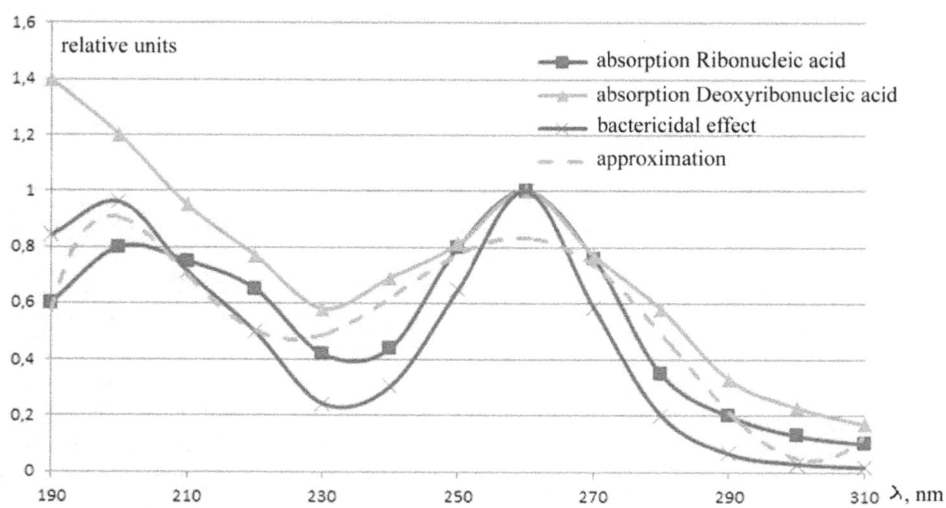

Figure 3. Spectrums of optical radiation

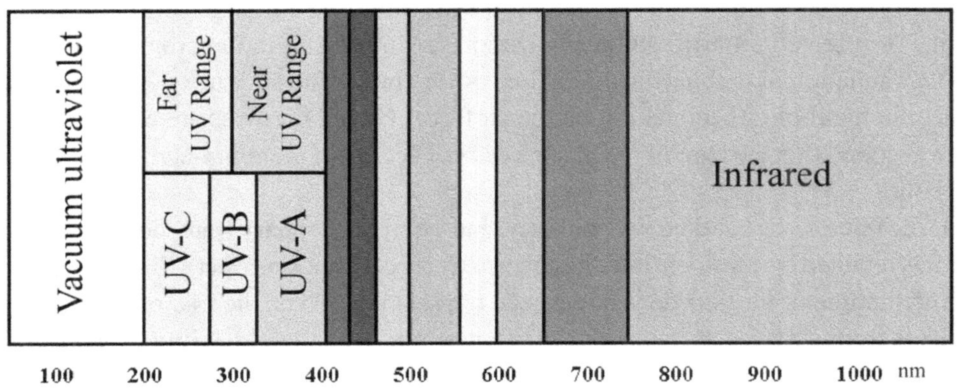

BACKGROUND

Electrophysical Methods of Air Disinfection in Livestock and Poultry Houses

The peculiarity of air disinfection in cattle-breeding conditions is the necessity of using it in the presence of animals. To disinfectants in this case, the following basic makes demands: they must have a strong bactericidal effect; to be harmless to humans and animals, even with prolonged use; do not pollute the environment, do not cause corrosion of metal, do not disable equipment; application of them must be cost-effective and technologically

Methods of air disinfection are usually classified according to the principle based on bactericidal action. Mechanical, chemical and physical methods and means of air disinfection of cattle-breeding premises are known presently.

In medicine and veterinary medicine, aerosol dispersion of chemical solutions (formaldehyde, caustic soda, calcium hypochlorite, etc.) is used to eliminate airborne infections. Lactic acid, resorcinol and other most used substances have a bactericidal effect on the pathogenic microflora. With periodically use in small doses, they are harmless to humans and animals.

However, the wide use of chemical aerosols in enclosed spaces in the presence of animals is hampered by a number of technical and biological difficulties: the need for uniform disinfectant spraying in the air, maintaining a constant concentration for a sufficiently long time, the need to exclude damage to process equipment, the negative effect of toxic substances on humans and animals with their regular application. In addition, during the ventilation of the room, after treatment, a large part of the disinfectant is carried to the atmosphere and pollutes the environment

The physical methods of disinfection include short-wave UV radiation, ozone and artificial air ionization, that is, nanoelectrotechnologies-biological and physiological processes carried out with the help of nanoelectrons (electrons, protons, ions, photons) at the level of living cells and microorganisms.

For the disinfection of poultry and pigs, the use of ozone, which is an environmentally friendly oxidant, is recommended.

The sterilizing advantages of ozone include a wide range of its biocidal action at low concentration, the possibility of using for disinfection of hard-to-reach surfaces, a shorter half-life, compared to other gases, and the presence of a deodorizing effect.

However, due to the high toxicity of ozone, disinfection is possible in the presence of animals and birds only at ozone concentrations not exceeding the maximum permissible concentration (MPC).

The results of research on artificial air ionization for improving the microclimate of premises are presented. Accumulation of pathogenic microflora occurs inside the air ionizer, on the surfaces of parts and equipment in the room. Precipitating on the surface, charged particles are eliminated from air circulation. It was found that horizontal metal surfaces collect much more bio-aerosols and dust particles than vertical ones.

However, the microflora continues to multiply, and often a secondary transition occurs from the surfaces where it is retained by aeroions, back into the air (reaerozolization effect). Since after disconnecting the device, the lifetime of air ions does not exceed ten seconds. Thus, the risk of secondary transfer of microorganisms from surfaces to the aerodisperse state excludes the possibility of wide application of devices in rooms with a standardized level of semination (Alferova 2002)

To carry out the sterilization process, air ionization is often combined with ozonation or ultraviolet irradiation. Such a combination has low productivity, requires constant monitoring and frequent maintenance of generators, and also adds the following negative consequences:

- excessive ionization of purified air;
- emission of ozone and nitric oxides.

In such cases, it is necessary to control the air parameters: the concentration, mobility and sign of air ions in the air, as well as the gas composition of the air in the room.

When air ionizers operate in clean rooms, the concentration of microbial bodies in the air decreases by 30-60%, and in rooms with strong dust formation and dirty floors - by 10-30%.

However, the use of ultraviolet germicidal plants requires strict implementation of safety measures that exclude possible harmful effects on humans and animals of ultraviolet bactericidal radiation, ozone and mercury vapors.

Compliance with these requirements when installing and operating bactericidal UV plants is a reliable protection against their damaging effects.

Studies of combined methods of air disinfection, using the positive qualities of different methods, are being conducted.

Sources of Ultraviolet Radiation

At present, there are the following types of discharge lamps for obtaining short-wave (bactericidal) ultraviolet radiation:

- Mercury quartz lamps of high pressure are made of quartz glass. The range of optical radiation of lamps is wide. The share of ultra-violet radiation is 17-20%, visible light - up to 5%, the rest of the radiation refers to the infrared spectrum. These lamps have a high self-temperature ≈ 950 degrees, so the ambient temperature has little effect on the operating parameters.
- **Scope:** Sources of ultraviolet radiation, installations for disinfecting air, water, food products, containers, household items, medical instruments. Features - low service life, low efficiency in the field of bactericidal radiation (9-12%), high specific power (100-3000 W).

Figure 4. The radiation spectrum of high-pressure UV lamps

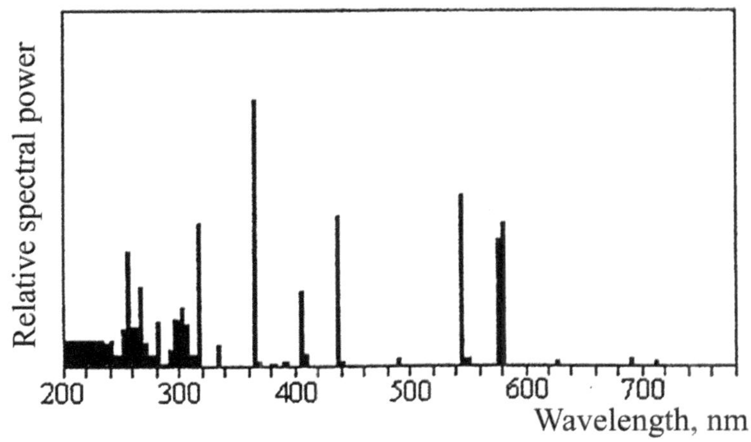

An example of high-pressure lamps are lamps PHILIPS of the HOK series with a power of 400W, an efficiency in the field of bactericidal radiation of 13%, or Heraeus Amba high-pressure mercury lamps.

The advantage of these lamps is that they have a unit power from 100 to 1000 W for small dimensions, which allows to reduce the number of lamps in the room, but they have a low bactericidal efficiency and a short service life of 500-1000 h. In addition, the normal combustion mode comes in 5-10 minutes after their ignition.

Low-pressure mercury lamps are structurally and electrically almost indistinguishable from ordinary fluorescent lamps, except that their flask is made of a special quartz or uviol glass with a high transmission coefficient of UV, the inner surface of which does not have a phosphor layer. These lamps are available in a wide range of capacities from 4 to 60 watts. The advantage of low-pressure mercury lamps is that the main radiation falls on a line with a wavelength of 254 nm lying in the spectral region of the maximum bactericidal action. They have a long service life and instant ability to work after their ignition. However, the parameters of low-pressure lamps depend on the ambient temperature (Yuferev 2010).

Bactericidal lamps of low pressure are divided into ozone-generating and ozone-free.

Ozone-generating lamps have a spectral line with a wavelength of 185 nm (Fig.5) in the radiation spectrum, which, as a result of interaction with the oxygen molecules of the air medium, forms ozone. High concentrations of ozone can have an adverse effect on human health. The use of these lamps requires monitoring of ozone in the air and careful ventilation of the room (Alferova 2000).

To exclude the possibility of ozone generation, so-called bactericidal "ozone-free" lamps have been developed. Such lamps, due to the production of a bulb from a special material (coated quartz glass or uviol glass), eliminates the emission of the 185 nm line.

Low-pressure discharge lamps are made in a flask made of uviol glass. Lamps are used in irradiators for disinfecting air, water, food, containers, household items, medical instruments and materials. For example, a lamp series TUV Phillips.

The low-pressure bactericidal discharge lamps have a wavelength of 185 nm, which allows the production of ozone from the oxygen of the air by photochemical method. Such lamps are used in irradiators for disinfecting air, water, food, containers, household items, medical instruments and materials.

Figure 5. Spectrum of lamps made of quartz glass

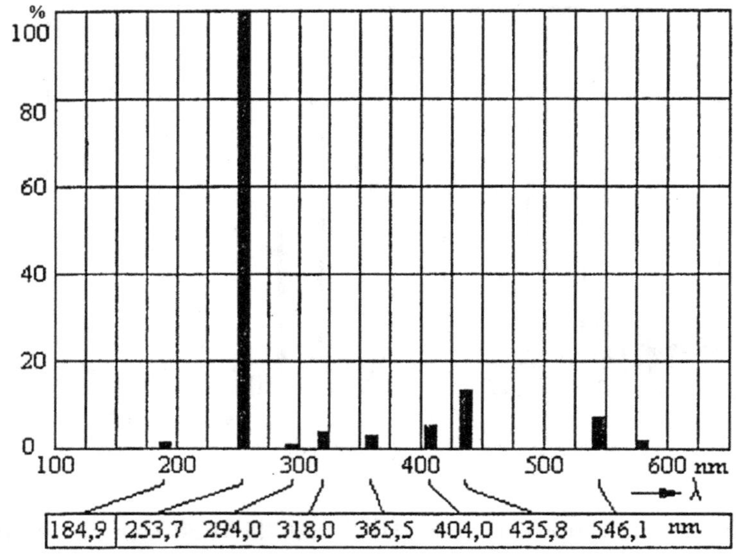

The main advantages of low-pressure mercury lamps are a long service life (8000-12000 hours) and high efficiency of bactericidal radiation (30-40%).

A comparison of the spectrum of high-pressure lamps and low-pressure lamps is shown in Fig. 6

As can be seen from Fig. 6, in high-pressure lamps, the highest peak in the ultraviolet range is at a wavelength of 365 nm, and the wavelength of 254 nm is emitted 3 times weaker.

For low-pressure lamps, the maximum radiation in the ultraviolet range is at a wavelength of 254 nm, which coincides with the maximum bactericidal effect by 87%, which makes it possible to create economical bactericidal irradiators. Therefore, we will not consider high-pressure lamps in the future.

Analyzing the characteristics of low-pressure lamps, it is established that the lamps of Phillips TUV-36, 36 W (Fig. 7), have the highest efficiency of bactericidal radiation (η up to 42%). But since they have

Figure 6. The bactericidal effect of high and low pressure lamps

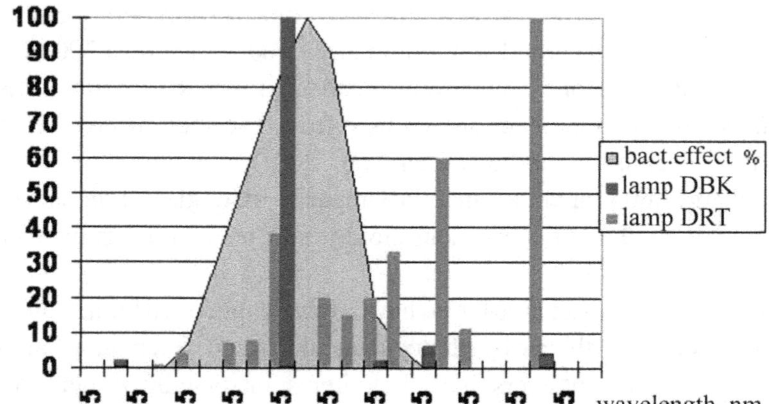

Figure 7. The appearance of UV LED LEUV

a special coating and are made of uviol glass, then these lamps do not produce ozone. Therefore, we chose low-pressure bactericidal lamps of the DBK series for the manufacture of irradiators-ozonizers. In such lamps, tubular flasks are made of quartz glass, transparent both for short-wave UV radiation of 205-315 nm, and for vacuum ultraviolet 185 nm.

A significant disadvantage of all mercury lamps is the possibility of mercury vapor contamination of the environment when the lamp is destroyed. In the event of a violation of the integrity of bactericidal lamps and the entry of mercury into the room, a thorough demercurization of the contaminated room must be carried out.

Impulse ultraviolet radiation sources are also known. The principle of their operation is based on high-intensity pulsed irradiation of air and ultraviolet radiation surfaces of the continuous spectrum.

Pulsed ultraviolet radiation is obtained with the help of xenon lamps. The spectrum of the xenon lamp contains a region with a wavelength of 260 nm (15% of the maximum) corresponding to the maximum bactericidal effect. Data on the difference between the biocidal effect of pulsed ultraviolet radiation and that of traditional UV are not found at present.

Also known are high-pressure lamps based on lead. The spectrum of these lamps contains a bactericidal component of 250-260 nm (18% of the maximum). The main application of these lamps is the hardening of paint coatings reacting to long-wave UV (350-400 nm). These lamps do not pollute the environment in case of damage.

At present, the production of UV LEDs with wavelengths from 250 to 390 nm, which have a huge lifetime.

Bactericidal Irradiators and Equipments

For irradiation of animals and birds, as well as air disinfection in the premises, ultraviolet irradiators and equipments were used.

Disinfection of surfaces, walls and floors of premises can be carried out with the help of open, combined and mobile irradiators, only in the absence of people.

A bactericidal irradiator is an electrical device in which a bactericidal lamp, a reflector and other auxiliary elements are placed, as well as devices for attaching it. Bactericidal irradiators redistribute

Table 4. Specifications of UV LEDs

Brand	Wavelength, nm	Power Consumption, mW	Output Power, mW	Efficiency,%
LEUVK11H00	278	850	10	1,1
LEUVA66G00K	305	700	10	1,5
NCSU276A	365	2590	800	31
LEUVA35T01	385	1700	875	51
LEVUA66X00T	385	5400	2230	41

the radiation flux into the surrounding space in a given direction and are divided into two groups - open and closed.

Open irradiators use direct bactericidal flow from lamps and reflector (or without it), which covers a wide area of space around them. Installed on the ceiling or wall.

Open irradiators are intended for disinfection of premises only in the absence of people. Open combined irradiators can be used during short-term stay of people. Closed irradiators can be used in the presence of people.

Ultraviolet radiation acting on open areas of the human skin and the retina of the eyes can cause burns of I-II degree, exacerbation of cardiovascular diseases, and in some cases lead to the appearance of malignant tumors.

When operating irradiators using irradiance from reflected flows from the ceiling and walls on a conditional surface at a height of 1.5 m from the floor, the irradiation power should not exceed 0.001 W / m^2.

In closed irradiators (recirculators), the bactericidal flux from the lamps is distributed in a limited small enclosed space and has no outlet to the outside, while air is disinfected during its pumping through the ventilation holes of the recirculation system. Closed-circuit irradiators (recirculators) should be placed indoors on the walls, along the main air flows (in particular, near the heating devices), at a height of at least 2 m from the floor.

When used in supply and exhaust ventilation, bactericidal lamps are placed in the outlet chamber. The speed of the air flow is provided either by natural convection or forced by a fan. The most common devices for cleaning and disinfecting air and surfaces with bactericidal UV rays were open-type irradiators.

Also known irradiating the installation of mechanized and self-propelled installation for irradiation of chickens and Chicks multi-tiered cellular accommodation.

MAIN FOCUS OF THE CHAPTER

Investigation of the Process of Disinfection of the Air and Surfaces in Poultry Farming by Exposure to UV Radiation

Technical means of disinfection with ultraviolet bactericidal radiation of air and indoor surfaces include: sources of ultraviolet bactericidal radiation, irradiators (modules) that distribute the radiation flux to the surrounding space in a given volume; bactericidal equipments, representing a group of modules installed in the room to ensure a given level of decrease in microbial semination.

The main characteristic of technical means for disinfection with UV radiation is bactericidal efficiency, i.e. the ratio of the number of dead microorganisms to their initial number before irradiation. The main working organ of the disinfection equipment is bactericidal lamps.

In low-pressure discharge lamps, UV radiation is generated most strongly at a certain mercury vapor pressure in the discharge tube. This pressure is determined by the operating temperature and reaches an optimum at the wall temperature of the lamp + 400 ° C. The temperature difference between the ambient air and the lamp wall is 200 ° C. This must be taken into account when operating the irradiators (ozonizers). The relationship between the operating temperature of the lamp and the air temperature with immobile air is shown in Fig. 8.

The dependence of the UV generation efficiency of the lamp on the ambient air temperature and its motion is shown in Figure 9.

For the operation of the irradiator-ozonator in the recirculation mode, the optimum speed of air movement near the lamp is 1.5 m / s. At this air speed, the range of air temperatures at an efficiency of 85% of the maximum is + 17 ... + 40 °C. Maximum efficiency is provided at an air temperature of + 28 °C (Yuferev 2007).

The Effectiveness of Air Disinfection With UV Radiation in Rooms

The main characteristic of technical means for air disinfection by UV radiation is bactericidal efficiency, i.e. the ratio of the number of dead microorganisms to their initial number before irradiation.

Different types of microorganisms of common and pathogenic microflora that are in the air have different doses of lethal effect, which depend on humidity, temperature of the air environment, and the amount of dust in it.

Figure 8. The value of the flow of UV radiation (ordinate axis) depending on the temperature of the tube wall (the upper scale of the abscissa axis) and the ambient temperature (the lower scale of the abscissa axis)

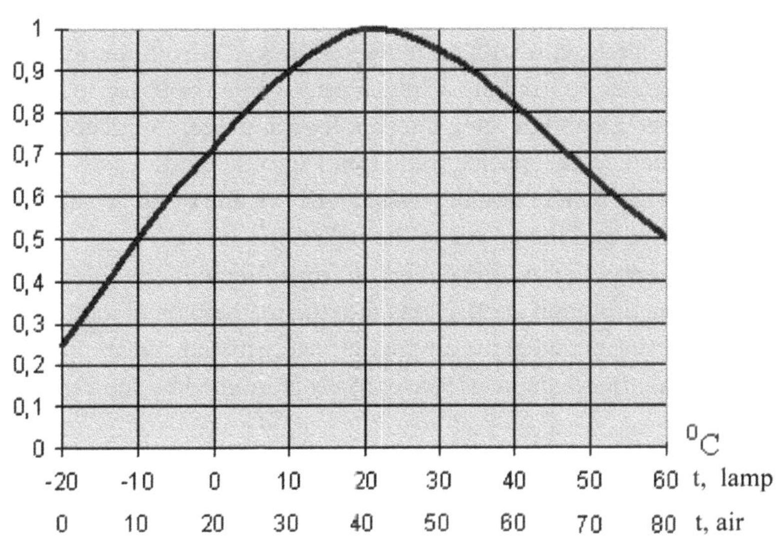

Figure 9. The dependence of the efficiency of generation of UV radiation by the lamp on the temperature of the motionless ambient air (the upper curve) and at a speed of 1.5 m / s (lower curve)

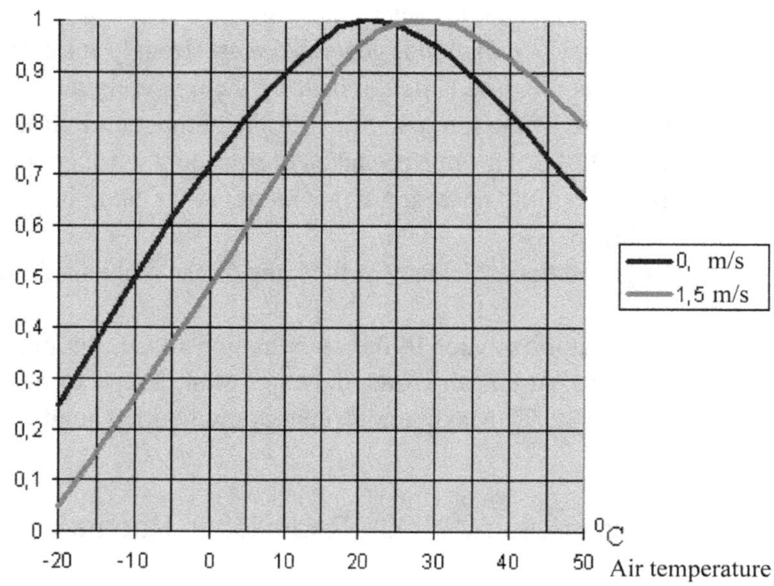

The parameters of the air in the room and the distribution of microflora in them depend on many particular characteristics of the microclimate of the room, including the nature and mode of ventilation, the speeds and directions of the movement of air currents, temperature fields, and technological features of production. Therefore, we have developed a general mathematical model of air disinfection with UV radiation, which makes it possible to calculate the power of UV bactericidal irradiation units taking into account the parameters of the disinfected air.

Most of the produced irradiators with a ceiling and wall mount are designed to irradiate the upper part of the room (ceiling) or their radiation is directed to the wall. Based on the developed mathematical model, it can be assumed that in order to increase the energy efficiency of UV bactericidal radiation during air disinfection, an input system with an increased distance from the irradiators to the absorption surface can be proposed (i.e., to irradiate as much as possible the space located above the level of view of animals and birds) and also to develop a technological process of decontamination that allows to regulate the direction of the UV-bactericidal radiation flow.

Thus, the effectiveness of air disinfection depends on the volume of the irradiated space, the absorption coefficient of UV radiation and the power of the radiation source.

We introduce a number of restrictions, the deviation from which, in the future, can be characterized by introducing additional functions and coefficients into the mathematical model.

Suppose that all particles suspended in the air are uniformly mixed and are in it an equal time. Then, the time of each colony of microorganisms in the room is determined by the expression:

$$t = \frac{V}{B}, \text{ [s]}$$

Figure 10. Irradiated space in a poultry house

where

V is the volume of the room, [m^3],
B - air flow in the room, [m^3 / s].

Then the time of the colony of microorganisms in the irradiated space is:

$$t_{obl} = \frac{V_{obl}}{B}, \text{[s]}$$

where V_{obl} is the volume of the irradiated space, [m^3].

The radiation dose "D_x" received by a colony of microorganisms depends on the irradiance of "E_x" in the place of its location and the duration of its exposure under the influence of this irradiance "t_x" (Fig.11)

For a colony moving through the irradiated space at each instant of time, the irradiance will vary according to its movement.

The radiation dose "D_x", obtained in an elementary volume dV, located at a distance "x" from the radiation source "A".

$$D_x = E_x \cdot t_x,$$

where

Figure 11. Model of irradiated space. (l-distance to the surface bounding the irradiated space)

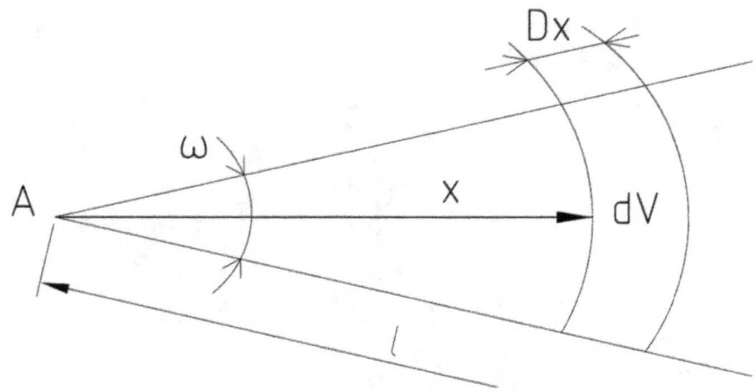

E_x is the irradiance at a distance "x" (W / m^2),
t_x is the time the particle is in the "V" volume.

The irradiance "E_x", taking into account the absorption and scattering of radiation by the medium, is given by

$$E_x = E_0 \cdot e^{-\sigma x},$$

where

σ – is the transmittance of radiation by the medium,
E_o – is the initial irradiance.

However, irradiation does not allow us to estimate the change in the radiation dose as a function of the distance to the irradiated object. Therefore, we convert the expression for calculating the dose to the form:

$$D_x = \frac{I_0}{x^2} \cdot e^{-\sigma x} t_x,$$

where I_0-bactericidal lamp flow (W).

Taking into account the previously introduced restriction, we can write down that the time of finding a particle in the elementary volume "dV" will be

$$t_x = \frac{dV}{B} \text{ [s]}.$$

The elementary volume "dV" (Fig. 11) is determined by the expression:

$$dV = \frac{\omega}{4\pi}\left(\frac{4\pi(x+dx)^3}{3} - \frac{4\pi}{3}x^3\right) = \omega x^2 dx\,.$$

Then the time of irradiation of aerosol particles in an elementary volume will be:

$$t_x = \frac{\omega x^2 dx}{B}$$

In view of the previous expressions, we can write:

$$D_x = \frac{\omega I_0 \cdot e^{-\sigma x}}{B} dx\,.$$

The expression obtained determines the dose of irradiation of colonies of microorganisms in the volume "dV" located at a distance "x" from the source "A". To determine the total dose received by the colonies of microorganisms in the air from the source "A" in the irradiation volume, bounded by the solid angle "ω" and the absorption surface located at a distance "l" from the source "A", it is necessary to integrate the obtained expression from 0 to l.

$$D = \int_0^l D_x(x)\,dx = \int_0^l \frac{\omega I_0}{B} \cdot e^{-\sigma x} dx\,;$$

or

$$D = \frac{\omega I_0}{B} \cdot \frac{1}{\sigma}\left(1 - e^{-\sigma l}\right).$$

The new expression shows the radiation dose received by colony microorganisms, depending on the parameters of the medium, characterized by the absorption coefficient "σ" and the location of the source, which causes the distance to the absorption surface "l". For the convenience of using the formula for calculating the dose in calculations, it can be converted to the form:

$$D = \frac{\frac{S}{l^2}\cos\alpha \cdot I_0}{B} \cdot \frac{1}{\sigma}\left(1 - e^{-\sigma l}\right),$$

where

S is the area of the absorption surface,
l – is the distance from the radiation source to the absorption surface,
α – the angle between the normal to the surface and the direction of the radiation force I_0.

For a point source, the radiation flux:

$$\omega I_0 = \Phi_0$$

Therefore, the expression for calculating the dose can be written as:

$$D = \frac{\Phi_0}{B} \cdot \frac{1}{\sigma}\left(1 - e^{-\sigma l}\right)$$

For the irradiator with the distribution of the radiation flux different from the point source, the dose will be different from that calculated by the formula obtained. This difference can be characterized by the introduction in the formula of the proportionality coefficient "*K*", which is determined experimentally. Then the dose of irradiation received from the real irradiator is determined by the formula:

$$D_p = K \cdot \frac{\Phi_0}{B} \cdot \frac{1}{\sigma}\left(1 - e^{-\sigma l}\right)$$

Similarly, the radiation dose formula for a linear radiation source is derived. For him, the elementary volume of irradiation "*dV*" is determined by the expression

$$dV = \left[\pi\left(x + dx\right)^2 - \pi x^2\right] \cdot b \cdot \frac{\alpha}{2\pi}$$

where

b – is the length of the considered source segment,
α – plane angle of radiation in the plane perpendicular to the axis of the source.

After the transformations, the expression has the form:

$$dV = \alpha \cdot b \cdot x dx$$

Then, the dose received by colonies of microorganisms from a length of a linear source "*b*" in the angle of radiation "α" at a distance perpendicular to it "*l*" will be:

$$D = \frac{\alpha \cdot b \cdot I_0}{B} \cdot \frac{1}{\sigma}\left(1 - e^{-\sigma l}\right)$$

For a linear radiation source:

$$\Phi_0 = \alpha \cdot b \cdot I_0$$

Therefore, considering this, dose is calculated:

$$D = \frac{\Phi_0}{B} \cdot \frac{1}{\sigma}\left(1 - e^{-\sigma l}\right).$$

From this expression it follows that with increasing distance "*l*" from the irradiator to the surface limiting the irradiated space, with the other parameters unchanged, the dose received by the microorganisms decreases, but the volume being processed is proportionally increased..

We assume that the increase in the irradiated space is achieved by reducing the height of the location of the ceiling bactericidal irradiators oriented with the lamp upward. As a result, the distance to the absorbing surface, in this case of the ceiling, increased from l_1 to l_2.

For reference: For atmospheric air, the main component of the transmittance "σ" is:

- σ_r is the component characterizing the scattering of radiation by air molecules (Rayleigh scattering)
- σ_0 is the component of the transmission coefficient due to the presence of ozone in the air;
- σ_{H2O} - component of the transmittance, due to the presence of water vapor in the air;
- σ_{aer} is the component of the transmission coefficient due to the presence of an aerosol in the air.

For atmospheric air at sea level under normal conditions, $\sigma_0 = 0.0003 m^{-1}$, in poultry houses approximately $0.015\ m^{-1}$.

Most of the produced irradiators with a ceiling and wall mount are designed to irradiate the upper part of the room (ceiling) or their radiation is directed to the wall. Based on the developed mathematical model, it can be assumed that in order to increase the energy efficiency of UV bactericidal radiation during air disinfection, an input system with an increased distance from the irradiators to the absorption surface can be proposed (i.e., to irradiate as much as possible the space located above the level of view of animals and birds) and also to develop a technological process of decontamination that allows to regulate the direction of UV bactericidal radiation.

Thus, the effectiveness of air disinfection depends on the volume of the irradiated space, the absorption coefficient of UV radiation and on the power of the radiation source.

Development of Structural Elements and Parameters of UV Irradiator

The industry produces bactericidal irradiators, designed to disinfect air and surfaces by direct exposure to ultraviolet radiation of the short-wave spectral range. There are a lot of modifications of irradiators, but they have drawbacks that limit the scope of application in poultry and farming.

We developed an irradiator, the known bactericidal properties of ultraviolet radiation are based on the disinfecting and deodorizing effects of the device. The irradiator can be used in the recirculation mode (disinfection takes place inside the housing of the device), or in the UV irradiator mode with open covers (disinfection occurs mainly from the UV lamp radiation).

Devices for UV irradiation are designed to influence the air environment and agricultural objects with ultraviolet radiation. They use a arc low pressure bactericidal lamps, located in a closed case with hinged covers and a fan.

Installations can be used in farms and other farms for disinfection and deodorization of the air, surface treatment for the destruction of microflora and harmful insects, for disinfection of cellars, cellars before laying products and during storage.

The main element of the bactericidal irradiator is a mercury lamp, since the wavelength of the mercury vapor almost coincides with the wavelength of the maximum bactericidal effect (265 nm).

The efficiency of the irradiator is influenced by the dimensions, the configuration of the hull, the reflective characteristics of the hull material, and the speed of air movement.

For optimal operation of the UV lamp and to achieve maximum radiation generation efficiency, an air velocity of about 1.5 m / s is required. For this purpose, an axial air fan is installed in the casing, providing forced convection of the ozone-air mixture, while simultaneously preventing accelerated destruction of ozone near the ultraviolet lamp.

The efficiency of the case of a square cross-section is less than that of a round casing by 10%. However, this decrease in efficiency can be neglected due to the more technological design of the saving metal and the fan of a widely distributed square section is installed in the housing (Fig. 12). In addition, the square section makes it possible to make the body dismountable (with an opening upper part), and to use the device as an irradiator, and also provides convenience of operation (cleaning, replacement of lamps).

The upper part of the body is made in the form of two hinged covers (3) on the hinges, fastened by locks (4). Under the lamp is installed a reflector (8), which increases the return of UV radiation in the irradiator mode. The internal arrangement of the irradiator-ozonizer is shown in Fig. 12.

To achieve the necessary air speed in the irradiator-ozonizer, a fan (1) is used. To improve the efficiency of the device in the irradiator mode, a reflector (8) is placed under the lamp.

The reflector surface should have a large reflection coefficient of UV radiation with a wavelength of 254 nm. In the experiments conducted by Phillips (Phillips 1979), the following spectral reflectance coefficients of various metal surfaces were determined (Fig. 13).

As can be seen from Fig. 13, the high reflectivity of the short-wave UV radiation is aluminum, therefore, the paint based on the aluminum powder was used for the reflector (8).

Figure 12. General view of the irradiator: 1-fan, 2-UV lamp,3-hinged covers,4 locks

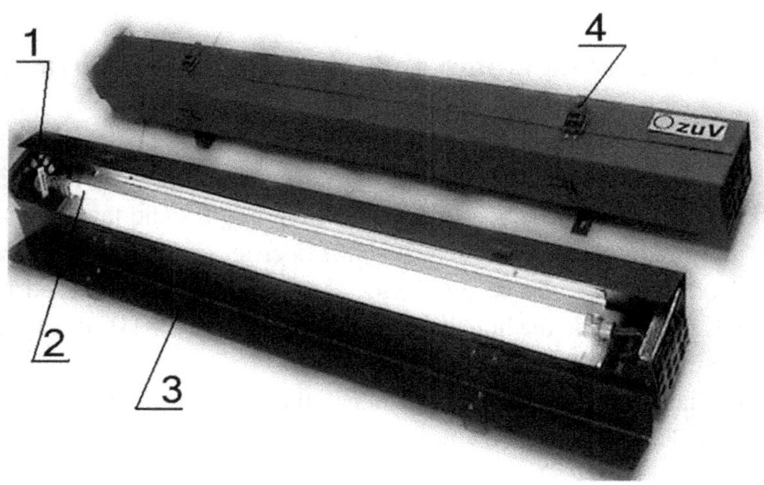

Figure 13. Spectral coefficients of reflection of UV radiation from different metal surfaces: a) evaporated aluminum, (b) aluminum foil, c) chromium, d) nickel, e) silver,f) stainless steel

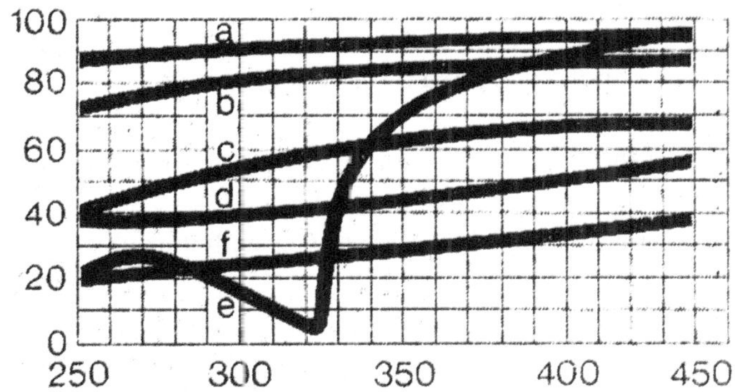

The case of the device is made of galvanized iron, the thickness of metal 0.3 mm provides the necessary strength and light weight. The device is powder coated with antistatic effect, which facilitates easy removal of dust. The surface of the reflector is coated with a paint with a high reflection coefficient of UV radiation with a wavelength of 254 nm.

The bactericidal disinfection plant is intended for operation in closed heated agricultural premises under the following conditions:

- **Air Temperature:** From 4 to 40 °C;
- **Relative Humidity of Air:** Up to 96%;
- Reactive environment;

Figure 14. Internal device of UV irradiator and aerodynamic tests of air flow

- Presence in the environment of aerosols and with veterinary-sanitary rules for cattle-breeding premises.
- **Working Position:** Floor, wall, suspended. The operating mode is continuous. Type of service - periodic.
- **Season of Work:** With technological need throughout the year.

The electrical connection of the "lamp socket" of the lamp is protected from the effects of aggressive media and moisture. The degree of protection must be at least 1P44.

The operation of the installation is controlled manually or with the help of a software device (Yuferev 2008)

The general view of the installation for ultraviolet air disinfection is shown in Fig. 15.

The use of compact mercury UV lamps of high power of the new generation allows to reduce the time of treatment of premises several times in comparison with irradiators on the basis of UV lamps of low power. At the same time, operating costs and energy consumption are reduced.

Despite the fact that the power consumption of new UV installations is higher (95 W) compared to standard installations (30 ... 40 W) collected on low-pressure UV lamps, the rate of reduction of the total microbial number, for example, up to 99.9% in rooms for sorting eggs, 3,2 times in comparison with bactericidal ozone-generating lamps with the power of 36 W and 15 times in comparison with the installations on bactericidal lamps with a power of 30 W. As a result, with the operation of new irradiation plants, considerable energy savings are achieved.

Figure 15. Equipment for UV disinfection of air with UV lamp

Table 5. Technical parameters and characteristics

Supply voltage, V	220
Power consumption, W	100
UV power, W	35
Air disinfection of industrial premises,%	99,8
Poultry house air disinfection,%	76,2
Overall dimensions, mm	750×120×140

Conducting Tests of Equipment

The essence of the tests lies in the fact that in the sealed chamber at a height of 2 m from the floor in the center of the chamber there was installed a irradiator-recirculator with an air divider to eliminate vortex flows and reduce the dustiness of UV lamps. In the second stage, a fan for drawing air with a capacity of 120 m^3 / h of air to the fan with an air capacity of 180 m^3 / h was replaced.

As a sanitary-indicative microorganism, Staphylococcus aureus strain 209P was used. In a chamber of 30 m^3 volume, a 2 billion suspension of culture was sprayed at a pressure of 3 atm in an amount of 1 ml / m^3. Air samples for bacteriological studies were taken before the irradiator-recirculator was switched on (initial background), and then after 10, 20, 30, 45 and 60 minutes of lamp operation with a Krotov device using 20 liters of air per Petri dishes with saline meat-peptone agar . The crops were grown in a thermostat at 37.5 ° C for 24 ... 48 h, and then the grown colonies were counted per 1 m^3 of air.

Doses of bactericidal UV rays were measured with an automatic universal dosimeter in J/m^2 in time.

The experiments were carried out in triplicate.

Results of Experiments

The average results of experiments on the efficiency of the irradiator-recirculation after its improvement for the decontamination of air contaminated with Staphylococcus aureus strain 209P are shown in the table.

It can be seen from the table that after the installation of the air diffuser, the efficiency of the irradiator-recirculator for air disinfection in the chamber increased by 2.5 ... 4.5%, and after installing a more powerful fan (P = 180 m^3 / h) and increasing the multiplicity of air treatment in hour (from 4 to 6 times) - by 7.48 ... 20%.

The use of an experimental energy-saving system of a new generation, consisting of UV irradiators with a power of 60 ... 100 W (Fig.16) and an automatic disinfection control unit, allows 1.5 ... 2 times to reduce capital and operating costs per unit area and increase the productivity of poultry.

These UV systems have also been tested for efficiency in various locations. A survey of a number of mayonnaise factory showed that in air the concentration of microorganisms is from 12 to 30 thousand / m^3. And on the surfaces of equipment and surfaces of premises - from 5 to 20 thousand bacteria per 100 cm^2 of area. High bacterial contamination of air and surfaces leads to a deterioration in the quality of manufactured products.

With this in mind, we conducted studies on the development of the regime and technology of air disinfection and surfaces of premises and equipment with the use of short-wave UV radiation.

Table 6. Efficiency of the irradiator-recirculator for air disinfection after its improvement

Exposition, Minutes	Dose of Ultraviolet Radiation, J / m³	Before the Installation of the Divider (Fan Capacity - 120 m³ / h) of air		After Installing the Air Divider		After Replacing the Fan (Fan Capacity - 180 m³ / h) of Air	
		Thousand Bacteria / m³	% of Disinfection	Thousand Bacteria / m³	% of Disinfection	Thousand Bacteria / m³	% of Disinfection
—	—	414,08	—	430,0	—	496,0	—
10	7700	155,28	62,5	150,5	65,0	77,4	85,0
20	15600	122,10	70,5	104,25	72,5	56,44	88,62
30	23500	60,86	85,3	90,3	89,0	37,10	92,52
45	39660	38,50	90,7	12,04	97,2	0	100,0
60	42200	20,70	95,0	2,15	99,5	0	100,0

(Prokopenko 2004)

Figure 16. Energy-resource-saving system of UV irradiation of premises for growing poultry

The experiments were conducted in a mayonnaise shop.

To this end, in a mayonnaise shop with a volume of 334 m³ on the side wall at the center at a height of 2.5 m, an irradiator with a UV lamp was installed. Two exposures of air and surface decontamination were tested: 3 and 4 hours. The effectiveness of the regime and technology was determined by the effect of bactericidal UV radiation on the air and surface contamination of the mayonnaise plant by microorganisms.

The total bacterial contamination of the air was determined with the help of the Krotov apparatus and Petri dishes with meat-peptone agar medium. The number of microorganisms on surfaces - by taking washings from 100 cm² of area and sowing by rubbing on Petri dishes with medium meat-peptone agar.

Doses of bactericidal UV radiation were determined by a dosimeter automatic universal DAU-81, and the amount of ozone - by iodometric method.

Approbation of the regime and technology of air disinfection and surfaces of premises and equipment of the workshops confirmed the high efficiency of the regime, which is recommended for wide application.

Tests of the Developed Disinfection System at the Poultry Farm

Production tests of irradiators were carried out at the poultry farm.

During the production tests of the system, the following are installed:

1. The effectiveness of the regime and technology of air disinfection by irradiators with UV lamps.
2. Influence of bactericidal UV radiation on the microclimate and bacterial composition of air in the premises for growing chickens.
3. Recommended modes and exposure exposures.

During the experiment, the dose of irradiation was repeatedly monitored and corrected. Work of irradiators with UV lamps was carried out at three exposures: 1, 2 and 3 hours.

During the tests, zooguinic parameters were determined (temperature, relative humidity, air speed, ammonia, carbon dioxide, ozone, bactericidal UV exposure dose), electrical network parameters (voltage), bacteriological parameters (bacterial air contamination in the test and control rooms), and zootechnical parameters (daily records were made of the mortality and safety of chickens, weighed the experimental and control chicks).

Determination of the physical properties of air - temperature, relative humidity was carried out with the help of the August psychrometer, the speed of movement in the ventilation ducts - a cup anemometer, in rooms - with a catameter.

Determination of chemical properties of air - the concentration of ammonia, hydrogen sulfide and carbon dioxide was carried out with a gas analyzer and corresponding indicator powders.

The electrical state of the air was determined from the concentration and type of air ions in it by the aeroion counter; light ions expressed a sign $n^{\pm}$, and heavy $N^{\pm}$.

Determination of the weight concentration of dust in the air of the chambers and rooms was made by a dust concentration meter.

The determination of the amount of ozone in the air was carried out by iodometric method and using an ozone gas analyzer.

Measurement of intensity and doses of bactericidal and erythemic UV radiation was carried out by an automatic universal dosimeter.

The effectiveness of the regime and technology of air disinfection by irradiators-ozonizers was determined by their effect on the total bacterial air contamination.

The quality of cleaning, disinfection and deodorization of air in livestock premises was determined by generally accepted methods.

For the sanitary and hygienic characteristics of bacterial air contamination, the total number of bacteria contained in 1 m3 of air (microbial number), the number of staphylococci and E. coli were determined.

Bacterial semination of air in the test and control rooms was determined at three points along the diagonal using the Krotov device. Air samples in an amount of 20 g were taken on petri dishes with medium meat-peptone agar and grown in a thermostat at 37 C for 24-48 hours, after which the grown colonies were counted per 1 m^3 of air.

A total of 12 experiments, 144 analyzes.

Control over the change in live weight and determination of the average daily weight gain of chickens was carried out by weighing them every 7-15 days of cultivation. Based on the results of the weights, the indicators for the growth of live weight were established.

On the preservation of young birds and calves were judged according to the records of the diseased and fallen.

To study the meat production of chickens, we took into account the mass of the carcasses and the subsequent determination of the slaughter yield.

Tests of the electrophysical two-component disinfection system for poultry facilities were carried out in the poultry house for the production of repair young birds aged 1 to 140 days in the poultry factory.

The tests were carried out in a room divided into 10 sections for rearing young animals for floor maintenance, 9x9 m in size with separate supply and exhaust ventilation.

The essence of the work was that three sections of the house were isolated from the rest with a polyethylene film (Figure 17)

In the first three sections at an altitude of 1.8-2 m from the floor, six irradiators with UV lamps were installed at a rate of one irradiator per 45 m^2 of room in one row at equal distances from each other. The stream of rays was directed to the upper zone of the room (Figure 18).

In the middle section, two irradiators were equipped with an uninterruptible power supply system having an automatic charger. In these sections there was an experimental group of chickens.

After the installation of the irradiators and the preparation of the premises for the experimental and control sections were completed, daily chickens were planted in the number of 3849 and 7500 heads, respectively.

Figure 17. General view of the test chamber

Figure 18. Arrangement of irradiators in the test chamber

In the experimental sections, the irradiators were switched on. In the first month, the cultivation of chickens irradiators with a low air volume worked according to the mode of 1 h of work and 2 h of break, and then on the regime of 2 h of work and 2 h of break during the daylight hours.

The experiments lasted for 2 months.

The need to clean the lamp and reflector was determined with the help of the ultraviolet radiation indicator developed by us.

All the studies were conducted according to generally accepted methods.

Economic Efficiency

Calculations of the economic effect are made on the basis of the conducted industrial tests of irradiators at the poultry farm. In calculations of economic efficiency, the method of growing young birds of egg-bearing breeds is compared with the same method in which irradiation regimes were included.

The irradiation period for each batch of chickens is 2 months. Chickens were irradiated every day 2 times for 3 hours.

The tests gave the following results - see Table 8.

The total increase in the live weight of commercial chickens of the experimental batch per 1 irradiator = 54.5 631 = 34381 g or 34.38 kg.

SOLUTIONS AND RECOMMENDATIONS

Methods of air disinfection with ultraviolet radiation are an effective, preventive, sanitary and anti-epidemic method of suppressing the vital activity of harmful microorganisms in the air and on the surfaces of objects of premises. Technical means for disinfection with ultraviolet germicidal radiation include:

Table 7. Test conditions

Indicators	Unit. Meas.	Maximum Permissible Concentration	Groups of Poultry	
			Experienced	**Controlled**
Temperature	°C	Мин.18	24,9	23,5
Relative humidity	%	64	64,7	64,2
Ammonia	mg / m^3	20	12	23
Hydrogen sulfide	mg / m^3		0	0
Carbon dioxide	%	0,03	0,02	0,05
Ozone	mg / m^3	0,1	0,03	–
Air velocity	m/s	0,2-0,4	0,2	0,25
Number of microorganisms in the air	thousands of microorganisms/m^3	240	272	1008
Ionic Composition				
light n +	ions/cm^3		556	465
n –	ions/cm^3		300	295
heavy N +	ions/cm^3		12500	72600
N –	ions/cm^3		2500	20500

Table 8. Results of irradiation chickens

Indicators	Inspection Lot of Chickens	Experimental Party of Chickens
Chickens, heads	7500	3849
Died, heads	224	61
Percent of died	2,99%	1,58%
Commodity number of chickens, heads	7276	3788
Live weight at the beginning, g / heads.	48	48
Live weight at the age of 2 months, g / heads.	864	899
Growth of live weight of 1 chicken of experimental lot, g		**35**
Increase in the live weight of the average commercial chicken of the experimental batch due to the reduction of the case, g.		**19,5** *1
The total increase in the live weight of the average commercial chicken of the experimental batch, g .		**54,5** *2
The total increase in the live weight of commercial chickens of the pilot lot per 1 irradiator, kg		**34,3 кг.** *3

* 1 Increase in the live weight of the average commercial chicken of the experimental batch due to the reduction of the mortality = 899 (224-61) / 7500/3788 = 19.5 g.

* 2 The total increase in the live weight of the average commercial chicken of the experimental batch = 35 19.5 = 54.5 g.

* 3 Number of chicks per irradiator = 3788/6 = 631 pcs.

sources of ultraviolet bactericidal radiation and irradiators that redistribute the radiation flux into the surrounding space in a given direction.

Ozonation, as an environmentally friendly oxidizer, has a strong disinfecting effect and is used in sanitation as a means of combating pathogenic microflora. One way to get ozone from the air is the photochemical method of obtaining ozone.

In this regard, we proposed to conduct a study on the possibility of using ozone-forming UV lamps for disinfection, in the calculation that the disinfecting effect of UV radiation directly will be enhanced by the oxidizing properties of ozone.

The research task in this work was the creation of a system of two-component electrophysical air disinfection on the basis of simultaneous exposure to ultraviolet radiation and ozone. Poultry premises are selected for production approbation.

Conducted comparative tests have confirmed the assumption that the bactericidal effectiveness of ultraviolet irradiators with ozone forming lamps is several times higher than that of irradiators with conventional bactericidal lamps due to the use of ozone as an additional disinfectant.

A mathematical model of the effectiveness of air disinfection with ultraviolet radiation is theoretically justified and suggested, which allows to substantiate the technological decontamination process that regulates the direction of the UV bactericidal radiation flow in the poultry house, under which the maximum disinfection takes into account the characteristics of the disinfected medium, including the transmission of UV germicide radiation to the decontaminated air and the distance between the radiation source and the absorbing surface.

The calculations for the installation were checked as a result of the experiment at the poultry farm, the reliability of the calculations was confirmed. As a result of approbation it was confirmed that for poultry people the amount of irradiated space is an important component of the energy efficiency of disinfection.

The experimental calculation of the maximum number of irradiators in a room for growing poultry was confirmed during production trials. Based on the results of the studies and calculations in Chapter 2, we substantiated and developed the design of the ozonizer's irradiation units, the materials of the body, the reflector, and also the lamps of the type DBK-8 and DBK-36 were chosen, on the basis of which the models of irradiators were developed.

The scheme of electronic high-frequency ballast developed by us and replacement of the old fan with the fan JF0825 reduced the energy consumption of the irradiator by 45 VA, extended the range of supply voltages to 170-245 V; increased lamp life. In addition, since the dimensions of the irradiator are determined by the cross section of the fan, the material consumption of the structure has decreased by 11% and the weight of the irradiator.

The modified model of the irradiator is optimal for the construction, the shape of the hull and the quality of disinfection for poultry facilities.

The laboratory tests of the UV indicator showed that the device is operational. The irradiators were tested and passed production tests in pilot farms and a poultry farm, which allowed setting the UV recession rate of lamps and adjusting the program of operation of the units.

However, design defects were also revealed, namely: the need for shielding from UV radiation from the eyes and exposed parts of the body during measurements.

To ensure the necessary exposures, we have developed a control system that makes it possible to substantially simplify the maintenance of irradiation facilities.

The use of developed installations at poultry farms will increase the storage conservation of livestock, increase the weight gain of poultry, improve the composition of air in the room, and prevent the entry of epidemic diseases into the poultry house.

REFERENCES

Alferova, L.K.,, Kozlov, A.I., & Yuferev, L.Y. (2000). Ul'trafioletovyy obluchatel'-ozonator. *Sel'skiy mekhanizator*, *6*, 35-36.

Alferova, L.K., & Yuferev, L.Y. (2002). Bakteritsidnaya ustanovka povyshennoy effektivnosti. *Tekhnika v sel'skom khozyaystve*, *2*, 29-31.

Band, S. (1901). *Die Eirkungen des Eichtes Sikreorganiscen*. Milt, Finsen.

Gates, F. L. (1928). On nuclear derivatives and the lethal action of ultraviolet light. *Science*, (68): 259–275. PMID:17758687

Mironyuk, S.S. (2015). The effect of antimicrobic treatment on the quality and storage duration of aubergine fruits. *Collection of scientific works of Vinnitsa National Agrarian University*, *1*, 111-116.

"OZUF" na ob'yektakh veterinarnogo nadzora. (n.d.). *Trudy mezhdunarodnoy nauchno-tekhnicheskoy konferentsii Energoobespecheniye i energosberezheniye v sel'skom khozyaystve*, *3*, 247-249.

Philips, N.V. (1979). Gloeilampenfabrieken. *Germicidal lamps and applications*.

Popova, N.V., Tokar, A.Y., & Mironyuk, S. S. (2015). Influence of treatment of antimicrobial agents with microbial damage of eggplant fruits. *Scientific works of the National University of Food Technologies*, *3*, 236-242.

Prokopenko, A. A. (1978). Influence of different doses of ultraviolet rays of unshielded bactericidal irradiators on the microclimate and productivity of the laying hens. *Proceedings of VNIIVSGE*, *61*, 139–141.

Prokopenko, A.A., & Yuferev, L.Y. (2004). *Effektivnost' primeneniya obluchateley-retsirkulyatorov*. Academic Press.

Tokar, A. Yu., & Mironyuk, S. S. (2015). Preservation of eggplant fruits provided processing of antimicrobial substances. *Progressive technique and technologies of food production of restaurant and trade*, *1*, 400–409.

Vashkov, V. I. (1973). *Means and methods of sterilization, used in medicine*. Moscow: Medicine.

Yuferev, L. Y. (2006). Increase the operational and energy characteristics of irradiators. *Proceedings of the International Scientific and Technical Conference Energy supply and energy saving in agriculture*, *3*, 270-275.

Yuferev, L. Y., & Alferova, L. K. (2016). *Svetotekhnika v sel'skom khozyaystve*. Moscow: FSAC VIM.

Yuferev, L.Y., Alferova, L.K., & Baranov, D.A. (2008). Avtomatizatsiya obezzarazhivaniya ul'trafioletovym obklucheniyem v pomeshcheniyakh dlya vyrashchivaniya molodnyaka ptitsy. *Sbornik nauchnykh dokladov VIM*, *1*, 448-456.

Yuferev, L.Y., Alferova, L.K., Baranov, D.A., & Prokopenko, A.A. (2010). Obluchatel'nyy pribor na baze amal'gamnykh UF lamp dlya ob'yektov veterinarnogo nadzora. *Trudy mezhdunarodnoy nauchno-tekhnicheskoy konferentsii Energoobespecheniye i energosberezheniye v sel'skom khozyaystve*, *3*, 292-296.

Yuferev, L.Y., Alferova, L.K., & Prokopenko, A.A. (2007). Elektrofizicheskoye obezzarazhivaniye i dezodoratsiya vozdukha v zhivotnovodcheskikh pomeshcheniyakh. *Nauchnyye trudy GNU VNIIMZH Rossel'khozakademii*, *17*(4), 63-68.

Chapter 4
Energy-Saving Electrical Installations for Heat Supply of Agricultural Objects

Dmitry Tikhomirov
Federal Scientific Agroengineering Center VIM, Russia

Alexey Vasiliev
https://orcid.org/0000-0002-7988-2338
Federal Scientific Agroengineering Center VIM, Russia

Stepan Dudin
Federal Scientific Agroengineering Center VIM, Russia

ABSTRACT

Electrical energy in the thermal processes of agricultural production has a number of technical and technological advantages in comparison with other energy carriers. At the same time, electricity is the most high-quality but expensive type of energy. The effective use of electrical energy in thermal processes is possible with the implementation of all its advantages over fuel energy. This requires a system of technical, economic, and energy analysis and the selection justification of the most effective systems and technical means of heat supply of stationary agricultural technological processes. The chapter presents functional and technological schemes, physical models, the experimentally obtained dependencies, parameters, and operating regimes of the developed energy-saving electrical equipment and systems for the main thermal technological processes.

INTRODUCTION

Agricultural production is a major consumer of fuel and energy resources (FER) in stationary processes in the amount of about 10 million tons of reference fuel. Over 65% of them are used in the heat supply systems of many production facilities: livestock farms, poultry farms, enterprises for drying, processing and storage of agricultural products, peasant and subsistence farming. Heat is used in the technological

DOI: 10.4018/978-1-5225-7573-3.ch004

processes of providing microclimate, hot water and steam supply, heating, heat treatment of feed, drying, storage, processing of various types of agricultural products and a number of other technological processes.

At present imperfect systems and technical means of heat supply of agricultural production facilities which are ineffective are still used without sufficient scientific justification. These include Central boiler-houses of various types. They are characterized by excessive capital intensity, large losses and significant energy consumption and do not provide normalized housing conditions. In addition, the state of existing fuel boilers, heat pipelines and heat power equipment are due to aging along with a significant decrease in technical and economic indicators.

There is a direct link between production costs and energy expenses. The share of energy expenses in the cost of agricultural products has increased from 5-8% to 15-20% and even more. This is primarily due to the rapid growth of prices for electricity and fuel compared to the prices of agricultural products as well as the use of morally and physically obsolete technical means and heat supply systems. There is an urgent need for livestock and other enterprises for more economical use of heat and electricity. The relatively inexpensive projects with a "cold" method of keeping animals in this regard are implemented without sufficient justification. At the same time the established zootechnical requirements and norms of technological design are not observed, especially for premises with young animals.

The justification of the directions of development and modernization of heat supply systems for livestock production facilities is an urgent task for the implementation of the country's energy policy on energy efficiency and energy expense reduction.

BACKGROUND

Many scientists have devoted their scientific works to separate issues of electric heat supply of agricultural facilities. The economic efficiency of the use of electrical energy in the thermal processes of agricultural production is justified in the works of domestic and foreign authors (Kanakin & Kogan, 1986; Rasstrigin, 1988; Morozov, 2011).

There should be noted scientific works and developments of scientists in the field of hot water and steam supply (Kagan et al., 1980). In the development of electric steam boilers and water heaters much attention is paid to the heat exchange between electric heater surface and water. The correct choice of heating elements depends on their service life and the reliability of electric heating equipment. The basis of the method of calculation of heat transfer between electric heater and water in bubble boiling mode are the works of well-known experts in the field of heat transfer (Kutateladze, 1990; Rozhenau, 1966; Tolubinskij, 1989). The works of Kuzma-Kichta (1994), Pis'mennyj (1999) and Tikhomirov (2002) are devoted to the intensification of heat exchange in bubbling boiling water. Currently much attention is paid to the creation of combined systems of hot water and steam supply with the use of hot water accumulation in a capacitive water heater-thermos (Tikhomirov, 2013a). The thermal accumulation occurs in the time when there is a reduced cost for electricity.

The creation of energy-saving ventilation and heating systems for livestock premises allows to realize the maximum genetic potential of livestock and poultry in order to ensure the maximum productivity with a significant reduction in the production energy intensity. A significant energy effect can be obtained by using the utilization of heat from the air removed from the room and its partial recirculation. (Kazantsev & Ignatkin, 2013). The recirculation system cannot be directly applied in livestock premises due to the high bacterial contamination of the circulating air and gaseous substances resulting from the animal life and high moisture. It is necessary to provide modern methods of air cleaning and disinfection. Cleaning and disinfecting the air is also necessary because with the ventilation air from the premises carried a large amount of dust, microorganisms, gases, which leads to pollution ecological environment. With high content of dust and microorganisms in the outside air ventilation it becomes ineffective.

A new method of creating a microclimate in livestock premises based on the principles of heat utilization, ozonation and air recirculation is proposed in the work of Markelova et al. (2013). Partial recirculation of warm air removed from the room is possible during its cleaning and disinfection, including using the ozonation method.

Another important area of energy saving in microclimate systems is the use of local heating using radiant (infrared) electric heaters. The most effective method is the use of radiant electric heating in rooms with young animals, where thermal zones with different temperatures can be built; as well as in livestock houses with periodic operating regimes (milking parlors, veterinary sanitary inspection, warehouses, isolation wards, box stalls etc.). The method of heat exchange calculation between the animal and the environment also is based on the methodological framework proposed by the works of Bystritsky et al. (1981) and Rastimeshin (1990). However, the equations for determining the convective heat transfer coefficients proposed in these works do not consider the age (or weight) of the animal. It makes a certain error in the calculations for different age groups of animals. Heat exchange of the animal with the environment under their artificial heating depending on the age of animals (calves, pigs) is considered in the scientific works of Kuzmichyov (2017) and Tikhomirov (2017).

Heat treatment of liquid foodstuffs (milk, juice etc.) refers to the number of common and energy-intensive processes of primary processing with the purpose of their preservation. At the same time the task is to preserve the maximum nutritional and taste qualities of the product. The effect of infrared radiation (IR) on the thin layer of liquid products is one of the energy-efficient ways of heat treatment. Short - term exposure to IR radiation with a high density simultaneously across the thickness and surface of the thin layer of a liquid creates the necessary conditions for the elimination of toxic and ballast microflora. It allows to keep useful biological and physic-chemical components (proteins, vitamins, enzymes) that determine the nutritional and organoleptic properties of the product (Spalart & Rumsey, 2007). The method of cold pasteurization is drawn attention due to the low energy intensity of the process. However, the effect of ultraviolet (UV) radiation on milk has a number of limitations (Smith, Lagunas-Solar & Cullor, 2002; Yong Sik Cho, Kyung Bin Song & Koji Yamda, 2010). For UV treatment of milk it is possible to use bactericidal lamps of low pressure without ozone formation with the power of 100-300 W. In the experimental equipment (installation) developed in the work of Kuzmichyov, Malyshev & Tikhomirov (2011) there was bactericidal amalgam lamp (DB-145, power is 145 W). The bactericidal flow of the lamp is 45 W. The area of irradiated surface S= 600 cm2. The effective effect of UV radiation for the bacteria destruction by this lamp is in the spectral region with λ max= 253,7 nm.

At the same time not much attention is paid to the problems of complex electrification of thermal processes for livestock facilities. Rastrigin V. N tried to solve this problem in the 80-90 and early 2000-ies (Rasstrigin, 1988). Among modern works the works of Barkov (2006) and Aldibekov (2010) should be noted. The purpose of their research was to create a scientific and methodological basis for the development of the resource and energy-saving systems of electric heat supply in animal husbandry on the basis of improved electrode thermal plants (boilers electrode type). More fully energy-saving electrical systems and technical means of heat supply in agro-industrial complex are considered in the work of Tikhomirov (2016a).

SUBSTANTIATION OF THE EFFECTIVE USE OF ELECTRIC HEATING SYSTEMS IN AGRICULTURE

The nature of thermal loads and the requirements most directly correspond to the electrical system of heat supply. It is known that the use of electric energy in the thermal processes of agricultural production in comparison with other energy carriers (solid and liquid fuels, gas, biomass, etc.) has a number of undeniable technical and technological advantages. The unique property of electric energy transmission and distribution allows to bring it directly to the place of heat use and make electrical heating systems more economical and easily controlled. The use of electric heating systems of a decentralized type reduces losses and capital investments due to not using such equipment as external heat lines and boiler houses. High efficiency and automation level of electrical systems allows more economical using of electricity and accurately maintaining the parameters of these processes with a high degree of readiness for commissioning. Electrical systems are environmentally friendly and create the best conditions for the staff. The main technological processes of agricultural production where thermal energy is used are shown in Figure 1.

Figure 1. Application of thermal energy in agricultural production

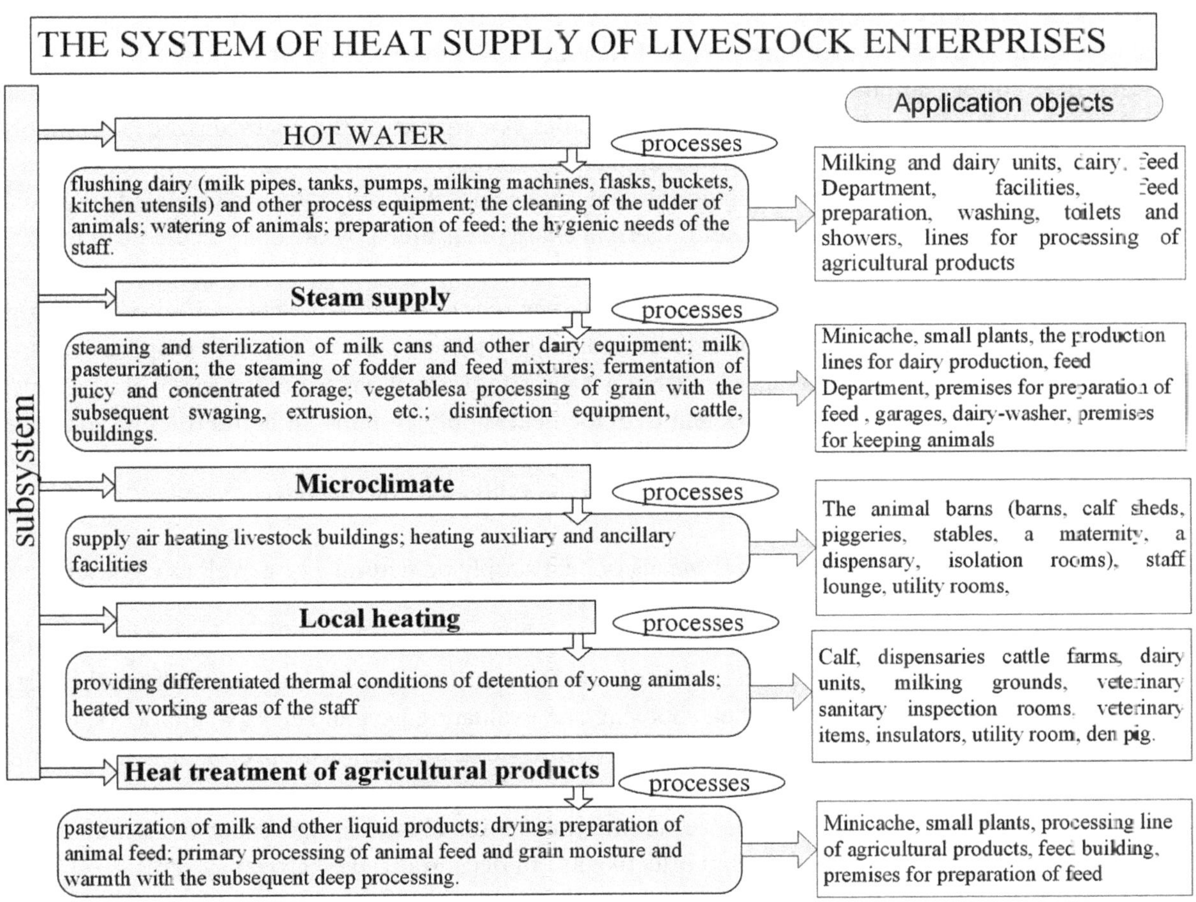

Thermal electrical power equipment is used on a limited scale due to high tariffs. The justification and development of modern, high-energy-saving systems and technical means of electrical heat supply for livestock production facilities therefore is the most important research task. This will lead to increased productivity, reducing total and energy expenses by 20-30%, and hence decreasing the energy intensity and agricultural product cost as well as improving work conditions and the environment.

The possibility of using different energy carriers in the technological processes of agricultural production, the variety of systems and technical solutions suggest a technical and economic comparison of different options for heat supply and choose the most economical option.

The calculation of the required power and annual energy consumption in the thermal processes of agricultural is the basis for the selection of thermal power equipment and the subsequent justification of the heat supply system.

The program project for a personal computer designed in works of Tikhomirov & Vasiliev (2016b) allows to carry out automated calculation of the heat and power indicators of the object. The project contains software modules, database and screen forms. The calculation is carried out for each premise where the animals are kept considering the technology and method of their maintenance. The calculation algorithm is based on the theoretical basis of well-known methodological recommendations with a number of new significant additions. All parameters and coefficients included in the project database are taken from the current recommendations and standards of technological design of livestock enterprises and the relevant sanitary norms and rules. At the same time a user can independently change and set the initial data in order to assess their impact on the final heat and power indicators of the object.

Serial heat power equipment was selected on the basis of the results of calculation for various and typical livestock facilities. The technical and economic assessment of the characteristic heat supply systems was carried out as well. Generally accepted both centralized systems (on solid, liquid, gaseous) and decentralized systems (on solid, liquid, gaseous fuels and electricity) were considered. Figure 2 shows the specific costs per 1 kWh of useful thermal energy consumed by the dairy cattle farm for 200 heads of cattle.

The graphs, shown in Figures 3 and 4, explain the dynamics of changes in the specific reduced costs per 1 kWh of useful thermal energy for various heat supply systems with a possible proportional and outstripping growth in the tariff for electricity. Even at a 1,5-fold rate of outstripping growth of electricity tariffs electric systems will be more efficient than the heat supply systems on liquid (diesel) and solid (coal) fuels but still inferior to natural gas systems.

Technical and economic analysis allowed to draw the following conclusions:

1. Decentralized systems and technical means of heat supply on natural gas as well as electricity are the most effective.
2. The cost of obtaining and using 1 kWh of useful thermal energy is 2 ... 4 times higher than just for its production. The specific reduced costs of obtaining and using 1 kWh of useful thermal energy therefore should be considered when choosing and evaluating heat and energy supply systems.
3. Capital investments and operating costs of electric systems are the lowest in comparison with fuel systems.
4. The less facility capacity is, the more total unit costs for obtaining and using 1 kWh of useful thermal energy are and thus the more effective use of decentralized electrical systems is.

Figure 2. The specific costs per 1 kWh of useful thermal energy
Centralized systems: 1- ordinary coal, 2- diesel fuel, 3 – natural gas;
Decentralized systems: 4 - ordinary coal, 5- diesel fuel, 6 - natural gas, 7- electric energy

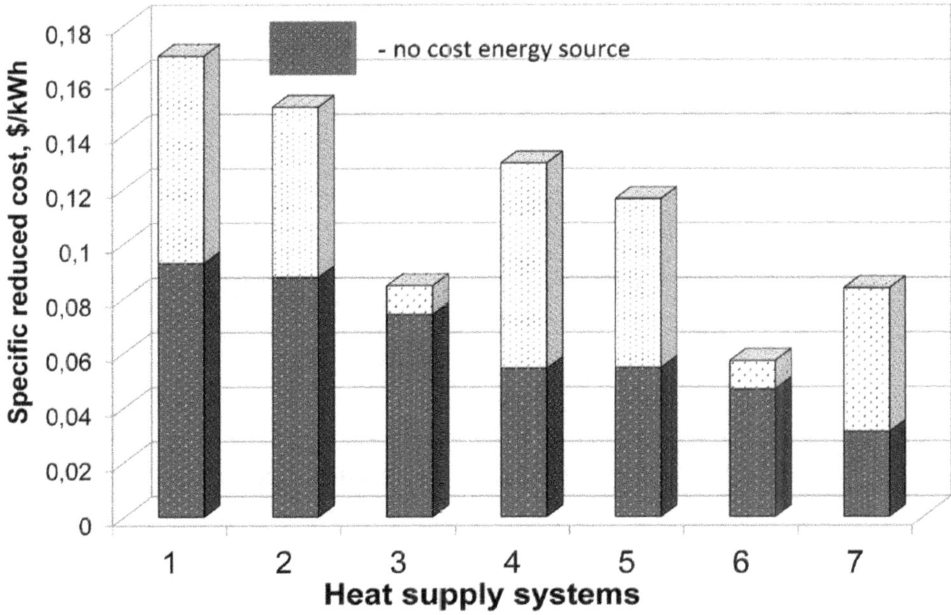

Figure 3. The proportional increase in energy prices

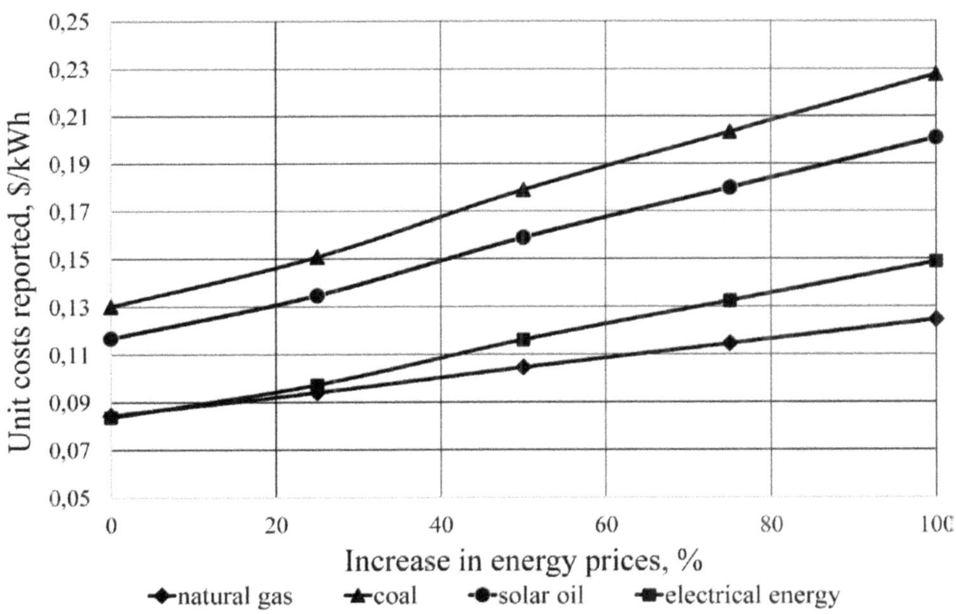

Figure 4. More rapid growth in electricity tariff compared with the price growth in fuel energy carriers

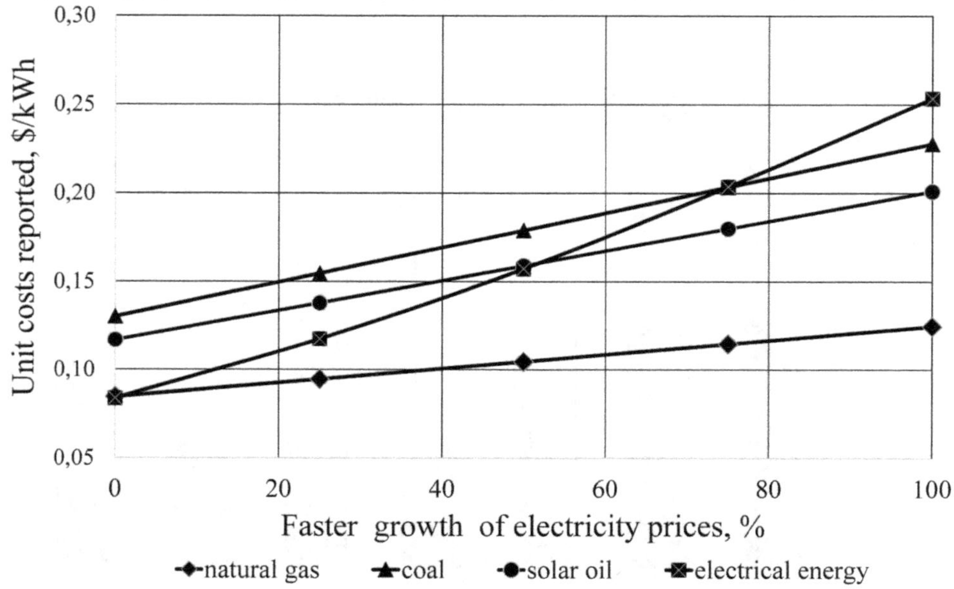

The analysis and calculations of heat supply systems of livestock facilities allowed to assess the most effective application area of electrical systems and their power equipment in order to identify areas for improving its energy efficiency and reducing energy costs. This determined the research on the development and improvement of power systems and equipment.

The basic conceptual provisions of development of heat electric systems were accordingly formulated. There are main directions among them:

- The application of the systems and equipment of decentralized type heat supply;
- The heat recovery of exhaust air in systems of supply air heating;
- The development of the technology able to work according to the differentiated electricity tariff; the wide use of local radiant (infrared) heating technology with the development of irradiators adapted for agricultural production;
- The reasonable use of autonomous cogeneration plants, heat pumps and more advanced heat-cooling equipment;
- The development of heat-power equipment that allows easy integration into modern technologies and processes of milking, feed preparation, primary processing of agricultural products;
- The use of renewable, local and non-traditional energy resources (Vasant & Voropai, 2016) etc.

RESEARCH AND DEVELOPMENT OF A COMBINED INSTALLATION FOR HOT WATER AND STEAM SUPPLY OF LIVESTOCK FACILITIES

Hot water and steam are widely used for washing and sterilization of dairy and other equipment, for the preparation of feed, for sanitary purposes, the needs of maintenance personnel as well as for processing agricultural products in numerous workshops etc. There is various serial equipment for agriculture. A significant part of them does not have autonomous devices for steam generation and water heating while these heat carriers are needed to carry out technological processes.

It is necessary to know the volumes and modes of consumption of these heat carriers to justify the parameters and operation modes of steam generation and hot water as well as possible modes of operation of the plants themselves. Daily schedules of thermal loads give the most complete picture about this. For the construction and analysis of daily schedules of steam and hot water consumption were considered typical projects of livestock facilities: milking units, dairy, feed preparation facilities of cattle farms as well as pig houses with facilities for the preparation of feed and hot water.

The consumption of hot water and steam in livestock farms is clearly uneven due to technological requirements. The processes of steam and hot water consumption are partially simultaneous and partially diverged within work time. The low load factor of the steam production unit (not exceeding 0,5) suggests its use for the preparation of hot water and confirms the possibility of developing a unit that allows separate production of steam and hot water in one device – an electric steam heater.

The installation can work for hot water preparation in 2 modes: in the flow mode with a free schedule and with full or partial accumulation of hot water in the storage tank or conjunction with another water heater, for example, accumulative type. Graphs comparing the two modes are shown on Figure 5. The mode of full accumulation of hot water can significantly reduce the peak power consumption and align the schedule of loads for the preparation of hot water and steam during the day. This mode allows consumers to reduce the cost of paying for electric energy when using a differentiated tariff during the day.

The block diagram of the electric steam and water heating installation contains a storage water heater, steam generator and an electric steam superheater (Figure 6).

Figure 5. Daily schedule of loads on hot water and steam supply of cattle farm for 200 heads

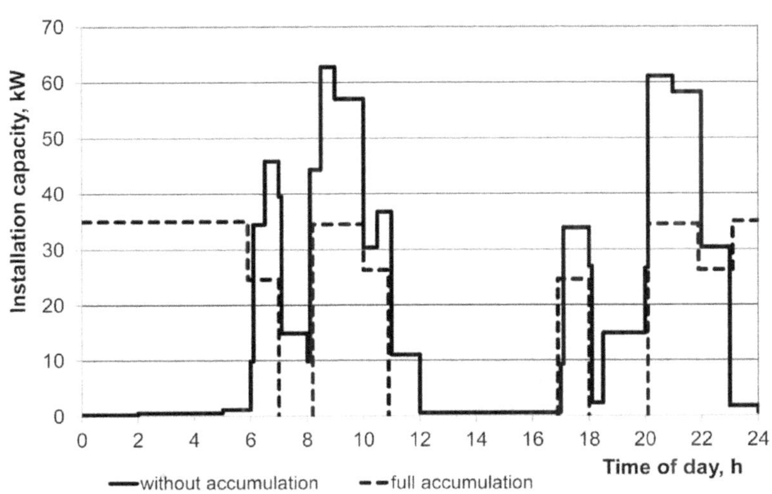

Figure 6. Block diagram of the electric heating installation

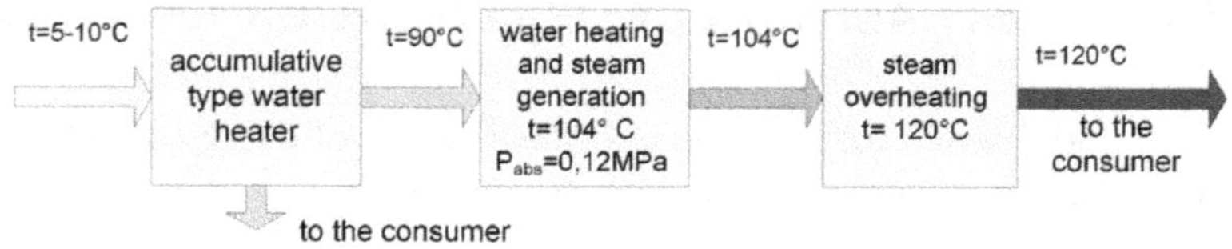

The original method of calculation of the electric heating installation determines the energy, thermal and structural parameters (Tikhomirov, 2015).

The power of the steam generator:

$$P_{sg} = \frac{c_w G_s (t_{bp} - t_a) + G_s r}{3600\eta}, \tag{1}$$

where c_w is the specific heat of water, J/kgK; G_s is the performance of steam, kg/h; t_{bp} is the boiling point of water in the steam generator at a pressure of P, °C; t_a is the temperature of water entering in the steam generator (90 °C); r is the specific heat of vaporization, J/kg; $\eta = 0{,}98$ is efficiency of the steam generator.

To estimate the heat transfer conditions, we should apply a known expression:

$$P_{sg} = \alpha F_h \Delta t, \tag{2}$$

where α is heat transfer coefficient from electric heaters to water, W/m^{2}K; F_h is the surface area of electric heaters, m^2; Δt is the difference between the surface temperature of the heaters and the average water temperature in the steam generator, °C.

Heat transfer coefficient α should be taken in the range of 9000…11000 W/m^{2}K.

The volume of vapor space V_v in the tank should be not less than one third of the volume of the tank V_t:

$$V_v = FH, \tag{3}$$

where H is the height of the vapor space, m;

$$H = \left(\frac{\omega}{0,54(0,65 + \omega)} \right)^2, \tag{4}$$

$$\omega = \frac{1,5 G_s}{3600 F \rho}, \tag{5}$$

where w is reduced steam speed, m/s; F is the area of evaporation, m^2; ρ is the saturated steam density, kg/m^3.

The electric superheater is used to overheat steam and its power and is determined by the expression:

$$P_{sh} = \frac{G_s(i - i'')}{3600\eta}, \tag{6}$$

where *i, i"* are enthalpy of superheated and saturated steam, kJ/kg; η is the superheater efficiency.

An expression for the evaluation of the heat exchange conditions of the electric heater and steam in the superheater is obtained.

$$P_{sh} = \varepsilon_1 \varepsilon_2 \alpha F_h \Delta t, \tag{7}$$

where ε_1=1,12…1,15 is the coefficient taking into account the radiation component of the heat flow from the heater to the steam; ε_2 = 1,2…1,4 is the coefficient taking account the convective component from the inner surface of the superheater body to the steam; α is the heat transfer coefficient (convective component) from the heater to the steam, (26…28 W/mK); F_h is the surface area of the heater, m^2; Δt is the difference between the surface temperature of the heater and the average temperature of steam in the superheater (150…160 K).

The surface temperature of the electric heater can be determined by the dependence:

$$t_h = t_s \Big/ \left(1 - 0,6e^{-\frac{\alpha \cdot F_h}{c_w \cdot G_s}}\right). \tag{8}$$

When using an accumulative electric water heater its power can be calculated from the expression:

$$P_w = \frac{1,05(V_w + V_s)c_w(t_h - t_c)}{3600\tau_w \eta_w}, \tag{9}$$

where V_w, V_s are the daily consumption of hot water and steam, kg; t_c, t_h are the temperature of cold and hot water, °C; τ_w is the time of water heating, h; η_w is the efficiency of the device.

$$V_s = G_s \tau_{gs} / \eta_{sp}, \tag{10}$$

where τ_{gs} is the time of steam generator operation, h; η_{sp} is the coefficient considering the loss of steam in the steam pipe.

$$V_w = \frac{1}{t_h - t_c} \sum_{i=1}^{n} \left[G_i(t_i - t_c)\right], \tag{11}$$

where G_i, t_i are flow rate, and water temperature for the *i*-th process.

The technique is applicable for the standard size range of installations and their working bodies of similar design, arrangement of working bodies and separate devices in accordance with the design and technological scheme (Figure 7).

Figure 7. Design and technological scheme of energy-saving hot water and steam supply installation
1 is accumulative water heater; 2, 9 are heater units; 3 is hot water intake pipe; 4 is electric steam boiler; 5 is separator; 6 is steam overheating section; 7 is steam inlet; 8 is water level sensors; 10 is automatic water supply valve

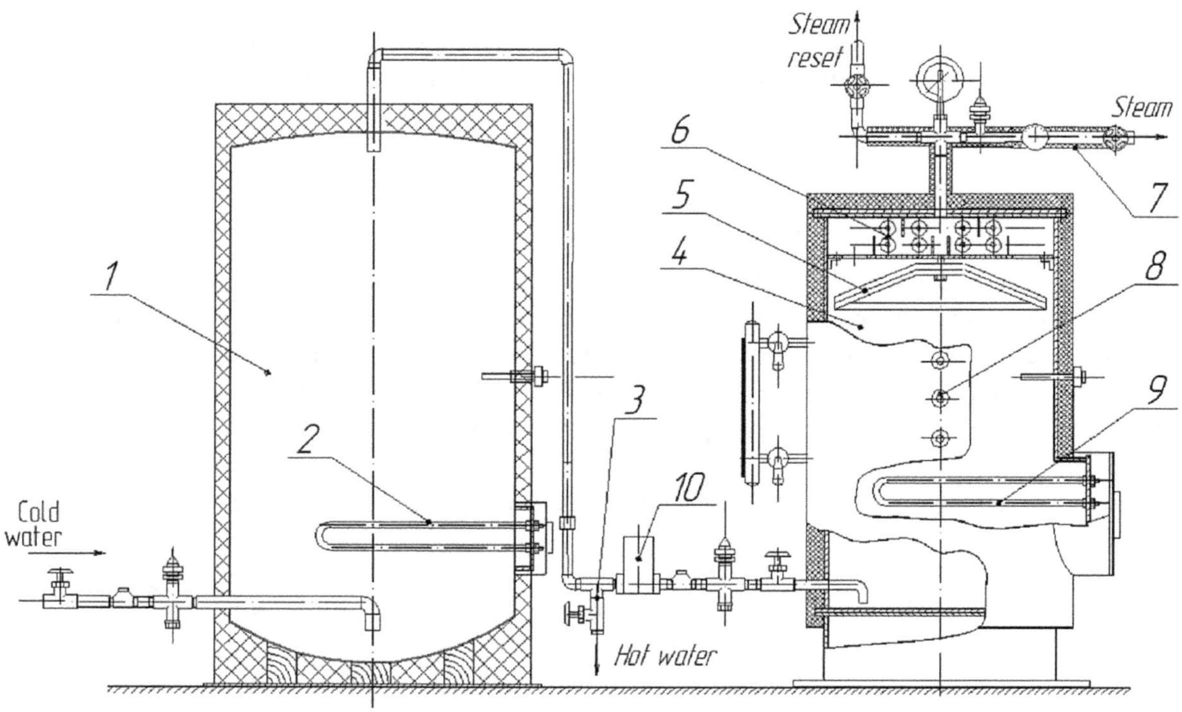

The physical modeling of the heat exchange process in each individual installation and its mathematical description allowed:

- To determine the heat transfer coefficient and the heat flow from the heaters to the heat carrier;
- To make the equation of the heat balance;
- To calculate the efficiency;
- To analyze the operating conditions;
- To check the correctness of the choice of energy and structural parameters of heating elements;
- To identify methods to intensify heat exchange.

Issues related to boiling fluid have been studied in the works of scientists known in the field of thermal engineering: M. A. Mikheev, V. M. Borishanskii, V. I. Tolubinskij, G. N. Kruzhilin, S. S. Kutateladze, U. Roshenau, T., Hobler and others (Rozhenau, 1966, Tolubinskij, 1989, Kutateladze & Borishanskii, 2012). A number of empirical formulas describing the process of heat transfer in boiling liquids have been proposed to date, but all of them are universal and at the same time approximate. The heat transfer coefficient depends on a large number of different factors and the specific conditions of heat exchange including the type and condition of the surface which cannot be considered in a single generalized analytical dependence (a criterion equation).

Figure 8. Heat transfer coefficient at bubble boiling of water in the electric steam heater

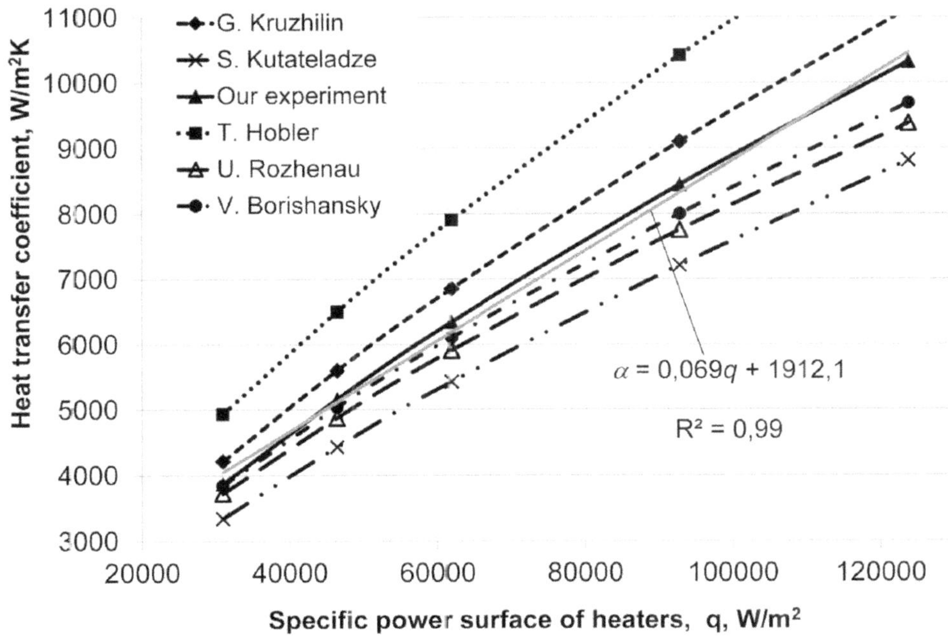

The difference between the obtained results of the heat transfer coefficient α in the problem under study and the known dependencies reaches 30% (Figure 8), which makes a significant error in the thermal and constructive calculations.

Specifying the task and conducting special studies for the electric steam heater became necessary in this regard.

The analysis of statistical processing of the experimental data obtained by means of the Maple mathematical system allowed to determine an interpolation polynomial enabling to calculate the heat transfer coefficient α from the specific surface power of the heaters q in the electrical steam boilers for the same type-size series (Lebedev, Rasstrigin & Tikhomirov, 2000):

$$\alpha = -1{,}6q^4 + 48{,}3q^3 - 513{,}8q^2 + 3000{,}1q - 1780{,}8 \quad (12)$$

where $q = 4{,}0\ldots13$ W/cm^2

The dependence in the interval $\alpha = f(q)$ is in fact linear. With the accuracy of the approximation $R^2 = 0{,}99$ the equation has the form:

$$\alpha = 0{,}069q + 1912{,}1 \quad (13)$$

where $q = (4{,}0\ldots13) \cdot 10^4$ W/m^2.

The maximum permissible electrical and thermal loads of tubular electric heaters (TEH) in boiling water are determined

$$\Delta t_{cr} = t_{sh} - t_s = (4,0 - 0,105 t_s)(q_{cr} \cdot 10^{-6})^{0,3} \tag{14}$$

where t_{sh} is the heater shell temperature, °C; t_s is the steam saturation temperature, °C; q_{cr} is the maximum permissible specific heat flow of the electric heater in the mode of nucleate boiling, W/m², q_{cr}=150000W/m².

The dynamics of convective heat exchange between an electric heater and the flow of steam allows the determination of the heat and energy parameters of the electric superheater and the parameters affecting the normal operation of the heating elements (Tikhomirov, 2014). The system of equations of thermal balance describing the change in temperature of the heater and the coolant (water vapor) for the dx element is defined as follows:

$$c_h F_s \rho_h dx dt_h = q dx d\tau - \frac{\alpha F}{L}(t_h - t_s) dx d\tau, \tag{15}$$

$$c_s F_s \rho_s dx dt_s = \frac{\alpha F}{L}(t_h - t_s) dx d\tau. \tag{16}$$

After the transformation it was received:

$$\begin{aligned} c_h F_h \rho_h dx \frac{\partial t_h}{\partial \tau} &= q dx - \frac{\alpha F}{L}(t_h - t_s) dx, \\ c_s G_s d_s &= \frac{\alpha F}{L}(t_h - t_s) dx, 0 \le x \le L, \end{aligned} \tag{17}$$

where c_h, c_s are the specific heat capacity of the material of the heater and steam, J/kgK; F_h, F_s are the sectional areas of the heater and steam m²; F is the heat transfer surface, m²; ρ_h, ρ_s are the density of the heater material and steam, kg/m³; q is amount of heat generated by the heater without taking into account heat loss, W/m; α is the coefficient of heat transfer from the heater to the steam, W/m²K; t_s, t_h are the steam and heater temperature, ºC; τ is the time; G_s is the steam consumption, kg/s; m_h is the mass of the heater, kg; L is the working part length of the superheater, m.

The solution of equation (15) on t_s/t_h:

$$\frac{t_s}{t_h} = 1 - C e^{\frac{-\alpha \cdot F \cdot X}{c_s \cdot G_s \cdot L}} \tag{18}$$

The partial solution of the equation (18) has the following form:

$$\frac{t_s}{t_h} = 1 - 0,62 e^{\frac{-\alpha \cdot F \cdot X}{c_s \cdot G_s \cdot L}} \tag{19}$$

For steam capacity G_s=45 kg/h:

$$\frac{t_s}{t_h} = 1 - 0,62e^{-0.22 \cdot X} \quad (20)$$

After jointly solving equations (17) and (20) the expression was obtained:

$$t_H = 289 - 185e^{-2,5 \cdot 10^{-3} \cdot \tau} \quad (21)$$

The graph of the temperature change of the heater surface in the superheater since the moment of switching on to the nominal mode of operation with the initial steam temperature t_s=104 °C is shown in Figure 9.

The radiation component of the heat flow from the heating elements to the steam was about 12%. The evaluation was carried out using the individual provisions of the method described by Volenberg.

Starting characteristics allow the estimation of the time spent on the device output to the nominal mode and the average temperature of the tubular heater surface with different steam capacity G_s and the constant parameters of the superheater.

This characteristic shows how the change in steam capacity affects the heater shell temperature and also makes it possible to predict the emergency and maximum permissible modes of its operation.

An energy-saving installation based on an electric steam generator and a storage water heater has been developed in order to provide cattle farms, pigsties and other facilities with hot water and steam.

The proposed design solution of the installation allows a reduction in energy costs up to 40% due to preliminary heating of water at a reduced tariff and its accumulation, the decrease in the peak power of the installation during the day hours up to 20% as well as the simultaneous and separate use of steam and hot water.

Figure 9. The surface temperature of the heater

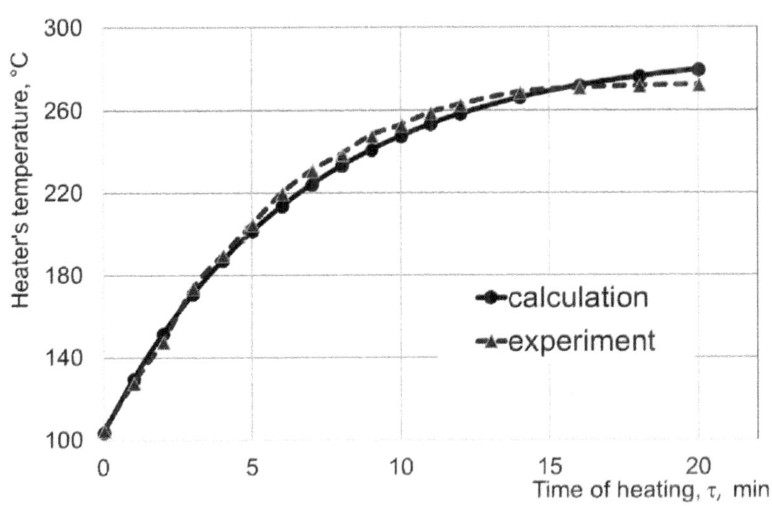

ENERGY SAVING INSTALLATION FOR MICROCLIMATE MAINTENANCE WITH UTILIZATION OF LOW-GRADE HEAT

The most energy intensive process in the heat supply systems of livestock farms is to support a microclimate. For this purpose, more than 60% of the heat energy is expended from the total cost of heating. The problems of the preservation of heat (energy) as well as the cleaning and disinfection of indoor air remain important for all livestock farms (Delyagin, 2004).

The subject of the study is thermal technological processes using electric energy when providing a microclimate to livestock facilities. A new highly effective method of decentralized heat supplying and creating a microclimate in livestock buildings based on the principles of heat recovery, ozonation, and deep air recirculation have been proposed in works of Markelova et al. (2013).

Structural diagram of the ventilation-heating installation implementing the described method (Figure 10) includes: a plastic film heat exchanger, a built-in peak electric heating coil, the fans of inlet and exhaust air, an ozone generator and a recirculation channel with the mixing chamber. The significant recirculation of the warm air removed from the premises is possible with its deep cleaning and disinfection including the use of the ozonization method.

The operation of installation showed in Figure 10 is as follows. The warm air in the room is taken up by means of the exhaust fan FAN2 and cleaned in the filter F2. Then this air drained during heat exchange in the heat exchanger HE is discharged to the street. Fresh outside air by means of the fan FAN1 is taken through the inlet and through the filter F1 is supplied to the heat exchanger HE where it is heated in the process of heat exchange. If the air temperature is not enough the electric heater EH can heat it additionally.

If the outside temperature lower than minus 12 ° C the partial recirculation (0 ... 75%) of the internal air is used to avoid frost and snow coat formation on the heat exchange surface from the exhaust air side. In this case the part of the inner warm air through the AAV flap is fed to the mixing chamber MC where it is purified and disinfected by the ozonization method. The partial recirculation of the internal air ensures the saving of the heat energy expended on heating the fresh supply air and the stable operation of the heat exchanger. In summer time an air cooler AC can be installed to cool the supply air.

Developed methodical principles and methods of calculating ventilation and the heating system include heat-power, hydro-mechanical and electrical calculations. The feature of thermal power calculation is the calculation of critical parameters - heat transfer coefficient when condensing the part of moist exhaust air into snow on the heat exchanger and the calculation of the possible thickness of this snow coat as well as its distribution along the length of the heat exchanger surface (Tikhomirov, 2013b).

The equation of heat balance in the heat exchanger in general is:

$$Q = G_1 c_1 (t_{12} - t_{11}) = G_2 c_2 (t_{21} - t_{22}) \eta \tag{22}$$

If the coolant changes its aggregate state (to condensate) the calculation should be carried out according to the expression:

$$Q = G_1 c_1 (t_{12} - t_{11}) = G_2 (i_{21} - i_{22}) \eta \tag{23}$$

Figure 10. Structural diagram of the ventilation-heating installation with heat recovery, ozonction and air recirculation

F1, F2 are filters; HE is a heat exchanger; FAN1, FAN2 are intake and exhaust fans; EH is electric heater; SAD is supply air distributor; OG is ozone generator; MC is mixing chamber; AAV is adjustable air valve; OG is ozone tube, AC is air cooler

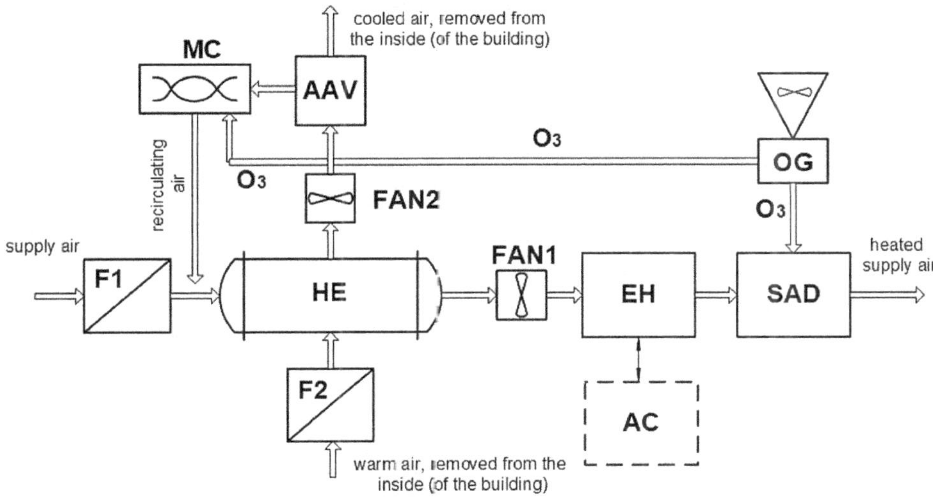

The heat transfer coefficient for film-type heat

$$k = \frac{1}{\frac{1}{\alpha_1} + \frac{\delta_{pf}}{\lambda_{pf}} + \frac{1}{\alpha_2} + R_p}, \tag{24}$$

where G_1, G_2 are the coolant flow rate, kg/s; c_1, c_2 are the average heat capacity of the coolant, J/kg·°C; t_{11}, t_{12}, t_{21}, t_{22} are the initial and final temperature of supply (cold) and exhaust (warm) air, °C, i_{21}, i_{22} are the enthalpy of the removed exhaust air before and after heat exchange, J/kg; η is the coefficient taking into account the heat loss of the installation; δ_{pf} is the thickness of the polymer film, m; λ_{pf} is the coefficient of thermal conductivity of the polymer film, W/m·°C; R_p is the thermal resistance of the layers of dirt on both sides of the polymer film, m²·°C/W; α_1, α_2 are the coefficients of heat transfer from the heat exchanger wall to the supply air and from the exhaust air to the wall of the heat exchanger, W/m²°C.

The calculated value of the heat transfer coefficient is 21…24 kW/m²°C.

The criterion equations using for determining the heat transfer coefficient α for supply and exhaust air are defined after evaluating the mode of air movement in the heat exchanger channels by the Reynolds criterion.

For dry air: $Nu = 0,021\,\mathrm{Re}^{0,8}\,\mathrm{Pr}^{0,43}(\mathrm{Pr}/\mathrm{Pr}_{wall})^{0,25}$. (25)

For wet air: $\alpha_w = \alpha_d + \beta r \Delta p / \Delta t$. (26)

$$Nu_m = \beta d_e / D_p ; \mathrm{Pr}_m = \upsilon / D_p R_s'' ; Nu_m = 0{,}023\,\mathrm{Re}^{0{,}8}\,\mathrm{Pr}_m^{0{,}4} . \tag{27}$$

where Nu, Pr, Re are the Nusselt, Prandtl and Reynolds numbers; Pr_{wall} is the Prandtl number at division wall temperature; β is the coefficient of mass transfer in vapor condensation from wet air related to the gradient of partial pressures, s/m; Δt is the temperature difference between the wet air and the wall, °C; Δp is the difference of partial pressures of steam in the core flow and the near wall, Pa; R_s is the gas constant of steam, J/kg·K; D_p is the diffusion coefficient; α_w, α_d are coefficients of heat transfer for wet and dry air, W/m²°C.

In the case of condensate freezing on the heat exchange surface from the side of the exhaust channel the average δ_{av} and maximum δ_{max} thickness of the snow coat are determined by the follow expressions:

$$\delta_{\max} = -\frac{(\alpha_w t_{22} + \alpha_1 t_{11})\lambda_s}{\alpha_w \alpha_1 t_{22}}, \tag{28}$$

$$\delta_{av} = -\frac{(\alpha_w t_{2av} + \alpha_1 t_{1av})\lambda_s}{\alpha_w \alpha_1 t_{2av}}, \tag{29}$$

where t_{1av}, t_{2av} are the average temperature of supply and exhaust air when passing it through the heat exchanger, °C; $\lambda_s = 0{,}116$ is the coefficient of thermal conductivity of snow coat, W/m°C.

The correctness of the mathematical description of heat and mass transfer processes in heat exchanger design and the calculation of the surface area of heat transfer are confirmed by the results of the experimental sample tests where the differences between parameters do not exceed 5%.

Hydro-mechanical calculation of heat exchanger has been made to determine the fan power at the minimum possible flow resistance of intake and exhaust tracts.

The design power of the additional supply air heater integrated in the electric air heater is calculated for the cold period of the year at a given ambient temperature t_o.

The heating element for the built-in electric heater is made from nichrome ribbon of rectangular cross section and small thickness. It is evenly spaced around the circumference. This design can significantly reduce the hydraulic resistance of moving air while providing high heat removal. Heat flow by convection from the built-in electric heater is determined by the formula:

$$P_c = \alpha_c F_h (t_h - t_a), \tag{30}$$

where α_c is the heat transfer coefficient by convection, W/m²°C; F_h is the area of the heat-transfer surface of the heater, m²; t_h is the temperature of a heater surface, °C; t_a is the temperature of air, °C.

The following expression was obtained using the mathematical method of experiment planning that relates the convection coefficient α_c to the air velocity ω through the heaters and the equivalent diameter d_e of the heating element.

$$a_c = 106{,}5 + 10{,}6\omega - 10{,}4d_e, \text{ at } 4 \le \omega \le 12;\ 2 \le d_e \le 6. \tag{31}$$

The electric power and the structural parameters of the electric air heater have been determined by means of the theory of heat exchange and the basics of electrical engineering. The method of ozonization is used for disinfection and purification of return and supply air.

The validation of the parameters was performed. The corona discharge ozonator for described ventilation and heating system was calculated and selected. The technological scheme of the ventilation-heating installation is shown in Figure 11.

The algorithm of controlling the air supply by changing the ratio of supply and exhaust air flows with the use of the partial recirculation of internal air was developed. The algorithm allows to improve the installation efficiency when significant negative ambient air temperatures (t ≤ -12 °C) and to prevent freezing of the heat transfer surfaces. The intelligent system of automatic control of the installation modes leads to a reduction in energy consumption for additional heating of the incoming air with an built-in electric heater. A circuit diagram for implementation of the above operation modes was developed. The experimental sample of the module the ventilation-heating installation with recycling, ozonation and recirculation of air for livestock buildings air with output up to 1500 m3/h was built based on the results of theoretical research and calculations.

The dependence of the heat flux Q_{hr} returned by the heat exchanger from the temperature difference of the external t_{21} and the internal t_{11} air was obtained on the basis of statistical processing of the experimental data and is shown in Figure 12.

Figure 11. Technological scheme of the ventilation-heating unit

1 is the exhaust fan; 2 is the polymer heat exchanger; 3 is the ozone generator; 4 is the intake air fan; 5 is the built-in electric heater; 6 is the air supply regulator; 7 is the supply air dispenser; 8 is the ozone feeding pipeline; 9, 11 are filters; 10 is the condensate trap; 12 is the supply duct; 13 is the return air duct with mixing chamber, T1...T6 are thermometers, P1...P4 are flow meters; AF1...AF3 are air dampers.

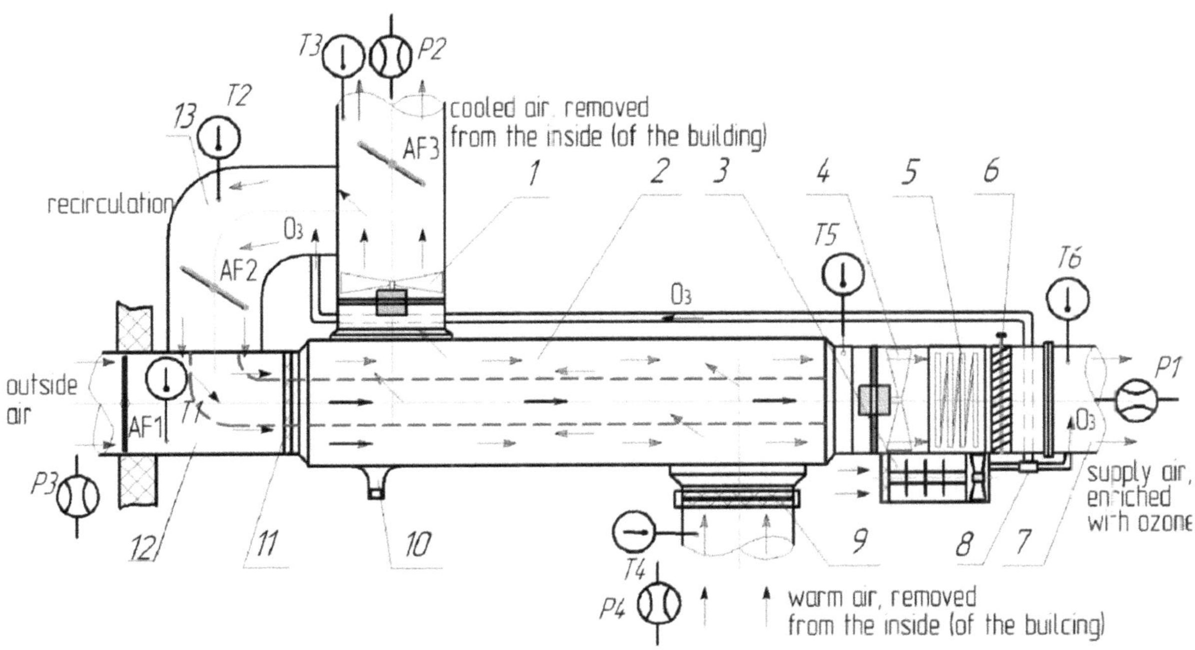

Figure 12. Heat flow returned by the heat exchanger

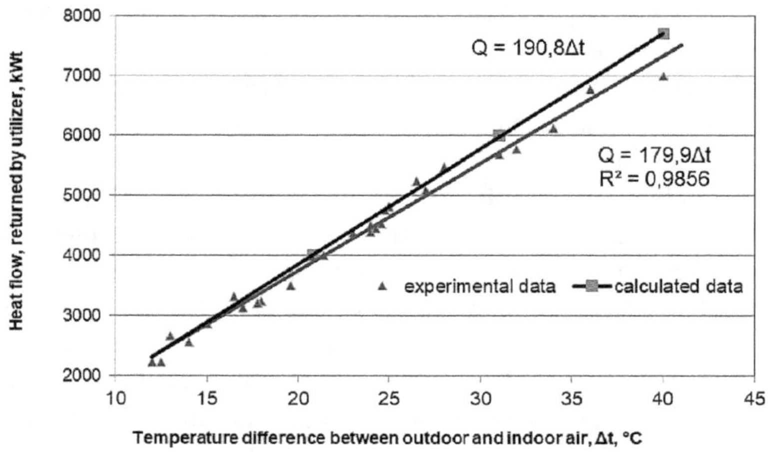

$$Q_{hr} = A_{hr}(t_{21} - t_{11})\,. \tag{32}$$

where A_{hr} is the temperature constant for the heat exchanger under study, J/kgK.

The researchers of heat and power operational parameters of the device under various environmental conditions including icing of heat exchange surfaces from the exhaust air were conducted out. The effect of the recirculation frequency of internal air on the heat transfer amount of the heat exchanger and the efficiency of the built-in electric heater capable of preheating the inlet air at a temperature of 20 ° C (Figure 13) have been confirmed by practice

The use of partial (30%) recirculation of internal air leads to further increase the return of warmth to 20%. The temperature of the air intake in the room reaches 3°C (Figure 5) when heating of the supply

Figure 13. Modes of ventilation-heating installation with the utilization and recirculation of internal air

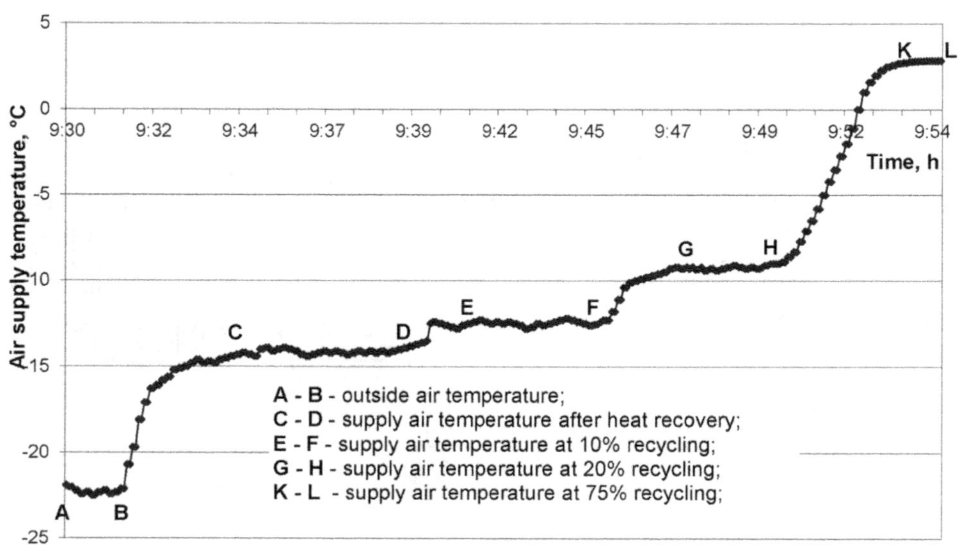

air in the heat exchanger mode and 75% of the recirculation of the indoor air. The energy expenses are reduced up to 60% in this case.

The methods of calculating new ventilation and heating system to create a necessary microclimate in livestock premises are confirmed by practical results. The design difference between the calculated and experimental data does not exceed 5%.

New dependencies of the coefficient of heat transfer from the heating elements of the built-in electric air heater to supply air and the heat-exchanger constant determining the amount of the return of warmth were discovered.

The exergy analysis of heat exchanger with the evaluating of the impact of each type of exergy losses in the heat exchanger on its efficiency ($\eta_e = 0{,}48$) were performed.

The installation has successfully passed the economic tests for supplying a predetermined atmosphere in the calf house (50 goals) of the dairy farm.

HEAT EXCHANGE BETWEEN YOUNG ANIMALS AND THE ENVIRONMENT WHEN LOCAL HEATING

Local heating using radiant (infrared) electric heaters is also an important area of energy saving in the systems of microclimate. Infrared local electric heating is particularly effective in premises with young animals demanding thermal zones with different temperatures as well as in the premises of periodic action (milking parlors, veterinary sanitary inspection, storage facilities etc.).

Most of the developed designs of radiant (infrared) heaters have the following disadvantages. The area of heating by the irradiator does not correspond to the standard size of the areas of young animals (sow stalls, calf boxes, cages etc.). This does not provide a uniform heat flow in a given area and reduces the efficiency of thermal radiation. The irradiators also are not adapted to the conditions of aggressive environment of livestock premises.

Irradiators with inexpensive infrared lamps have found the greatest use for heating young animals, despite the fact that they also have a number of disadvantages (Gobelev & Lyamcov, 2006). The "dark radiator" of plate type was developed in conjunction with the microelectronic company "Onyx". The irradiator has a uniform heat flow over a given area, high service life and reliability. Control scheme, a methodology, algorithm and program of calculation of radiant electric heaters of the plate type are developed. Their standard series are justified.

The radiant electric heater of young animals contains a metal heat-insulated rectangular body, a flat heater, a reflector, a protective grid. The metal casing inside is covered with fire-resistant, anti-corrosion protection and thermal insulation. The heater is a flat electrical resistance element installed by the method of burning into a steel plate thickness 1...2 mm and with size up to 200 x300 mm. It has low inertia, environmental safety, high reliability and durability (12,000 hours). The average surface temperature of the heater is 400 ... 450 ° C.

The method of heat exchange calculation between the animal and the environment as well as the electric heater is based on the methodological framework proposed by the works of Bystritsky et al. (1981) and Rastimeshin (1990).

IR irradiance E (W/m2) is considered to be the main parameter determining the choice of technical means and IR heating mode. It can be in general determined from the equation of the thermal balance between the animal and the environment.

$$Q_{sh} = EF_a a_{sur} = Q_{conv} + Q_{rad} + Q_{cond}, \tag{33}$$

$$Q_{con\,v} = \alpha_c F_a (t_{sk} - t_r), \tag{34}$$

$$Q_{rad} = C_0 F_a \varepsilon_{red} \left[\left(\frac{273 + t_{sk}}{100} \right)^4 - \left(\frac{273 + t_r}{100} \right)^4 \right]^4 \varphi r_r, \tag{35}$$

$$Q_{cond} = \frac{F}{R}\left(t_{sk} - t_f\right), \tag{36}$$

where Q_{sh} is the sensible heat of the animal corresponding to the optimal perceived temperature of the room, °C; Q_{conv}, Q_{rad}, Q_{cond} are the animal heat by convection, radiation and conduction, W; a_{sur} is the absorption coefficient of the animal surface infrared radiation; α_c is the heat transfer coefficient by convection of the animal surface, W/m^2°C; F_a is the surface area of the animal, m^2; t_{sk} is animal temperature °C; t_f is the floor temperature, °C; C_0 is the coefficient of the radiation of absolutely black body equal to 5,76 W/(m^2 K^4); ε_{red} is the emissivity factor of the animals skin and walling; φ is the coefficient of mutual irradiance of the model and fences (for the reduced volume of the climatic chamber φ =1); r_r is the coefficient taking into account the fraction of the radiating surface of the body (for all species of animals r_r = 0,65...0,70).

This calculation is associated with certain difficulties. The study of Rastimeshin (1991) allows its solution which is universal and can be used for different kinds of animals. This method serves as a basis for the calculation unification for the energy parameter choice of local heating of animals, but the use of these calculations for practical purpose is considered difficult. The author equations for determining the convective heat transfer coefficient αc do not consider the age (or weight) of the animal. It makes a certain error in the calculations for different age groups of animals.

The analytic expression for determining the coefficient of convective heat transfer of the animal surface depending on its mass (age) was obtained in works of Kuzmichyov, Lyamcov & Tikhomirov (2015) in order to complement the Rastimeshin provisions:

$$\alpha_c = 4{,}11 \left(\frac{t_{sk} - t_r}{k_a m + a} \right)^{0{,}25}, \tag{37}$$

where t_{sk} is skin temperature of the animal, °C; t_r is room temperature, °C; k_a is coefficient characterizing the species of the animal.

For calves: k_a = 0,2, a =16,3; for piglets: k_a = 0,35, a =12,0.

A number of researchers conducted their experiments in climatic chambers. We have carried out the analysis of these data and compiled the expression to calculate the sensible heat emitted by calves and piglets Q_{sh} depending on the weight of animals (calves m=30…80 kg, pigs m=7…40 kg) and an estimated room temperature 0...20 °C.

For calves: $Q_{sh}(m, t_r) = -56{,}9 + 5{,}5m + 5{,}3t_r - 0{,}087mt_r - 0{,}017m^2 - 0{,}23t_r^2$, (38)

For pigs: $Q_{sh}(m, t_r)=38{,}5+5{,}3m-1{,}1t_r-0{,}072mt_r-0{,}046m^2-0{,}011t_r^2$, (39)

where m is the animal weight, kg; t_r is the room temperature, °C.

The peculiarity and novelty of these expressions is that they allow the evaluation of the heat transfer change for different age groups of animals. The laboratory test bench was specially created for the study of IR heaters of various types and designs.

Studies have been carried out for uniform irradiation of a given area by various typical installations: based on infrared lamps (IRL), with a ceramic heater (CH) and with an electric planar IR radiator (EPIR).

The graphs of the heat flow distribution and analytical expressions for determining the value of the heat flux density (irradiation) at a given point depending on the height of the irradiator suspension were obtained within the study. When the height of the irradiator is 0,6 m and the installation power is 0,25 kW the greatest uniformity of the heat flow is achieved by plate-type irradiators.

The experimental data were subjected to statistical and mathematical processing using the mathematical environment of MathCAD and Maple. The analytical dependences of the heat flux density Q on the height of the suspension h (0,5…1,1 m) and the distance r (0…0,8 m) from the vertical axis of the suspension are obtained:

For IRL: $Q = 649{,}7 - 947{,}8r - 502{,}5h + 447rh + 301{,}5r^2 + 70{,}8h^2$

For CH: $Q = 785 - 603{,}8r - 948h + 431{,}5rh + 27{,}5r^2 + 303{,}5h^2$

For EPIR: $Q = 512{,}6 - 350{,}3r - 535{,}1h + 321rh - 0{,}3r^2 + 148h^2$ (40)

The dependence of the heat flow in the area of animals on the heater power and the height of the suspension is shown in Figure 14.

The proposed design provides: the reduction of energy consumption up to 25% in automatic mode providing the specified heat flux by regulation, the power of the heating element depending on the ani-

Figure 14. The dependence of the heat flow in the area of animals on the heater power and the height of the suspension

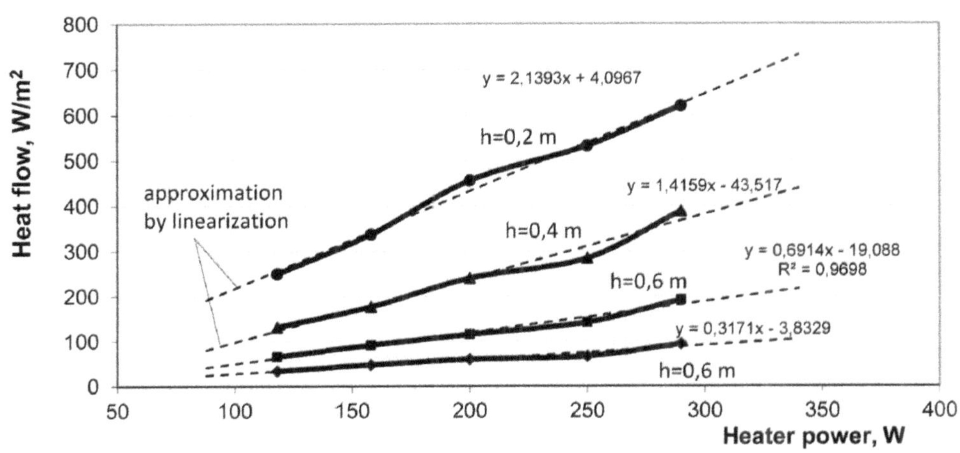

mal position (standing or lying); the possibility of using the heater with standard industrial equipment (individual cages, boxes) for breeding calves, shaping and directing the heat flow only in the area of the calf and ensuring uniform temperature field throughout the area of its content, thereby increasing the safety and productivity of animals.

FUTURE RESEARCH DIRECTIONS

Future researches can be directed to the development of intelligent energy-saving systems and technical means of heat supply, implementing operational control and decision-making over the normalization of microclimate parameters in agricultural premises. Process control should be based on the reactions of biological objects (animals, birds). The system should decide on the regulation of the microclimate parameters on the basis of information coming directly from the animal. This will create optimal living conditions for animals and birds in order to maximize their genetic potential and ensure high productivity.

Among the promising areas of research are new technologies and methods for processing biomass, agricultural waste (including manure) and wood processing for high-quality process fuel used in the installations producing electrical and thermal energy.

CONCLUSION

Technical and economic analysis showed that the most effective heat supply systems for livestock breeding are the decentralized ones with using a natural gas as well as electrical systems with a thermal capacity up to 200 kW. Systems on liquid and solid fuels are less efficient. It is established that capital investments and costs of the current operation of electric systems are the lowest in comparison with fuel systems. It found that the less facility capacity is, the more total unit costs for obtaining and using 1 kWh of useful thermal energy are. Under these conditions the efficiency of electrical systems is increased.

The required parameters and operating modes of energy-saving electrical systems and technical means for ensuring the main thermal technological processes are substantiated: hot steam and water supply, microclimate, local heating of young animals and workplaces, primary heat treatment of animal products and feed. A new method of providing a microclimate with the use of heat recovery, ozonation and air recirculation provides a reduction of energy consumption up to 60%. A new method of providing a microclimate using heat recovery, ozonation and air recycling reduces energy consumption up to 60% during the heating season.

The physical models of the heat exchange processes between an electric heater and the environment as well as their mathematical description make it possible to scientifically justify the electrical, structural and thermal parameters of the new energy-saving equipment for heat supply.

The proposed methods for calculating energy-efficient technical facilities are applicable to justify the parameters of the system-forming electrical heating equipment of various power and performance.

The efficiency of developed samples of electric heat supply facilities were tested in laboratory and production conditions including at the level of state acceptance tests.

Modern energy-efficient electrical heat supply systems are competitive and preferable for non-gasified agricultural production facilities.

REFERENCES

Aldibekov, I.T. (2010). *Vybor i primenenie ehnergosberegayushchih sistem teploobespecheniya molochnyh ferm i fermerskih hozyajstv na baze ehlektrodnyh teplovyh ustanovok (rekomendacii)* [Selection and application of energy-saving heat supply systems for dairy farms and farms based on electrode heat installations (recommendations)]. Almaty, Kazakhstan: Inzhu-Marzhan.

Barkov, V. I. (2006). *Vybor i postroenie sistem ehlektroteploobespecheniya v zhivotnovodstve* [Selection and construction of electric heating systems in animal husbandry] (Doctoral dissertation). Almaty, Kazakhstan.

Bystrickij, D.N., Kozhevnikova, N.F., Lyamtsov, A. K. & Murugov, V. P. (1981). *EHlektricheskie ustanovki infrakrasnoro izlucheniya v zhivotnovodstve* [Electrical installations of infrared radiation in animal husbandry]. Moscow, Russia: EHnergoizdat.

Cho, Y. S., Song, K. B., & Yamda, K. (2010). Effect of Ultraviolet Irradiation on Molecular Properties and Immunoglobulin Production-regulating Activity of β-Lactoglobulin. *Food Science and Biotechnology, 19*(3), 595–602. doi:10.100710068-010-0084-0

Delyagin, V.N. (2004). Obosnovanie racional'nyh temperaturno-vlazhnostnyh rezhimov zhivotnovodcheskih pomeshchenij [Substantiation for rational temperature and humidity conditions of livestock buildings]. *EHnergoobespechenie i ehnergosberezhenie v sel'skom hozyajstve, 3*, 250-255.

Gobelev, S.N. & Lyamcov A.K. (2006). Rezul'taty ispytanij infrakrasnyh lamp na srok sluzhby [The results of the tests of infrared lamp service life]. *Tekhnika v sel'skom hozyajstve, 6*, 39–41.

Kagan, N.B., Kaufman, V.G., Pron'ko, M.G., & YAnevskij G.D. (1980). *EHlektrotermicheskoe oborudovanie dlya sel'skohozyajstvennogo proizvodstva* [Electrothermal equipment for agricultural production]. Moscow, Russia: EHnergiya.

Kanakin, N.S. & Kogan, Y. M. (1986). *Tekhniko-ehkonomicheskie voprosy ehlektrifikacii sel'skogo hozyajstva* [Technical and economic issues of agriculture]. Moscow, Russia: EHnergoatomizdat.

Kazancev S.P. & Ignatkin I.Y. (2013). Rekuperator teploty dlya svinovodcheskogo kompleksa [The heat exchanger of the heat for a pig-breeding complex]. *Mekhanizaciya i ehlektrifikaciya sel'skogo hozyajstva, 4*, 17-18.

Kutateladze S.S. (1990). *Teploperedacha i gidrodinamicheskoe soprotivlenie.*[Heat transfer and hydrodynamic resistance]. Moscow, Russia: EHnergoatomizdat.

Kutateladze, S. S., & Borishanskij, V. M. (2012). Spravochnik po teploperedache [Handbook of heat transfer]. Moscow, Russia: Kniga po Trebovaniyu.

Kuzma-Kichta, Yu. A. (1994). *Metody intensifikacii teploobmena* [Methods of heat transfer intensification]. Moscow, Russia: MPEI.

Kuzmichyov, A. V. (2017). Opredelenie i ocenka teplovyh uslovij v zone obogreva infrakrasnymi obogrevatelyami molodnyaka zhivotnyh [Determination and evaluation of thermal conditions in the heating zone by infrared heaters of young animals]. *Vestnik VIEHSKH, 4*(29), 26–32.

Kuzmichyov, A. V., Lyamcov, A. K., & Tikhomirov, D. A. (2015). Teplotekhnicheskie pokazateli IK obluchatelej dlya molodnyaka zhivotnyh [Thermal performance of IR irradiators for young animals]. *Svetotekhnika, 3*, 57–58.

Kuzmichyov, A. V., Malyshev, V. V., & Tikhomirov, D. A. (2011). Efficiency of the combined pasteurization of milk using UV and IR irradiation. *Light & Engineering, 19*(1), 74–78.

Lebedev, D.P., Rasstrigin, V.N., & Tikhomirov D.A. (2000). Teploobmen v ehlektricheskih parovodonagrevatelyah [Heat exchange in electric steam and water heaters]. *Promyshlennaya teplotekhnika, 4*(22), 34–39.

Markelova, E. K., Petrova, L. N., Uhanova, V. Yu., Tikhomirov, A. V., Tikhomirov, D. A., & Pershin, A. F. (2013). Pat. RF MPK A01K1/03. Sposob i ustrojstvo ochistki vozdushnoj sredy zhivotnovodcheskih i pticevodcheskih pomeshchenij [Method and device of air cleaning of livestock and poultry premises]. Moscow, Russia: Academic Press.

Morozov, N. M. (2011). Organizacionno-ehkonomicheskie i tekhnologicheskie osnovy mekhanizacii i avtomatizacii zhivotnovodstva [Organizational, economic and technological basis of mechanization and automation of livestock]. Moscow, Russia: FGBNU «Rosinformagrotekh».

Panova, T. V. (2011). *Razrabotka i obosnovanie parametrov pritochno-vytyazhnogo utilizatora tepla dlya uluchsheniya mikroklimaticheskih uslovij v zhivotnovodcheskih pomeshcheniyah* [Development and justification of parameters of the heat supply and exhaust heat exchanger to improve the microclimate conditions in livestock buildings] (Doctoral dissertation).

Pis'mennyj, E.N. (1999). Konstruktivnye metody povysheniya teploaehrodinamicheskoj ehffektivnosti trubchatyh poperechno-orebrennyh poverhnostej teploobmena [Design methods for increasing heat and aerodynamic efficiency of tubular transversely finned heat transfer surfaces]. *Promyshlennaya teplotekhnika, 2-3*(21).

Rasstrigin, V. N. (1988). *Osnovy ehlektrifikacii teplovyh processov v sel'skohozyajstvennom proizvodstve* [Fundamentals of electrification of thermal processes in agricultural production]. Moscow, Russia: Agropromizdat.

Rastimeshin, S. A. (1990). *Tekhnicheskie sredstva lokal'nogo obogreva* [Technical means for local heating]. Moscow, Russia: Rosagropromizdat.

Rastimeshin, S. A. (1991). *Lokal'nyj obogrev molodnyaka zhivotnyh (teoriya i tekhnicheskie sredstva)* [Local heating of young animals (theory and technical means)]. Moscow, Russia: Agropromizdat.

Rozhenau, U.M. (1966). *Teploobmen pri kipenii* [The boiling heat transfer]. Moscow, Russia: Sb. Sovremennye problemy teploobmena, EHnergiya.

Smith, W. L., Lagunas-Solar, M. C., & Cullor, J. S. (2002). Use of pulsed ultraviolet laser light for the cold pasteurization of bovine milk. U.S. *Journal of Food Protection, 65*(9), 1480–1482. doi:10.4315/0362-028X-65.9.1480 PMID:12233862

Spalart, P. R., & Rumsey, C. L. (2007). Effective inflow conditions for turbulence models in aerodynamic calculations. *AIAA Journal, 45*(10), 2544–2553. doi:10.2514/1.29373

Tikhomirov, D. A. (2002). Povyshenie ehffektivnosti teploobmena v ehlektricheskih parovodonagrevatelyah [Improving the efficiency of heat transfer in electric steam heaters]. In Sovremennye ehnergosberegayushchie teplovye tekhnologii (sushka i termo-vlazhnostnaya obrabotka materialov): Trudy 1-j mezhdunarodnoj nauchno-prakticheskoj konferencii (pp. 213–216) Moscow, Russia: MGAU.

Tikhomirov D.A. (2013a). Teploehnergeticheskie parametry sistemy goryachego paro- i vodoobespecheniya ob"ektov zhivotnovodstva [Heat and power parameters of the system of hot steam and water supply of livestock]. *Mekhanizaciya i ehlektrifikaciya sel'skogo hozyajstva, 3*, 17–19.

Tikhomirov D.A. (2013b). Metodika teploehnergeticheskogo rascheta ehnergosberegayushchej ventilyacionno-otopitel'noj ustanovki dlya zhivotnovodcheskih ferm [Method of heat power calculation of energy-saving ventilation and heating plant for livestock farms]. *Al'ternativnaya ehnergetika i ehkologiya, 2*(1), 125–131.

Tikhomirov D.A. (2014). Fizicheskaya model' dinamiki teploobmena v ehlektricheskom paroperegrevatele [The physical model of the dynamics of heat transfer in electrical superheater]. *Tekhnika v sel'skom hozyajstve, 2*, 13–15.

Tikhomirov D.A. (2015). Opredelenie koehfficienta teplootdachi pri kipenii vody v ehlektricheskom parovodonagrevatele ehksperimental'nym putem [Determination of the heat transfer coefficient at boiling water in the electric steam heater by experiment]. *Innovacii v sel'skom hozyajstve, 3*(13), 85-89.

Tikhomirov, D. A. (2016). EHnergoehffektivnye ehlektricheskie sredstva i sistemy teploobespecheniya tekhno-logicheskih processov v zhivotnovodstve [Energy-efficient electric means and systems of heat supply of technological processes in animal husbandry]. *Vestnik VNIIMZH, 4*(24), 15–23.

Tikhomirov, D.A. (2017). EHlektricheskie IK-obogrevateli: sostoyanie i perspektivy innovacionnogo razvitiya [Electric IR heaters: status and prospects of innovative development]. *Tekhnika i oborudovanie dlya sela, 1*, 2-7.

Tikhomirov D.A. & Vasiliev A.N. (2016). Programmnyj proekt rascheta teploehnergeticheskih parametrov ob"ektov zhivotnovodstva [Program project for calculation of heat and power parameters of livestock facilities]. *Mezhdunarodnyj nauchno-issledovatel'skij zhurnal, 8*(50), 95-98.

Tolubinskij, V. I. (1989). Teploobmen pri kipenii [The boiling heat transfer]. Kiev, Ukraine: Naukova dumka.

Vasant, P., & Voropai, N. (2016). *Sustaining Power Resources through Energy Optimization and Engineering*. IGI Global. doi:10.4018/978-1-4666-9755-3

KEY TERMS AND DEFINITIONS

Heat Accumulation: (for electrical installations) Accumulation of heat energy during a reduced tariff for electric energy.

Heat Recovery: (in the case of ventilation) Return of heat from the exhaust air. Warm air exchanges heat with cold supply air in the heat exchanger. The cooled air is emitted outside, and the fresh and heated air is supplied to the room.

Heat Supply: Activity on production, transfer, distribution, and use of heat energy without its sale (for own needs).

Microclimate: Complex of physical factors of the internal environment of rooms influencing a thermal exchange of an organism and health of both animals and people.

Ozonation: Purification technology based on the use of ozone gas, which is a strong oxidizer.

Steam and Water Heater: Device for generating steam and heating water in one unit.

Superheater: Device designed to overheat steam, that is, increase its temperature above the saturation point. Superheated steam has higher heat content.

Chapter 5
On Efficiency Rising of the Technological LED Lighting for the Floor Poultry Keeping

Tatyana R. Gallyamova
Izhevsk State Agricultural Academy, Russia

Aleksey L. Vozmischev
OOO "Udmurt Poultry Farm", Russia

Nicolai V. Khokhriakov
Izhevsk State Agricultural Academy, Russia

Vladimir G. Baltachev
Izhevsk State Agricultural Academy, Russia

Svetlana Y. Ponomaryova
Izhevsk State Agricultural Academy, Russia

ABSTRACT

A method for solving the problem of the uniform LED illumination for the floor maintenance of birds is proposed. On the basis of physical modeling, the dependences have been obtained between the geometric dimensions of the poultry house and the coefficient of uniformity of illumination with the purpose of optimization of the parameters on which the energy consumption and quality of illumination depend. The method has been used to estimate the values of the optimal parameters of the functioning system. Thus, in the poultry house of 78m x 18m, four lines of LED lamps have been installed for illumination, 40 pieces per each line. With the computer calculations of illuminance, estimates of the optimal values have been obtained: the total number of LEDs is equal to 1920 and allocated over 120 lamps instead of 160 previously used, which allows predicting a fall in actual energy consumption by 25% with an irregular illumination coefficient being 1.13. Based on the developed and manufactured LED lamp, the applicability of the mathematical model has been experimentally justified.

DOI: 10.4018/978-1-5225-7573-3.ch005

INTRODUCTION

Poultry farming is the most dynamic branch of agricultural business that is characterized by rapid paces in reproduction of the poultry population, high productivity and low cost per a unit of output. Dietary production of poultry farming is significantly cheaper compared to other food products containing animals' protein. Therefore, from the point of view of public and state interests, the development of the poultry farming holds a special priority for solution of the feed problem.

Development of modern science and technology is aimed at investigation, development and implementation of energy efficient and energy saving techniques for goods and services production. One of the available methods to increase power efficiency in poultry farming is optimization of power consumption of technological illumination being one of the important microclimate factors for the aviary. In recent times, modernization of the aviary's lighting systems has been completed by fluorescent lamps replacement by the LED lamps. These innovations have led to prominent saving of the electric power. However, the authors of the present paper are confident of the threshold for power saving of the LED illumination that has not been reached yet. Respectively, problems of energy efficiency and saving increase in the field remain to be actual, and require further investigation.

Presently, despite of the high costing value of the land, in most European countries there is still floor poultry farming being actively cultivated. To rise the efficiency of a uniform distribution of the led illumination in the floor poultry farming it is necessary to work out the lowest LED power mode at the given value of the set up poultry farm illumination, and not interfering with the illumination quality requirements. In the sense, it is admissible to qualify power consumption as optimal provided that the difference will be minimum between the average lightness with the light emitting diodes, and that of the specified value of the normalized lightness. The parameters, upon which power consumption and the led illumination quality depend, thus appear to be the principle parameters for technological illumination.

Calculation of optimal values of the led lighting principle parameters (in the terms of each led bulb) is rather a complicated problem analytical solution of which has not been known yet. Some of the authors suggest the problem to be solved by the numerical method. However, their calculations made are intended for local lighting thus applying just one LED module for relatively a small space lighting capacity. Along with that, there are no researches existing on optimization of the LED lighting principle parameters for a poultry house of a prominently larger area (up to 1000 square meters, and more) respectively applying to a significant number of led bulbs (up to 100) hang up on several electric wirelines (two, and more).

The aim of the present research is to estimate optimal values for the LED lighting principal parameters based on the worked out model. Those values are subjected to optimization upon which the quality of illumination and energy consumption will evidently be depending on the floor poultry maintenance.

To achieve the goal the following tasks have been defined:

1. To substantiate the relevance of the research and of the capacity to increase the led lighting efficiency for the floor keeping of the poultry.
2. Basing on the optimization method, to develop a mathematical model thus allowing estimating the poultry house's power consumption with its most uniform led illumination.
3. To work out methodology of computer researches for estimating poultry house's optimal electric power consumption.
4. To justify experimentally utility of the created mathematical model and its computerized utilization.

1. RELEVANCE OF THE RESEARCHES

Technological illumination is one of the principle factors of microclimate in the poultry house and it plays the key role in chick breeding. Utilization of the illumination admits management of physiological growth of the poultry, thus providing more comfortable conditions for their keeping, and achieving substantial productivity growth of the poultry flock. Investigations held in many countries in the run of the last century (Hammer, 1963; King, 1958; Ralph, 1963; Wilson, Abplanalp & Atrington, 1962) had proved great influence of the illumination upon health, fertility, metabolism and productivity of the poultry. For today, grades of the light impact on the chickens have been fairly well investigated. Hence, the principle parameters inducing the chicks' vital activities are illumination, the emission spectrum of the lighting, daylight hours and modes of illumination. Automatic regulation of the daylight hours is possible to be realized with the electronic devices installed, and thus artificial illumination provided in the morning and evening hours ("dawn-to-dusk" modes), and that would bring in supplementary advantages to the farm.

In the works below (Kavtarashvili, Novorotov & Kolokol'nikova, 2014; Kavtarashvili & Ridzhal, 2001; Kavtarashvili, Novorotov, Gladin & Kolokol'nikova, 2012; Kondratieva, Baranov & Kondratiev, 2012) the priority is given to the lamps-based LEDs with warm–and-white colour with the regard of the best sources of artificial illumination. The led lighting in poultry keeping can outrightly be called innovative since its advantages over other well-known sources of illumination are unquestionable, and its usage actually is devoid of any drawbacks known with its predecessors (Yozef, 2011; Zyudkamp & Bakalets, 2008). Despite relatively high cost of LEDs, innovative illumination is intensively spreading thus gradually superseding all presently functioning systems from today's market (Kovalenko, 2009; Mikhalyov, 2014; Novosyolov, 2011; Pil'schikova, 2014; Yuferev, 2016). Progress making in the field matches the avalanche-like computerization of today's society. Already now, the advance-pacing farms acquire LED lamps in the terms of their energy management, longevity and fast recoupment.

One of the factors of successful participating at the meat market is the principle of humanity towards animal and poultry farming. Cage breeding of broilers contradicts public norms of humanity, whereas a public view is an essential term for successful utilization of food products, and particularly of meat. In many countries, the Green party and other public organizations monitoring the environment and wildlife are seeking for banning the cage manufacturing, even be it a single table egg. Presently, majority of the states have made their sure choice in favour of the floor poultry keeping.

Thus, to market the poultry products successfully, the floor poultry keeping looks the most promising and actual. For the purpose of uniformity of livestock by weight, it is necessary to create uniform conditions of poultry keeping in every corner of the aviary. Since the illumination management is one of the primary factors of sustaining the microclimate within the poultry house, it is of no less important for the aviary with the floor poultry keeping to provide the most uniform illumination of the big square building, thus the required lightness reached (some 100 lx) and monitored. Increase of durability and intensiveness of illumination, especially within the windowless poultry house, leads to overconsumption of energy, and the increasing the productivity and production costs. Therefore, creating a regime of optimal energy consumption for the house is a primary task when solving the problem of qualitative (most uniform) aviary illumination with the floor poultry keeping.

Under the floor keeping the parent flock, the hatching egg is a production output. Its quality and quantity depends on a number of parameters. For instance, an egg laid down on the floor-covering bedding (not into the nest), it is discarded. The standard of illumination E_0 with the coefficient of light reserve k_0 is maintained over the feeder and a drinking bowl, and, as a rule, is not maintained at the walls of the poultry house. This is the main reason for eggs' casting off, since a chicken tends to carry the egg in a darken place. If the illumination is the same everywhere, the bird tends to the only darken place-nest. The literature presents various algorithms of optimization for engineering issues and economics (Vasant, Weber, & Dieu, 2016). The authors propose a new approach to optimizing energy consumption by calculating the technical parameters of the lamp built in into the lighting system of a poultry house.

Increase of illumination quality meaning to improve aviary's microclimate is available by lessening the coefficient of light non-uniformity *Z* (within the boundary up to one: $Z \to 1$). Problems of the light uniformity and calculations of the *Z* coefficient when keeping the egg-laying chickens under cage conditions are revised in (Novosyolov, 2011). The object of illumination is normally set up in the form of a direct line of consequently installed troughs, though all that restricts utilization of the results of the above-mentioned activities during further developing a system of illumination of a busied area for the floor kept poultry. In our view, a conclusion for the need to account for LEDs over the neighbouring lines should be expanded and taken into account while calculation for lighting is being made for some particular point. Furthermore, it would be also necessary to consider the impact upon lighting of that particular point of all *N*-placed point-LEDs spread all over the LED system of lighting in the poultry house.

In numerical method (Guttsait, Krasnopol'skij, & Milyutin, 2007) for calculation of an illuminated area is suggested. It is utilized for local illumination (comparatively small space if compared to that of an area in a standard poultry house) with the help of a single 2-dimentional LED module. And the principle of superposition is considered as the basic to determine the influence of all *N* LEDs spread over the module upon the lightness of one particular point. As a result, the better illumination uniformity is obtained when the amount of *N*-LEDs has been increased, whereas the longitudinal in-between-LED-space has been lessened. However, the present article does not concern the problem of optimal energy consumption and the increase of illumination worthiness.

To solve the problem it is necessary to use certain calculations to obtain the estimations of those optimal parameters, upon which the illumination worthiness and the energy consumption (utilized power) depend during the floor poultry farming.

According to what has been mentioned above, a determining parameter for the worthy illumination is the coefficient of non-uniformity of the illumination *Z*. The very parameter, upon which the consumed energy *P* of the lighting system (in terms of consumed power p_1 for one led) depends is a total number of leds distributed among all the lamps, since

$$P = N \cdot p_1 . \tag{1}$$

To calculate lightness at a particular point of working space by applying to the principle of superposition, the parameter N is also required. Thus, the coefficient of non-uniformity of the lightness Z, and the total amount of LEDs N are the principle parameters for the led illumination of the aviary. To obtain the justified assessment for all those parameters (needed calculations applied) a certain mathematical model needs to be developed.

2. MATHEMATICAL MODEL FOR LIGHTING THE AVIARY FOLLOWED BY THE ESTIMATE OF AN OPTIMAL ENERGY CONSUMPTION

2.1. Setting Up the Task and Allowances Adopted for Mathematical Modeling

Coefficient of lightness non-uniformity Z (Baranov & Zakharov, 2006; Izenberg, 2006)

$$Z = \frac{E_{\max} + E_{\min}}{2 \cdot E_{\min}} \tag{2}$$

is an important parameter defining the quality of illumination, whereas according to (Kozinski, 1991; Vodyanikov, 2002) its recommended value is set within the interval

$$1.1 \leq Z \leq 1.15 \,. \tag{3}$$

The lightness produced by entire led lamps $E(x,y)$ at the point coordinates of which are x and y, in longitudinal and cross-section directions towards the plane is determined as per the principle of superposition, used in (Guttsait et al., 2007):

$$E\left(x,y\right) = \sum_{s=1}^{N} E_s\left(x,y,s\right). \tag{4}$$

The energy consumed P for the poultry house's lighting system depends on N as per (1). Therefore, number of led lamps N and the coefficient of the lightness non-uniformity Z are evident to determine the energy consumption and the quality of illumination. To raise the effectiveness of energy consumption and of the quality of lighting, let us solve the following task: to estimate optimal values for the principle parameters under the needed level of poultry house's illumination E_0 with the surplus coefficient k_0.

Lightness E_s produced by the point led number s, as per the principle law of lighting engineering:

$$E_s\left(x,y,s\right) = \frac{I_\alpha \cdot \cos_{\,}\left(x,y,s\right)}{r_s^2}\,. \tag{5}$$

Let's suppose, a led admits a round-symmetrical light spreading, (Knorring, Fadin & Sidorov, 1992) and the luminous intensity as per

$$I_\alpha = I_0 \cdot \cos^m \theta\left(x,y,s\right), \tag{6}$$

where θ (x,y,s) is the angle between the axial line of the curving luminous intensity of the led and the direction at the point with the coordinates x and y; I_0 - axial luminous intensity of the led number s. A degree indicator m implies the opening angle of the luminous flux angle of the led lamp α_0 as per

$$m = \frac{\ln 0.5}{\ln \cos\left(0.5\pm\right)}. \tag{7}$$

The distance r_s in (5) is explicated through the led lamp's coordinates x_s, y_s, z_s:

$$r_s = \sqrt{\left(x - x_s\right)^2 + \left(y - y_s\right)^2 + \left(z - z_s\right)^2}.$$

If the led lamps are placed linearly with the λ-pace, and the lamps being hanged at the H- height with the line distance indicated by L_0, then the values x_s, y_s and z_s are available through λ, L_0 and *H*. Whereas values y_s and z_s to be tasked, it is necessary to determine the angle between the lines y, and the line number from that of given number of lines *j* in the lamp. Normally, the *H* value is to be known beforehand, thus to calculate lightness as per (4) and (5) it is necessary to be aware of the parameters, *j*, λ, L_0, γ and the led parameters as per (6) and (7), provided that the led lamp is defined, I_0, α_0 are given.

When the number of lines is defined, the in-between –line angle under uniform distribution within the angle of illumination limits would be calculated as per

$$\gamma = \frac{\pi}{2\left(j-1\right)}.$$

For the *m* consequence 100 as in (Gallyamova, Novosyolov & Shirobokova, 2012; Gallyamova, Shirobokova & Kochetkov, 2015; Shirobokova, Kochetkov & Gallyamova, 2014) the number of lines proposed is *j*=7. Enlargement of the number of the lines *j* by exceeding the seven has significantly complicated the lamp construction. However, at the same time their lessening restricts the feasibility of controlling uniformity of lighting. Besides these parameters needed to calculate illumination by (4) it is necessary to know the *N* value, its distribution over the lamps, and also the lamp's length and the distance between them in the line.

Dependence between the parameters λ and *N* is available when the following tolerance is admitted: the lamp's length and the in-between distance makes 1m. Analysis of the information in (Yozef, 2011) indicating the led lamps' measurements does not contradict the tolerance accepted. Then, the number of the lamps S_0 in the poultry house would reach

$$S_0 = \frac{a \cdot N_L}{2}, \tag{8}$$

where a is the house's length, N_L – number of lines, interlaced with the house's width *b*, and the distance of the first and of the last line from the house's walls y_b, that runs as follows:

$$b = \left(N_L - 1\right) \cdot L_0 + 2y_b. \tag{9}$$

Let us admit that N_1, j and n_1 implicate the respective number of the leds for one lamp, the number of lines and leds in a line with the length of 1m. Then,

$$N = S_0 \cdot N_1;\ N_1 = j \cdot n_1;\ \lambda = \frac{1}{n_1}. \tag{10}$$

Using (8), (9) and (10), we achieve the λ and N relation as:

$$\lambda = \frac{a \cdot j}{2N} \cdot \left(1 + \frac{b - y_b}{L_0}\right) = \frac{a \cdot j \cdot N_L}{2N}. \tag{11}$$

To determine the interrelation between major parameters and constructive parameters in (11) determining the geometry of the poultry house and its lighting system let us turn to the method of physical modelling.

2.2 Physical Modeling of the Poultry House Lighting

It is well known that under physical modeling involves mathematical analogy of different by their physical nature processes occurring according to the laws involving in their turn similar equations (formulas). Thus, different physical processes characterized by similar quantitative relations are the basis for further physical modeling (Feynman, Leighton & Sands, 1977).

In the case, we apply to analogy between lightness created by a point led and by the electric field intensity created the point charge from the electrostatics. Analogy is based on the fact that lightness and intensity of an electric field change inversely proportional to the square of the distance from the point respectively till the point led lamp and the point charge. In the direction of normal towards the plane this proportionality is explicated for two cases:

$$E_s\left(x, y, s\right) \sim \frac{\cos\theta}{r^2},$$

where θ is the angle between the normal towards the plane and the direction of the vector r.

Notice also the principle of superposition (4) to follow for these two cases.

Suppose, the distance between the lines is equal to L_0, whereas the lines indicate parallel charged threads of infinite length. Field intensity of the charged infinite thread is inversely proportional to the distance R from the thread:

$$E\left(y, z\right) \sim \frac{1}{R}, \tag{12}$$

which means under any x within the plane yOz the charged threads become visible as the point charges (e.g. positive) at the distance L_0 from each other (figure 1).

Figure 1. Distribution of potential under the parallel charged threads of infinite length

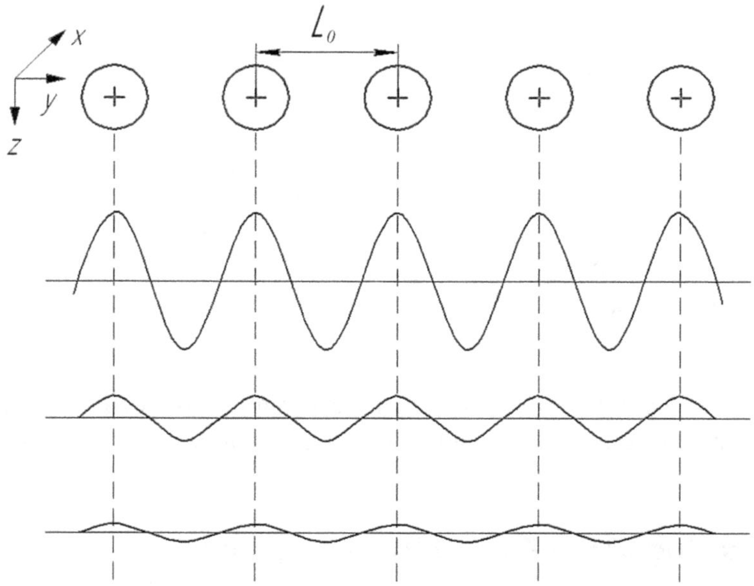

Potential φ of the point charge is inversely proportional to the distance *R* from the charge similarly to the intensity of the field of a charged infinite thread as in (12). According to the analogy by means of the potential φ we simulate the intensity, i.e. the lightness. Therefore, the results achieved for the potential can be used for the lightness we are seeking for.

Potential produced by the point charges at the point of the plane *y*O*z*, according to Feynman can be explicated as a periodical function (Feynman et al., 1977):

$$\varphi\left(y,z\right)=F_n\left(z\right)\cdot\cos\left(\frac{2\pi\cdot n\cdot y}{L_0}\right), \tag{13}$$

where n =1,2,3… - number of oscillations. In the thread area, the potential should meet the Laplas's equation and be the only solution under the given boundary conditions. After introducing the function φ(*y*, *z*) into the Laplas's equation Δφ = 0 we obtain Fourier's component of the $F_n(z)$ field n – harmonic at the distance $z=H$ from the charged threads:

$$F_n\left(H\right)=A_n\cdot e^{-\frac{2\pi\cdot n\cdot H}{L_0}}.$$

It is known that the Fourier coefficients of A_n decline, and when n=1, and n=2

$$e^{-\frac{2\pi\cdot H}{L_0}}>e^{-\frac{4\pi\cdot H}{L_0}}$$

then for assessment it would be sufficient to consider the case when n=1: $F_1 = A_1 \cdot e^{-\frac{2\pi \cdot H}{L_0}}$, under A_1 equal to the average value of the field (of potential, intensity or lightness). In the present case, A_1 is also equal to normalized illumination E_0 with the reservation coefficient k_0:

$$A_1 = \frac{E_{max} + E_{min}}{2} = k_0 \cdot E_0 .$$

Applying to illumination E_1 (H) oscillation amplitude as

$$E_1\left(H\right) = \frac{E_{max} - E_{min}}{2},$$

and coefficient of non-uniformity of illumination Z as in (2), we would obtain correlation between the height H of the lamps hung over and that of the distance between the lines L_0 in the form of the function $f(Z)$:

$$\frac{H}{L_0} = f\left(Z\right) = \frac{1}{2\pi} \ln\left(\frac{Z}{Z-1}\right). \quad (14)$$

According to the function (figure 2), for Z <1.15 it should be

$$\frac{H}{L_0} > 0.324 \text{ or } L_0 < 3.085H .$$

Let us study the illumination on the working surface of the poultry house (under z=H) in the direction of the axis y, i.e. in transverse direction. According to (13), the potential φ(y, z) is a periodical function. Therefore, the intensity of the field, meaning illumination as well, would be also a periodical function

Figure 2. The graph of dependence f(Z) as per (14). The condition of uniformity of illumination Z <1.15 is performed at (H / L_0)> 0.324

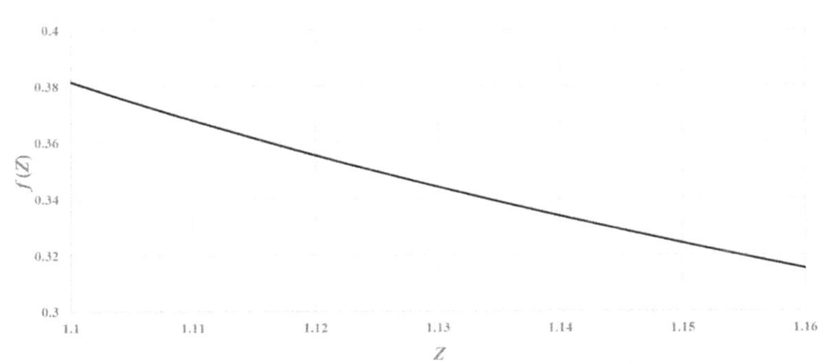

in transverse direction with the period L_0. When shown from the representation of an arbitrary periodic function as a Fourier series with the number of oscillations n=1, we write the illumination $E(y)$ in the transverse direction as

$$E(y) = \frac{a_0}{2} + A \cdot \cos\left(\frac{2\pi \cdot y}{L_0} + \theta_1\right), \tag{15}$$

where $a_0/2 = (E_{max}+E_{min})/2$ is the mean value of illumination, $A = (E_{max} - E_{min})/2$ is the amplitude of oscillations of illumination in the transverse direction relative to k_0E_0 – normalized illumination with a reserve factor k_0 (figure 3). Expression (15) must satisfy the following boundary condition,

$$E(0) = E(b) = \frac{E_{max} + E_{min}}{2} = k_0 \cdot E_0, \tag{16}$$

which means that the floor and the wall boundary of the poultry house illumination should be iniform, and close to normalized E_0 with the safety coefficient k_0.

Function (15) satisfies boundary condition (16) at zero at phase $\theta_1 = \frac{\pi}{2}$, then

$$E(y) = \frac{a_0}{2} + A \cdot \sin\left(\frac{2\pi \cdot y}{L_0}\right).$$

To perform (16) at the point b it must be $\frac{2\pi \cdot b}{L_0} = \pi \cdot n$, where n = 0, 1, 2… since $b \geq L_0$, then

$$b = \frac{n \cdot L_0}{2} \text{ under } n \geq 2. \tag{17}$$

Thus, it follows from (14) and (17), that there is a link between the values H, b and n as per

Figure 3. Illumination E(y) in the transverse direction for a periodic function with period L_0 (for the number of lines N_L =2 perpendicular to the plane of the figure)

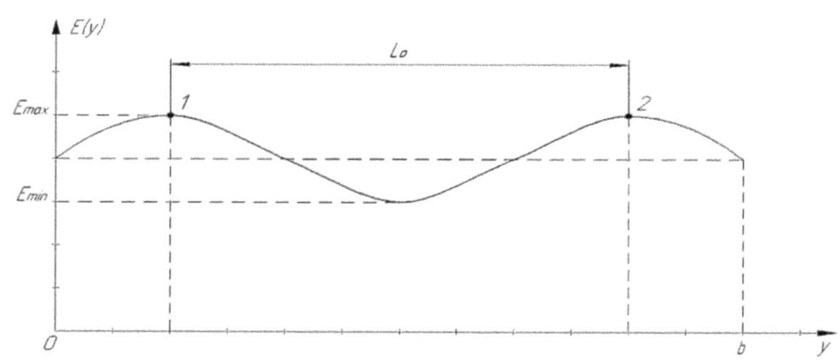

$$\frac{n \cdot H}{2b} = f(Z).$$

Therefore to calculate $Z<1.15$, it must be

$$n > 0.648\frac{b}{H}. \quad (18)$$

Thus, for the b/H known we receive rounded to an integer value n as in (18), and calculate the distance between the lines L_0 as in

$$L_0 = \frac{2b}{n} \quad (19)$$

that is to be used as an initial parameter necessary to calculate illumination for the poultry house.

Regarding the (9) and (19), we put it down as $b-\left(N_L-1\right)\cdot L_0 > 0$. To define the lamp lines by rounded to an integer value from the inequality

$$N_L < \frac{b}{L_0} + 1. \quad (20)$$

If the number of led lamps N_L is specified, the top estimate of the distance between the lamp lines is calculated from the inequality (20)

$$L_0 < \frac{b}{N_L - 1}.$$

When uniformly the lamp lines are spread across the poultry house's width (if $y_b = L_0$), we receive the bottom estimate of the distance between the lamp lines

$$L_0 = \frac{b}{N_L + 1}.$$

Thus, at the known number of lamp lines N_L, an interval for the L_0 optimization is calculated as

$$\frac{b}{N_L+1} \le L_0 < \frac{b}{N_L-1}. \quad (21)$$

The estimates of the L_0 parameters received can be used as initial (entry) parameters needed to calculate the poultry house's illumination.

2.3. Estimation of Optimal Values of the Principle Parameters Using the Target Function

Illumination (5) with the account of (6) is equal to

$$E_s\left(x,y,s\right)=\frac{I_0\cdot\cos^{m+1}\theta\left(x,y,s\right)}{r_s^2}. \tag{22}$$

Equation (4) subject to (22) can be written as and denote the sum on the right by S:

$$\frac{E\left(x,y\right)}{I_0}=S=\sum_{s=1}^{N}\frac{\cos^{m+1}\theta\left(x,y,s\right)}{r_s^2}. \tag{23}$$

If the uniformity of lighting is ideal (Z=1), the value $E(x,y)$ is constant and remains the same at every point. Then, for a particular led lamp with I_0 and α_0 the sum S must be also constant, the numerical value of which depends on N. For now, there is no analytical method of calculation existing yet; respectively, there is no chance to designate obviously N from (23).

However, applying to computer multiple calculations there is a chance to achieve minor difference I_0S from the illumination $E(x, y)$ =const required at some N. To provide uniform illumination of the horizontal surface in the transverse direction in the poultry house is more problematic, that is why computer calculations N must be conducted at transverse surface under x=const at the uniformity of illumination approved.

Calculation of illumination at point P is done with the account of every led lamp's contribution to the system of illumination (figure 4).

If

$$y_s=R_j\cdot\sin\left(j\cdot\gamma\right);\ z_s=H_0+D_0-R_j\cdot\cos\left(j\cdot\gamma\right),$$

values

$$R_j=\frac{D_0}{2}\sqrt{2\cdot\left(1+\cos\left(2j\cdot\gamma\right)\right)};$$

$$\cos\theta=\frac{r_s^2+c_{\beta j}^2+d^2}{2r_s^2\cdot c_{\beta j}}$$

are expressed by theorem of cosines when

Figure 4. For calculation of illumination at point P by a led lamp. Axis «x» and «y» form longitudinal and transverse directions as well as the working area of the poultry house

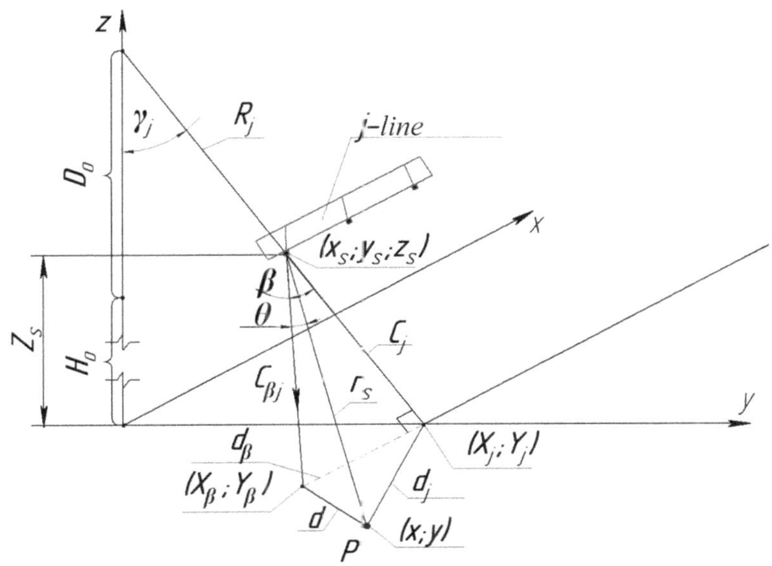

$c_{\beta j} = \dfrac{c_j}{\cos\beta}$ and $d = \sqrt{\left(y - Y_\beta\right)^2 + \left(x - X_\beta\right)^2}$.

The rest of the values can be determined by *Figure 4.*

$$Y_j = \left(H_0 + D_0\right) \cdot tg\left(j \cdot \gamma\right);$$

$$Y_\beta = Y_j;\ Y_\beta = y_s;\ X_\beta = x_s - c_j \cdot tg\beta = x_s - d_\beta;$$

$$c_j = \sqrt{X_j^2 + \left(H_0 + D_0\right)^2} - R_j;$$

$$d_j = \sqrt{\left(y - Y_j\right)^2 + \left(x - X_j\right)^2}\text{, if } X_j = x_s;$$

$$d_\beta = \sqrt{\left(Y_j - Y_\beta\right)^2 + \left(X_j - X_\beta\right)^2}.$$

After calculating of illumination at the points of transverse direction and if the value of the number of the led lamps at one line is n_0, we will get the array $E(x$=const, $y)$ to determine E_{min} and E_{max} the average meaning of which must not be lower than the level of illumination required. Therefore, to provide a normalized illumination level of E_0 with a reserve factor of k_0, it is necessary to introduce and investigate the target function in the form of

$$f(t) = k_0 E_0 - \frac{E_{min} + E_{max}}{2} \tag{24}$$

at parameters of led lighting of the poultry house $t = t_1, t_2, ...t_n$ through which $f(t)$ is not explicitly expressed. Note that the parameters $t = t_1, t_2, ...t_n$ are the number of leds N, the design parameters that define the geometry of the poultry house and lighting systems: the length a and width b of the poultry house, the height of the suspension of the lamps H, the distance between the lines L_0, the number of lines in the lamp j, the angle between the led lines γ, etc.

To estimate the optimal value of N, the target function $f(t)$ is checked against the condition

$$f(t) = k_0 \cdot E_0 - \frac{E_{min} + E_{max}}{2} \leq 0, \tag{25}$$

whereas it is a calculation of the coefficient of irregularity of illumination (2), and checking the conditions of uniformity of illumination (3). Failure of (25) will denote that illumination is insufficient, so then one led should be supplied additionally into each led lamp line. Note that increasing the number of leds in one line by a unit will lead to the increase of total number of leds N for the whole illumination system by

$$\Delta N = j \cdot S_0,$$

where j, S_0 and N_L were calculated above. (Thus, if $j = 7$, $S_0 = 66$, then, evidently, $\Delta N = 462$), which means that the value of the parameter N is a multiple of ΔN, and the minimum of the target function (24) is determined when it passes through zero, that is, when the condition (25) is satisfied.

Thus, calculations are repeated when increasing n_i by one led until the conditions (25) and (3) are satisfied. Satisfying the condition (25) results, according to (11), in value N, designated by N_0 and called the intermediate number of leds in the lighting system of the poultry house. The L_0 value chosen for physical modeling can also be called the intermediate distance between the luminaire lines.

One can see from (23) that the left part of $E(x, y)/I_0$, so then a sum of parts in the right part S can remain unchanged even when the number of parts N_0 is less, if each part is larger. Then L_0 can be reduced, and that allows to receive a new value but less than intermediate N_0 with some L_{opt} (figure 5).

However, the intermediate value of L_0 may be lower in advance than L_{opt} (figure 5 not shown). Therefore, calculations must be proceed in small increments h_L and with increasing L_0. This means that the L_{opt} should be searched for near the L_0 value in the form

$$L_{opt} = L_0 \pm \Delta L_0, \tag{26}$$

where ΔL_0 is a multiple of the step size h_L.

At some ΔL_0 a L_{opt}, is achieved, which gives the best values of N_{opt}, as well as an estimate of the optimal power consumption

Figure 5. When the distance between the lines L_0 decreases, the angle θ (x,y,s) and r decreases

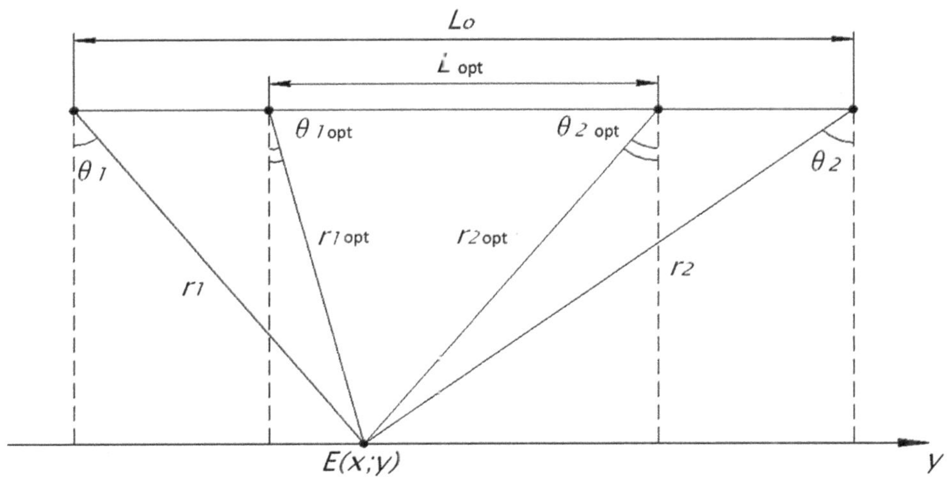

$$P_{opt} = p_1 \cdot N_{opt}.$$

From the point of view of optimization theory, we implemented a method of zero order (Bandi, 1988; Mzhel'ski & Mzhel'skaya, 2001), in which it is not necessary to calculate derivatives, and the direction of minimization is determined by successive calculations of the values of the target function

3. ESTIMATION OF THE POULTRY HOUSE'S OPTIMAL ENERGY CONSUMPTION BASED ON THE COMPUTER PROGRAMME

3.1 Algorithm for the Computer Programme

To calculate the illumination and afterwards estimation of optimal consumed power applying to the target function (24) a computer programme has been worked out suggesting conversion of the illumination array into an Excel spreadsheet database (Gallyamova & Shirobokova, 2015). Computer calculations of illumination created by means of N of point leds are based on the mathematical model proposed in paragraph 2.

As a result of calculations done, it should result in obtaining estimates of the optimal number of N_{opt} leds (with a limit of $Z<1.15$) as an output parameter which (as mention above) determines the consumed energy P of the technological illumination system of the poultry house. The output parameters also include the numerical value of the estimation of the optimal distance between the luminaire lines L_{opt} in the poultry house, for obtaining of which it is necessary to set the initial value L_0 according to the physical modeling.

The output parameter N_{opt} is connected with the number of lines j in one lamp, and with the number of leds in the line n_1, which, in turn, is related by the formula (11) with the distance (step) between the leds in the λ line. Therefore, all four parameters become dependent output parameters, and to obtain them one parameter is sufficient. Easier to search for assessment of the optimal value n_{1opt}, because the calculation of illumination E must be repeated with $n_i = n_i + 1$. All the rest of the parameters, needed for calculated estimation of the parameters, are distinguished by their meaning as input parameters and variable ones. We will write down all three types of parameters used in calculations.

Entry parameters:

a, b – length and width of the poultry house, [m];
l_c – lamp length, [m];
H – the height of the luminaire above the illuminated surface, [m];
h_x – step in the longitudinal direction, [m];
h_y – step in the transverse direction, [m];
N_L – number of lines of lamps;
L_0 – the initial distance between the lines of the luminaires, [m];
S_0 – number of lamps in the lighting system;
k_0 – coefficient of light reserve;
I_0 – the axial light power of the led, [cd];
E_0 – normalized illumination [lx];
α – the angle of disclosure of the light flux of the led, [rad];
β_0 – the initial angle (of the first led in the line between the axial line of the curve intensity of the light and the transverse direction), [rad];
γ – the angle between the led lines in the transverse plane of the lamp, [rad];
j – number of lines in the transverse plane;
n_0 – the initial number of leds per 1 m of the luminaire line;
I, U – current and voltage for one led, respectively [A], [B];
$p_1 = I \cdot U$ – power consumption by one led [W].

Output parameters:

n_{1opt} – estimation of the optimal value of the number of leds for one line;
Z_{opt} – the optimal value estimation of the coefficient of irregularity of illumination;
L_{opt} – estimation of the optimal distance value between the lines, [m];
$P_{opt} = p_1 N_{opt}$ – the optimal value estimation of the power consumption by the illumination system of the poultry house, [W];

Variables:

x, y – coordinates in longitudinal and transverse directions, [m];
x_s, y_s, z_s – coordinates of the led at the height z_k, [m];
n_1 – number of leds per 1 m of the lamp line;
$\lambda = 1/n_1$ – the distance (step) between the leds on the board, [m];
β – the angle between the axial line of the led's light intensity curve and the transverse direction in the plane of the board, [rad];
r_s – the distance between the led and the point with coordinates x and y, [m];
$E(x, y)$ – the illuminance at the point with coordinates x and y, [lx];

Z – the coefficient of uneven lighting;

θ – the angle between the axial line of the led's light curve intensity and the direction to the point with coordinates x and y, [rad].

The algorithm of the program provides for calculation of the number of leds in the line n_1 for a given number of lamps S_0, designed for lighting from the same height H of the working area of the poultry house with the length a and the width b. According to the accepted tolerances at the length of the lamps $l_c = 1$ m, suspended from each other at a similar distance of 1 m, the number of lamps for one line will be numerically equal to $a/2$, then S_0 will be by N_L times greater than (8). The number of lines of N_L lamps is defined according to physical modeling from the inequalities (20), or (21). After entering the input parameters, the program starts to calculate the illumination with the creation of an array $E\ (x, y)$, from which E_{min} and E_{max} are defined.

Estimation of the optimal value of the main parameter is carried out as per two cycles with a postterm:

- in the first cycle, when the leds' number in one line per a unit increases, the condition for the output from the cycle is the transition of the target function through the zero: exceeding the average (median) value of illumination in the transverse direction E_y=0.5(E_{max}+E_{min}) over the value k_0E_0. However, thus the current ratio of the uneven lighting, Z (3) is calculated. If after the exit from the first cycle Z >1.15 (more than the admitted unevenness of lighting), then the second cycle is performed;
- in the second cycle, the condition of exit from the cycle is to follow the condition Z <1.15, also with an increase in the number of leds in one line per a unit. After performing the two cycles, we obtain estimates of the intermediate value of the parameter N_0 at a constant value of the structural parameter L_0, selected in physical modeling.

To estimate the optimal value of the structural parameter L_0, calculations are made with a small step h_L about L_0. The condition to exit the cycle is a deterioration (increase) of the values of N calculated, and compared to the previous values (starting with the intermediate value of N_0). Then we obtain the estimation of the optimal value of the L_{opt} as per (26).

This means that at some value ΔL_0 (as mentioned above), such a L_{opt}, is obtained which provides the best values $N_{opt} \leq N_0$. The resulting values of the array $E(x, y)$ with the values of N_{opt} are converted into the database of Excel spreadsheets for further processing and plotting. When comparing the above method with the classical one, it should be noted that the latter is an approximate (inaccurate) method of calculation of lighting, since while making calculations:

- essential graphical information is needed (isolux depending on altitude and distance from the point to the projection of the lamp onto the plane);
- only the influence of the nearest luminaires is taken into account;
- coefficients are used, upon the choice of which depends on the accuracy of the lighting calculation.

These shortcomings in the proposed method are excepted, and that allows us to consider the method much more accurate than the classical one.

3.2 Computer Research of the Real System in Order to Improve the Efficiency of Lighting

Based on the proposed algorithm, methods of computer-aided study of illumination with an assessment of the optimal energy consumption of poultry houses have been created. This technique allows computer studies of the real led lighting system to improve its power consumption and the quality of lighting. Thus, at the floor maintenance of birds in the poultry house 78m x 18m for illumination, four lines of led lamps per 40 lamps in each line have been used. The total power consumption of 160 luminaires of "Gelan" LLC manufacturing makes 1280 w (power consumption of one led lamp is 16 W). With this, for one lamp of GL036 -D016.0N brand 16 leds of Z5M2-WN-C8 brand with axial intensity I_0 = 35.8 cd and the opening angle of the luminous flux of the led α = 120° are used. Power consumption of a led makes 1 W with the current equal to 120 mA. By experimental measurements of illumination in the working area the coefficient of unevenness of lighting in the range from 1.15 to 1.43 has been determined which exceeds the recommended limit Z=1.15.

By means of computer calculations of illumination in the transverse direction the estimates of optimal values have been obtained: those of the total number of leds equal to N_{opt} =1920, spread across 120 lamps, instead of 160 previously used, with the coefficient of irregularity of illumination of the Z=1.13, and if the distance between the lines in L_{opt} =5.44 m. In this case the normalized luminance level E_0 =100 lx with the illumination coefficient of the reserve k_0 =1.2 (figure 6). Thus, computer modeling of illuminance predicts decline in real consumption on 640Bt (or 25%) when improving the quality of lighting to Z=1.13 is reached.

Figure 6. Calculated illumination of the poultry house78m x 18m in transverse direction with x=39m; E_{min}=107.14 lx; E_{max}=134.86 lx; Z_{opt}=1.130 with L_{opt} =5.44m. Abscissa axis - distance in the transverse direction of the poultry house, [m]; ordinate axis – the amount of light at the points E, [lx]

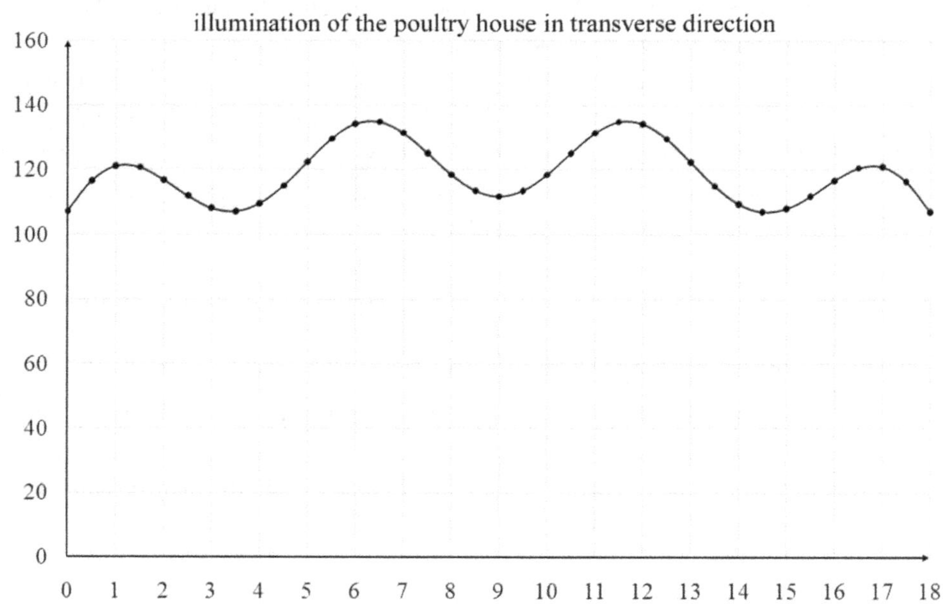

4. EXPERIMENTAL SUBSTANTIATION OF THE APPLICABILITY OF THE DEVELOPED MATHEMATICAL MODEL AND ITS COMPUTER IMPLEMENTATION

4.1. Experimental Research Methods

For laboratory experiments two prototypes of the led lamp have been designed and manufactured, the design of which is protected by a patent of the Russian Federation (Gallyamova & Shirobokova, 2016). The novelty of the design of the lamp is that the leds are fixed from the edge of the board for a more compact arrangement of boards (figure 7). This allows to increase their number in the transverse plane by 2-3 times and, accordingly, to increase the uniformity of the illumination of the working surface.

According to the proposed design, a prototype of a led lamp has been manufactured. Experiments on measuring illumination were carried out on a laboratory bench (figure 8) to check the adequacy of the designed mathematical model and its software implementation by comparing the calculation results of illuminance with the experimental values. To do this, the computer program introduced the parameters of the prototype led lamp: length 600 mm, the number of led lines - 7, the distance between the leds in one line 25mm, the angle between the lines 15°.

In the lamp, there are used round-symmetric leds of brand FYL – 3014WWC/s (warm-white) with technological data for axial light intensity I_0=15cd and for the radiation angle α_0 =30°. Measurements (from a sample of 20 leds) of the axial light intensity I_0 and of the radiation angle α_0 gave the following results:

$$I_0 = \left(14.9 \pm 0.6\right) cd\,,\ \alpha_0 = \left(29.4 \pm 1.5\right)^\circ .$$

The outfit of the stand consisted of the studied led lamp, of the power source "Agilent" E3634A, and of a testified luxmeter-pulsemeter Argus-07 with the manually admitted error 8%.

Figure 7. Design of the led lamp: a) front view; b) side view; 1 – tubular light-transmitting bulb; 2 – leds; 3 – double-sided boards (led line); 4 – led findings; 5 – grooves; 6 – flanges; 7 – tightening flanges stud; 8 – screws; 9,10 – end caps; 11 – a sealing gasket; 12 – power cord; 13 – suspension eyebolt

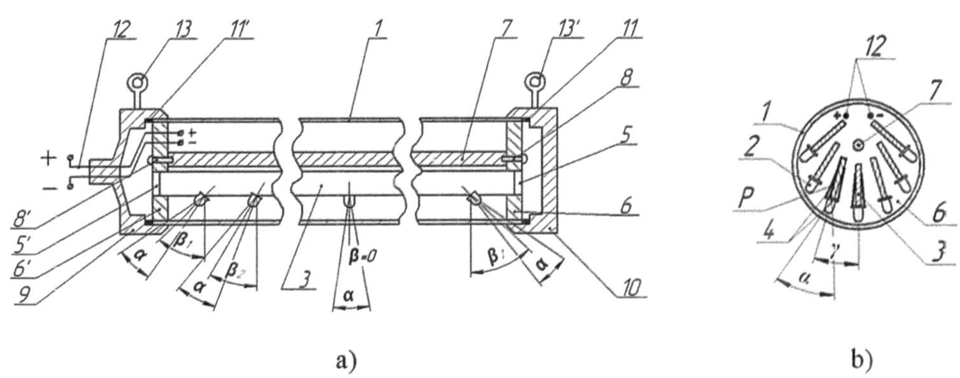

Figure 8. Fixing LED lamps on the frame construction of the laboratory stand above the coordinate grid

4.2 Study of Illumination Produced by One LED Lamp

The lamp was fixed on a framework metal construction (figure 8), at a certain height above the coordinate grid with the working area of 3m x 3m. At the grid nodes (number of nodes: 16x16=256) with a pace of 20 cm, measurements of illumination were made using a sensor of a luxmeter. The results of measurements of the illumination created by the manufactured prototype were obtained in the view of a spatial spread over the working plane (figure 9).

Figure 9. Spatial distribution of illumination produced by the sample of the LED lamp

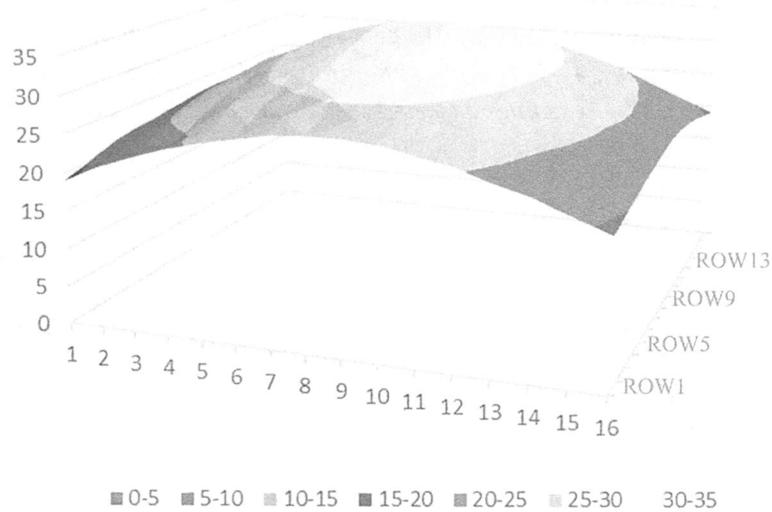

4.3 Study of Illumination Produced by Two Lamps

To further verify the adequacy of the developed mathematical model and its software implementation, a second copy of the prototype of the led lamp was made. The luminaires were fixed on the frame structure of the stand at a height h=2.80 m above the grid in parallel and in series in respect of each other (figure 10). With a parallel arrangement of lamps, the distance between them (parameter L_0) was being changed in the range up to 1.5 m. With the sequential arrangement of lamps, the space between them was equal to a_x=0.38m.

The results of experimental measurements and the results of computer calculations of the illumination created by two experimental samples were obtained for parallel arrangement of lamps on the working plane with dimensions of 3m x 3m.

A comparative analysis of the illumination of experimental measurements and computer calculation was carried out for the transverse planes x=1.6 m (figure 11).

The results of experimental measurements and the results of computer calculations of illumination created by two consecutive lamps have been also obtained as in the above paragraphs on the working plane with the dimensions of 3m x 3m.

For the comparative analysis of illumination of experimental measurements and computer calculation, the transverse planes x =3.0 m and x=1.6 m are also considered (figure 12) at a distance between the lamps a_x = 0.38 m.

Figure 10. Location of the two LED lamps on the frame structure of the stand above the grid at a height H=2.80 m: a) - in parallel (L_0=1.5 m), b) - in series (a_x=0.38 m)

Figure 11. Results of laboratory measurements of illumination (solid line) in the transverse plane of two parallel led lamps in comparison with computer calculation (dashed line). The height of the lamp above the sensor of the liuxmeter H=2.80 m. The plane x=1.6 m close to the central one. Illumination E (y) in lc for different distances L_0 between lamps: a) - L_0=0.5 m; b) - L_0=1 m; c) - L_0=1.5 m

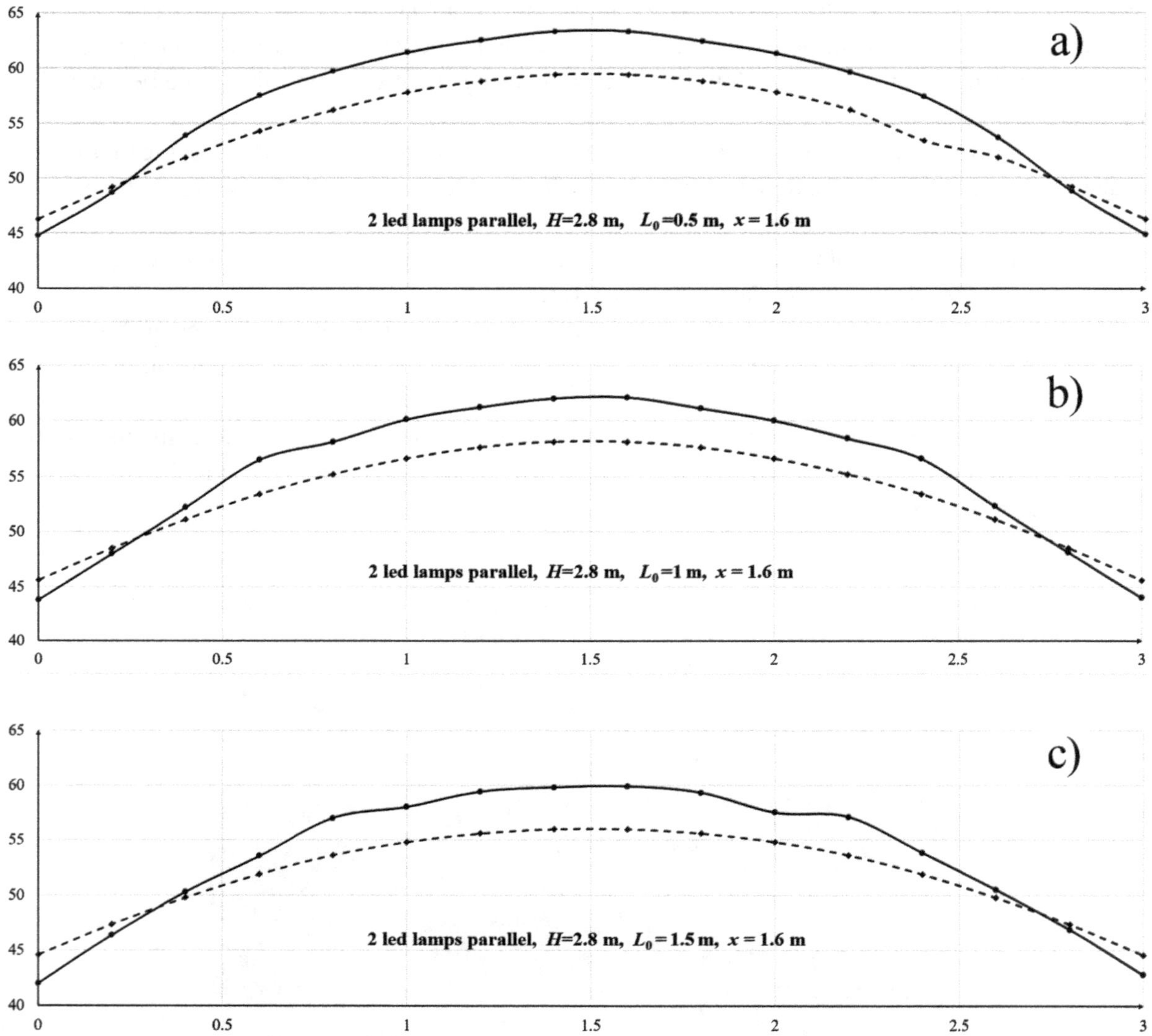

Processing of the measurement results has shown that the standard (mean-square) deviation of the measured values of illumination from the calculated ones with reliability of 95% does not exceed the manual error of the luxmeter being 8%. That confirms the applicability of the proposed mathematical model and its software implementation.

Figure 12. Results of laboratory measurements of illumination (solid line) in the transverse plane of two consecutive led lamps, in comparison with computer calculation (dashed line). The height of the lamp above the sensor of the luxmeter H=2.80 m. Illuminance E(y) lc for the distance a_x=0.38 m between the lamps. The plane x=1.6 m close to the central one

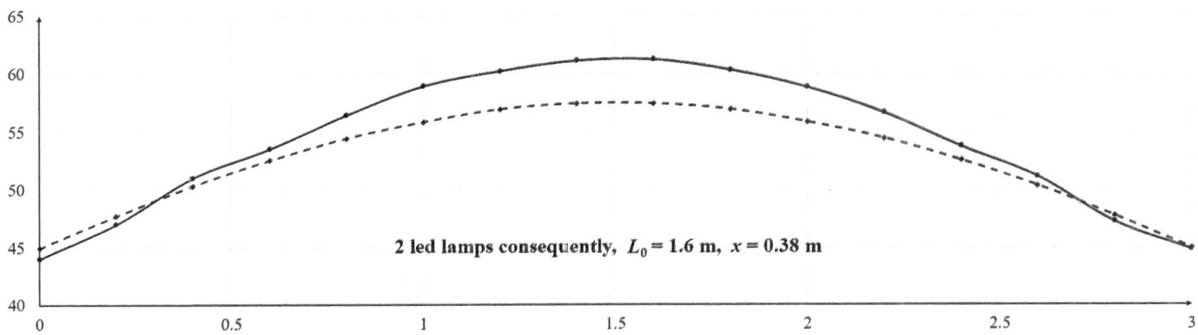

CONCLUSION

Solution to the problem of estimating the optimal values of the principle parameters of led lighting is based on a mathematical model with a computer implementation of the illumination calculations done.

While developing the mathematical model, the physical modeling of uniform led lighting of a large area of the poultry house had been used. Modernization of technological led lighting of poultry houses based on the proposed method allows obtaining additional energy savings. Thus providing normalized illumination of approximate 100 lx, and illumination uniformity ratio $Z < 1.15$.

The proposed method can find its application in the development and design of a system of technological led lighting of a poultry house of a fairly large area, as well as to develop recommendations for reducing the energy consumption for existing led lighting systems for poultry houses. Mathematical model and methodology of calculation of the illumination systems with the Leds used can be used at any agricultural object where it is necessary to illuminate uniformly a large horizontal area.

ACKNOWLEDGMENT

The reported study was supported by the Supercomputing Center of Lomonosov Moscow State University (Voyevodin et al., 2012)

REFERENCES

Bandi, B. (1988) *Metody optimizacii. Vvodnij kurs* [Methods of optimization. Introductory course]. Radio i svyaz'. (in Russian)

Baranov, L.A. & Zakharov, V.A. (2006) *Svetotechnika i electrotechniologia* [Lighting engineering and electrotechnology]. Kolos. (in Russian)

Feynman, R., Leighton, R., & Sands, M. (1977) Feinmanovskie lekcii po fizike v 8 t [Lectures on physics by R. Faiyman: in 8 vol.], Vol. 5. Elektrichestvo i magnetism [Electricity and magnetism]. Mir. (in Russian)

Gallyamova, T. R., Novosyolov, I. M., & Shirobokova, T. A. (2012) Opredelenie optimal'nogo svetoraspredelenia dlya obespechenia ravnomernosti osveschenia pticevodcheskih pomeschenij [Determination of optimal light spreading for providing uniformity of illumination of the poultry facilities]. *FGBOU VPO Izhevskaya gosudarstvennaya sel'skohozaistvennaya academia, 3,* 271-274. (in Russian)

Gallyamova, T.R. & Shirobokova, T.A. (2015). Svidetel'stvo o gosudarstvennoi registratsii programmy dlya EVM 2015617923 [Certificate of registration 2015617923 for the computer programme]. In *Yevribrid svetodiodov v sisteme osveschenia ptichnika* [Led eurobreed in the system of the poultry house illumination]. Zayavitel' i patentoobladatel' FGBOU VPO Izhevskaya GSHA, 2013116823/07. Publ. 24.07.2015. (in Russian)

Gallyamova, T. R., & Shirobokova, T. A. (2016). *Svetodiodny svetil'nik dlya ravnomernogo osveschenia rabochei poverhnosti ptichnika* [A led lamp for the uniform illumination of the working plane in the poultry house]. Pat. 159705 RF, MPK F21S 13/00. Publ. 20.02.2016. Bulletin 5. (in Russian)

Gallyamova, T. R., Shirobokova, T. A., & Kochetkov, N. P. (2015). Experimental'nye issledovania osveschyonnosty, sozdavaemoi opytnym obrazcom svetodiodnogo svetil'nika s lineikamy kruglosimmetrichnyh svetodiodov [Experimental studies of illumination created by an experimental sample of a led lamp with the lines of round symmetric leds]. In *Mezhdunarodnaya prakticheskaya konferencia "Teoria i praktika - ustoichivomu razvitiu agropromyshlennogo kompleksa"* [International practical conference "Sustainable development of agroindustrial complex: theory and practice]. FGBOU VPO Izhevskaya GSHA. (in Russian)

Guttsait, E. M., Krasnopol'skij, A. E., & Milyutin, D. V. (2007). Paschyoty svetodiodnyh modulei dlya mestnogo osveschenia [Calculations of led modules for local illumination] [in Russian]. *Svetotehnika, 4*, 52–56.

Hammer, W.M. (n.d.). Diurnal rhythm and photoperiodism in testicular recrundes sense of the house finch. Poultry Science, 42(5), 1294–1295.

Izenberg, Y. B. (2006). Spravochnaya kniga po svetotehnike [Reference book on lighting engineering] (3rd ed.). Znak. (in Russian)

Kavtarashvili, A. S., Novorotov, E. N., Gladin, D., & Kolokol'nikova, T. N. (2012). Svetodiodnoye osveschenie pri soderzhanii roditel'skogo stada [Led illumination during the parents flock keeping]. *Pticevodstvo, 5*, 15–17.

Kavtarashvili, A. Sh., Novorotov, E. N., & Kolokol'nikova, T. N. (2014). Racional'ny vozrast kur sovremennyh yaichnyh krossov pri nachale svetovoi stimulyacii [Rational age of chickens at the launch of illuminating stimulation]. *Pticevodstvo, 2*, 43–46.

Kavtarashvili, A.S. & Ridzhal, P. (2001). Preryvistoye osveschenie kur vo vtorom cikle produktivnosti [Intermittent illumination of the chickens during the second cycle of their productivity] *Peredovoi nauch.-proizv. opyt v ptitsevodstve: Ekspress-inform. Vseros. N.-i. i tehnol. in-t ptitsevodstva, 2,* 10-12. (in Russian)

King, D. (1958). Get more eggs with new lighting plan. *Poultry Tribune*, *64*, 26–28.

Knorring, G.M., Fadin, I.M., & Sidorov, V.N. (1992). *Spravochnaya kniga dlya proyektirovschikov elektricheskogo osveschenia* [Reference book for the designers of electrical lighting]. Sankt Petersburg: Energoatomizdat. (in Russian)

Kondratieva, N. P., Baranov, S. A., & Kondratiev, R. G. (2012). Energoresursosberegayuschaya systema osveschenia myasnogo krossa "Smena 7" [Energy resources conservation system of illumination of the meat cross "Smena 7"]. *Trudy mezhdunarodnoi nauchno-tehynicheskoi konferencii "Energoobespechenie i energosberezhenie v sel'skom hoziastve, 3,* 177-179. (in Russian)

Kovalenko, O. Y. (2009). *Svetotehnicheskie ustanovki dlya povyshenia produktivnosti sel'skohozyaystvennyh zhivotnyh* [Light engineering installations for increasing the productivity of farm animals] (Unpublished doctoral dissertation). Saransk. (in Russian)

Kozinski, V.A. (1991). *Elektricheskoye osveschenie i obluchenie* [Electrical lighting and radiation]. Agropromizdat. (in Russian)

Mikhalyov, A. A. (2014). *Razrabotka sistemy osvescheniya pticevodcheskih pomeshchenij na osnove svetodiodnyh svetil'nikov i rezonansnogo istochnika pitania* [Development of a lighting system for poultry houses based on LED lamps and a resonant power source]. Dissertation Abstracts PhD in Engineering. (in Russian)

Mzhel'ski, B.I., & Mzhel'skaya, E.B. (2001). *Ocnovy teorii optimizatsii. Uchebnoye posobie* [Fundamentals of the theory of optimization]. MEI. (in Russian)

Novosyolov, I. M. (2011). *Obosnovanie effektivnosti tehnologicheskogo svetodiodnogo osveschenia promyshlennogo stada kur-nesushek krossa "Highsex – Brown"* [Substantiation of the effectiveness of an industrial flock of laying hens of the cross "Highsex – Brown"]. Dissertation Abstracts PhD in Engineering. Izhevsk. (in Russian)

Pil'schikova, Y. A. (2014). *Povyshenie effektivnosti obluchatel'noy ustanovki dlya pticevodstva* [Increasing the efficiency of the irradiation plant for poultry farming]. Dissertation Abstracts PhD in Engineering. Saransk. (in Russian)

Ralph, A. (1963). *Lighting programmes for replacement pullets. In Extension Poultry Specialist* (p. 95616). Davis, CA: Animal Science Department, University of California.

Shirobokova, T.A., Kochetkov, N.P. & Gallyamova, T.R. (2014). Modelirovanie svetodiodnogo svetil'nika dlya osveschenia ptichnikapri pri napol'nom soderzhanii roditel'skogo stada [Modeling of a led lamp for illumination of the poultry house with floor maintenance of the parent flock]. *Agrarny naychny zhurnal, 9,* 52-54. (in Russian)

Vasant, P., Weber, G.-W., & Dieu, V. N. (2016). *Handbook of Research on Modern Optimization Algorithms and Applications in Engineering and Economics*. IGI Global. doi:10.4018/978-1-4666-9644-0

Vedenyapin, G. V. (1967). Obschaya metodika eksperimental'nogo issledovania i obrabotki opytnyh dannyh [General methodology of experimental studies and of processing the experimental data] (2nd ed.). Kolos. (in Russian)

Vodyanikov, V.T. (2002). *Organizacionno- ekonomicheskie osnovy sel'skoi elektroenergetiki. Uchebnoye posobie dlya vuzov po agroinzhenernym special'nostyam* [Organizational and economic fundamentals of rural electroenergineering. A tutorial for colleges of agroengineering specialties]. Academic Press. (in Russian)

Voyevodin, V. V., Zhumatij, S.A., Sobolev, S.I., Antonov, A.S., Bryzgalov, P.A., Nikitenko, D.A., & Stefanov, K.S. (2012). Practice of "Lomonosov" Supercomputer. *Open Systems J., 7,* 36-39. Retrieved from http://www.osp.ru/os/2012/07/13017641 (in Russian)

Wilson, W., Abplanalp, H., & Atrington, L. (1962). Sexual development of coturmix as affected by changes in photoperiods. *Poultry Science*, *41*(1), 17–22. doi:10.3382/ps.0410017

Yozef, M. (2011). OSRAM o buduschem biznesa v svetotehnicheskoi otrasli [OSRAM about the future of business in lighting industry]. *Svetotehnika*, *1*, 59–61.

Yuferev, L. Y. (2016). *Energoresursosberegayuschie osvetitel'nye i obluchatel'nye systemy i ustanovki sel'skohozyastvennogo naznachenia na osnove rezonansnoi systemy elektropitania* [Energy resources saving illumination and radiation systems, and agricultural installations on the bases of a resonant power supply system] (Unpublished doctoral dissertation). Moskva. (in Russian)

Zyudkamp, H. & Bakalets, A. (n.d.). Viper-kompiuter dlya pticefabrik. [A Viper-computer for the poultry houses]. *Zhivotnovodstvo Rossii, 2,* 74-75. (in Russian)

Chapter 6
Use of an Elastic-Damping Mechanism in the Tractor Transmission of a Small Class of Traction (14 kN):
Theoretical and Experimental Substantiation

Sergey Evgenevich Senkevich
https://orcid.org/0000-0001-6354-7220
Federal Scientific Agroengineering Center VIM, Russia

Nikolay Viktorovich Sergeev
Don State Agrarian University, Russia

Evgeniy Konstantinovich Vasilev
Federal Scientific Agroengineering Center VIM, Russia

Zakhid Adygezalovich Godzhaev
Federal Scientific Agroengineering Center VIM, Russia

Vugar Babayev
Agrarian Science and Innovation Center of the Ministry of Agriculture of Azerbaijan, Azerbaijan

ABSTRACT

Each modern tractor should be adapted to perform an increasing number of various traction, transportation, and other works. Therefore, their designs are becoming more complex. To improve tractor productivity, the power-to-weight ratio and the speed of its units are constantly increasing. This inevitably improves dynamic loading of the chassis and the transmission parts, the level of the generated oscillations, which increases the vibration load of the transmission. One of the promising ways to solve this problem is to introduce elastic links into the tractor transmission closer to an engine or a clutch. Installed closer to the engine, the elastic element in the transmission of the tractor absorbs the impact of the shocks, protecting the tractor engine from large dynamic loads. The material of the chapter contains studies confirming the efficiency of the use of an elastic mechanism (closer to the engine) in the tractor transmission of a small class of traction (14 kN).

DOI: 10.4018/978-1-5225-7573-3.ch006

INTRODUCTION

The universality of modern tractors is constantly increasing. Each tractor should be adapted to perform an increasing number of different traction, transport and other works, therefore their designs become more complex. To increase labor productivity, the power-to-weight ratio of tractors and speed of their units are increasing. But this leads to an increase in the dynamic loading of the chassis and transmission parts and to an increase in the level of generated oscillations, which increase the vibration load of the transmission.

Academician V.N. Boltinsky started the research work on the effect of unsteady loading conditions on the economic and operational performance of the tractor, and the engine in particular. Various studies have shown that the source of engine speed fluctuations is not only the oscillation of the traction resistance, but also the oscillations of the transmission elements. Therefore, it is necessary to increase the transmission compliance to reduce the dynamic loads on the engine.

But there is no consensus among the researchers on the installation location in the transmission of elastic elements. One of the ways to solve this problem is to introduce elastic links into the tractor transmission closer to the engine or clutch. Mounted closer to the engine, the elastic element in the tractor transmission absorbs the shocks impact, thereby protecting the tractor engine from large dynamic loads. The development of tractors with such more advanced transmissions acquires a high scientific and practical significance. Based on the above mentioned, we come to the conclusion that to increase the tractor operational efficiency it is necessary to install special devices damping the load and protecting the engine from fluctuations in the external traction load.

GENERAL CONTENTS

Theoretical Substantiation

Achievement of permanent growth of agricultural production, the population can be provided with food and agricultural raw materials due to the improvement of technical level, quality and durability of agricultural machinery and the widespread introduction of scientific and technological achievements aimed at increasing the efficiency of its use (Denisov, 2005;Nekhoroshev, 2014; Hansson, 2002).The improvement of mobile agricultural machinery, its efficiency, technological efficiency and reliability, the traction-dynamic and economic characteristics of machine-tractor units (MTAs) in their set-making and operation in production are of great importance in the solution of these problems (Iofinov & Mikuberg, 1971; Zoz &Grisso, 2003).

The most technological operations result in overexpenditure of fuel, energy and money costs per unit of received products and processed area. This occurs because of the inconsistency of traction characteristics of the tractor and the aggregated machines at operating speeds and, as a consequence, the lack of an optimum load of the engine by traction (Kutkov, 1980).

The introduction of the latest achievements of science, technology and advanced technologies is of great importance in the development of agricultural production efficiency. The main task in the development of a new system of machines for agricultural production is to increase their productivity, to reduce fuel consumption in various types of work, to decrease the negative consequences of the interaction of the engine with the soil to a minimum. Each modern tractor must be adapted to perform an increasing number of different traction, transport and other works, therefore their designs become more complex. To increase labor productivity, the power-to-weight ratio of tractors and speed of their parts are constantly increasing. But this inevitably leads to an increase in the dynamic loading of the chassis and transmission parts and to an increase in the level of generated oscillations, which increase the vibration load of the working place of an operator.(Antonov, 1978; Hansson, 1994; Gurhan, &Cay, 2008; Duke & Goss, 2007; Marjoram at al, 2008; Zheng at al, 2016; Deboli at al, 2017; Pobedin & Shekhovtsov, 2016).

The negative consequences of these impacts can and should be decreased by installing additional damping devices in the tractor transmission. The effectiveness of their installation has been proved by many researchers (Polifke, 2017; Polifke, 2016; Feuerhuber at al 2013; Hansson, 2002; Fritz & Elsner, 2002).

We have made an attempt to prove the efficiency of introducing a traction mechanism of a damping mechanism into a tractor transmission of a small class. Its theoretical and experimental parameters have been justified. Its effectiveness has been proved.

Analysis of the Tractor Operating Conditions in the Machine-Tractor Unit (MTU)

The regime of maximum productivity, the most complete use of the traction capabilities of the machine and the most economical fuel consumption corresponds to the work with steady constant nominal traction power and nominal (maximum) traction power.However, it is practically impossible to achieve a constant value of operating resistances during the working process. The vibrations of the load on the working element affect on the operation regime of the engine, transmission and running gear, which cause a deviation in the traction power of the tractor (Kravchenko V.A., Oberemok V.A. & Kravchenko L.V., 2014; Kravchenko V.A., Senkevich A.A., &Senkevich S.E., 2003).

According to the researchers (Zoz, 2003), in order to carry out agricultural work with maximum productivity it is necessary:

1. to maximize the efficiency of fuel consumption by an engine and a chassis;
2. to maximize traction power of a tractor;
3. to select the optimal speed for this MTU.

The Russian researcher S.A. Iofinov distinguishes three groups of factors that affect the oscillations of traction power of the tractor (Iofinov & Mikuberg, 1971; Iofinov, 1974):

1. natural and climatic factors;
2. design characteristics;
3. operating conditions.

Natural and climatic factors include the type and condition of the soil, the presence of stones in them, the meteorological conditions and properties of the treated materials. As the physical and mechanical properties of the soil considerably vary due to the moisture content, function of the traction resistance of soil cultivating and sowing machines also considerably varies because of soil moisture (Kirtbai, 1976).

In Mochunova's research work (Mochunov, 2015) it was experimentally proved that the soil background could affect on the oscillation frequency of traction power and other tractor parameters.

Design characteristics affecting the specific resistance of machines include a type, a shape and a number of working elements, machine weight, availability of auxiliary devices, a type and arrangement of the running gear.

The last group of factors, *operating conditions*, comprise the technical condition and operational regimes of the machine-tractor unit, such as speed of movement, depth of processing, utilization degree of the transmission capacity, etc.

All mentioned factors cause large fluctuations of resistance moment to MTU movement even on a small part of the field.

Thus, when studying the working processes of machine-tractor units and selecting the optimal parameters and operating regimes, it is necessary to take into account the stochastic (random in the probability-statistical sense) character of the resistance to movement, and to evaluate the changing processes of the resistance forces by statistical indicators.

Studying the diagrams of resistance forces of machines and implements, academician V.N. Boltinsky came to the conclusion that it is possible "... to present a diagram of the resistance force (or resistance moment) as a chain of consecutive sinusoids with different maximum values of the function ... and periods ..." with a reasonable degree of accuracy (BoltinskiyV.N., 1949).

On the basis of the analyzed experimental data professor Yu.K. Kirtbai distinguished the following harmonic components of traction resistance (Kirtbai, 1976):

1. Micro oscillations are the load fluctuations up to 2 meters of the path, caused by the peculiarities of the field micro relief and the technological process of its tillage.
2. Meso oscillations are the load fluctuations up to 12 meters of the path, caused by the field unevenness and by the changeable physical-mechanical properties of the soils.
3. Macro oscillations are the loadfluctuations associated with a change of the field gradient, etc.

Other authors, for example G.M. Kutkov (Kutkov, Puchkov & Kholin, 1975; Kutkov, 1980) distinguished the other five main components of load oscillations on the hook:

- the formation of the first component mainly occurs due to the field microrelief and to the change in the physical and mechanical properties of the soil;
- the presence of the second component is connected with the longitudinal-angular fluctuations of the tractor;
- oscillations frequency of the third component coincides with its own frequency of vertical fluctuations of the tractor's body;
- the fourth component can be generated in the running system of the track tractor when the track rollers move through the same unevenness;
- the fifth component is also generated in the running system of the tractor, due to the track chain links.

Such division of the traction resistance into components is convenient to identify and analyze the causes of the load oscillation on the hook.

However, the authors themselves indicate the conventionality of such division of traction resistance because the components are not strictly harmonic functions, as their fluctuation periods vary within certain limits.

Effect of Dynamic Loads on the Tractor Operation

The work of the machine-tractor unit (MTU), accompanied by oscillations in the moment of resistance to movement, has its own peculiarities. These features distinguish it from work with a stable load. The change in resistance to the movement of agricultural machines connected with continuous changes in the resistance to the tractor movement deteriorates the operating conditions of the unit, decreases the coefficient of useof engine power and tractor traction (Karabanitsky, 2014; Babanin, 2015; Babanin & Polivaev, 2015).

In his works academician V.P. Goryachkin wrote thatso-called agricultural machines provided fertile field for all kinds of fluctuations. He pointed to the variable nature of the traction resistance of machines and implements used in agricultural production. The further studies have confirmed the validity of these provisions (Goryachkin, 1965).

V.N. Boltinsky started studying the effect of unsteady loading conditions on the economic and operational indexes of the tractor (Boltinskiy, 1949; Boltinskiy, 1959; Gabdrafikov, 2017; Sinitsky, 2005). Academician V.N. Boltinsky identified the main two factors that affect the energy losses of engine:

- load oscillations lead to rotational speed variations of the engine crankshaft, and this negatively affects the organization and working process of the engine, reduces the coefficients of filling, therefore the engine develops a lower effective power;
- fluctuations in the angular velocity of the engine crankshaft cause significant disturbances in the regulating system, namely phase shifts between the input and output elements of the control system coordinates.

The engine power decrease when working with unsteady loads can be characterized by the power utilization factor for unsteady loads (Svistula & Matievsky, 2011).

Due to large load fluctuations up to 30 ... 40% of the average value, tractor traction power reduces on 20 ... 30%, and the fuel consumption increases on 15 ... 25% (Boltinsky, 1959). Unsteady load conditions have a significant effect on the value of the engine load coefficient, and therefore the load of the tractor engine cannot be brought to power more than 75-80% of the standard value (Denisov, 2005; Okhotnikov, 2014).According to the conducted theoretical and experimental studies (Boltinsky, 1959), the degree of utilization of the engine power of 14 kN class tractors is: 88…96% in plowing, 75% in harrowing, 92% in shallow plowing, 75 ... 83% in cultivation. The average annual load of engines of various tractorsis 58...81% when machine-tractor units carry out the whole range of works (Boltinsky, 1959).

Studying the dynamics of agricultural units, A.B. Lurie noted the stochastic nature of external effects on the units in the form of random sequences, which reveals in the uneven load regime, significantly affects the indices that determine the level of machine use during technological operations (Lurie, 1969; Bespamyatnova, 2002).

Recently, due to speed increase and introduction of advanced technologies (with the combined operations), the engine power use has slightly increased. However, even now it is 75...78%. This is because the external load increases faster with increasing speed per kilometer (Kravchenko, Duryagina & Gamolina, 2017).

It was found out that under unsteady load the tractor-dynamic indicators of the tractor deteriorate. The decrease of unit productivity in such conditions reached 20...30%, and fuel consumption increased on 15...25% (Boltinsky, 1959; Shekhovtsov, 2002a; Shekhovtsov, 2002b). In the works of the above mentioned researchers, information has been presented on the change in the indicator, power, and economic parameters of engines operating in the unsteady load regime. The data is not enough to assess the entire tractor traction. A significant number of elements and links (inertial masses and elastic elements of the transmission) including engineparticipate in machine movement. They significantly affect the machine movement and if the parameters of the elements of the tractor dynamic systemaren't taken into account, it'll lead to fairly rough approximations.

At present, the operating speed increasing is one of the main ways to increase the productivity of machine-transport units. When the working speeds increase, irregularity of the traction power and the resistance to movement also increases. It negatively affects the indexes of the engine work, leads to a significant reduction in the load coefficient, and contributes to an increase in the wear of all machine-tractor mechanisms (Shekhovtsov, 2014; Nekhoroshev, D.D., &Nekhoroshev D.A., 2015).

Therefore, the problem of dynamic loading decrease and the fluctuations in the loading regime of the MTU, which significantly influences productivity and other technical and economic indicators, is of great urgency.

The Ways to Increase Operation Efficiency of a Tractor as a Part of the Machine and Tractor Unit

To improve the operating conditions of the machine-tractor unit, it is necessary to know the factors that determine its productivity.

MTU efficiency is described by the following formula (1):

$$W = C \cdot \frac{N_{en}}{K_a} \cdot \eta_t \cdot \lambda_N \cdot \nabla_w \cdot \tau \tag{1}$$

where:

C is coefficient of proportionality;
N_{en} is nominal engine power;
K_a is resistivity of the unit;
η_t is traction coefficient of a tractor;
λ_N is degree of engine power use;
T_w is time of MTU work per a shift;
τ is degree of useful time per a shift.

Analyzing the formula (1), we can assume that the MTU productivity largely depends on the degree of engine power use functionally connected with the engine load coefficient:

$$\lambda_N = \frac{M_e}{M_{en}}, \tag{2}$$

where M_e, M_{en} are engine torque effect, respectively, developing and nominal developing.

Obviously, in order to increase the engine load coefficient and thereby its productivity, it is necessary to carry out the optimum degree of engine loading. The works of many researcher deal with the problem (Bychkov, 2003; Vedenyapin at al, 1968; Parshin,1978; Labetskas G.S., 2002; Kuznetsov, 2006; Kuznetsov et al, 2013; Polivaev, 2017).

As V.A. Filin's research showed (Filin et al, 2016), the deviations of the engine energy parameters from their nominal values greatly depend on the loading degree, which is determined by the load profile and controlled during the whole period of operation. Modern engines with high torque effect and optimized fuel consumption significantly increase the requirements for modern drives. Figure 1 shows the load profile of a modern tractor.

The coefficient of engine power use, and, consequently, the machine and tractor unit productivity can be increased by selecting the operating parameters of the unit (working width, speed, etc.) (Zimagulov &Nurulin, 2003; Iofinov, 1974; et al).

From the university books it is known that agricultural implements and machines need to be combined with the tractor so that the unit resistance is equal to the tractor traction power in the transmission corresponding to the largest traction power.

However, it is known that traction resistance of machine-tractor units does not remain constant, it varies in fairly wide limits, which results indecreaseof average engine power, engine load factor and the planned MTU's productivity.

At the present the task of the engine operational load improvement is successfully being solved by the development of constant power engines (CPE). Scientists are engaged in research and optimization of the MTU's speed and load regimes and determine the optimum frequency of crankshaftrotation (Svistula, 2011; Eviev, at al, 2014).

The most rational way to optimize the engine loading regimes of a machine-tractor unit is to compensate the fluctuations of resistance.If the unsteady loads of the engine are damped and the angular velocity of the engine crankshaft is removed, the amount of used power will increase and approach to the nominal values (Tarasik, 1984).

Comparative tests of caterpillar and wheel-type tractors of class 3traction power showed that the caterpillar tractor is satisfactorily aggregated with high-speed soil-cultivating and sowing machines to the wheel-type tractors, but it results in4...11% increase of specific fuel consumption.

Figure 1. Load profile of a high power tractor
(Filin, 2016)

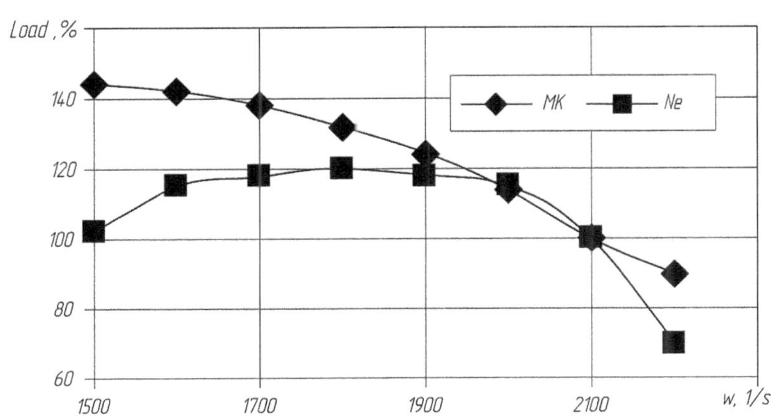

A step mechanical transmission is spread on agricultural tractors of a small class of traction (14 kN). Such a transmission usually consists of a clutch, gearbox, transfer case (for four-wheel drive tractors) and ground drive axle (one or two). The work on development of transmissions of other types is constantly being done. But back in 1966, A.M. Lysov analytically proved the inexpediency of using hydromechanical type of transmission on tractors of a small class of traction (Lysov, 1966).

One interesting solution of the problem is the power-split transmission developed by Fendt's firm (Janulevičius &Giedra, 2008) shown in Figure 2.

In such a transmission, some of the power is transmitted through a mechanical transmission, and the other part is transmitted through a hydrostatic transmission. This arrangement together with various intelligent control systems provides the best power distribution at the optimum fuel consumption. But we should take into account that the presence of a large number of elements increases the dimensions of the tractor and the complexity of the design.

In his works Shekhovtsov V.V. (Shekhovtsov at al, 2002a; Shekhovtsov at al, 2002b; Shekhovtsov at al, 2016; Shekhovtsov K.V., 2014) proposed to install torsional vibration dampers in the tractor transmission, which help to reduce the intensity of torsional oscillations in it and to decrease the impact forces in the spline or gearing units.

Since 2011 the tractor transmission testing showed that the elastically friction spring elements (dampers) installed in the driven clutch disc allow to reduce significantly the amplitude of the oscillations in the engine. Besides, the compliance of such a disk results in a smoother interaction of the friction material with the flywheel and the pressure plate (Grebnev V.P., 2011).

Figure 2. Kinematic scheme of transmission with the power-split transmission: a,b. adjustable hydraulic pumps; c.a shaft; h. transmission mechanism; 1,2. solar and hollow gears; 3. satellite; 4, 5, 6, 7. gears; 8. two-stage gearbox (Fendt Vario)

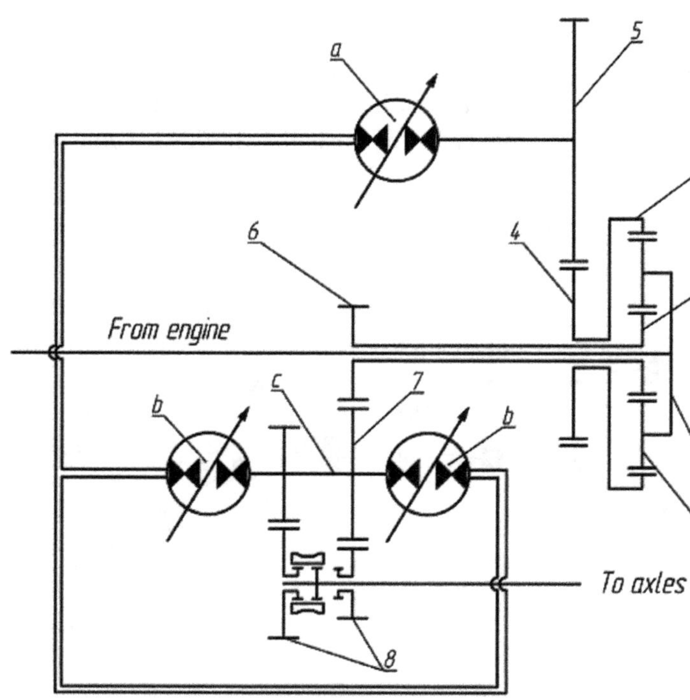

In 2016 the German explorer G. Polifke (Polifke, 2016) proposed Voith Hydrodamp, a hydraulic shock absorber of torsional oscillations, which is mainly used in thetractor drive between the engine and transmission. The main advantages of hydraulic damping combined with insulation results in a significant reduction in peak torque moments and oscillations of a drive. In 2017 G. Polifke (Polifke, 2017) continued designing a drive. The optimal Hydrodamp parameters were found by changing parameters of spring characteristic and a damping system.

Russian researcher O.I. Polivaev continued the investigation of damper springs. He found out that the introduction of elastic elements in the driven disk results in a shift of the resonant oscillations towards the reduced speed of the diesel engine, contributes to a smoother engagement of the clutch, decreases a drive skidding on 12...20% andreduces the wear on 29% in comparison with serial parts (Polivaev, Vedrinsky &Derkanosova, 2016). The influence of elastic elements in the initial link of the mechanical transmission on the operation of the machine-transport units is of great urgency (Babanin &Polivaev, 2015). This raises an important problem of combining drives with vibration dampers, i.e. the use of drives with additional damping. This allows to improve significantly the reliability and durability of machines and to increase the uniformity of MTU speed. But here we should solve the following problems: to what extent the pneumohydraulic elastic-damping elements of the power transmission extinguish the oscillations on the transmission shafts; to what extent they restrict the vibration transmission to the engine; how this affects the MTU technological processes.

Many researchers suggest a number of ways to reduce the transmissionrigidity, which results in a reduction in peak loads, dynamic forces in the drive, and in an increase in MTU speed (Kuznetsov, 2006; Kuznetsov, Gapich & Shiryaeva, 2013).

One of the promising ways to solve this problem is to introduce elastic links in the tractor transmission.

A huge contribution to the study of the elastic-damping mechanism (EDM) makes O.I. Polivaev. In his works (Polivaev & Zhidenko, 2017; Polivaev &Kutkov, 2010; Polivaev O.I., Pilyaev S.N.& Kutkov, 2011; Polivaev & Vedrinsky, 2013; Polivaev & Vedrinsky O.S., 2014, etc.), he repeatedly proved the efficiency of its application in wheeled tractors.For example, the researchers (Babanin & Polivaev, 2015) demonstrated the efficiency of the developed EDM application for smooth work. Due to its elastic and dissipative properties, the core oscillation amplitudes decrease till 60% on average; the amplitude maxima reduce their frequency ranges (0.2-1.8 Hz), skidding is reduced on 10-12% on average, the fuel economy of the tractor increases on 8-14% and productivity improves on 6-15%.

D.A. Nekhoroshev believes that in order to achieve the necessary long-eternity and MTU productivity increase while reducing fuel consumption and improving the working conditions of the machine operator, it is necessary to introduce a special clutch with pneumatic-hydraulic elastic element into the tractor transmission (Nekhoroshev, 2004; Nekhoroshev, 2014; Nekhoroshev, D.A., & Nekhoroshev D.D., 2009). According to these experimental studies, the installation of additional pneumatic-hydraulic batteries with a damping system increases the realized capacity on 15.6%. This occurs due to the fluctuation decrease of hook load.The achieved effect is determined by the stabilization of traction resistance and the ability of the engine to realize high power in the changed conditions. This provides productivity increase on 18% and fuel consumption decrease on 14.5% (Nekhoroshev, 2009).

As a result of experimental and computational study (Kalmykov, 2014) there has been developed and substantiated a method to reduce the dynamic loading of the tractor power transmission by changing the torsional rigidity of the reactive link.It has been established that due to the change it is possible to reduce the ratio of the maximum torque to the average on 4% with rectilinear movement, and on 35% on the regimes high dynamic loads.

To reduce the effect of fluctuations of the external load on the engine the elastic-damping mechanism (EDM), which structure is shown in Figure 3, was developed under V.A. Kravchenko supervision (RU Patent 2222440). Later it was improved (RU Patent 2398147). The use of EDM in the tractor transmission during the acceleration of the MTU reduces clutch friction on 22% and reduces the dynamic loads in 1.2...1.5 times.

In working conditions of MTU, the use of EDM increases the capacity of the unit on more than 10%, reduces the fluctuations of the external load on 15-40% and reduces the fuel consumption on 9%. (Kravchenko, 2014; Kravchenko, 2017). V.A. Kravchenko supervises a group of scientists who contribute a lot to the research of EDM in the tractor transmission. He has a lot of personal works, which show the positive influence of the mechanism on the parameters of the MTU (Kravchenko, at al, 2003; Kravchenko, at al, 2008; Kravchenko, at al, 2014; Kravchenko, at al, 2017).

In the work of Senkevich S.E. and Kravchenko V.A. (2007) it was proved that EDM installation in the tractor transmission eliminates the resonant regimes of the MTU work in the area of oscillations frequencies of the external load. As a result, the amplitude of the translational velocity fluctuations is reduced on 8%, and the unit productivity increases on more than 5%).

Figure 3. Scheme of the elastic-damping mechanism in the transmission of the class 14 kN tractor (patent 2222440 RU, IPC B60K 17/10): 1. a planetary reduction gear; 2. a hollow gear; 3. a solar gear; 4. a gear of an oil pump drive; 5. an oil pump; 6. movable loads for changing inertia moments; 7. satellites; 8.a regulator of cargo position; 9. a suction port (passage); 10.a delivery port (passage); 11. a carrier shaft (a primary shaft of a gearbox); 12.a gear change box; 13.a clutch; 14. an engine; 15.an adjustable throttle; 16. a throttle valve; 17.a safety valve; 18.a throttle control valve; 19. a damper valve; 20. ahydro tank; 21. compressed air; 22.a pneumatic hydro accumulator; 23.a movable piston; 24. an oil tube(patent 2222440 RU)

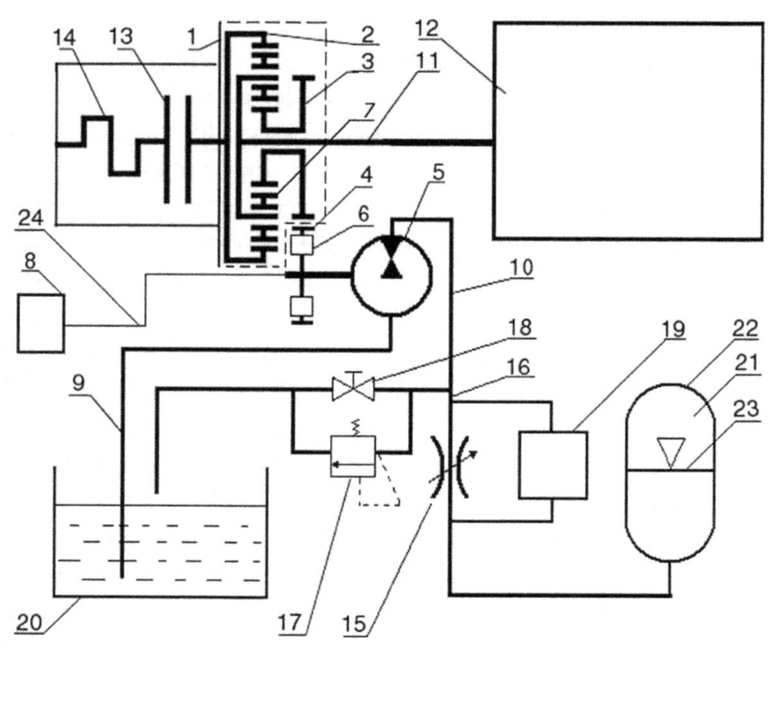

The department of tractors and cars of the Donskoy State Agrarian University (Russia, the Rostov region) continues the study and proves the effectiveness of the pliable links used in tractor transmissions. Now the Federal Research Agro-Engineering Center VIM (Russia, Moscow) has joined the research. The elastic damping mechanism (Figure 3) is designed for the MTU smooth starting during acceleration, for dynamic loads reduction in the transmission, and also for protection against external load fluctuations due to variable rigidity and automatic change of the gear ratio and for increase in the engine load.

The mechanism is designed to reduce transmission rigidity (Figure3). It comprises a planetary reduction gear (1), a solar gear (3) of which is connected with a gear of an oil pump drive (4). An oil pump (5) is connected witha suction passage (9) and with a delivery port (10). A planetary reduction gear (1) is connected with a gear change box (12) by a carrier shaft (11) and is started by an engine (14). In a delivery port there is a two-stage adjustable throttle (15) with a safety valve (17) and a throttle control valve (18) connected to it.A damper valve (19) is parallel connected with a throttle (15). There is also a hydro tank (20) and a pneumatic hydro accumulator (PHA) (22). A pneumatic hydro accumulator is a final element of a delivery passage. It consists of a hydro cylinder and a movable piston (23). In one of the parts of PHA there is pre-compressed air, which plays the role of vibration damper. The proposed device differs from the already existing one in a gear of an oil pump drive made with changing inertia moments. Its necessity is explained by the fact that in different agricultural operations the traction load fluctuates in different ranges of frequencies. By changing inertia moment of the drive, it is possible to adjust the transparency degree of the entire mechanism, thereby setting it up to a certain range of vibration frequencies of the external traction load at a given operation. As the mechanism itself is a complex system, it is necessary to find the optimal values of the system parameters in order to obtain the best result of the optimization parameter.

EXPERIMENTAL SUBSTANTIATION: STUDY OF A TRACTOR WITH AN EXPERIMENTAL

Transmission Loaded With a Traction Laboratory

For conducting the study there was prepared a machine-tractor unit comprised of a class 14 kN tractor with an elastic-damping mechanism in the transmission and a traction laboratory TL-3 (Figure 4).

Figure 5 gives a general view of the pneumatic hydro accumulator, a throttle, a safety valve and an oil pump, installed in the tractor.

The experiment was done to find optimal parameters (these factors are listed in Table 1) that will ensure the best working characteristics. As a criterion, there was chosen a “degree of transmission transparency” (Kravchenko, 2010; Kravchenko, Duryagina & Gamolina, 2017), which characterizes the damping properties of the tractor transmission with EDM in comparison with transmissions of the serial tractor. The studies were carried out on a specially prepared concrete foundation (a track). External disturbances were done by the traction laboratory.

For regression analysis, there was adopted a plan for four factors, which comprised 27 experiments. In the center of the plan, i.e. when all factors at the zero levels ($x1 = 0$, $x2 = 0$, $x3 = 0$, $x4 = 0$) are found, three experiments are to conduct. Values of factors in coded and physical forms are presented in Table 1.

Figure 4. General view of the tractor and the traction laboratory

Figure 5. General view of the pneumatic hydro accumulator, the throttle, the safety valve and the oil pump. a pneumatic hydro accumulator; 2. a throttle; 3. a safety valve; 4. an oil pump; 5. a pressure gauge

Table 1. Factors and their levels

Factors	Physical Symbol	Code Symbol	Natural Values, Corresponding to the Levels of Coded Factors		
			Upper (+1)	Basic(0)	Low (-1)
Section area of a throttle	S_t, m^2	X_1	$4.624 \cdot 10^{-4}$	$3.494 \cdot 10^{-4}$	$2.364 \cdot 10^{-4}$
Volume of pneumatic hydro accumulator (PHA)	V_{PHA}, m^3	X_2	$6.05 \cdot 10^{-3}$	$4.44 \cdot 10^{-3}$	$2.83 \cdot 10^{-3}$
Air pressure in PHA	P_{PHA}, $10^5 \cdot Pa$	X_3	5	4	3
Inertia moment of the gear of an oil pump drive	J_g, $kg{\cdot}m^2$	X_4	0.05780	0.0563	0.05481

Regression analysis and graphical representation of the results were carried out using the MathCAD program (Part 'Vector and matrix operations'). The application of this simple and convenient program helped to solvethe systems of equations and increased the visibility of the presented results.

After conducting the regression analysis and evaluating the significance of the coefficients, there has been obtained a dependence of the transparency degree of the transmission on the considered parameters of the UDM in the following equation:

$$y = 0.578 - 0.081x_4 + 0.148x_1^2 + 0.078x_2^2 + 0.073x_3^2 + 0.101x_4^2 \tag{3}$$

To find the optimal values of the factors, it is necessary to take the partial derivatives of equation (3). As result, we obtained a system of four equations (4) – (7):

$$\frac{dy}{dx_1} = 2 \cdot 0.148 \cdot x_1 \tag{4}$$

$$\frac{dy}{dx_2} = 2 \cdot 0.078 \cdot x_2 \tag{5}$$

$$\frac{dy}{dx_3} = 2 \cdot 0.073 \cdot x_3 \tag{6}$$

$$\frac{dy}{dx_4} = -0.081 \cdot 1 + 2 \cdot 0.101 \cdot x_4 \tag{7}$$

Solving the equations (4-7) there has been established optimal values of each factor. They are:

is section area of a throttle (x_1=0), $S_t = 3.494\times10^{-4} m^2$;
is volume of pneumatic hydro accumulator (x_2=0), $V_{PHA} = 4.44\times10^{-3} m^3$;
is preliminary air pressure in PHA (x_3=0), $P_{PHA} = 4\times10^5 Pa$;
is inertia moment of the gear of an oil pump drive (x_4=0.401) $J_g = 56.90\times10^{-3} kg\times m^2$;

The value of the function "degree of transmission transparency" P = 0.561.

Analyzing the received equation, we received surfaces of response ("degree of transmission transparency") and level lines (for the MTU cultivator on the basis of the class 14 kN tractor, they are presented in Figures 6, 7, 8).

The carried out researches have shown expediency of EDM application ina class 14 kN tractor transmission, as the value of damping properties of the "degree of transmission transparency" mechanism is less than one.

The found optimal parameters of the mechanism allowed us to adjust the mechanism to the working conditions for transport units. Then it was decided to conduct research for the other agricultural operations. In particular, this chapter presents research on the cultivating unit.

Figure 6. Surface response (a) and its cross section (b) for the factors x_1 and x_4.

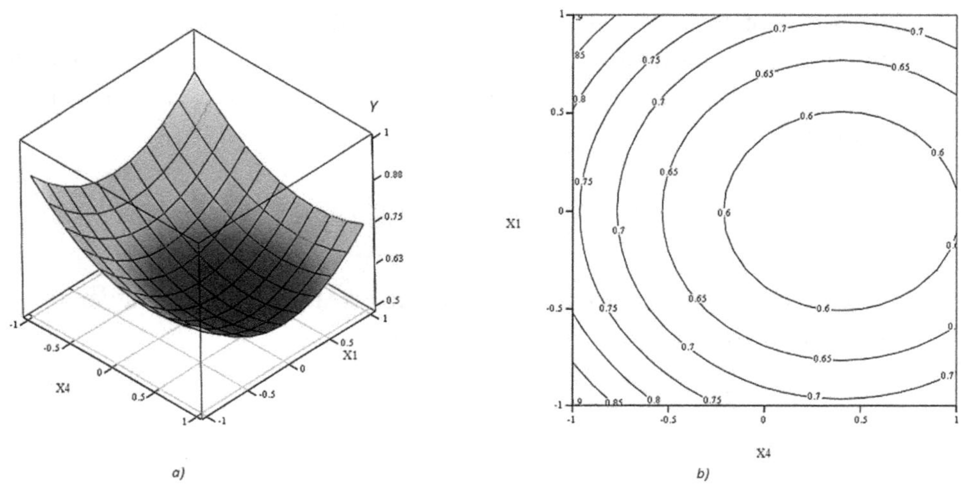

Figure 7. Surface response (a) and its cross section (b) for the factors x_2 and x_4.

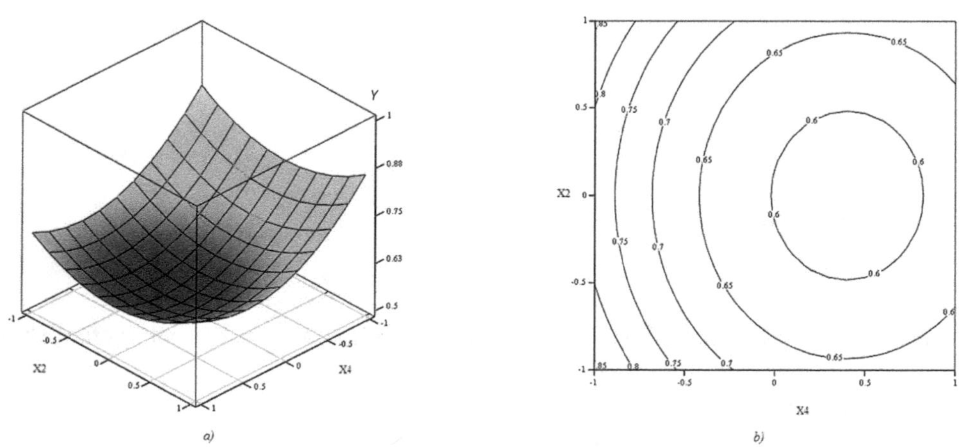

Figure 8. Surface response (a) and its cross section (b) for the factors x_3 and x_4.

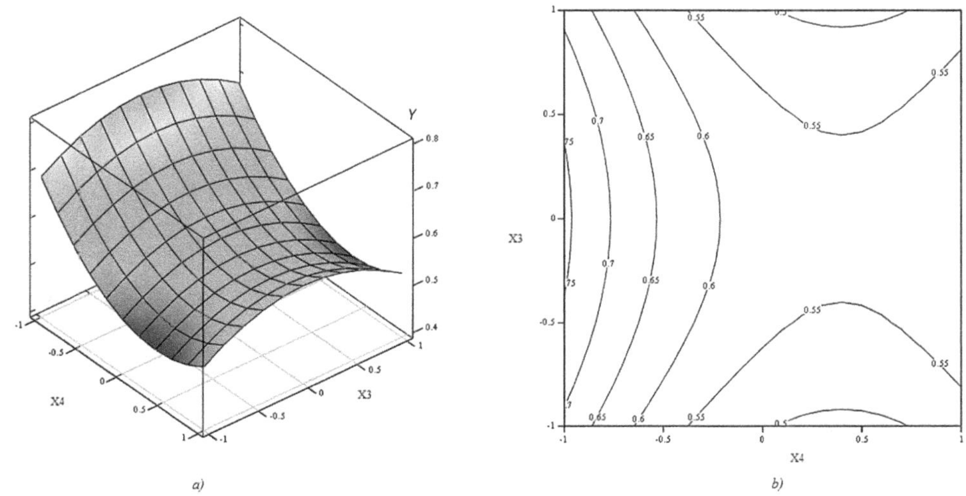

Testing of a Cultivating Unit

The research was carried out using the system of automatic accumulation and processing of metrological mobile information (SANO), which was developed by the Federal Agricultural Research Center "Donskoy" (Zernograd, Russia).

The system consists of the hardware and software, namely an on-board computer, ananalog-digital converter (ADC), an adapter board, an instrument amplifier. All recording equipment was installed in the cabin of the movable strain gauge laboratory TL-2 on the basis of the all-wheel drive truck (Figure 9).

A set of sensors (primary converters) was installed on the unit. This set allows measuring the momentary values of the following energy parameters: traction resistance; torque on the driving wheel axis; rotation frequency of the tractor generator shaft; rotation impulses of a drive wheel; revolution impulses of a tractor wheel; oil pressure in the hydro line (up to the throttle); rotation frequency of a gear of an oil pump drive; fuel consumption.

The experiments with the cultivator unit took place in the fields of the Zernograd district of the Rostov region (Russia). When the unit was operating in the field, its main working parameters were recorded. The time of experiment and the covered distance were registered with the help of SANO. The results of the experiment were processed on PC by the special software. The study of the cultivating machine-tractor unit with an elastic-damping mechanism was carried out in two regimes: acceleration and steady-state process.

The experiments were carried out on a horizontal section of the field with a slope of not more than 2 degrees with strictly rectilinear movement of the unit on the 7th speed of the tractor. The fuel controlling lever was set to the maximum value.

Before the experiments, there was conducted a technical examination of the tractor, cultivator, measuring system and the vehicle. The experiments started only after warming up the engine and transmission. After preparation (i.e. charging the accumulator, calibrating the sensors and adjusting the instruments),

Figure 9. General view of the unit and the mobile laboratory

the unit was sent to the field, where a further check of the working abilities of its devices was made. One second before the unit starting, the recording equipment (SANO) was switched on. When carrying out the measurements, the strain gauge laboratory moved along with the machine-tractor unit.The cultivation was carried out on a specially prepared field on a 10...12 cm depth. During the experiments, the cultivation depth and the grasp width were controlled. The fluctuations in the cultivation depth and the grasp width were within the limits permitted by the agro-requirements.As a standard reference of the machine-tractor unit, the same machine-tractor unit but with a locked elastic-damping mechanism was used in the comparative experiments. The tractor control in both variants was carried out by the same class 1 tractor driver.

During field testing of the cultivating unit, qualitative working characteristics were determined. The depth of the working bodies was calculated for the established soil depth of 10...12 cm. It was estimated by the mathematical expectation m, the mean square deviation δ and the coefficient of variation v.Thef igurecharacteristicsofthetillage are given in table 2.

An analysis of the obtained data shows that the tillage depth with the lowest coefficient of its variance can be seen when the unit works with an elastic-damping mechanism in the tractor transmission.

The installation of an elastic-damping mechanism in the transmission, as it can be seen in Table 2, improves the stability of the technological process.

Table 2. Figures of the working quality of the cultivating unit

Name of the Parameters	MTU With an EDM	MTU Without an EDM
Average tillage depth *m*, cm	11.2	10.88
Variance of tillage depth D, cm^2	4.58	5.86
Average quadratic deviation δ, cm	2.14	2.42
Coefficient of variance *v*	0.19	0.22

THE RESULTS OF ASSESSMENT OF THE CULTIVATOR POWER

There has been done an assessment of the cultivator power. The power characteristics are given in Table 3.

The figures of power assessment indicate that the cultivating machine with an elastic-damping mechanism in the tractor transmission has better power figures (speed, productivity and fuel consumption per hectare) than a similar machine-tractor unit with a rigid transmission. Analyzing Table 3, we see traction resistance decrease. For an MTU the traction resistance is the internal force between the connected links: the tractor and the cultivator. The amount of the force is determined by the external influence of the soil on the cultivator working parts and the interaction of the links through the elastic and deforming connection between them. The installation of an elastic-damping mechanism in the transmission creates more favorable conditions for the load character formation of the tractor units, both under unsteady and steady driving regimes. This results in an engine and drive wheels' load decrease, which causes the increase of angular shaft velocity and the progressive tractor speed in comparison with the standard tractor with a high rigidity of the power transmission.

For a more detailed analysis of the running gear skidding, there has been conducted an estimation of skidding deviation over time. The skidding variation is shown in Figure 10. The data on the running gear skidding are presented in Table 4.

Statistical analysis of experimental skidding figures showed that the use of an elastic damping mechanism makes it possible to reduce the slippage deviation on 37.6%.in this case the skidding unevenness reduces on 21%, and the decrease in the mathematical expectation of skidding is 2%.

Table 3. Power figures of the cultivating MTU work

N°	Power Figures	MTU With an EDM	MTU Without an EDM
1	Traction resistance P_{tr}, N	9287.77	9203.84
2	Speed of the unit *V*, m/s (km/h)	1.81 (6.52)	1.96 (7.06)
3	Rotation frequency of a crankshaft n_c, rad/s	235.32	244.75
4	Fuel consumption per hour *G*, kg/h	9.972	9.108
5	Skidding of a running gear *δ*, %	14.96	14.65
6	Unit productivity *W*, ha/h	2.61	2.82
7	Fuel consumption per hectare G_{za}, kg/ha	3.82	3.23

Table 4. Statistic figures of skidding

Figures	MTU With an EDM	MTU Without an EDM
Mathematical expectation $M_δ$, %	14.96	14.65
Deviation D, %	51.72	32.28
Root-mean-square Deviation $σ_δ$, %	7.19	5.68

Figure 10. Skidding deviation of the cultivating unit: --------- skidding of the unit with an EDM; ——— skidding of the unit without an EDM

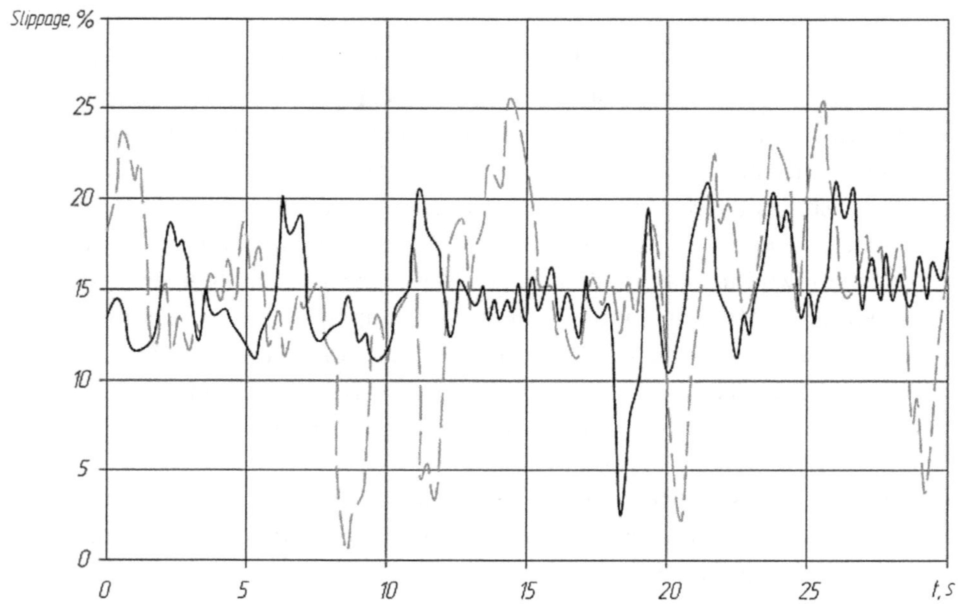

Spectral Analysis of the Experimental Data

The work of the machine-tractor unit was recorded in the serial and experimental variants in three replications. One mathematical expectation was derived out of the three realizations of one process. The processing of the obtained results (electronic oscillograms) was carried out according to a special program for conversion developed by the Federal Agricultural Research Center "Donskoy" (Rostov Region, Russia). After analyzing of the cultivator unit movement, a correlation-spectral analysis of the obtained experimental data has been carried out. Correlation functions and spectral densities characterizing a random process were obtained. The analysis showed that the functions are different. Many graphs show a slower adaptation to the variable work of the unit with a blocked EDM than that of the unit with EDM. On the graphs characterizing the random process in the frequency domain, the prevailing frequencies are clearly expressed, which account for the maximum of the variances.

The normed autocorrelation function of the engine torque shows that the decay time of MTUwork without EDM is 0.55 s, compared with MTU with an elastic-damping mechanism, whose decay time of work is 0.35 s (Figure 11). Rapid decay is a sign of an ergodic property. Ergodicity of the process in MTA with an EDM is higher.

Figure 11. Standard autocorrelation (serial correlation) function of engine torque: --------is the unit without an EDM; ———is the unit with an EDM

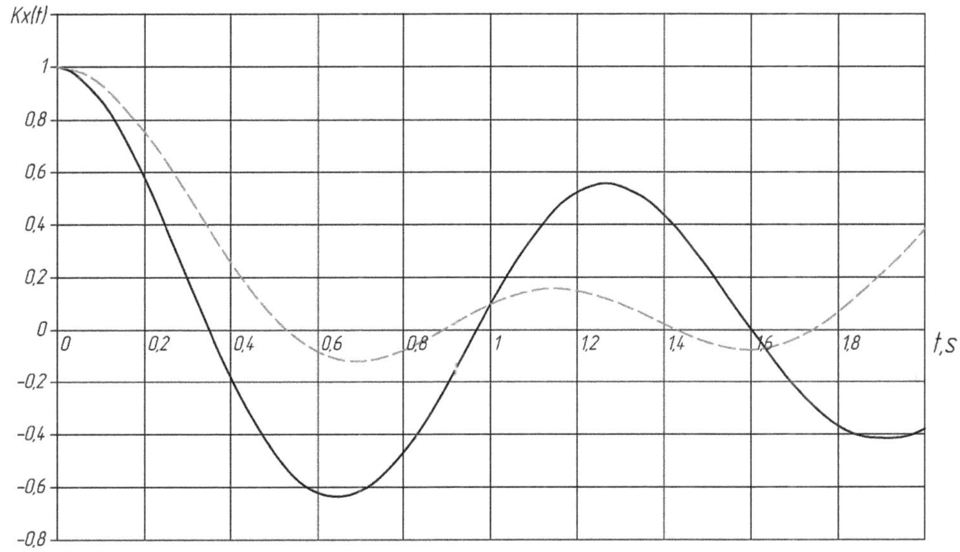

The graph of spectral densities (Figure 12) shows that the number of prevailing frequencies in MTU-without an EDM differs from it in MTU with an EDM. For MTU with an EDM there is one prevailing frequency of 1.0 s^{-1}, and for MTUs without an EDM there are not revealed any prevailing frequencies. The frequencies range from 0.5 s^{-1} to 7.0 s^{-1} and 12.0 s^{-1}as waves. There are small increases between 20.0 $s^{-1}$and 22.0 s^{-1}.

Figure 12. Standard spectral density of the engine torque: -------- is the unit without an EDM; ——— is the unit with an EDM

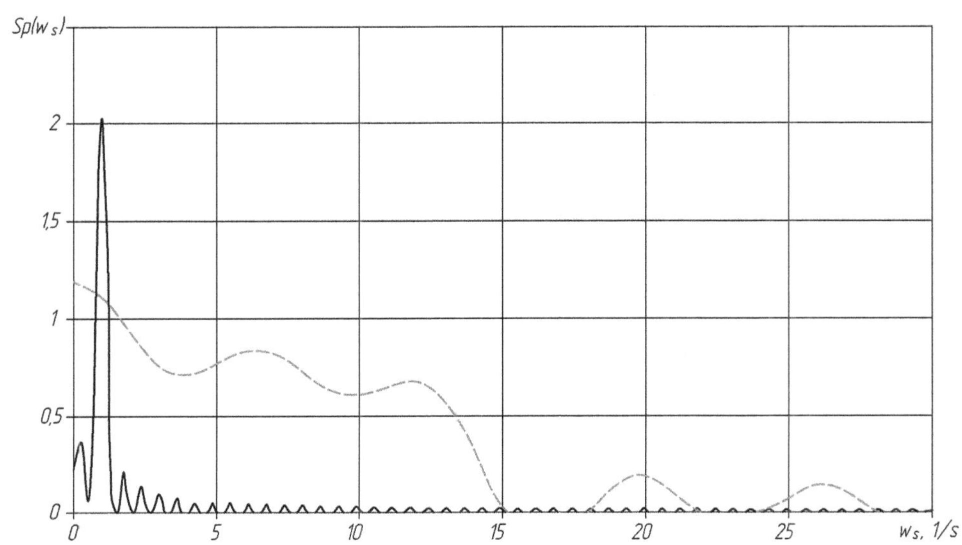

Cutoff frequency in the unit with an EDM is $\omega_s = 4\ s^{-1}$, in the unit with an EDM it is $\omega_s = 28\ s^{-1}$. The graphs show that for MTUs without an EDM, the limit of variation of the prevailing frequencies ranges wider. This indicates that the transmission of the unit without an EDM is less adaptable to external factors.

The standard autocorrelation function for crankshaft rotational speed shows that the decay time for the unit without an EDM is 1.1 s (Figure 13). For an experienced tractor with an EDM in the transmission, the function decreases on 0.2 s faster (the value is 0.9 s), which indicates greater adaptability of the process in the experiment, as well as greater ergodicity of the process.

The graph (Figure 14) of the spectral densities shows that the unit without an EDM has two prevailing frequencies, the so-called zero and equal to 4.0 s^{-1}.

For the unit with an EDM, the number of prevailing frequencies is also two: 1.0 s^{-1} and 3.0 s^{-1}. The predominant frequencies are shifted to a zone of lower frequencies. For the unit without an EDM in the range from 7.0 s^{-1} and more, there is a similar characteristics of the spectral densities. The spectral densities of the studied process have a fairly extended character; the cutoff frequency is about 7.0 s^{-1}. The maximum values of the spectral density have peak characteristics at frequencies of 4.0 $s^{-1}$and 6.0 s^{-1}. The graph shows a fairly wide process, as the correlation time is very short. The periodic components are more revealed, that indicates its exposure to fluctuations of the external traction load.

The standard autocorrelation function of the traction load (Figure 15) shows that the decay time of the unit with an EDM is 0.1 s. For the unit without an EDM, the function decreases more slowly and does not reach zero values at the obtained length (zero value is achieved from 1.5 to 2.0 s). This indicates a greater ergodicity of the process, which is realized by the unit with an EDM.

A similar behavior of the spectral densities ranged from 0 to 12.0 s^{-1} is observed in the graph (Figure 16). It can be seen that the unit without an EDM and the unit with an EDM has several prevailing frequencies: 0.5 s^{-1}; 1.0 s^{-1}; 0.5 s^{-1}; 4.9 s^{-1}; 6.8 s^{-1}. The cutoff frequency in both variants has approximately

Figure 13. Standard autocorrelation (serial correlation) function for crankshaft rotational speed:- - - - - is the unit without an EDM; —— is the unit with an EDM

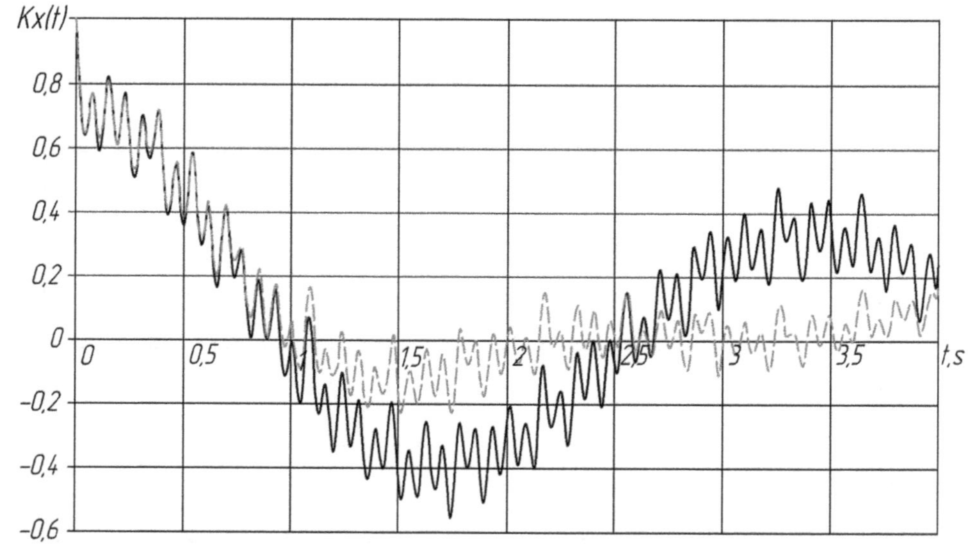

Figure 14. Standard spectral density of crankshaft rotational speed: - - - - - - is the unit without an EDM; ——— is the unit with an EDM

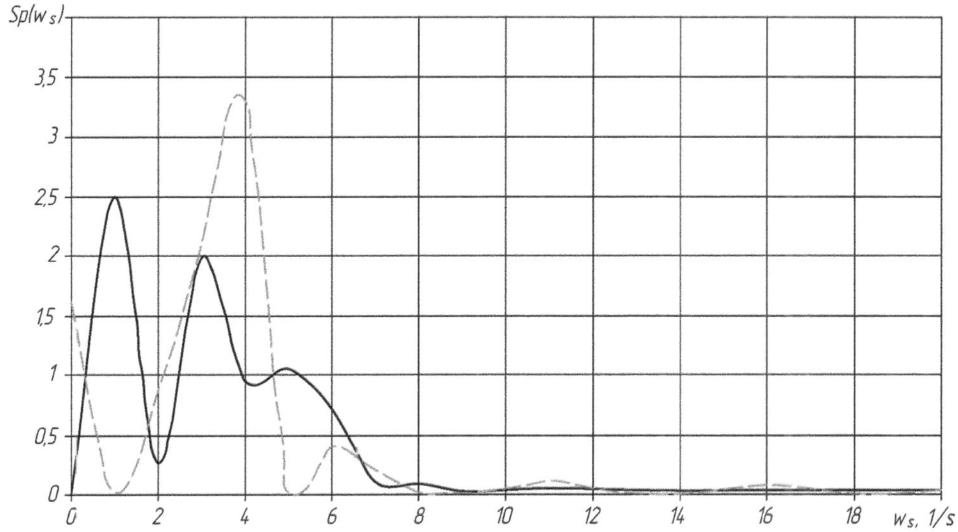

Figure 15. Standard autocorrelation (serial correlation)function of the traction load: - - - - - - is the unit without an EDM; ——— is the unit with an EDM

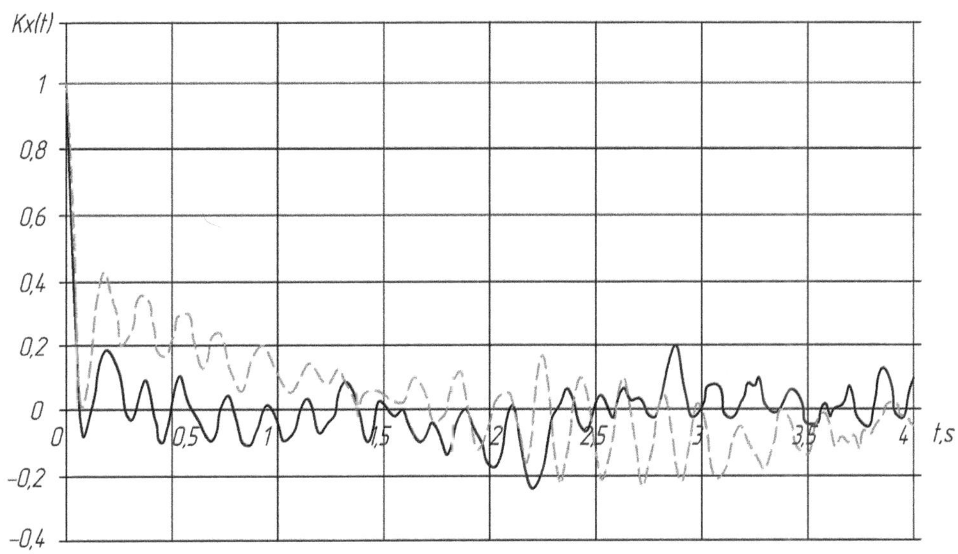

the same value and it is $\omega_s = 12.0\ s^{-1}$. The first component of the spectral density is about the oscillations of the implement. The second component is about the fluctuations in the transmission.

The graph of the standard autocorrelation (serial correlation) function and spectral density of the traction load shows that the oscillations of the cultivator's paws are superimposed on the mentioned above low-frequent fluctuations of the cultivating unit. The so-called "oscillations seizure" takes place.

Figure 16. Standard spectral density of traction load: - - - - - - is the unit without an EDM; ——is the unit with an EDM

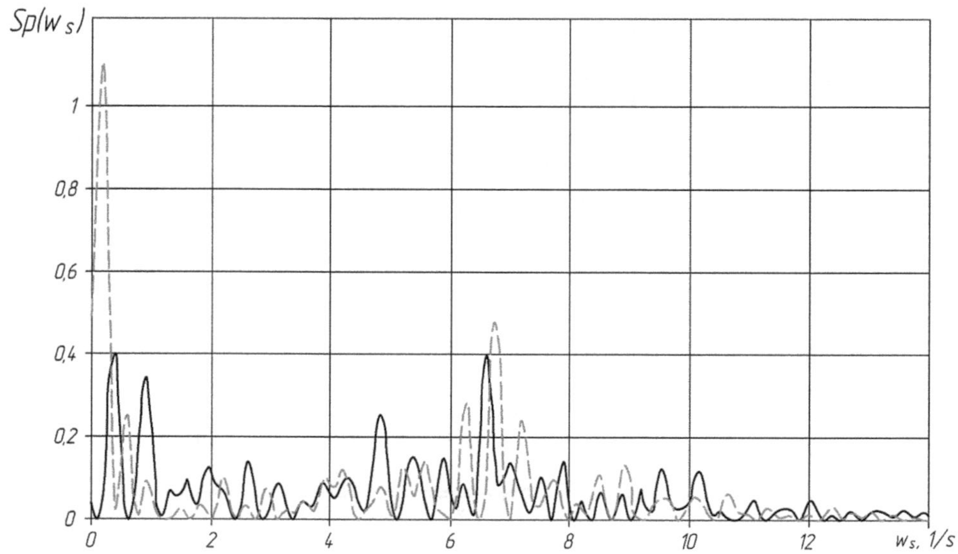

This phenomenon can be clearly seen on the spectral density curves. Only when the unitworks with an EDM, the maximum values of frequencies are a little lower than when the unit is without an EDM.

The graph shows that the tractor with an EDM is more adapted to external changes in the traction load. The mechanism installed in the transmission absorbs vibration energy and protects the entire unit.

The normed autocorrelation (serial correlation) function of the driving wheel torque (Figure 17) shows that the processes proceed almost identically. The decay time for a unit without an EDM is 0.4 s, in contrast to the unit with an elastic-damping mechanism in the transmission, in which the decay time of the correlation function is 0.3 s.

The graph of spectral densities (Figure 18) shows that the number of prevailing frequencies for both units is approximately the same: for the MTU with an EDM it is 0.4 s^{-1} and 1.1 s^{-1}; for the MTU withoutan EDM it is 0.2 s^{-1} and 1.1 s^{-1}. The cut-off frequencyfor both unitsis in about $\omega_s = 7.0$ s^{-1}. Spectral density, as an indicator of power expenditures, describes the distribution of small amplitudes in frequency. In other words, this is the frequency distribution of the power process. A visual comparison of the curves shows that the MTU with an EDM has less energy expenditure on the oscillation process than the MTU without an EDM.

The conducted correlative and spectral analysis shows that the tractor with an elastic-damping mechanism in the transmission (the unit with an EDM) has smaller values of the spectral densities for the traction load, smaller angular velocity of the engine shaft and the engine torque, and the frequencies predominant in the random process are shifted to the area of lower frequencies than the unit without an EDM. This indicates a decrease of the dynamic loading on the unit's links in the tractor transmission with an installed elastic-damping mechanism.

Figure 17. Standard autocorrelation function of the driving wheel torque- - - - - - is the unit without an EDM; ——is the unit with an EDM

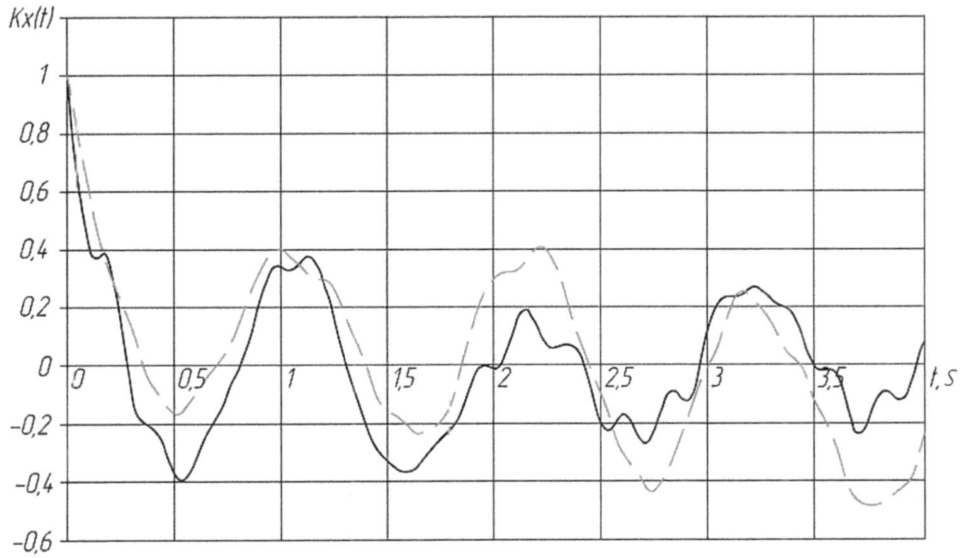

Figure 18. Standard spectral density of the driving wheel torque: - - - - - - is the unit without an EDM; —— is the unit with an EDM

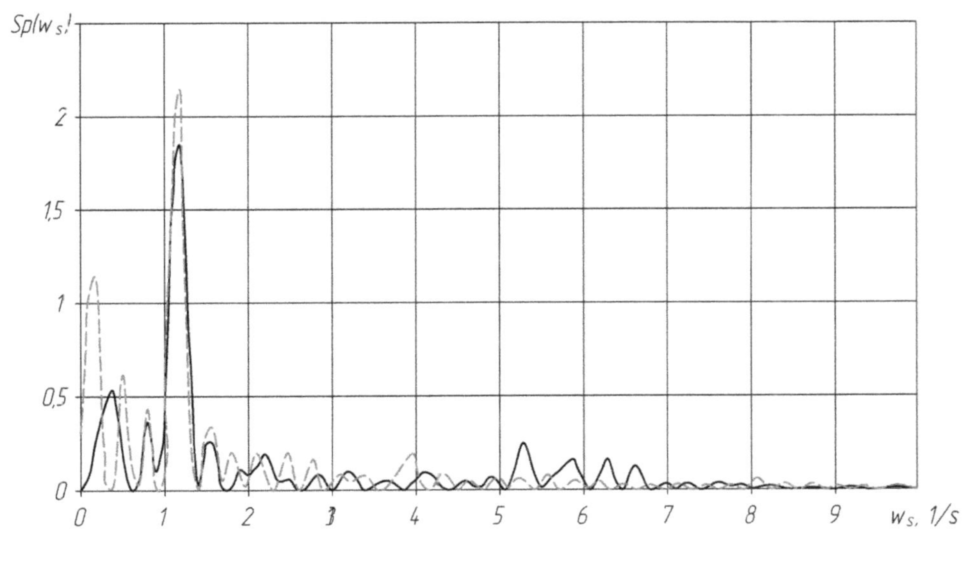

CONCLUSION AND FURTHER DIRECTIONS OF RESEARCH

In general, the following conclusions can be drawn based on the results of experimental studies:

1. The obtained data show that the use of an EDM can reduce the oscillations of the external traction load on15...20% (on average)in the field and on 30...40% on solid stable backgrounds (concrete coating, asphalt coating, etc.).
2. The use of an elastic-damping mechanism in the transmission improvesagro technical characteristics during cultivation, i.e. uniformity of treatment depth, on 12%.
3. The deviation of skidding decreased on 37.6%, which favorably influences on the work of the machine-tractor unit.
4. Due to more stable engine work,the per-hour fuel consumption decreased on 9%, and the per-hectare consumption reduced on 18%.
5. The angular velocity of the engine crankshaft with the use of an elastic damping mechanism increases on 4%. This indicates engine work improvement.

Thus, the installation of an elastic-damping mechanism in the tractor transmission is an effective constructive measure that improves the output parameters of the unit under real working conditions. It should be noted that the study of an EDM effect on the MTU work is going on till nowadays. Future research work and further practical application of an EDM are aimed at studying its impact on the MTU work during various agricultural operations, such as plowing, sowing and transporting.

REFERENCES

Antonov, N. M. (1978). *Issledovanie nagruzhennosti transmissii traktora Kirovets na pahote* [Investigation of the loading of the transmission of the tractor "Kirovets" on the plowing] (PhD dissertation). Saint Petersburg, Russia: SPBGAU

Babanin, N. V. (2015). Uluchshenie plavnosti hoda mashinno-traktoronogo agregata na baze traktora klassa 1.4 s uprugodempfiruyushchim privodom na vedushchih kolesah [Improvement of the smoothness of the machine and tractor unit on the basis of a class 1.4 tractor with an elastic damping mechanism on driving wheels]. *Current Problems of Science and Education, 22*, 46.

Babanin, N. V., & Polivaev, O. I. (2015). Eksperimentalnie issledovaniya na plavnost hoda, proizvoditelnost i toplivnuyu ekonomichnost mashinno-traktornogo agregata na baze traktora klassa 1.4, oborudovannogo gazogidravlicheskim uprugodempfiruyushchim privodom [Experimental researches on smoothness of a course, productivity and fuel economy of the machine-tractor unit on the basis of a class 1.4 tractor, equipped with a gas-hydraulic elastic-damping drive]. *Bulletin of the Voronezh State Agrarian University, 3*(46), 112–118.

Bespamyatnova, N.M. (2002). *Nauchno-metodicheskie osnovi adaptatsii pochvoobrabativayushchih mashin* [Scientific and methodological fundamentals of adaptation of soil cultivating and sowing machines]. Rostov-on-Don, Russia: LLC "Terra".

Boltinskiy, V. N. (1949). *Rabota dvigatelya tractora pod nagruzkoy* [The work of the tractor engine under unsteady load]. Moscow, Russia: OGIZ-SELHOZGIZ.

Boltinsky, V. N. (1959). Moshcnost traktornogo dvigatelya pri rabote s neustanovivsheysya nagruzkoy i eye opredelenie [Tractor engine power when working with unsteady load and its determination]. *Mechanization of Socialist Agriculture*, *4*, 13–16.

Bychkov, N. I., & Yakulin, V. G. (2003). Vibor tractor dlya vspashki [Choice of a tractor for plowing]. *Mechanization and Electrification of Agriculture*, *3*, 11–12.

Deboli, R., Calvo, A., & Preti, C. (2017). Whole-body vibration: Measurement of horizontal and vertical transmissibility of an agricultural tractor seat. *International Journal of Industrial Ergonomics*, *58*, 69–78. doi:10.1016/j.ergon.2017.02.002

Denisov, V. P. (2005). *Optimizatsiya rabochego protsessa zemleroyno-transportnyh mashin s uchetom sluchaynogo haraktera nagruzok* [Optimization of the working process of earth-moving machines taking into account the random nature of the loads] (PhD dissertation). SibADI, Omsk, Russia.

Duke, M., & Goss, G. (2007). Investigation of tractor driver seat performance with non-linear stiffness and on-off damper. *Biosystems Engineering*, *94*(4), 477–486. doi:10.1016/j.biosystemseng.2007.01.005

Eviev, V. N., Jabborov, N. I., & Ochirov, N. G. (2014). Povishenie effektivnosti gusenichnyh traktorov klassa 4, osnashchennyh dvigatelyami postoyannoy moshchnosty [Increase of efficiency of caterpillar tractors of class 4, equipped with constant power engines]. *Technology of Wheeled and Caterpillar Vehicles: Research Institute of Automotive Design*, *6*, 51–56.

Feuerhuber, K., Lindert, S. O., & Schlacher, K. (2013). *Vibration attenuation by semi-active dampers.* Paper presented at 14th International Conference on Computer Aided Systems Theory (EUROCAST), Las Palmas, Spain.

Filin, V. A., Pronin, V. Y., Kryuchkov, S. V., & Filin, A. V. (2016). Obosnovanie stepeni zagruzki traktornogo dvigatelya pri vypolnenii selskokhozyastvennykh rabot [Substantiation of the degree of loading of the tractor engine in the performance of agricultural works]. In Proceedings of Energy-Efficient and Resource-Saving Technologies and Systems. Saransk, Russia: FSBEU of Higher Professional Education of the Moscow State University named after N.P. Ogaryova.

Fritz, N., & Elsner, W. (2002). *Non linear numerical torsional vibration analysis of the JD 6920 tractor drive train.* Paper presented at Conference on Agricultural Engineering, Halle, Germany.

Gabdrafikov, F. Z., & Safina, R. R. (2017). Povyshenie tehniko-ekonomicheskih pokazateley traktornyh dizeley na chastichnyh i neustanovivshihsya rezhimah raboty [Increase of technical and economic indicators of tractor diesel engines on partial and unsteady operating modes]. *Russian Electronic Scientific Journal*, *2*(24), 95–109.

Goryachkin, V. P. (1965). *Zemledel'cheskaya mekhanika. Sobraniye sochineniy. Tom 2* [Agricultural mechanics. Collected Works: Volume 2]. Moscow, Russia: Kolos.

Grebnev, V. P., Poliaev, O. I., & Vorokhobin, A. V. (2011). *Traktory i avtomobili: teoriya i ekspluatatsionniye svoystva: uchebnoye posobie dlya studentov vuzov, obuchaushchihsya po napravleniyu «Agroingeneriya»* [Tractors and automobiles: theory and operational properties: a teaching aid for university students studying in the direction of "Agroengineering"]. Voronezh, Russia: KnoRus.

Gurhan, R., & Cay, C. I. (2008). Performances of tractor's driver seats with passive damping system under simulated test conditions. *Tarim Bilimleri Dergisi-Journal of Agricultural Sciences, 14*(4), 401–408.

Hansson, P. A. (1994). *Gain scheduling based controller for active cab suspension on agricultural tractors.* Paper presented at 2nd IFAC/ISHS Workshop on Mathematical and Control Applications in Agriculture and Horticulture, Silsoe, England.

Hansson, P. A. (2002). Working space requirement for an agricultural tractor axle suspension. *Biosystems Engineering, 81*(1), 57–71. doi:10.1006/bioe.2001.0007

Iofinov, S. A. (1974). *Ekpluatatsiya mashinno-traktornyh agregatov* [Operation of machine-tractor units]. Moscow, Russia: Kolos.

Iofinov, S. A., & Mikuberg, B. K. (1971). Opredelenie ekspluatatsionnih parametrov i pokazateley raboty agregatov pri veroyatnostnom haraktere issleduemyh velichin [Determination of operational parameters and performance indicators of units for the probabilistic nature of the investigated quantities]. *Mechanization and Electrification of Social Agriculture, 12*, 42–46.

Janulevičius, A., & Giedra, K. (2008). Analysis of the main dynamic parameters of a split power transmission. *Transport, 23*(2), 112–118. doi:10.3846/1648-4142.2008.23.112-118

Kalmykov, A. V. (2014). *Snizheniye dinamicheskoy nagruzhennosty silovoy peredachi traktora za schet izmeneniya krutilnoy zhestkosti reaktivnogo zvena* [Reduction of the dynamic loading of the power transmission of the tractor due to the change in the torsional stiffness of the reactive link] (PhD dissertation). Volgograd State Agrarian University, Volgograd, Russia.

Karabanitsky, A.P., &Levshukova, O.A. (2014). *Teoreticheskoe obosnovanie parametrov energosberegaushchikh mashinno-traktornykh agregatov: uchebnoe posobie* [Theoretical justification of the parameters of energy-saving machine-tractor units: manual]. Krasnodar, Russia: KubSAU.

Kirtbai, Yu. K. (1976). *Rezervy v ispolzovanii mashinno-traktornogo parka* [Reserves in the use of machine and tractor fleet]. Moscow, Russia: Kolos.

Kravchenko, V. A. (2010). *Povishenie dinamicheskih i ekspluatatsionnyh pokazateley selskohozyastvennyh mashinno-traktornyh agregatov* [Increase of dynamic and operational parameters of agricultural machine and tractor units] (PhD dissertation). FSEE HE ABSSAEA, Zernograd Russia.

Kravchenko, V. A., Duryagina, V. V., & Gamolina, I. E. (2014). Matematicheskoe modelirovanie tyagovoy nagruzki MTA [Mathematical modeling of traction load MTU] [Kuban State University Scientific Journal]. *Polytematic Network Electronic Scientific Journal of the Kuban State University, 100*, 459–472.

Kravchenko, V. A., Duryagina, V. V., & Gamolina, I. E. (2017). Analiz resultatov analiticheskih issledovaniy pahotnogo agregata na baze traktora klassa 1.4 s UDM v transmissii [Analysis of the results of analytical studies of arable unit on the basis of a class 1.4 tractor with EDM in transmission]. *Bulletin of the Don Agrarian Science*, *37*, 36–42.

Kravchenko, V. A., Oberemok, V. A., & Kravchenko, L. V. (2014). Povishenie effektivnosti mashinno-traktornyh agregatov na baze kolesnyh traktorov [Increase of efficiency of machine-tractor units based on wheeled tractors]. *Technology of Wheeled and Caterpillar Machines*, *6*, 45–50.

Kravchenko, V. A., Senkevich, A. A., & Senkevich, S. E. (2003). *Povysheni eeffektivnosti funktsionirovaniya mashinno-traktornogo agregata na baze traktora klassa 1.4 putem vvedeniya v transmissiyu uprugo-dempfiruyushchego elementa* [Increase of efficiency of functioning of the machine-tractor unit on the basis of a tractor of a class 1.4 by introduction in transmission of an elastic-damping element]. Paper presented at Materials of the XLII Scientific and Technical Conference of the Chelyabinsk SAU, Chelyabinsk, Russia.

Kravchenko, V. A., Senkevich, S. E., & Senkevich, A. A. (2007). Issledovanie vliyaniya uprugodempfiruyushchego mehanizma na pokazateli raboti posevnogo mashinno-traktornogo agregata [Investigation of the influence of an elastic-damping mechanism on the performance of a seed machine-tractor unit]. *Mechanization and Electrification of Agriculture*, *11*, 21–23.

Kravchenko, V. A., Senkevich, S. E., Senkevich, A. A., Goncharov, D. A., & Duryagina, V. V. (2008). *Ustroystvo dlya snizheniya zhestkosti transmissii mashinno-traktornogo agregata* [*Device for reducing the stiffness of the transmission of the machine-tractor unit*]. RU Patent 2398147 F16H47/04.

Kravchenko, V. A., Senkevich, S. E., Senkevich, A. A., Yarovoy, V. G., & Tolstoukhov, Y. S. (2004). *Ustroystvo dlya snizheniya zhestkosti transmissii transportnogo sredstva* [Device to reduce the rigidity of the vehicle's transmission]. RU Patent 2222440, B60K 17/10

Kutkov, G. M. (1980). *Tyagovaya dinamika traktorov* [Traction dynamics of tractors]. Moscow, Russia: *Mechanical Engineering (New York, N.Y.)*.

Kutkov, G. M., Puchkov, V. S., & Kholin, A. I. (1975). Analiz istochnikov genaratsii kolebaniy nagruzki na dvigatel selskohozyastvennyh traktorov [Analysis of the sources of oscillation of the load on the engine of agricultural tractors]. *Tractors and Agricultural Machinery*, *6*, 9–10.

Kuznetsov, N. G. (2006). *Stabilizatsiya rezhimov raboty skorostnyh mashinno-traktornyh agregatov* [Stabilization of the operating modes for high-speed machine-tractor units] (PhD dissertation). Volgograd State Agricultural Academy, Volgograd, Russia.

Kuznetsov, N. G., Gapich, D. S., & Shiryaeva, E. V. (2013). Vliyanie neustanovivshegosya haraktera nagruzhenuya kolesnogo traktora krukovim usiliem na koeffitsient buksovaniya [Influence of the unsteady nature of loading of the wheeled tractor by hook force on the coefficient of slipping]. *Tractors and Agricultural Machinery*, *10*, 25–26.

Labetskas, G. S., & Slavinkas, S. S. (2002). Kak ekonomit pri ekspluatatsii traktora MTZ-82 [How to save while operating the tractor MTZ-82]. *Tractors and Agricultural Machinery*, *9*, 9–10.

Lurie, A. B. (1969). *Dinamika regulirovaniya navesnykh selskokhozyaystvennykh agregatov* [The dynamics of regulation of mounted agricultural units]. Leningrad, Russia: Mechanical Engineering.

Lysov, A. M. (1966). Issledovanie vliyaniya tipa transmisii na proizvoditelnost selskohozyaystvennykh traktorov, provedennoe na analogovoy vychislitelnoy mashine [Investigation of the effect of transmission type on the performance of agricultural tractors, conducted on an analog computer]. In Proceedings of Hydrostatic and Hydromechanical Tractor Transmissions. Moscow, Russia: Works of SATI.

Marjoram, R., St-Clair, K., McMahon, B., Goelz, A., & Achen, A. (2008). *Integrated MR cabin suspension for reduction of human exposure to whole-body vibration in heavy-duty vehicles*. Bremen, Germany: Messebremen-hvg hanseatische veranstaltungs-gmbh.

Mochunova, N. A. (2015). Vliyanie chastoti kolebaniy tyagovoy nagruzki na soprotivleniye perekativaniyu kolesnogo trakrora [Influence of the oscillation frequency of the traction load on the rolling resistance of a wheeled tractor]. *Nature Management*, *4*, 85–87.

Nekhoroshev, D. A. (2014). *Stabilizatsiya rezhimov MTA izpolzovaniem pnevmogidravlicheskoy mufty stsepleniya* [Stabilization of MTU modes using pneumatic hydraulic clutch] (PhD dissertation). Volgograd State Agrarian University, Volgograd, Russia.

Nekhoroshev, D. A., & Nekhoroshev, D. D. (2009). Planetarnaya uprugaya mufta v transmissii kolesnogo tractor klassa 1.4 [Planetary elastic coupling in the transmission of class 1.4 wheeled tractor]. *Achievements of Science and Technology of Agro-Industrial Complex*, *12*, 60–61.

Nekhoroshev, D. D. (2004). *Povishenie ekspluatatsionnykh pokazateley MTA putem ispolzovaniya traktorov s pnevmogidravlicheskoy muftoy* [Increase of operational indicators of MTU by using tractors with pneumatic hydraulic clutch] (PhD dissertation). Volgograd State Agrarian University, Volgograd, Russia.

Nekhoroshev, D. D., & Nekhoroshev, D. A. (2015). Energeticheskoe sredstvo v transmissii kolesnogo traktora [Power means in the transmission of the wheeled tractor]. *Unique Research of the XXI Century*, *12*(12), 40–41.

Okhotnikov, B.L. (2014). *Ekspluatatsiya dvigateley vnutrennego sgoraniya: uchebnoye posobie* [Operation of internal combustion engines: a training manual]. Ekaterinburg, Russia: Publishing house Ural.

Parshin, I. V. (1978). *Issledovanie raboty traktornogo dvigatelya s garoturbinnym nadduvom v usloviyah selskokhozyastvennogo mashinno-traktornogo agregata* [Investigation of the operation of a tractor engine with gas turbine supercharging in the conditions of an agricultural machine-tractor unit] (PhD dissertation). FSEE HE ABSSAEA, Zernograd Russia.

Pobedin, A. V., Dolotov, A. A., & Shekhovtsov, V. V. (2016). *Decrease of the vibration load level on the tractor operator working place by means of using of vibrations dynamic dampers in the cabin suspension.* Paper presented at 2nd International Conference on Industrial Engineering (ICIE), Chelyabinsk, Russia. 10.1016/j.proeng.2016.07.136

Polifke, G. (2016). *Hydrodamp - hydraulic torsional vibration damper for tractors*. Paper presented at Conference on Agricultural Engineering, Koln, Germany.

Polifke, G. (2017). *Hydrodamp - hydraulic torsional vibration damper for tractors and construction machines design for a race-sensitive drive train*. Paper presented at VDI Conference on Couplings and Clutch Systems in Drives 2017 / 2nd VDI Conference on Vibration Reduction in Mobile Systems, Ettlingen, Germany.

Polivaev, O.I., & Kutkov, A. Y. (2010). Effektivnost ispolzovaniya transportnykh sredstv s uprugodempfiruyushchim privodom vedushchikh koles pri tormozhenii [Efficiency of the use of transport units with an elastic-damping drive of driving wheels during braking]. *Machinery in Agriculture*, *5*, 30-32.

Polivaev, O. I., Pilayev, S. N., & Kutkov, A. Y. (2011). Obosnovanie parametrov i effektivnosti ispolzovaniya uprugodempfiruyushchego privoda vedushchikh koles traktorno-transportnogo agregata v rezhime tormozheniya [Substantiation of parameters and efficiency of using the elastic-damping drive of the driving wheels of the tractor-transport unit in braking mode]. *Bulletin of the Voronezh State Agrarian University*, *3*, 48–52.

Polivaev, O. I., & Vedrinsky, O. S. (2013). *Vliyanie uprugodempfiruyushchego privoda vedushchikh koles traktora na dinamicheskie nagruzki v transmissii pri razgone MTA* [Influence of the elastic-damping drive of the tractor's driving wheels on the dynamic loads in the transmission when accelerating the MTU]. Innovative Technologies and Technical Means for Agroindustrial Complex.

Polivaev, O. I., & Vedrinsky, O. S. (2014). Analiz vliyaniya uprugodempfiruyushchego privoda koles na dinamicheskie nagruzki v transmissii traktora pri razgone [Analysis of the effect of the elastic-damping drive of the wheels on the dynamic loads in the tractor's transmission during acceleration]. *The Bulletin of the Don Agrarian Science*, *28*, 5–9.

Polivaev, O. I., Vedrinsky, O. S., & Derkanosova, N. M. (2016). Povishenie dolgovechnosti stsepleniya traktorov za schet uprugofriktsionnogo dempfera [Increasing the durability of tractor traction due to elastic-friction damper]. *Science and Education in Modern Conditions*, *12*, 226–230.

Polivaev, O. I., Zhidenko, K. S., Maksimov, I. I., & Levkin, I. G. (2017). *Uprugo-dempfiruyushchiy privod vedushchih koles traktora* [Elastic-damping drive of the tractor's driving wheels]. Paper presented at Modern Trends in the Development of Technology and Technical Means in Agriculture, Voronezh, Russia.

Shekhovtsov, K. V. (2014). *Snizhenie urovnya vibronagruzhennosti rabochego mesta operatora traktora za schet primeneniya dinamicheskikh gasiteley kolebaniy v sisteme podressorivaniya kabiny* [Reduction in the vibration level of the operator's workplace due to the use of dynamic vibration dampers in the cabin suspension system] (PhD dissertation). Volgograd State Technical University, Volgograd, Russia.

Shekhovtsov, V. V., Shevchuk, V. P., & Zelenko, S. V. (2002a). Rasprostranenie krutilnyh kolebaniy v valoprovode silovoy peredachi traktora VT-100 [Dissemination of torsional oscillations in the power transmission line of the power transmission of the tractor VT-100]. *Tractors and Agricultural Machinery*, *8*, 10–11.

Shekhovtsov, V. V., Shevchuk, V. P., & Zelenko, S. V. (2002b). Issledovaniya rezonansnykh rezhimov silovoy peredachi traktora VT-100 [Investigation of the resonance modes of the power transmission of the tractor VT-100]. *Tractors and Agricultural Machinery*, *7*, 5–6.

Shekhovtsov, V. V., Sokolov-Dobrev, N. S., & Potapov, P. V. (2016). Decreasing of the Dynamic Loading of Tractor Transmission by means of Change of the Reactive Element Torsional Stiffness. *Procedia Engineering*, *150*, 1239–1244. doi:10.1016/j.proeng.2016.07.129

Sinitsky, S. A. (2005). *Vliyanie nagruzki mashinno-traktornogo agregata na pokazateli dvigatelya v usloviyakh ekspluatatsii* [Influence of the load of the machine-tractor unit on the engine performance under operating conditions] (PhD dissertation). Kazan State Agricultural Academy, Kazan, Russia.

Spiridonov, A. A. (1981). *Planirovanie eksperimenta pri issledovanii tehnologicheskikh protsessov* [Planning an experiment in the study of technological processes]. Moscow, Russia: Mechanical Engineering.

Svistula, A. E., & Matievsky, G. D. (2011). Issledovanie optimizatsionnoy skorostnoy kharakteristiki dvigatelya postoyannoy moshchnosty [Investigation of the optimizing speed characteristic of a constant-power motor]. Internal Combustion Engines: National Technical University "Kharkov Polytechnic Institute, 2, 46-49.

Tarasik, V. P. (1984). *Proektirovanie kolesnykh tyagovo-transportnykh mashin* [Design of wheel traction-transport machines]. Minsk, Belarus: High. Sc.

Vedenyapin, G. V., Kirtbaya, Yu. K., & Sergeev, M. P. (1968). *Ekspluatatsiya mashinno-traktornogo parka* [Operation of the machine and tractor fleet]. Moscow, Russia: Kolos.

Zheng, E. L., Fan, Y. D., Zhu, R., Zhu, Y., & Xian, J. Y. (2016). Prediction of the vibration characteristics for wheeled tractor with suspended driver seat including air spring and MR damper. *Journal of Mechanical Science and Technology*, *30*(9), 4143–4156. doi:10.100712206-016-0826-x

Zimagulov, A. Kh., & Nurulin, R. G. (2003). Trebovaniya k tyagovym MES dlya vypolneniya tehnologicheskih operatsiy [Requirements for traction MEM for the execution of technological operations]. *Tractors and Agricultural Machinery*, *4*, 8–9.

Zoz, F., & Grisso, R. (2003). Traction and Tractor Performance. Paper presented at *Agricultural Equipment Technology Conference*, Louisville, KY.

KEY TERMS AND DEFINITIONS

Correlation Function: Is a function of time and spatial coordinates, which determines the correlation in systems with random processes.

Degree of Transmission Transparency (Transmission Transparency): Is the degree of filtration of oscillations from an external load to a power unit (engine). A very rigid transmission is called transparent, that is, one that passes a reactive external load through itself without any changes.

Elastic-Damping Mechanism (EDM): Is a mechanism that damps vibrations and dissipates energy inside the system under the effect of cyclic loads. In this case, the mechanical energy of the oscillations is converted into the thermal energy.

Gearboxis (GB): A change speed gearbox for changing the torque and for transfer it from engine to wheels.

Machine-Tractor Unit (MTU): Is a tractor connected to a working machine or machines. A tractor in the unit is a source of mechanical energy which helps the machine to move around the field, for example when a tractor pulls a cultivator.

Multifactorial Experiment: Is an experiment that involves varying several factors, possibly influencing upon the studied phenomenon.

Regression Analysis: Is a statistical study of the effect of one or several independent variables on a dependent variable.

Spectral Density Function (or Power Spectrum): Refers to the characteristics by which the basic properties of stationary random processes are analyzed. The spectral density function characterizes the harmonic composition of the studied process and determines the energy spectrum.

Chapter 7
Electric Robotized Organic Technology for Livestock Production on a Pasture Field

Valentin Krausp
Federal Scietific Agro-Engineering Center VIM, Russia

Grigoryan Bagdasar Ovik
Vanadzor Branch of the Yerevan Natonal Polytechnic University, Armenia

ABSTRACT

The concept and principle solutions create breakthrough electro-robotized technology organic food ECO-1m used in plants for the production of ecologically clean meat and dairy food. Breakthrough technology overcomes competition and international sanctions, providing export expansion due to the novelty of domestic solutions, supported by patents for new robotic machines and equipment. ECO-1m for the first time used highly reliable, technically advanced serial technique of the sectors of the defense industrial complex, aerospace, and electrified railways. The stages and results of fundamental research on the improvement of technical solutions and a list of implemented automated farms, where the parameters of breakthrough technology ECO-1m. Technology is a modular plant consisting of 12 departments placed on 2500ga of arable land, and 3000 head of mixed aged cattle, which contains on cultural pastures. According to the indicators of the developed business plan, the plant has 3-3.5 times higher productivity, with a profitability of 30-40%.

INTRODUCTION

The parameters of a breakthrough organic biotechnology have been validated. The feature and novelty of the approach for development of biotechnologies includes sustainable prospects, cost effectiveness, simplicity, and intellectuality of the solutions proposed. The advantages of industrial modular robotized production have been demonstrated featuring high level of labor efficiency and the use of reliable and mass-produced domestically manufactured machines for electrified railway transport, railway carriage manufacture, and in digital management of electric power systems with distributed plants. In order to

DOI: 10.4018/978-1-5225-7573-3.ch007

achieve this modular youth innovative project has been developed for meat and dairy electric and robotized organic plant in one module (ECO-1m) which utilizes the defensible biotechnology. The modular ECO-1m plant features 2,500 hectares of pasture fields and contains 3,000 cattle animals and 12 shops. It processes milk into production-grade milk, (cattle) butchering for manufacture of food products. Parameters of the breakthrough organic biotechnology have been explained and validated. One feature and novelty of the approach to developing the biotechnology includes sustainable prospects, cost effectiveness, simplicity, and intellectuality of the solutions proposed. The trends for future development of industrial modular robotized manufacture have been shown. They feature high level of labor efficiency and the use of reliable, mass-produced and domestically manufactured machines in electrified railway transport, railway carriage manufacture, and digital management of electric power systems with distributed plants.

The breakthrough industrial biotechnology, under market conditions, is aimed at preemptive competition, expansion on international markets, winning global markets and legitimately driving international companies and corporations that supply eco-friendly food products from the Russian market. In order to achieve this modular youth innovative project has been developed for meat and dairy electric and robotized organic plant in one module (ECO-1m) which utilizes the organic biotechnology.

The plant counts 30 farms with milk cow, calves, bred heifers, breeding and battening of bull calves and culled cows. All the farms synchronously once every 10-12 are moved using an electrified railway to new cultivated pasture fields. A system of electric fences is used in order to gather an animal stock in one group and in order to migrate the animal stock to another cultivated pasture field, as well as to daily standardize drafting of cultivated pasture fields.

A system of processing of organic waste into fertilizers has been designed. The control station provides digital, Internet-based management of electric and robotized production along with quality certification of the organic food products. Water delivery and filling of farms, cleaning and transportation of manure pump into fertilizer processing shop, cleaning and delivery of harvest into storage tanks, transportation of livestock, delivery of machines to fields and other transport-related operations are performed by electric robots. There are 9 types of such robots in the plant. A high level of mechanization, electrification, automation, and robotization allow for increasing of the labor efficiency 3-3.5 times, while retaining the level of profitability at 30%-40%.

DEVELOPMENT OF DIGITAL AND BREAKTHROUGH ELECTRIC ROBOTIZED ORGANIC TECHNOLOGIES

Organic technology, by definition is means a biotechnology which is developed on empirical, scientific methods, art, and mastery of processing of raw materials in the process of food manufacture. It has been shown that the raw materials in the technology developed include the earth, standing grass crop, animal stock, milk produced, weight of animals, processing of meat and milk into food products, electric energy and heat energy, and water. The earth is evaluated in terms of its fertility and contamination; it is processed, fertilized, and grassed. The grass is eaten by animals on cultivated pasture fields for fattening and weight gaining for subsequent milking and butchering. The meat of animals and milk are processed into various food products and delivered to stores. There are many technologies of food production, and the skills and mastery of food technicians and process engineers help choose the best option. Such option depends heavily on a specific location in Russia.

Electrified roads. Electrified roads are used for solving the ages-old Russian problem of bad roads. This is resolved by way of building a network of rural-area railways for ECO-1m typical modular plant. The railways are designed to securely provide for migration of electric robotized platforms along the process flows in order to transport harvest from the fields to the warehouse and in order to distribute animal stock by farms, to deliver water to the farms and to the fields to the irrigation devices, for transportation of milk, manure pump, and fertilizers. As compared with other competitive solutions, highways, unsurfaced roads, and packways, railway-tracked municipal roads and electrified railways, tramway lines, and subway lines have significant advantages since they enable for the performance of seasonal works. They are endurable (up to 40 years of service life), reliable, they have high lift capacity (up to 60 tons per one railway car), they are easily automated, mass-produced and therefore they are relatively inexpensive. Railway cars, railway tracks, and contact power lines are standardized. The routes are programmable at the stage of laying of railway tracks. The rolling stock and electric robots, which move along strictly and accurately laid routes, reach their process points en route by the estimated time. Low sped rural electrified railways have been designed. Their speed is up to 10 km/h and load lifting capacity of a railcar of up to 60 tons. With this carrying load it is possible to transport manure pump, liquid organic fertilizers, water for farms, and plant watering in rail cisterns. Trains from 15 farms migrate to new cultivated pasture fields. Importantly, such railcars travel with equipment for animal stock servicing, while animals being constrained by electric fences and confined to herds, together with the railcars smoothly at the speed of 0.2 km/h – 0.4 km/h migrate to their new pasture field. Inside the pasture field and surrounded by a 500-m long electric fence, from the railcar with equipment, moves the transversal fence with the width of 24.5 m in the form of a wheeled electric fence, which frees up a day-long area of the pasture field for pasturing, for instance, with the size of 25 m x 20 m = 500 m^2. Animals eat the grass from that area of the cultivated pasture field within half a day or one day. The next morning a farm operator once again moves the transversal fence by another 20 m. If the animals eat all the grass by midday, then in the afternoon the farm operator once again moves the wall by 20 m. This is how a cultivated pasture field gets gradually and systematically pastured. Daily movement of the fence is calculated based on the density of plant formation and the number of animals in a given group.

Initial requirements and restrictions for the development of organic biotechnology which is part of ECO-1m plant:

- Environmental soil cleanup must be ensured in order to achieve the required initial environmental level of cleanness, as it has a dramatic impact on output cost parameters of meat and dairy food products;
- A closed loop of production waste processing must be ensured, so as to produce organic fertilizers which improve soil fertility;
- Ensure soil- and environment biocenosis in the climatic zone where the plant is to be built;
- Organic biotechnology must function in a closed-type production cycle ‘field-store’ while excluding intermediaries;
- Grass growing for cultivation of pasture fields and ensuring a winter reserve of columnar fodder and small grains as used for concentrated fodder must be produced using own, area-specific seeds in a 4-field fodder crop rotation;
- Animals during the pasture period must be housed in groups (stocks) with few movements in order to retain the hierarchy of relations between animal units inside the animal stock;
- Apply daily and standardized pasturing of grass at cultivated pasture fields;

- In organic biotechnology they should use local cattle stock breeds for the production of meat and milk. Such breeds must be fit for cold winter conditions with partial air heating in farms of up to +5°C;
- A closed-type cycle must be provided for extended reproduction of an animal stock of up to 20% with information accumulation in databases;
- It is planned to apply group diet for feeding of all the age-sex groups of animal stock as balanced by nutritive substances;
- Animals must be practically healthy due to high percentage of culling of sick animal units;
- Organic biotechnology required general and mechanical fencing from outside impact, agronomical and vegetarian preventive protection from infestation of plants, animals, and support personnel;
- Artificial cow mating must be used, as well as natural mating of calves with bulls;
- Upon completion of pasture period animals must not migrate to new farms; they should stay in the same farms for wintering;
- Transport operations must be performed using rural tracked electrified railways with standardized intertrack distances and overhead systems.

Electric robotized organic plants (ECO-1m) with one module means as developed using the organic biotechnology are based on engineering and interdisciplinary solutions for mechanization, automation, electrification, as well as construction, transportation, and use of heat- and electric energy and intellectual automated control systems.

Electric robotization, by definition it means electrification to include generation, distribution, and utilization of electric energy; robotization and automation of facilities aimed at replacement and optimization of human labor and mental work.

Breakthrough technology enables people to overcome competition and international economic sanctions and provides for export-oriented expansion of Russia due to the novelty of domestic solutions backed up by the relevant patents for the new systems, robotized machines and equipment and the application of reliable high-quality mass-produced equipment and machinery.

Concept for building ECO-1m plants. It reflects the fundamental solutions for developing a breakthrough electric- and robotized organic technology, which is at the heart of building ECO-1m modular plant for the production of organic meat and dairy products. The stages for this fundamental research have been shown to reflect improvements in the technical solutions and to list the commissioned automated farms which had defined the parameters of the breakthrough technology implemented in ECO-1m plant that consists of 12 shops.

Plant module is performed on 2,500 hectares of pasture fields, 3,000 cattle animals of mixed-age animal stock stored and pastured at cultivated pasture fields. In ECO-1m plant there is one module. If there are 2 modules in the plant it is marked as ECO-2m and it is housed on 5,000 hectares of pasture fields with 6,000 animal units. In case of 3 combined modules they use the designation ECO-3m… and 7,500 hectares of pasture fields and 9,000 cattle animals. Regardless of the number of such combined modules the central facility and feed preparation shop with feed storage, as well as milk processing shops, livestock butchering shop, and meat processing shop, finished product shop, and production control station remain common and unified. This allows to significantly improve the economic performance of ECO-3m.

Production of grain and grass seeds. At the time of the plant's launch it is proposed to purchase pedigree seeds of cereals and grasses. In subsequent years it is planned to grow grain seeds and grass seeds with 1-5 reproductions. In-house grown seeds must be produced at the plant's fields and processed in seed-cleaning and drying shop, equipped with vented hoppers, driers, and specialized machines for seed processing, sorting, and storage. This guarantees their availability and quality.

Business plan. The final indicators as shown in the business plan must reflect growth of labor efficiency 3-3.5 times with the level of profitability of 30%-40%. Along with the task of building farm for technical and technological imports substitution, the key trend in the transformation and development of meat and dairy organic production in Russia until 2025, considering the economic sanctions, should be the proposed development of unused lands, and organic food production in Central, North-Western, and Eastern regions of the Russian Federation under the auspices of the young innovative project ECO-1m project. "*Development and implementation of city-forming, gasified and electro- robotized plants for the production of organic food in Virgin Lands and unused pasture lands of the Russian agricultural complex until 2018- 2025*"

METHODOLOGY OF BUILDING ECO-1M DIGITAL PLANTS

"Science is a Productive Force, Mother Land." Using this principle it is proposed to develop ECO-1m breakthrough technology of the future. It is designed to export this technology. It is also proposed to use the same for export-oriented expansion, i.e. in order to drive suppliers of industrial livestock technologies from the Russian market (notably, such suppliers from Sweden, the Netherlands, the UK, and other countries. ECO-m1 plant designed to locate animals on cultivated pasture fields aims at the production of meat and dairy organic products with a full cycle 'field-store' in order to produce eco-friendly meat and dairy food products. Electric robotization of the plant will ensure higher labor efficiency in terms of final meat and dairy organic products. The organic technology of a modular plant (ECO-m1) is to be developed in a typical modular way and it will be adaptable to weather and climatic conditions of the Russian Far East, Central, North-Wester, Non-Black Earth areas, and Black-Earth area of the Russian Federation, where they had for centuries traditionally applied and cultivated grassland farming. However, many villages have been desolated and young people have been migrating to urban areas, there are no more herders, Russian cows, while pasture lands are now overgrown with agrestal and hogweed it is only possible to build modern agrocities, attract young families to live in rural areas if there are new, high-tech jobs with decent pay, with interesting and appealing jobs involving digital technologies, online management, and eco-friendly organic food manufactured in a plant close to home. In case of ECO-1m projects land cultivation is adapted to soils of a particular climatic area. Biocenosis is supported by the plant's technology. Local livestock breeds are used as they are adapted to zonal climatic conditions. The plants are proposed to be primarily implemented in unused pasture lands with new infrastructure, and the new master plan. First, a natural gas pipeline must be connected to the construction site. Next, a standard-type local gas-turbine power station must be built, electrified railways must be erected. Ahead of time, young support staff must be recruited and trained to include cyber-engineers, livestock experts, veterinarians, and managers, i.e. the ones who will be operating the plant. The master plan identifies the boundaries of a place for the agrocity construction. In this agrocity it is proposed to create favorable living conditions for modern and young professionals and their families. Such agrocity will grow as young families move into the city for employment as support personnel at ECO-1m plants.

The methodology of development of ECO-1m plants is based on using domestic developments and material- and technical, as well as manufacturing *reserves of the Russian Federation: The first reserve is* 34 million hectares of unused and mismanaged croplands; *the second reserve is* the possibility of connecting pipeline natural gas to the plant construction sites in rural areas; *the third reserve is* availability of local breeds of unpretentious Simmental, Switz, and other types of livestock which is adaptable to weather and climatic conditions of Russian regions and summertime pasturage at cultivated pasture fields; *the fourth reserve is* electrified railways which are reliable and which are irreplaceable in rural areas in order to ensure operation of high-capacity robotized technologies applied in the plants featuring year-round, all-weather operation cycle; *the fifth reserve is* the achievements of the Russian enterprises in the area of building electrified railway platforms and locomotives with electric drives for wheels or caterpillars and distributed cybernetic object management systems; *the sixth reserve is* the possibility of powering gas-turbine power stations with natural gas and to efficiently use electric and heat energy for the plant's operation with the efficiency factor of nearly 90%. And, finally, *the seventh reserve is* the creation of thousands of jobs for young professionals in breakthrough industries, at railroad enterprises, defense industry, and aerospace industry, in order to boost the national economy and eradicate the current 'brain drain.' No other country in the world has comparable reserves for the development of robotized enterprises. The only thing left is to set the task of producing organic food products in order to supply sufficient quantity of high-quality and eco-friendly food products and to ensure that this task is fulfilled.

Additional factors to be taken into account include that the breakthrough technology is designed to overcome competition from foreign companies and officials who are interested in imports by the Russian entities; an additional program is required for the training of students and retraining of instructors and professor of the interested universities for the purpose of implementing the ECO-1m plant project; the state program must be approved, in accordance with which funds are allocated for the first draft of the first modular plant along with georeferencing, training of young professionals and, in future, implementation of such plants using public-private investment schemes.

STAGES FOR THE DEVELOPMENT OF ECO-1M PLANTS PROJECT

The science applied for the development of a breakthrough technology is classified as an interdisciplinary one, as it combines farming agriculture, stock farming, arable farming, energy, management, cybernetics, and economics. This interdisciplinary science must ensure Russia's technological leadership drawing on the achievements of fundamental sciences of the Russian Academy of Sciences and the leading branches of Russian industry. This was the case in the past, when the country had to tackle its major economic issues. This is what is envisioned for the near future.

Stage 1: 1920-1930, the GOELRO Plan (the state plan for electrification of Soviet Russia) was aimed at building and developing the people's economics; it was designed for 10-15 years and the plan had been achieved by 1931; the plan provided for fundamental reconstruction of national economy based on electrification, i.e. construction of large-scale enterprises, erection of 30 district-level electric power stations including 10 hydro power stations with the total capacity of 1.75 GW and annual generation of 8.8 billion kWh.

Stage 2: 1954- 1960, the Virgin Lands Campaign, 43.8 million hectares of lands were plowed up, a number of cities have been built from scratch including Tselinograd, with the participation of the institute for electrification of agriculture VIESH, by the resolution of the Government of the USSR in order to increase daily performance of machines and devices they installed a system of night-time lighting of agricultural work and set up post-harvesting processing of grain in the Virgin Land farms.

Stage 3: 1966- 1985, development and implementation of livestock industry. Under the contract awarded by the Ministry of Agriculture of the USSR together with the Italian company Juza G&G and jointly with Giproniselkhoz (projects) and VIESH (development of automation systems under the leadership and management of V.R. Krausp, 1975 [3, pages 13-15]), over 60 large-scale industrial swine production complexes, poultry farms, dairy farms, and fattening complexes for young animal stock have been developed and commissioned.

Stage 4: 1970-1988, the institute for electrification of agriculture VIESH (under the leadership of V.R. Krausp) developed and implemented Automated process control systems (SCADA) at 6 industrial-grade dairy farms [3, page 180, Fig. 10.11 and the website: www.viesh.narod.ru, SCADA system for dairy farms, views 1-4, top down];

Type 1: AISU-400 with an in-house electronic system for identification of animals, farms designed for 400 animals, yard housing with Tandem milking, individual feeding in self-feeders, a database involving 53 indicators as implemented in Nikiforovo and Put' Ilyicha in Moscow Region and Okinino in Nizhny Novgorod Region;

Type 2: a mono-block designed for 1,100 cows and a conveyor for tailor-made standardization and feeding twice a day during 40-60 minutes with complete feed and milking using a carousel as implemented in Gorkovskaya Oblast (now Nizhny Novgorod Region), the city of Kholyazin, Kolkhoz im. Lenina. The automated process control system was developed by V.R. Krausp and various entities in the city of Nizhny Novgorod. Implemented were an electronic system for identification of animal tags (numbers), unique fodder conveyor – a mechanized pasture field with tailor-made standardization and dispensing of complete feed inside the mono-block building of designed for 1,100 cows.

Type 3: Petrilovo farm in Ivanovo Region located in the territory of Shuiskaya Poultry Farm. Secured pasture management of cows has been implemented featuring automated tie system and placement of cows in 3 rows of standing stalls, dispensing of group diet onto a conveyor belt of the feed table, individual dispensing of concentrated fodder using a track-mounted suspended dispenser. An automated system for calf weighing has been developed and implemented using automatic control room scales. This system has proven reliability after over 10 years of operation.

Type 4: Bred livestock farm for 100 cows, State Natural Reserve Petrovskoye, Moscow Region. Milk cow productivity is 10 thousand kg/year, stable system, individual dispensing of complete feed, mobile dispensing of coarse fodder, milking on Tandem, automated process control system with a database, diet calculation, livestock accounting, plans, and reporting. At this site, a multiyear international cooperation project has been carried out with the aim of automation of livestock industry. Together with the institute of radioelectronics / Bulgaria / an automated farm called Srebrena has been established and implemented. Two inventors certificates have been jointly obtained for the control system of a milking parlor, as well as for a transponder for the system of mobile object identification.

Stage 5: 2016 - 2025, VIESH – FNATs VIM. At stage 5 it is planned to develop and implement electric and robotized organic plants with one ECO-1m module and two ECO-2m modules featuring keeping the animal stock at cultivated pasture fields and reproduction of the animal stock and in-house fodder production. It is designed to build and commission by 2025 in various areas of the Russian Federation 160 ECO-1m plants with adjacent youth agrocities for operating personnel. It is planned to develop over 1 million hectares of unused lands and create 300,000 high-tech jobs for young professionals including engineers, livestock experts, veterinary physicians, managers, as well as to produce over 48,000 tons of meat and 800,000 tons of dairy organic products. The

Figure 1. Master plan of the ECO-1m beef and dairy plant

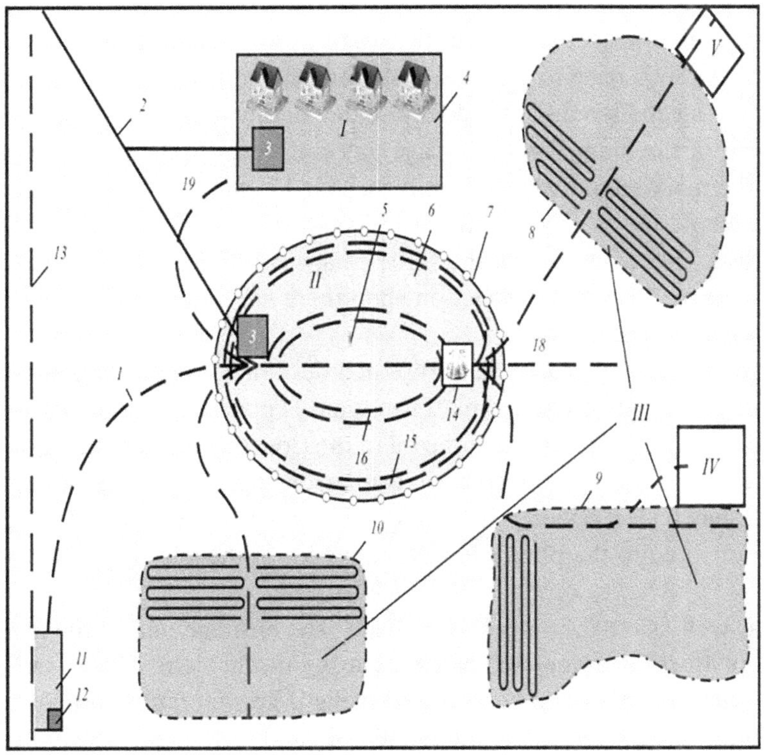

author and chief structural engineer of ECO-1m was V.R. Krausp /1, 2/ drawing on the experience gained during stages 1-4. Fundamental and applied research has been performed, mathematical models have been built: Probability chart for the delivery of grains to the processing and storage shop /3, page 13/; Mathematical model and the method and algorithm for optimization of process line parameters /3, pages 187 - 190/; a model for hierarchy-type automated process control system for livestock farms / 9, page 43/; an intellectual system for the management of farms at the current, operative, and annual stages. A mathematical method for evaluation of the quantity, quality, and value of information has been developed /4, pages 99-112/.

Master Plan and Composition of ECO-1m Plant Shops

The master plan for ECO-1m modular plant is shown in Figure 1. The plant consists of the following five architectural and process facilities: I – agrocity where operating personnel lives; II - livestock production shops; III – fodder production fields; IV – swine production shops; V - poultry production shops. All the shops are interconnected with electrified railways and, thus, they form an integrated production chain. In close proximity to the plant there is a main 13 intercity railway, from which there is one (1) railway spur stretching from the plant.

Electrification and heat supply for the plant and the agrocity is designed to be performed using two natural gas-fired mini CHP plants 3 and the natural gas comes from gas pipeline 2 branching off to the agrocity and the plant. The pivotal facility of the plant is 11 the livestock production shop. Territorially it is surrounded by two 15 and 16 ellipses consisting of two railway tracks.

In the center of 5 small ellipse, there are utility rooms, process shops and the feed preparation shop 14 with storage of winter feed reserve, a mixing robot for the preparation and dispensing of a balanced diet, and a robot for laying and loading of fodder into the mixing unit. Switching tracks 18 are provided for, as well as fencing of the plant 7 with passage gates. Railway spur 1 stretches from the plant to a railway junction of the mainline railway. The railway spur 19 stretches from livestock production shop to the agrocity. The wholesale store 12 with a warehouse and refrigerator chambers 11 is adjacent to 13 the intercity railway line, using which the finished agricultural products are delivered to stores of different cities. Around the livestock production shop there are fodder production fields 8,9,10 which are connected with railway spurs.

On fodder cropping fields they grow cereal crops in order to achieve feeder grain suitable for livestock fodder. After threshing on a self-propelled platform grain is poured into a detachable hopper, which is delivered to the fodder shop and fed onto a dump pit of the flow line for post-harvest grain processing. In the flow line, they perform a full cycle of coarse cleaning, drying, sorting, and storage in ventilated bins.

Grass growing for hay which is then rolled using self-propelled platforms. The rolls are laid onto the platform body and delivered to the plant's honeycomb silo. In the silo they store the rolls and dry the hay with warm air.

Corn is grown, as well as other croppers, for silage. The ensilage during the harvesting with a self-propelled hoppered platform is granulated using a cutting mechanism. The granulated mass is placed into airtight plastic container and then compressed using digestion tanks and covered with a pressure cap. Filled up containers are dispatched to the plant's silo and loaded into the silo cells.

At headlands and ends of radial lines of the electrified railways they install swine production shops and poultry production shops. Their products are delivered to the meat processing shop for the production of food-grade meat and production of modern grades of sausages and ready-to-eat products.

Fodder preparation and diet preparation is performed in the feed processing room. Fodder is dispensed by versatile robotized feed dispensers. Robotized water carriers deliver and pour water into process tanks water. Robots – shuttles take care of the transportation of support personnel. Power supply is performed using 10 kV power lines and using an overhead line of the electrified railway. In the ECO-1m plant there are 30 movable farms which are transported along cultivated pasture fields by railway and 34 2-directional railways. The plant technology is serviced by 9 models of electric robots which automatically move along railways. The 3-layer hierarchal Automated Process Control System is responsible for managing the plant. It is located in a unified control room. Organic food products are made without GMOs, exciters, and conservants. They only add organic fertilizers at fodder cropping fields. No pesticides are allowed. The organic quality certificate is issued for meat and daily food products.

Process shops, which provide for straight-line productivity of the plant, are controlled in real time from the central control room using a smart Automated Process Control System and using dispatcher's Internet connection and video surveillance. The following are part of this shop group: Energy shop; water supply shop; fodder cropping shop, and cultivated pastures; feed processing shop with a honeycomb silo; livestock shop; transportation shop; dairy processing shop; livestock butchering shop with a final-stage meat processing.

Support shops operated based on the time schedules preset in the control center using both operative mode and annual mode of operation. They include organic waste processing shop which produces fertilizers; movable farm- and railway collection, storage, and repairs shop. In the depot they service locomo-

Figure 2. Shop structure of the plant

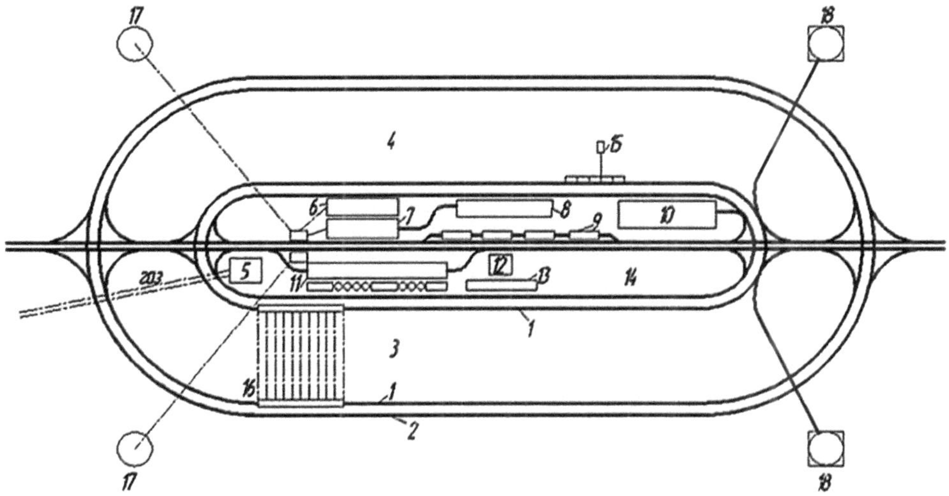

tives, self-propelled platforms, electric robots, and snow breakers. Finished product storage shop ensures storage of cooled food products and certification of such food products as per organic-grade quality.

The total production capacity of organic food products is determined based on the sum of yields of meat production shops and dairy production shops. The labor costs (man-hours) are allocated to the sum of meat and dairy production (t/h). As a result we get the labor costs (man-hours) per one ton of food products produced. In a similar way, we calculate the unit costs of electricity, water, waste, and organic fertilizers.

Energy Shop

An autonomous, small capacity bimodular gas turbine thermal power plant (mini CHP plant) with the capacity of 2x2 mV with electric energy distribution device by a group of feeders of 10 kV power lines and heat energy distributor as generated by heat losses of the turbine for heating of the shops and utility rooms. As a result of a combined usage of heat- and electric energy the efficiency ratio of the mini CHP plant is designed to reach 90%.

Power of supply of the plant facilities is performed via power lines designed for 10 kV, 0.4 kV and 0.8 kV of direct current. The energy generated by the mini CHP plant is, most often, transmitted to significant distances from the power plant to the relevant consumers via the power line 8 (0.8 kW direct current power line) and 10 kW power line which feeds railway substations.

In order to reduce energy loss the power supplied to power lines is amplified in the transformer substation 2 to the level of 35; 110; 220 kV and more depending on the remoteness of consumers. Close to the consumption point on the stepdown substation 4 the voltage level is reduced to 6 kV and 10 kV. From this point on the power is transmitted to consumers.

Power supplies to railway substations 6 of municipal electrified transportation is performed using cable (in rare cases, using airborne) three-phase lines 5. On the railway substation the voltage is reduced to 600 V and alternate current is converted by a rectifier unit into direct current. Using the feed lines of

Figure 3. Wiring diagram for power supply of the ECO-1m plant facilities.
1- Gas-turbine mini CHP plant. 2- 10 kV step-up transformer. 3 – 10 kV three-phase electric lines. 4 – 10 kV alternate current and 0.8 kV direct current railway substation; 5 - 0.8 kV direct current power lines. 6 - 0.8 kV direct current switchgear to 2 feeders. 7 - 0.8 kV direct current wires. 8- Overhead power lines of 0.8 kV direct current. 9 – Neutral conductor, track; 10 – 0.8 kV second overhead line and neutral conductor – track.

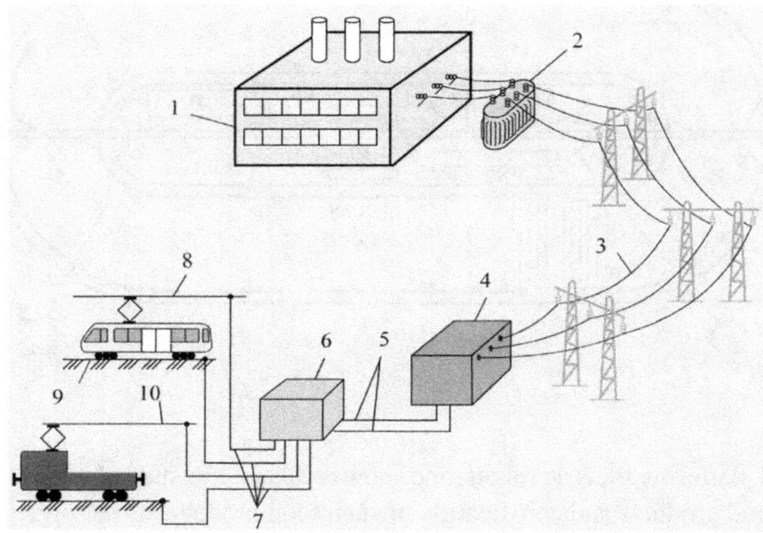

positive and negative polarity 7 electric power is supplied to the overhead 8 lines. The overhead line is fixed above tracks of the electrified railways with a pantograph sliding along them.

DEVELOPMENT OF WHEELED, MOVABLE FARMS INSTEAD OF FOUNDATION-TYPE FARMS

All the 30 ECO-1m farms are movable and in terms of their design they are based on 25 meter-long electrified railways. Instead of substructure, they are on wheels and tracks and they move along tracks.

Farm groups are interconnected as railcars, thereby forming trains. Figure 1 shows 2 trains made up of 15 farms, which are housed on inner- and small outer ellipses of railways.

The area of cultivated pasture field covered with electric fences for these 30 farms is 37.5 hectares (1.25 hectares per one farm). The size of a field with artificial plant formation of 25 m x 500 m = 1.25 hectares. On this area, 75 cows are run on in the summer season. Depending on the intensity of plant formation they calculate the number of days of pasture in this area. In one day of pasture of 75 cows and the standard of eating 50 kg/unit - 60 kg/unit per day and the known level of harvest of a cultivated pasture field, they determine the preset daily size of pasture field /13/.

The modular dairy farm developed, which moves along railways, is designed for 75 cows. The farm design allows to house animals in natural conditions of a self-organized animal stock. Cows freely move inside the farm and across the cultivated pasture field. Animals are treated in two modes: The summer mode which provides for movement between cultured pasture fields (Figure 4) and winter period (Figure 5) – a stationary mode. A modular farm for 75 cows and 20 calves (under 2 months old) has been designed.

Figure 4. Farm layout for the summer period
1 – Milking unit; 2 – Diagnostics point; 3 – Automated gate; 4 – Exit to pasture fields; 5 – Stalls; 6 – Electrified railways; 7 – Water tank; 8 – Milk tank ; 9 – Complete feed tank; 10 – Operator room; 11 – Control room and laboratory. 12 – Stalls for calf feeding; 13 – Incline; 14 – Pasture field; 17 – Feed table; 19 – Identification gate

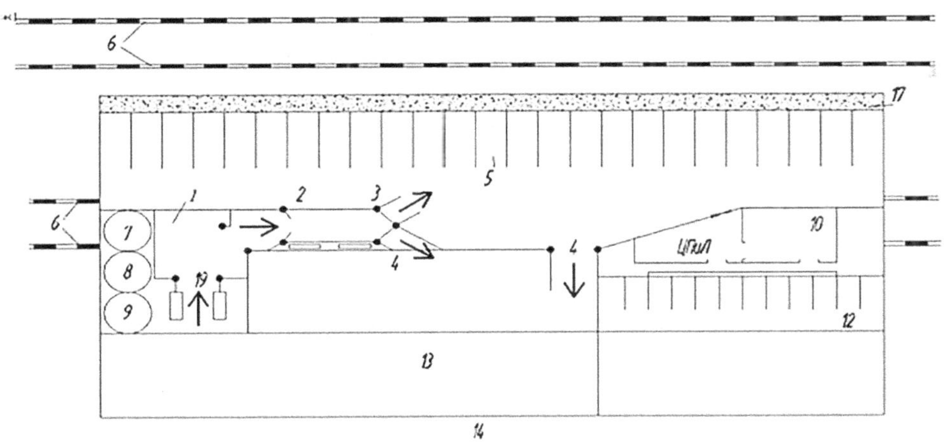

In the summer period animals eat green feed and fattening feed in the form of concentrate depending on the productivity of a milking unit. In this case for a 75-cow farm they only use one railcar platform which houses a robotized milker or a milking unit, identification point, diagnostics point, and veterinary service point, production center and laboratory, stalls for calf feeding, tanks for water, milk, concentrated fodder, and other equipment. After milking and diagnostics animals are re-identified and then they are either let onto the pasture field or they stay on the farm for preventing veterinary service. Severely diseased animals are either rejected or replaced with bred heifers. First calf-heifers are serviced at milk cow farms.

There are 25 stalls, of which 10-15 stalls are free. They are designed for placement of calved animals undergoing preventing treatment or rejected cows to be picked up by a cattle truck. Feeding problematic animals with group-based diet is performed using a feeding pan. In the summer period, microclimate is naturally sustained. Animals are protected from rain, wind, and sun by a hangar adjacent to the farm. In the hangar, they rest at daytime and nighttime.

In the winter period animals receive group-based complete feed in a pan and tailor-made fattening with concentrate fodder, depending on the productivity of milking robot or milking device. During this period the farm consists of 3 railcar platforms which are interconnected to create a uniform structure for 75 animal units (Figure 5).

For each cow there are stalls designed with the width of 0.8 m long and 1.8 m length where cows eat their group diet from the common tray or where they lay in their stalls after feeding, where they chew the cud, rest, and lactate. In order to penetrate its stall a cow goes past the milking robot (a milking device) and identification gate. Cows pass through the identification point and enter the milking robot (milking device) where milk cows receive their tailor-made complete feed diet. Dry, diseased, and calf-bearing heifers go through the milking device without milking and they get used to the machine milking technology. After milking each animal goes through the diagnostics point. Diseased, problematic, and blissom animals are detected using a set of tools and, while going through sorting gate, with feeding. An operator, who is a veterinary physician by profession, determines the numbers of such problematic

Figure 5. Plan of the modular farm during the housing season
1 – Milking unit, 2 – Diagnostics point, 3 – Automated gate, 4 – Exit to pasture fields, 5 – Animal stock stalls, 6 – Electrified railways, 7, 8 and 9 – Tanks for water, milk, and complete feed, correspondingly, 10 – Operator room, 12 – Stalls for calf feeding, 13 – Incline, 15 – Pasture field, 17 – Feed table, 19 – Identification system.

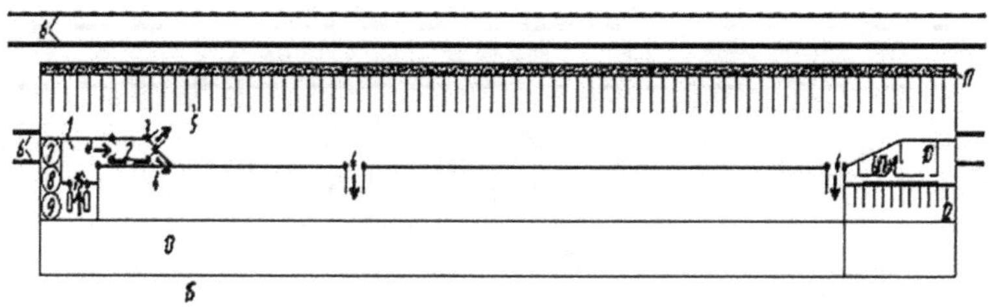

animals and dispatches them to stalls for zoo-technical or veterinary servicing and performs surgeries as necessary to include sowing, preventive treatment, etc.

Daily Regimen During the Housing Season

At 6 am an operator sends animals to pasturage (except for problematic ones) via the identification gate to the area in the hangar. Now cows are able to penetrate the stall for feeding only if they pass the milking device and the preventive inspection area. From 6 am until 7 am the mixing and feed dispensing robot dispenses feed to all the farms. During this time slot the first animal, most often, the animal stock leader passes the identification point and enters the milking area. The milking process takes on average 3-7 minutes, and the relevant information is logged into the database. Information about animal diagnostics is also logged in the database. 2.5÷3 hours later, the last animal passes through the milking area and the diagnostics area. When the animal enters the feeding stall it will have around 6-7 hours for eating the feed and resting. Milking and feeding is performed twice a day. The evening cycle of milking and feeding is completely identical to the morning cycle, and animals can stay in their stalls until 6 am. In order to create the necessary temperature- and moisture conditions in the winter period they use an automated system for sustaining the required microclimate. Air temperature in the farm and the hangar is controlled by this system. In the winter period, the temperature in the farm and the hangar is kept at the level of +5°C. This guarantees that there is no freezing of process equipment, as well as the pallet with liquid manure and the conditions for natural cold housing of animals. After birth, calves are treated with hay, cloth to dry condition and then they are fed with colostral milk. Beef-producing animals are housed in hangars and are laid on rubber mats at the temperature of +5°C. Beef-producing calves are grown with chock. The farms and the hangar are equipped with plastic banded curtains and animals have access to the feed table and are able to freely exit the hangar for pasturage, guarded with an electric fence. At the end of pasturage, via metal guards, they lay rolls (batches) of hay which animals eat in addition to their group diet (twice per day). In order to sustain above-zero temperatures in the farm and the hangar they use warm air fans. Air is warmed by heat exchangers of the gas-turbine mini CHP plant and pumped inside with the fan. The excess air from the farms and the hangars naturally exits the buildings via the banded curtains.

Figure 6. The plan for placing the pallet and spiral mixer for animal manure

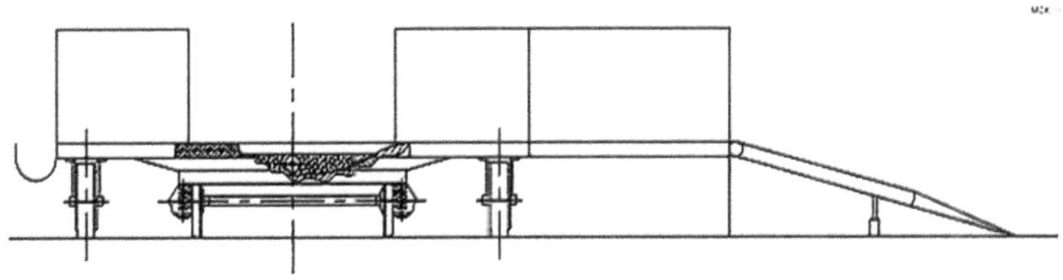

Technology and Equipment for a 75-Cow Farm for Manure Removal

The process of removing manure from the stalls and farm premises is performed using slatted floors and washing off with a water jet. Liquid manure is accumulated in a special heated pallet under the farm floor, from where is it periodically removed (pumped out) with an electric robot and transported to biologic treatment tanks prior to implanting into soil. The following information is taken into account during the design of pallets and devices for mechanical removal of animal manure:

- Adult animal stock yields 20 kg/day-30 kg/day of manure and 10 kg-15 кг of stale,
- Young animal stock yields 10 kg-15 kg of manure per day and 5 kg-8 kg of stale.

In the course of milking around 30% of daily manure volume is produced. The highest volume of manure takes place after 8 to 14 hours of the day (25%). In winter the volume of manure extracted is higher than in the pasture season, since most part of manure stays at the pasture field. During the housing season, for 75 cows they create 75 stalls and 3 pallets for manure pump. The waste generated by calves on the incline is washed off by a water jet and drained onto the field. The housing season (215 days):

Excrements $m_{K,CT} = 30 \cdot 75 \cdot 215 = 48$, $m_{K,CT} = 30 \cdot 75 \cdot 215 = 48\ 3\ t$;

Stale $m_{M,CT} = 15 \cdot 75 \cdot 215 = 24$, $m_{M,CT} = 15 \cdot 75 \cdot 215 = 24\ 1\ t$;

The pasture season (150 days):

Excrements $m_{K,\text{П}} = 30 \cdot 75 \cdot 150 = 3$, $m_{K,\text{П}} = 30 \cdot 75 \cdot 150 = 3\ 37\ t$;

Stale $m_{M,\text{П}} = 15 \cdot 75 \cdot 150 = 16$, $m_{M,\text{П}} = 15 \cdot 75 \cdot 150 = 16\ 8\ t$.

In the shop designed they use a rubber blanket as underlay. Floors are slatted, and manure penetrates the slats and is stored in the pallet for one week. It is mixed there using a spiral mixer (Figure 5).

The pallet's volume for manure storage is calculated using the formula (1) for a 25-m farm's one week's reserves:

$$m_n = 30 \cdot 25 \cdot 7 + 15 \cdot 25 \cdot 7 = 7875 \text{ kg}$$

$$m_H = 30 \cdot 25 \cdot 7 + 15 \cdot 25 \cdot 7 = 7875 \text{ kg}$$

$$V_б = \frac{m_П}{\rho} = \frac{7875000}{(1,0\ldots1,2)} = 6,6\ldots7,9 \text{ м}^3.$$

Let us assume that the pallet volume is 8 m^3. Then the pallet size can be determined as the volume of one half of the cylinder:

$$V_б = \frac{\pi r^2 l}{2};$$

The length of the semitube equals the platform's length l = 25 m, then the width:

$$2r = 2\sqrt{\frac{2V_б}{\pi l}} = 2\sqrt{\frac{2 \cdot 8}{3,14 \cdot 25}} = 0,9 \text{ м}$$

The spiral mixer must be able to continuously mix animal manure with the weight of 7.9 tons in order for the electric robot to remove waste from the pallet using the pump.

In order to determine the maximum speed of rotation of a shaft of the low-speed mixer they use formulas for the kinematic mode:

$$k = \frac{w^2 r}{g} < 30;$$

$$w = \sqrt{\frac{kg}{r}} = \sqrt{\frac{30 \cdot 9,81}{0,45}} = 25,6\text{с}^{-1} => n = 244,6\frac{об}{мін}$$

The power capacity of a continuous-action electric drive:

$$N = 0,01 \cdot k_{ПР} \cdot Q \cdot L = 0,01 \cdot 0,95 \cdot 8 \cdot 25 = 1,9 \text{ кВт},$$

where $k_{ПР}$ reduced factor of resistance to the movement of manure along the auger tube $k_{ПР} = 0,93\ldots;$

Q – hourly production output of the auger tube, t/h;
L– length of the auger tube, m.

Since a farm in a housing season consists of 3 interlinked platforms it is necessary to take into account the number of spiral mixers. There are 3 1.9 kW mixers per farm during a housing season.

We chose a rotary-field motor for the S1 –AIR80V2 mode with a synchronous speed of 2,850 rpm and a nominal power P = 2.2 kW; $\eta = 0.83$.

Cow Milking in Case of a Loose Housing System

During cow milking they use a robot milker by DeLaval or Yolochka-12 milking unit. The said DeLaval robot milker consumes 15 kWh to 25 kWh per one ton of milk produced. The average daily milk yield is 30 – 40 liters.

In this case a 75-animal unit farm is able to produce $75 \cdot (30 \ldots 40) = 2250 \ldots 3{,}000$ liters per day.

$$P = (15 \ldots 25) \cdot (2.25..3) = 33{,}75 \ldots 75 \text{ kWh/day}$$

Milk Cooking Tanks

Milk cooling by DeLaval equipment is performed by a heat recuperation system. For each liter of cooled milk it is possible to yield 0.7 liters of warm water. The process of water heating starts immediately, as soon as milk starts to cool. The water passing through the plate heat exchanger is heated to 50°C÷55°C. Then this temperature is sustained in a special isolated vessel. A 75-animal unit farm on a daily basis consumes 50 liters of hot water and 150 liters of heated water. In order to ensure such volume of heated water on an annual basis there must be 7,200 kWh of electric power. At the same time, a separate electric fence is provided for each farm.

Technical parameters: 1) Current amplitude of the pulse generator, which impacts an animal, must be at least 100 mA – 200 mA, and this must be monitored closely, i.e. the intervals between pulses must be around 1 second. 2) In order to overcome electric resistance of the wool bearing and skin cover, the pulse must have high voltage of around 5 kV-7 kV. 3) The estimated energy value in the pulse that goes through the animal's body should be 6 mA.sec. The range of up to 30 mA. sec is considered to be absolutely safe for both people and animals.

Intervals between the pulses enable an animal to move away from the electric fence after the first electric shock. The electric resistance of an animal's body is estimated at 500 Ohm for the calculations. The main elements of an electric fence include a pulse generator, electric power source, earthing, and an electric fence. The generator continuously feeds the electric fence connected to it with current pulses of a specific power, length, and peal voltage at the preset frequency of repetition of pulses (normally, 50-70 per minute). The most reliable and inexpensive power supply for the pulse generator is a power grid. The pulse generator is installed and earthed immediately from a low-voltage network, while a high-voltage pulse is fed to the object via electric wires. They install a standalone earth circuit close to the pulse generator. In the option developed they use railway tracks for earthing.

In order to create tension of the wires along the electric fence they install tension legs and angular towers once every 150÷200 m on direct areas, as well as at all the angular points and gates.

In order to nurture the reflect of electric fence fear with animals it is proposed to use additional irritants – repellents (deterrent agents) or attractants (attracting devices) which may be visual, scent-based, acoustic, or tactile.

Figure 7. Design of an electric fence

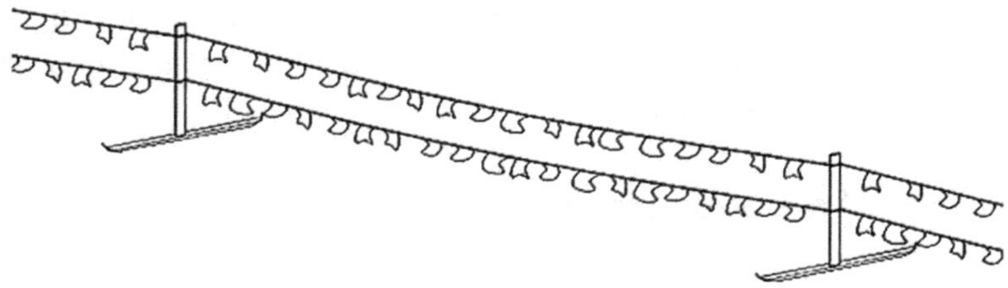

Application of such agents or devices is due to the need to draw attention of an animal to the electric fence. In case of electric and robotized farms such additional visual stimuli may include flags made of textiles, synthetic film, foil, etc., fixed using a thing synthetic cord once every 0.7÷1 m of the fence. A good material for such flags is a metallized (mirror-like) 15÷25 mm-thick lavsan film with a protection coating. It is sufficiently lightweight, durable, and weatherproof. IE-1-1 and IE-1-2 pulse generator blocks have been chosen for electric and robotized farms.

Migration of Farms with Electric Fences

Farms are connected to the train. Each farm is placed in a transit position. Thrust rollers, which sustain and fix the farm in a horizontal position, are lifted. Ramps are lifted and removed.

Electric fences for each farm are shortened and each animal stock is gathered into a compacted group. Railcars with farms are interconnected into a train, to which an electromotive is attached. Then they connect an operating railcar to the electromotive. The operating railcar has a self-propelled, mobile platform – a crop farming robot. The farm train begins to travel at the speed of 0.2 km/h to 0.4 km/h. Electric fences of all the farms move on their metal skids at the same speed as the speed of farms, due to the crop farming robot. As an electric fence constantly moves at low speeds, as an animal approaches the electric fence the animal moves away from the wires with flags due to their fear reflex, and together with the farm the animal moves to the new pasture field.

Figure 8 The scheme of migration of a farm with an electric fence is shown. All the legs and shoes are interconnected with cable wires. In order to eliminate rolling up of the fence the area 9-10 is manufactured using rigid galvanized pipes which are mounted on 4 wheels. They connect long cable wires to the crop farming robot 8 which moves the first electric fence. The crop farming self-propelled platform, just as the electric robot, is remotely controlled by an operator from their control railcar. This enables the operator to move the electric fence, which guards the animal stock, smoothly and synchronously with the farm.

CONCLUSION

The parameters of the organic technology have been validated for use in an industrial-grade plant for the manufacture of organic milk and meat products. It has been shown that eco-friendliness of food products depends on the preparation/treatment of lands and improving its fertility using organic fertilizers. In order to ensure sustainable operation of a farm it is necessary to solve the road/access problem in a reliable and long-term way. It has been proposed to use electrified railways for moving farms along cultivated

Figure 8. Scheme of migration of the farm with an electric fence
1 – Cultivated pasture fields for cattle stock, 2 – Cultivated pasture fields for calves, 3 – Electric fence, 4 – Support stands, 5 – Farm, 6 – Electrified railways for farms and robots, 7 – Opening between fences, 8 – Crop farming robot, 9-10 – Area with galvanized pipes.

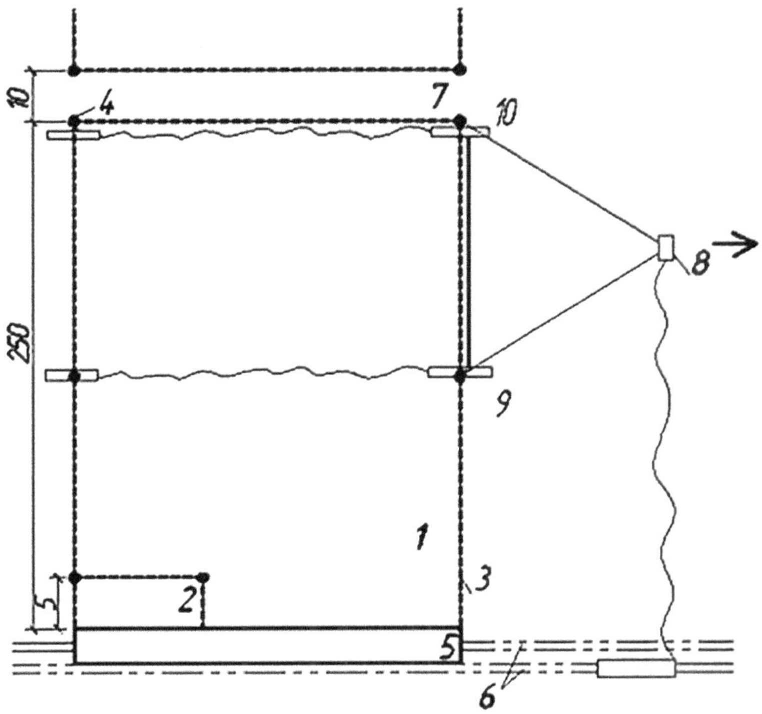

pasture fields. It is expedient to use own seeds of grasses and to standardize grass pasturing on a daily basis. The plant's electric power supply and heat supply is designed to be performed by an in-house, autonomous, natural gas-fired mini CHP plant.

REFERENCES

Electric, R. C. T. (2016). Utility Patent No. 2576209 registered in the State Register of Inventions of the Russian Federation 02.02.2016.

Goroshkov, Y. I., & Bondarev, N. A. (1981). Contact Network, Moscow. *Transport*, 397.

Intellectual Robotized Agricultural System for Food Production. (2012). Utility Patent No. 2462853 registered in the State Register of Inventions of the Russian Federation 10.10.2012.

Kalyanina. (2017). We Just Have To Stop And Think: What Is Our National Interest. *Expert Magazine, 49*, 22-27.

Kotelnikov, A. V. (1986). Earth Currents in Electric Railway Transport. Moscow. *Transport*, 279.

Krausp, V. R. (1975). *Automation of Grain Postharvest Treatment*. Moscow: Machine Building.

Krausp, V. R. (1980). *Complex Automation in Livestock Industry*. Moscow: Machine Building.

Krausp, V. R. (2008). *Strategy for Automation and Informatization of the Agricultural Production Management*. Moscow: Theory. The Internet- and Nanotechnologies. Practice.

Krausp, V. R. (2010). Scientific Methods and Experience in Computerization of Innovative Project Management in Industrial Agriculture until 2020. Youth Scientific School. The Internet- and Nanotechnologies. Automated Enterprises.

Krausp, V. R. (2014). Industrialization and Electric Robotization of Industrial Agriculture for Organic Food Production in Full Cycle "Field-Store" under the slogan "Science is a Productive Force and Mother Earth." In *1ST All-Russian Forum for Russia's Technological Leadership. Book of Reports*. "Technological Doctrine 2014," MOO RUSTO.

Reference Book for Hay Meadows and Pasture Fields. (1986). Moscow: Rosselkhozizdat.

System of Management of Robotized Mixer Feed Dispenser. (2016). Utility Patent No. 2585043 registered in the State Register of Inventions of the Russian Federation 28.04.

Tarnizhevsky, M. V., & Tomlyanovich, D. K. (1986). Design of Electric Power Supply Devices for Trams and Trolleybuses. Moscow. *Transport*, 376.

Chapter 8
Control and Optimized Management of Grain Drying in Forced-Aerated Bins

Alexey Nicolaevich Vasiliev
https://orcid.org/0000-0002-7988-2338
Federal Scientific Agroengineering Center VIM, Russia

Alexey Alexeevich Vasiliev
Federal Scientific Agroengineering Center VIM, Russia

Dmitry Budnikov
https://orcid.org/0000-0002-8729-1300
Federal Scientific Agroengineering Center VIM, Russia

Dmitry Tikhomirov
Federal Scientific Agroengineering Center VIM, Russia

Asan Bekeshovich Ospanov
Kazakh SRI MEA, Kazakhstan

ABSTRACT

In the chapter, both informational and system-based approaches have been applied to the analysis of grain drying process as an interaction of two systems, namely, "drying agent" and "grain layer." It made it possible to evaluate the information content of the process. Analytic expression has been obtained for grain layer informational entropy variations which enabled to make the conclusion that an adequate physical description of grain drying process has to include ambient air parameters control, as well as that of drying agent's parameters in its exit out of a grain layer, as well as parameters of both drying agent and grain in one point of grain layer. A mathematical model of heat-and-moisture exchange in dense grain layer has to be applied. As a result of modeling, it has been proved that the sensors of both drying agent and grain parameters have to be located at a distance of 10 to 11 cm from the central cylinder of the forced-aerated drying bin.

DOI: 10.4018/978-1-5225-7573-3.ch008

INTRODUCTION

Grain drying by forced aeration is a process of long continuance in which ambient air is used as a drying agent (Vanckul, Goryacheva, Garyaev & Sorochinskij, 2017 (Onhonova & Skryabina, 2017; Sabiev & Shonov, 2017; Vasilyev, 2008). Air is heated only in case that its relative humidity exceeds 70%. Air is normally heated with the use of electric fan heater. Drying ability of ambient air is not high. That is why it becomes wet soon while blowing through a grain layer. Grain layers located in the input of the air flux get dry faster. Within the day period, temperature and humidity of ambient air may change in a wide range. Therefore, a substantial irregularity of thick grain layer drying process may occur. Those grain layers that air enters first may appear overdryed. At the same time, grain layers located in the output of the air flux may absorb moisture (Vasilyev, Budnikov, Gracheva & Severinov, 2016; Filatov & Paulys, 2017; Lauva, Aboltins, Palabinskis & Karpova-Sadigova, 2006; Petraq & Francois, 2015).

In homogeneity in grain drying can be reduced and the productivity of drying process can be enhanced owing to the improvement of process control. It was suggested to control both fan performance and capacity of electric heating elements (Gulyaev, 1990; Mashukov & Safonov, 2017). It has to be done with the account of specifics of grain layer moisture content distribution in the bulk of layer. What has to be the set of drying agent and grain layer parameters under control that will provide the necessary and sufficient conditions of effective drying process? Certain studies have to be performed to answer this question. In these studies, it is essentially important to evaluate informational content of heat-and-moisture exchange process in grain layers. Therefore, informational contents for both drying agent and grain layer have to be first separately estimated and, only after that, it is possible to evaluate that of their interaction process.

Evaluation of Informational Content for Grain Layer Status

Within the system-based concept (Petrik, 2011; Yakovlev, 2005), drying process can be regarded as interaction of two systems, i.e. that of 'drying agent' (ambient air) and that of 'grain layer'. Interaction of these two systems results in heat-and-moisture exchange processes in grain layer. The volume of information obtained in the course of grain layer parameters measurements does not remain constant. Besides, the value of this information changes, as well, and it is dependent on the drying technology specifics.

Seeds of different crops may have different grain moisture content values in the range of 8 to 40%, in the stage of postharvest handling. In the assumption that all values in this range (with increment of 1%) are equally probable the maximal informational content of the entire value range will equal to (Gurov, 2000):

$$H(W) = log_2 n = log_2 32 = 5 \text{ data bit}, \quad (1)$$

where n is number of equi-probable events.

Factually, the whole range of grain moisture content values is not that informative since, from the technological point of view, it is more important to know only the value range in which grain moisture content actually falls. Particularly, for wheat, these are the following intervals: values within 14% correspond to standard moisture content, those of 14 to 18% correspond to the long-term storage conditions, provided that grain has been precooled, grain having moisture content in the range of 18 to 22% can be safely stored within the period of 2 to 3 days at temperature not exceeding 15 °C, grain having moisture content exceeding 22% shall not be subject of long-term storage even if it is cooled and, therefore, it has to be treated immediately (Verbickij, Maratov, Timoshenko & Novoselov, 1989; Melynik, 1986). For this particular case, the maximal value of informational content for the entire grain humidity range will not exceed $H(W) = log_2 4 = 2\ bit$. Postharvest temperature of grain may deviate in a range of 15 to 25°C. For the abovementioned grain temperature range, two such intervals can be specified: the interval of 15 to 20°C corresponds to the case when the rate of biological processes is low, while the interval of 20 to 25°C corresponds to conditions when grain respiration rate increases sharply (Egorov, 1973; Dimitrov, 2014). For these two temperature intervals, the maximal informational content equals to $H(W) = log_2 2 = 1$ bit.

In the initial stage just after harvest completion, temperature and humidity of grain can be assumed to be uncorrelated parameters. Therefore, the aggregate maximal entropy of temperature and humidity can be estimated as $H(W_1 \Theta) = H(W) + H(\Theta) = 3$ bit.

The other component of system 'grain–drying agent' is ambient air.

Evaluation of Ambient Air Informational Content

Conditions of ambient air change during the day, and they may substantially vary for different environmental and climatic zones of the country, during the harvest period. That is why studies of such changes have been made (Vasilyev, 1986). The accuracy of ambient air parameters control will make an essential effect on the choice of drying modes and on the effectiveness of automated control system operation. To define ambient air parameters and their changeability, records of one meteorological station in one of the administrative areas in Rostov Region have been analyzed for the time period from the 10th of June to the 10th of September.

The process of ambient air parameters change has an irregular character. That is why probabilistic-stochastic methods were applied in studies (Gulyaev, 1990; Vershinin, 2011). Estimations of mathematical expectation for air relative humidity on the base of sampling for July, August and September yielded the following values 66.1%, 62.8% and 61%, respectively, while the values of mathematical expectation for air temperature are 22.3 °C, 22.7°C and 19.7 °C, for these months. Confidential interval values for air humidity in July, August and September, were equal to ±1.23%, ±1.01% and ±1.84%, respectively, while for air temperature they were estimated as ±0.35°C, ±0.31°C and ±0.5 °C, respectively.

The possibility to employ ambient air for force-aerated drying of grain is defined by its relative humidity and temperature (Anisimova, 1998). Therefore, studies were performed focused on these two parameters.

Five-year sampling bar diagrams for months were drawn in order to define the distribution law for random values (Gracheva, Rudenko, Kononenko & Litvinov, 2017). The entire statistical array for each month was divided in 10 grades, namely from 0% to 100%, for humidity, and from 0°C to 35°C, for temperature. The whole range of ambient air relative humidity variation was divided in intervals. For grain drying by forced aeration, ambient air of relative humidity to 65% is used. In the assumption that air temperature after having passed through a fan increases by, at least, one degree ambient air humidity values up to 70% can be considered as acceptable. If this humidity value limit is exceeded the forced heating has to be applied. Heating is normally performed using two-step mode (Vasilyev, 2009). When relative humidity of air attains the level of 70 to 75% one of the heater sections is activated while the second section is switched on when air humidity attains the value of F = 85%. Thus, the whole range of air relative humidity change appears to be divided in three intervals: 0 to 70%, 70 to 80% and 80 to 100% (Vasilyev & Severinov, 2016). The probability of conditions when ambient air relative humidity falls within each of these three intervals, for any hour in the day, was determined with the use of bar diagrams. Particularly, for the sampling corresponding to July probability values for these intervals were equal to: P_{F1}= 0.564, P_{F2} = 0.252 and P_{F3} = 0,184, respectively, where P_{F1} is probability of conditions when ambient air relative humidity is in the range of 0 to 70%.

Therefore, the aggregate informational content related to the air humidity can be found as:

$$H(F) = H(F_1) + H(F_2) + H(F_3), \quad (2)$$

$$H(F) = 0.564\ log_2\ 0.564 + 0.252\ log_2\ 0.252 + 0.184\ log_2\ 0.184 = 0.14 \text{ bit}.$$

Specific features of drying technology have to be taken account of while dividing ambient air temperature range into intervals, as well. Dividing the entire temperature range into intervals was carried out with the account of the following conditions. For zero temperature, grain respiration is practically negligible, at 10 °C it is of medium intensity while at 20 °C it increases sharply. Grain temperature after forced-aerated drying affects the period of its safe storage (Makushin, 2012; Machihina, Ushakov, Lvova & Denisova, 2012). That is why the entire air temperature range has to be divided in three intervals: 0 to 10 °C, 10 to 20 °C and 20 to 35 °C, and the probability values corresponding to conditions when temperature falls within a particular interval are equal to each other, for sampling over July: P_{T1} = 0, P_{T2} = 0.443, P_{T3} = 0.557, where P_{T1} is probability of ambient air temperature value falling into the range of 0 to 10 °C.

Thus the aggregate informational content related to air temperature is:

$$H(T) = 0.443\ log_2\ 0.443 + 0.557\ log_2\ 0.557 = 0.99 \text{ bit},$$

where $H(T)$ is informational content for ambient air temperature measurements.

Integrated entropy of temperature and humidity can not be calculated as a simple sum of corresponding entropies since temperature and humidity are linked to each other by correlation dependence (Tarasenko, 1963). Therefore, integrated entropy of complex system 'temperature-humidity' equals to the entropy for one of its components plus conditioned entropy for the second component related to the first one. The obtained value of conditioned entropy equals to $H(T/F)$=0.546 bit. Integrated entropy of complex system 'temperature-humidity' can be found from the following equation:

$$H(F_1T)= H(F)+ H(T/F), \quad (3)$$

$$H(F_1T)= 1.43+0.546=1.976 \text{ bit}.$$

Evaluation of Informational Content of the Entire System

In accordance with the theory of information (Hilmi, 2010; Mehdi, 2017), when the status of one system (drying agent, A) entirely defines that of the other one (grain layer. B) it corresponds to this case $H(B/A)$ = 0, and the aggregate entropy of system $H(A,B)$ 'drying agent–grain layer' equals to entropy of the key component. Thus,

$$H(A,B) = H(A) = H(F,T) = 2 \text{ bit}.$$

For constant values of air input parameter, temperature and humidity of grain layer can be calculated, for any point of time. In this case, control can be limited by measurements of only input parameters of drying agent and maintaining them on a specified level. Calculated duration of grain drying process has to be adhered to. System entropy equals to $H(A,B) = H(A)$. Information related to the input parameters of drying agent is quite sufficient for drying process controlling.

For random variations of drying agent (ambient air) parameters, situation is different. In this case, the aggregate entropy of system 'drying agent–grain layer' can be represented in the following form:

$$H(A,B) = H(B/A) + H(A),$$

where $H(B/A)$ is conditioned entropy of system B for changing system A.

The aggregate entropy of system increases making its status undefined for any time period. This status uncertainty will oscillate with changing probability of interaction states of systems B and A. Thus, control systems will permanently suffer changeable deficit of information equal to the excess of entropy $H(B/A)$.

In this case, control of only input parameters of drying agent is not sufficient because the lack of information related to changes in moisture content and temperature of the grain layer has to be compensated. It can be done by either installation of additional sensors within the grain layer or by application of mathematical models for heat-and-moisture exchange processes in grain layer (Vasilyev, Severinov & Makarova, 2016; Vasilyev & Severinov, 2014; Severinov, 2016; Jalil, Nejadi, Ali Mohammad & Nikbakht, 2017).

It has to be noted that entropy $H(B/A)$ is distributed over all elementary layers within the grain mass:

$$H(B/A)= H(B_1/A)+ H(B_2/C_1)+ H(B_3/C_2)+\ldots+ H(B_n/C_{nis\ 1}), \tag{4}$$

where $H(B_1/A)$ is conditional entropy of the first (in the direction of air flow) grain layer B_1, for ambient air parameters variation, $H(B_2/C_1)$ is conditional entropy of the second grain layer B_2, for ambient air parameters variation C_1 within the inner grain area between the first and the second grain layers, $H(B_n/C_{nis\ 1})$ is conditional entropy of the nth, in the direction of air flux, grain layer B_n, for ambient air parameters variation $C_{nis\ 1}$, within the inner grain area between the last and the next-to-last grain layers.

Equation (4) can be presented in the following reduced form:

$$H(B/A)= H(B_1/A)+ \sum H(B_i/C_{iis\ 1}),\ 2 \le i \le n. \tag{5}$$

Since, in this case, entropy of elementary grain layer is equal to its conditional entropy $H(B_i)= H(B_i/C_{iis\ 1})$ entropy of the entire grain layer can be calculated as:

$$\mathrm{H(B)= H(B_1/A) + \sum H(B_i)}.$$

Therefore, entropy of system 'drying agent–grain layer' can be found from the following expression:

$$H(A,B)= H(B_1/A) + \sum H(B_i) + H(A). \tag{6}$$

In the process of drying, each elementary layer has entropy that can be defined using the following expression:

$$H(B_i)=[H(B_i/C_{iis\ 1})\cdot K + H(C_{iis\ 1})],\ 0 \le K \le 1, \tag{7}$$

where K is relative coefficient whose value changes depending on the amplitude of grain layer's state deviation from the equilibrium, in the system comprising grain layer and drying agent. Let us suppose that drying agent enters the ith elementary grain layer with its temperature and relative humidity values equal to those of the ith grain layer after the time moment when drying agent has just passed this elementary grain layer. In this case, coefficient $K = 0$, and

$$H(B_i)=[H(B_i/C_{iis\ 1})\cdot 0 + H(C_{iis\ 1})] = H(C_{iis\ 1}).$$

Values of coefficient K are the closer to zero the closer is the thermodynamic state of elementary grain layer to the equilibrium in the system 'drying agent–grain layer', during drying process.

Taking equation (7) into consideration we can modify equation (6) as follows:

$$H(A,B)= H(A) + H(B_i/A) + \Sigma\ [\ H(B_i/C_{iis\ 1})\cdot K_i + H(C_{iis\ 1})]. \tag{8}$$

This equation shows the change of 'drying agent–grain layer' system entropy, in the process of drying by ambient air whose parameters change in time.

The value of $H(A)$ was defined earlier while the values of other components will not remain constant during the drying process. In the course of drying, entropies of elementary layers change permanently. These changes are determined by the random-wise character of drying agent's input parameters oscillation. If we try to compensate for the information deficiency by installing additional sensors into the grain layer the number of such sensors shall be equal to that of elementary layers.

In the course of oscillation of drying agent's parameters in its exit from the grain layer, there is a possibility to define those of grain layer. In this case, the value of entropy of measurements depends on conditional entropy of drying agent $H(C/B)$, during grain layer parameters measuring. Informational content of measurements also depends on the number of elementary grain layers in the output of drying agent from the grain layer that are in the state of hygrothermal equilibrium with the drying agent:

$$\mathrm{H(C_n)= \Sigma H(C_i/B_i)\cdot n_k},$$

where $H(C_n)$ is entropy of drying agent in its exit from the grain layer comprising n elementary layers, $H(C_i/B_i)$ is conditional entropy of air between elementary layers, n_k is number of elementary layers in the state of hygrothermal equilibrium with the drying agent.

If we decide to control both grain layer's and drying agent's parameters in the input into the grain layer, within the layer and in its exit, the aggregate entropy of system 'drying agent–grain layer' will be defined from the following equation:

$$H(A,\ B)= H(A) + H(B_i/A) + \Sigma\ [\ H(B_i/C_{iis\ 1})\cdot K_i + H(C_{iis\ 1})] + \Sigma H(C_i/B_i)\cdot n_k, \tag{9}$$

Values of the third and the fourth components in this equation change in the process of drying.

Drying process is considered as having been successfully completed when the state of equilibrium in the system comprising drying agent in the output of grain layer and 'output' grain layers will correspond to a specified target value of standard moisture content of grain. Thus, controlling parameters of drying agent in its exit from grain layer reduces the uncertainty of drying process and makes it possible to indicate the moment of its completion.

Permanent control of drying agent's parameters and those of grain within the entire grain layer is associated with unacceptable capital expenditure. Therefore, it is expedient to use results of drying process modeling accompanied by monitoring process parameters in one or more points of grain layer. In this case, the rate of change of ambient air parameters and time required to reduce relative humidity of grain layers located upstream down to its value corresponding to the equilibrium state has to be taken in account (Vasilyev, 2005; Chang, Converse & Steele, 1993). Relevant means of monitoring have to be installed on the border of these areas. Let us consider example when ambient air parameters remain unchanged during two hours. For ambient air temperature of 20 °C and relative humidity of 65%, grain having initial humidity 20% will attain its equilibrium humidity point of 14% within a layer of 20 cm thick. This width of dry layer will vary depending on initial parameter values of both grain and ambient air. Therefore, it is more advisable to focus on statistical expectation for corresponding values.

The specifics of water removal progress in dense grain layers in conditions of varying ambient air parameters can be studied with the help of computer-assisted modeling process. This requires an adequate mathematical model of heat-and-moisture exchange in grain layer.

Installation points of sensors for monitoring parameters of drying agent and grain layer can be found with the use of mathematical model for optimal control of grain drying by forced aeration.

Model for Optimal Control of Grain Drying by Forced Aeration

This model was developed based on thermal balance equation for elementary layer (Ginzburg, 1973):

$$L(I_1 - I_2) = Q_n + Q_i + Q_p,$$

where L is drying agent mass flow ratio, kg/s, $L = SV\rho\varepsilon$, S is cross-section area of fragment within elementary layer (see Figure 1), m^2, V is velocity of drying agent, m/s, ρ_v is volumetric density of drying agent, kg/m^3, is fractional void volume of grain layer, a.u., I_1 is enthalpy of drying agent entering the grain layer, kJ/kg, I_2 is enthalpy of drying agent in the output of grain layer, kJ/kg, Q_n is quantity of heat consumed in the process of grain heating the, $Q_n = \gamma_z c_z (d\Theta/d\tau)$ kJ, Q_i is quantity of heat consumed for water evaporation, $Q_8 = \frac{\gamma_7 r'}{100}\frac{dW}{d\tau}$ kJ, Q_p is loss of heat in the process, kJ, γ_z is volumetric density of grain, kg/m^3, c_z is specific heat capacity of grain, kJ·kg^{-1}·deg^{-1}, r' is latent heat of vaporization for water, kJ/kg, d_{ef} is effective diameter of single seed.

Enthalpy of drying agent is defined as follows (Ginzburg, 1973):

$$I_1 = \left(0.597D_1 + c_v T_1\right)\cdot 4.19,\ I_2 = \left(0.597D_2 + c_v T_2\right)\cdot 4.19,$$

Figure 1. Schematic representation of elementary grain layer fragment having cross-section area of 1m×1m used in calculations

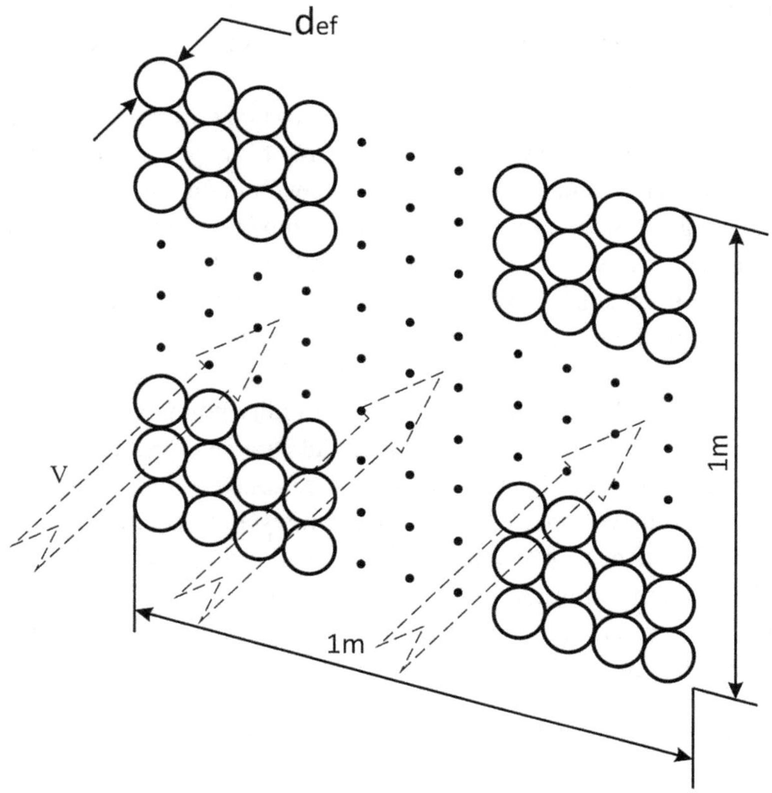

where $c_v = \left(0.24 + 0.00047D\right) \cdot 4.19$ is specific heat capacity of air, kJ·kg^{-1}·deg^{-1}, T_1 and T_2 are temperatures of drying agent in the input and output of grain layer, respectively, °C, D_1 and D_2 are moisture content values of drying agent in the input and output of grain layer, respectively, g/kg.

In the assumption that drying agent flux is controlled so that the heat energy of agent's mass is used efficiently the following base equation of drying process has been obtained that takes energy potential of drying agent into account:

$$SV\rho_v\varepsilon \cdot \left(I_1 - I_2\right) = n \cdot d_{ef} S\gamma_z c_z \frac{d\theta}{d\tau} + n \cdot d_{ef} S \frac{\gamma_z r'}{100} \frac{dW}{d\tau}, \tag{9}$$

where n is number of elementary layers.

The main controlled parameter during grain drying by forced aeration is its moisture content W. Temperature of grain is not critical parameter, in this process. That is why equation (9) can be simplified by excluding its first component. As a result, we will obtain:

$$V\rho_v\varepsilon \cdot \left(I_1 - I_2\right) = n \cdot d_{ef} \frac{\gamma_z r'}{100} \frac{dW}{d\tau}.$$

Transformation of this equation in relation to the rate of variation of grain moisture content yields the first variant of expression representing criterion of optimal control for drying process of elementary grain layer:

$$\frac{dW}{d\tau} = V\left(I_1 - I_2\right)\frac{\rho_v \varepsilon 100}{n \cdot d_{ef}\gamma_z r'} \to \max . \tag{10}$$

Optimization criterion can be expressed in relation to the required time of drying. For this purpose, we have to solve differential equation (10) to obtain the following dependence:

$$W = W_0 + \tau V \frac{\rho_v \varepsilon 100\left(I_1 - I_2\right)}{\gamma_z r' n \cdot d_{ef}},$$

where W_0 is initial value of grain moisture content, %.

Another expression for optimization criterion can be found with the use of the above equation:

$$\tau = \frac{n \cdot d_{ef}}{VK\left(I_1 - I_2\right)}\left(W - W_0\right) \to \min . \tag{11}$$

Unlike equation (10), this one makes it possible to minimize duration of grain drying process over a specified interval of moisture content from W_0 to W.

Let us assume that the velocity of drying agent V is controlled variable, in equation (11). In order to solve criterial equation (10) we have to present it in a detailed form by substituting enthalpy by drying agent's parameters. Here we have to consider certain specific features of the drying process by forced aeration. Relative humidity in the input of grain layer is permanently maintained at the level of 65%. To calculate value T_1 we have to define the difference between ambient air relative humidity F_a and that in the output of the fan heater F_1 (65%). Taking into account that heating air by 1°C reduces its relative humidity by 5% we can derive T_1, from the following expression:

$$T_1 = \frac{F_1 - F_a}{5} + T_a,$$

where T_a is ambient air temperature, °C.

To define humidity of air in the input of grain layer we will use the following expression:

$$D_1 = -\frac{622 \cdot F_1}{F_1 - 765\exp\left(\frac{5345 \cdot T_1 + 311}{500 \cdot \left(T_1 + 236\right)}\right)},$$

while the following equation yields air humidity in the output of grain layer:

$$D_2 = -\frac{622 \cdot \left(1.0 - \exp\left(W_p^{2.4752}\left(5,47 \cdot 10^{-6} T_a + 1.493 \cdot 10^{-3}\right)\right)\right)}{1.0 - \exp\left(W_p^{2.4752}\left(5,47 \cdot 10^{-6} T_a + 1.493 \cdot 10^{-3}\right)\right) - 765 \exp\left(\frac{5345 \cdot T_a + 311}{500 \cdot \left(T_a + 236\right)}\right)},$$

where W_p is equilibrium moisture content of grain for drying agent parameter values in the output of grain layer. For this particular case $W_p = W$, and if drying is made with the use of ambient air without preheating temperature T_2 in the output of grain layer is equal to the ambient air temperature T_a. In case when air is heated $T_2 = T_1$.

For the first case, humidity of air in the input of grain layer has to be calculated with the use of the following dependence:

$$D_1 = \frac{-622 \cdot F_a}{F_a - 765 \exp\left(\frac{5345 \cdot T_a + 311}{500 \cdot \left(T_a + 236\right)}\right)}. \tag{12}$$

For the second case, when air is preheated, its humidity in the input of grain layer will be determined as follows:

$$D_1 = \frac{-622 \cdot 65}{65 - 765 \exp\left(\frac{5345 \cdot \left(\frac{65 - F_a}{5} + T_a\right) + 311}{500 \cdot \left(\frac{65 - F_a}{5} + T_a + 236\right)}\right)}. \tag{13}$$

Accordingly, air humidity in the output of the grain layer has to be calculated with the help of the following two equations:

$$D_2 = \frac{-622 \cdot \left(1.0 - \exp\left(W_p^{2.4752}\left(5.47 \cdot 10^{-6} T_a + 1.493 \cdot 10^{-3}\right)\right)\right)}{1.0 - \exp\left(W_p^{2.4752}\left(5.47 \cdot 10^{-6} T_a + 1.493 \cdot 10^{-3}\right)\right) - 765 \exp\left(\frac{5345 \cdot T_a + 311}{500 \cdot \left(T_a + 236\right)}\right)}, \tag{14}$$

$$D_2 = \frac{-622 \cdot \left(1.0 - \exp\left(W_p^{2.4752}\left(5.47 \cdot 10^{-6}\left(\frac{65 - F_a}{5} + T_a\right) + 1.493 \cdot 10^{-3}\right)\right)\right)}{1.0 - \exp\left(W_p^{2.4752}\left(5.47 \cdot 10^{-6}\left(\frac{65 - F_a}{5} + T_a\right) + 1.493 \cdot 10^{-3}\right)\right)} \cdot \tag{15}$$

$$-765 \exp\left(\frac{5345 \cdot \left(\frac{65 - F_a}{5} + T_a\right) + 311}{500 \cdot \left(\left(\frac{65 - F_a}{5} + T_a\right) + 236\right)}\right)$$

Since conditions of drying process change with varying parameters of ambient air there should be two criterial equations. After sequential substitution by expressions (12) and (13) in equation (11) we obtain two criterial equations that make it possible to define optimal performance of fan of the heater Q. This is the value of fan performance required to perform drying of immovable grain layer comprising n elementary layer from initial moisture content value W_0 to specified moisture content W by air having ambient temperature T_a and relative humidity F_a, within the minimal time period τ.

In case when electric fan heater is used for air heating, criterial equation will have the following form:

$$\tau = \frac{n \cdot d_{ef}}{k_v Q K \left(\left(0.597 \cdot \frac{-622 \cdot 65}{65 - 765 \exp\left(\frac{5345 \cdot \left(\frac{65 - F_a}{5} + T_a\right) + 311}{500 \cdot \left(\left(\frac{65 - F_a}{5} + T_a\right) + 236\right)}\right)} + c_v\left(\frac{65 - F_a}{5} + T_a\right)\right) \cdot 4.19 - \left(0.597 D_2 + c_v T_a\right) \cdot 4.19\right)}, \tag{16}$$

$$\cdot \left(W - W_0\right) \rightarrow \min$$

where k_v is correlation coefficient between drying agent velocity V and fan performance Q.

In case when forced aeration is used without heating, criterial equation will be represented as follows:

$$\tau = \frac{n \cdot d_{ef}}{k_v QK \left(\left(0,597 \left(\frac{-622 \cdot F_a}{F_a - 765 \exp\left(\frac{5345 \cdot T_a + 311}{500 \cdot \left(T_a + 236 \right)} \right)} \right) + c_v T_a \right) \cdot 4,19 - \left(0,597 D_2 + c_v T_a \right) \cdot 4,19 \right)} \cdot \left(W - W_0 \right) \to \min \qquad (17)$$

To calculate air humidity in the output from grain layer D_2 we have to apply equations (14) and (15).

While solving criterial equation it is necessary to consider limitations of available technological equipment. For example, frequency-regulated fan drive enables performance control but it has to be kept in mind that electric motor capacity is limited and it is equal to 5.5 kWt (for BV-25 bunker). It is stipulated in limiting condition where the fan head is expressed via grain layer resistivity (Vasilyev, 2006).

$$\frac{2.778 \cdot 10^{-7} Q \left(0.0064 Q + 1.089 \cdot 10^{-6} Q^2 \right) n \cdot d_{ef}}{\eta} \le 5,5 . \qquad (18)$$

Let us define the two mathematical models.

Mathematical model for optimal control of drying by ambient air:

$$\begin{cases} D_2 = \dfrac{-622 \cdot \left(1.0 - \exp\left(W^{2.4752} \left(5.47 \cdot 10^{-6} T_a + 1.493 \cdot 10^{-3} \right) \right) \right)}{1.0 - \exp\left(W^{2.4752} \left(5.47 \cdot 10^{-6} T_a + 1.493 \cdot 10^{-3} \right) \right) - 765 \exp\left(\frac{5345 \cdot T_a + 311}{500 \cdot (T_a + 236)} \right)}, \\ \tau = \dfrac{n \cdot d_{ef}}{k_v QK \left(\left(0,597 \left(\frac{-622 \cdot F_a}{F_a - 765 \exp\left(\frac{5345 \cdot T_a + 311}{500 \cdot (T_a + 236)} \right)} \right) + c_v T_a \right) \cdot 4.19 - (0.597 D_2 + c_v T_a) \cdot 4.19 \right)} (W - W_0) \to min \\ \dfrac{2.778 \cdot 10^{-7} Q \cdot (0.0064 Q + 1.089 \cdot 10^{-6} Q^2) n \cdot d_{ef}}{\eta} \le 5.5. \end{cases}$$

Mathematical model for optimal control of drying by preheated air:

$$\begin{cases} D_2 = \dfrac{-622 \cdot \left(1.0 - \exp\left(W_p^{2.4752} \left(5,47 \cdot 10^{-6} \left(\frac{65 - F_a}{5} + T_a \right) + 1.493 \cdot 10^{-3} \right) \right) \right)}{1.0 - \exp\left(W_p^{2.4752} \left(5.47 \cdot 10^{-6} \left(\frac{65 - F_a}{5} + T_a \right) + 1.493 \cdot 10^{-3} \right) \right) - 765 \exp\left(\frac{5345 \cdot \left(\frac{65 - F_a}{5} + T_a \right) + 311}{500 \cdot \left(\left(\frac{65 - F_a}{5} + T_a \right) + 236 \right)} \right)} \\ \tau = \dfrac{n \cdot d_{ef}}{k_v QK \left(\left(0,597 \cdot \frac{-622 \cdot 65}{65 - 765 \exp\left(\frac{5345 \cdot \left(\frac{65 - F_a}{5} + T_a \right) + 311}{500 \cdot \left(\left(\frac{65 - F_a}{5} + T_a \right) + 236 \right)} \right)} + c_v \left(\frac{65 - F_a}{5} + T_a \right) \right) \cdot 4.19 - (0.597 D_2 + c_v T_a) \cdot 4.19 \right)} (W - W_0) \to min, \\ \dfrac{2.778 \cdot 10^{-7} Q \cdot (0.0064 Q + 1.089 \cdot 10^{-6} Q^2) n \cdot d_{ef}}{\eta} \le 5.5 \end{cases}$$

Combined solution of criterial equation and equation with limiting conditions for fan's electric motor capacity will yield the value of fan performance Q that ensures the minimal time of drying.

Use of product $n \cdot d_{ef}$ in criterial equation, in this model, makes it a multifunctional tool. Thus, using standard algorithm of grain drying process control for forced aeration when fan performance remains constant ($Q = const$) the following calculations can be made:

for a specified kind of crop (d_{ef}), the number of elementary layers and time required for moisture extraction to ensure specified final value of relative humidity can be determined,

- for a specified kind of crop, process time and final humidity value can be calculated for a specified number of grain layers.

In order to define the points for installation of drying agent and grain layer parameter sensors, calculations were made with application of the following initial data. Initial water content in grain is 20%. Fan performance is 11000 m^3/h. Ambient air temperature is 25 °C and its relative humidity is 65%.

System of equations was solved with the use of optimization command *fmincon* of the set of software package MATLAB (Dyakonov, 2008). Calculation result show that 21 grain layers will become dry to standard moisture content of 14% within 3.3 hours. It means that sensors for parameter control can be installed at the distance of 0.10 to 0.11 m from the central air distribution duct of the forced-aerated drying bin.

Let us correct these parameters by application of computer model for grain drying process in dense layer.

Development of Computer Model for Heat-and-Moisture Exchange in Dense Grain Layer

In drying bins where air bows vertically the speed of air blowing through the grain can be considered to remain constant. For radial air distribution, cross-section area of grain layer changes in the direction of air stream. Therefore, the velocity of air decreases (Aniskin, 1972). This specifics of air velocity change influences essentially the rate of grain drying process in different points of grain layer, and it has to be taken into consideration while developing mathematical model for grain drying in dense layer. Step-by-step computation method was chosen as the basic one. Grain layer is divided into elementary layers. Width of elementary layer is an essential parameter that affects calculation precision. Thus, in methodical recommendations by VIESH (1997) layers are considered as elementary ones if one weight unit corresponds to consumption of 1.7 kg/min of dry drying agent. The total number of elementary layers is defined as:

$$n = \frac{1.7 l \gamma_{z.n}}{60 v \gamma_v},$$

where $\gamma_{z.n}$ is bulk weight of grain, kg/m^3 (for wheat, $\gamma_{z.n} = 784$), γ_v is relative (volumetric) density of air, kg/m^3.

Other points of view there also exist related to this issue. For example, in (Aniskin, Okun' & Vercman, 1988) elementary layer is defined as a layer whose thickness is equal to single seed diameter. Zelenko

(1988) made the assumption that elementary layer is that in which drying rate is mainly determined by gradient of moisture content. In our study, elementary layer was defined as that whose thickness equals to single seed diameter. With this assumption the maximal possible number of elementary layers can be considered for process description and heat-and-moisture exchange processes within a single seed can be studied.

In step-by-step computation method, parameters of drying agent that has passed through an elementary layer are taken as input parameter values for the next elementary layer. Therefore, values of moisture content and equilibrium humidity of air that has passed through an elementary layer are subject to calculations. The initial system of equations for drying process description has the following form (Zhidko, Rezchikov & Ukolov, 1982):

$$\frac{\partial T}{\partial \tau} + 3600V\frac{\partial T}{\partial x} = -\frac{\gamma_z c_z}{\gamma_v c_v \varepsilon} \cdot \frac{\partial \theta}{\partial \tau} - \frac{\gamma_z r'}{\varepsilon \gamma_v c_v \cdot 100} \cdot \frac{\partial W}{\partial \tau}, \tag{19}$$

$$\frac{\partial T}{\partial \tau} + 3600V\frac{\partial T}{\partial x} = -\frac{\alpha_q s_v}{\varepsilon \gamma_v c_v} \cdot \left(T - \theta\right), \tag{20}$$

$$\frac{\partial W}{\partial \tau} = -\frac{\varepsilon \gamma_v}{10^3 \gamma_z} \cdot \left(\frac{\partial D}{\partial \tau} + 3600V \cdot \frac{\partial D}{\partial x}\right), \tag{21}$$

$$\frac{dW}{d\tau} = -K\left(W - W_p\right), \tag{22}$$

$$W_p = \left[\frac{\ln\left(1-F\right)}{5.4710^{-6}\left(T+273\right)}\right]^{0.435}, \tag{23}$$

$$K = 7{,}110^{-2} e^{0.05\,T}, \tag{24}$$

$$F = \frac{745\,D}{\left(622+D\right)e^{\left(0.622+\frac{7.5\,T}{238+T}\right)}}, \tag{25}$$

where T is temperature of drying agent, °C, D is relative humidity of drying agent, g/kg, W is actual moisture content of grain, %, θ is temperature of grain, °C, V is velocity of drying agent, m/s, c_v, c_z are, respectively, values of specific heat capacity of air and grain, kJ·kg^{-1}·°C, μ is fractional void volume of grain layer, s_v is specific surface area of seeds, 1/m, r' is latent heat of vaporization for water, kJ/kg, α_q is heat-transfer coefficient, J·kg·h·°C, γ_z is volumetric density of grain, kg/m^3, γ_v is volumetric density of air, kg/m³, K is relative coefficient of drying, 1/h, W_p is equilibrium moisture content of grain, %, x is spatial coordinate, m, τ is time coordinate, h, F is air relative humidity, %.

Calculations are performed in the following sequence of operations. Equilibrium moisture content W_p is first defined with the use of parameters of air entering the grain layer (23). Then drying coefficient K is calculated (24). Thereafter, moisture content of grain W is found (22). Then temperature T that has drying agent after having passed elementary grain layer is calculated (19). Then grain temperature ϑ is defined (20). Thereafter, moisture content D (21) and relative humidity of drying agent F in the output from the grain layer are calculated (25). This calculation operations sequence carried out for each elementary layer makes it possible to build the entire picture of grain drying process dynamics.

Taking into account that in equations (19), (20), (21) we assumed the thickness of elementary layer to be equal to single seed diameter, spatial coordinate x can be substituted by either seed radius R or by its effective diameter d_{ef}.

As a result of performed transformations (Vasiliev et al., 2016) of initial system of partial differential equations, for step-by-step computation method applied to thick grain layer we obtained the following system of equations in symbolic form:

$$T(p) = T_0(p)\frac{1}{p}\left(1 + pe^{-p\tau_1} - e^{-p\tau_1}\right) - A_1\theta(p)p - A_2W(p)p, \tag{26}$$

$$\theta(p) = T(p)\frac{1}{A_4} - T_0(p)\frac{1}{pA_4}\left((1 - A_4)^{\circ}C\cdot\left(1 - e^{-p\tau_1}\right) + pe^{-p\tau_1}\right), \tag{27}$$

$$D(p) = D_0(p)\frac{1}{p}\left(1 + pe^{-p\tau_1} - e^{-p\tau_1}\right) - \frac{1}{A_3}W(p)p, \tag{28}$$

where $A_1 = \frac{\gamma_z c_z}{\varepsilon\gamma_v c_v}$, $A_2 = \frac{\gamma_z r'}{100\gamma_v c_v \varepsilon}$, $A_4 = \frac{\alpha_q s_v,}{\varepsilon\gamma_\iota c_v}$, $\frac{\varepsilon\gamma_v}{10^3\gamma_v} = A_3$, $\tau_1 = \frac{d_{ef}}{3600V}$.

In order to solve this system of equations we will convert differential equations in symbolic form into transfer functions (Ospanov, Vasilyev, Budnikov, Karmanov & Vasilyev, 2017). For example, for equation (19), functional diagram implementing the corresponding dependence is shown in Figure 2.

This functional diagram is represented in form of transfer functions of standard elementary components used in automation engineering where: $\frac{1}{p}$ is integrating factor, p is differential factor, $e^{-p\tau_1}$ is time lag element and A is amplifying element. In a similar way, the above conversion procedure can be applied to equations (20) and (21).

Drying process modeling was carried out with the use of *Simulink* software (Dyakonov, 2008) that provides a rather wide range of tools for computer-assisted modeling of various processes. The use of standard library units of *Simulink* application makes it possible to build a required computation model. In Figures 3 and 4, functional diagrams are presented for computer-assisted modeling of equations (19) and (21) solution.

Figure 2. Functional diagram implementing equation (19) in form of transfer functions

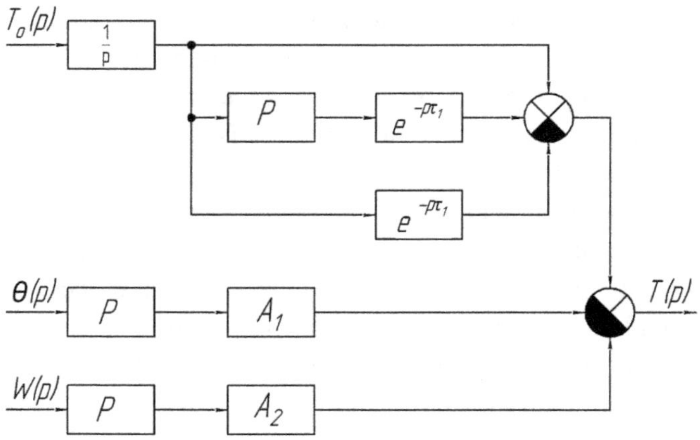

With the help of units $Q_0, T_0, W_{0,} D_0,$ initial values of, respectively, grain temperature, air temperature, moisture content in grain and air relative humidity are specified.

Units W, Q describe transfer of grain temperature and moisture content values into the computation module. Unit *Tau* implements dependence $\tau_1 = \frac{d_{ef}}{3600V}$ and it is used to specify drying agent velocity within the grain layer. To make the entire modeling diagram representation more compact, the concept of subsystem formation is applied in which a complicated network circuit can be represented as a single unit.

In conventional mathematical models, drying coefficient (24) depends on air temperature. Our mathematical model takes account of grain drying coefficient dependence on drying agent velocity and temperature. Experimental studies were carried out to define such dependencies. Experiments were made for convection-mode drying in temperature range from 20 to 50 °C. These values of temperature were chosen in order to estimate the change of grain drying process effectiveness using preheated air. Grain with initial moisture content of 16%, 20% and 24% was applied in experiments. In the course of experiments, drying agent velocity was controlled in the rangee of 0.2 to 2 m/s. Velocity of drying agent was measured in the output from the grain layer with the use of thermal mass flowmeter. Initial moisture content of grain was defined with the help of moisture tester 'Faunais M', and the quantity of grain loaded into the experimental installation was weighed. In the process of grain drying, its moisture content percentage was defined by application of the following expression:

$$W = \frac{m_1 - m_2}{m_1} 100, \quad (29)$$

where m_1 is grain mass prior to drying, kg, m_2 is grain mass just after drying, kg.

For all drying process modes, the value of drying coefficient was calculated with the use of the following well-known expression (Lykov, 1968):

Figure 3. Computational diagram for equation (19) solution modeling

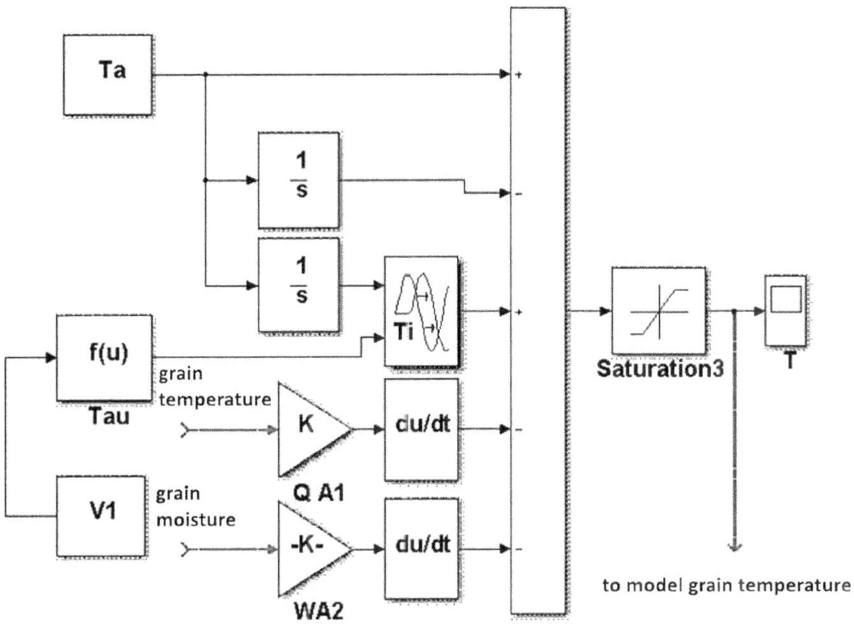

Figure 4. Computational diagram for equation (21) solution modeling

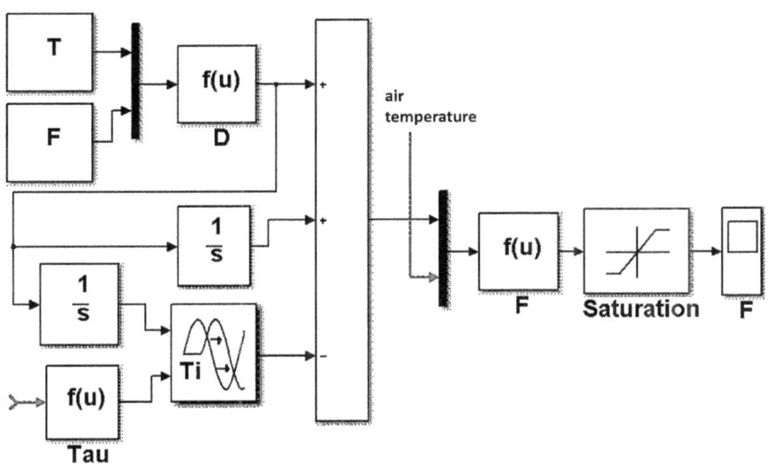

$$K = \frac{2,3\left(\lg\left(W_1 - W_p\right) - \lg\left(W_2 - W_p\right)\right)}{\tau_2 - \tau_1}, \tag{30}$$

where W_1, W_2 – moisture content of grain subject to drying, for time moments τ_1 *and* τ_2, respectively.

Based on the results of experimental studies, regression analysis has been performed and regression equation has been designed with the use of software package MATLAB. The obtained regression dependence was checked for validity, and its applicability has been proved. The adequacy of the mathematical model was determined by the Fisher criterion for degrees of freedom f1 = 30, f2 = 7, with a confidence probability of 95%. The calculated "lower" value of the Fisher test is Fp = 0.0007. Tabular, respectively, Fm = 2.84. The condition Fp <Fm is satisfied, therefore, the model adequately reflects the required dependence.

In research used the criterion R-square to evaluate the quality of the approximation. As a result of the calculations the following values of the criteria were obtained: R-square = 0.996; RMSE = 0.132, which confirm the high quality of the object description obtained by the regression equation.

For instance, the following dependence was found for convection-mode drying:

$$K = 0,009T^{1,026} \cdot V^{0,221} \cdot W^{0,440}. \tag{31}$$

This dependence was applied to the computer model to calculate drying coefficient in conditions of changing air temperature and velocity and moisture content of grain.

In order to perform modeling of drying process in thick layer on the basis of elementary layer concept, it is necessary to define the number of such elementary layers. Thickness of grain layer placed between the wall of air duct and that of outer cylinder of the forced-aerated drying bin is 1.2 m. For effective diameter of seed equal to 0.0049 m (Avdeev, Nachinov & Nachinova, 2009), the number of elementary layers is 245. Thus, complete model of thick layer has to comprise 245 elementary layer models.

While developing the model, it is necessary to consider cylindrical shape of forced-aerated drying bin that makes the velocity of drying agent change depending on the thickness of grain layer. Diameter of the inner air duct of grain drying bin 'BV-25' is 0.75 m while that of outer cylinder is 3.1 m. In the output of central air duct, drying agent has velocity of 0.8 to 0.7 m/s while in the output from grain layer close to the outer cylinder its velocity is only about 0.1 to 0.2. Similar dimensions has the grain layer in grain drying bins of type 'BVis 40' that are also often applied in farms. To develop a model, we have to define the required drying agent velocity in each point of grain layer.

Drying agent velocity within the grain layer can be calculated using the following expression:

$$V = \frac{Q}{S \cdot 3600},$$

where Q is air fan flow-rate, m^3/h, S is cross-section area of grain layer at a specified distance from the central air duct, m^2.

Assuming that, computational accuracy in the course of modeling will not be affected significantly if drying agent velocity is varied by increments of 0.05 m/s we can define the number of elementary layer in each subsystem of the model. Thus, drying agent velocity will be equal to 0.80 m/s, within the first 6 elementary grain layers. In the next 6 elementary layers its calculated value is 0.75 m/s. Within the next 7 layer, drying agent velocity is 0.70 m/s while its velocity is 0,65 m/s, in the next 9 layers. In 32 elementary layers, velocity of drying agent equals to 0.3 m/s, and within the next 43 layers it equals to 0.3 m/s, as well. Finally, within the last 65 elementary layers, in direction of drying agent flow, drying agent velocity is 0.25 m/s. Thus, in complete model of drying process in dense layer required values of drying agent velocity are specified for all elementary grain layers.

Computational diagram for process modeling in the first six elementary layers is shown in Figure 5.

Figure 5. Computational diagram for drying process modeling for a dense layer comprising the first six elementary layers using Simulink software

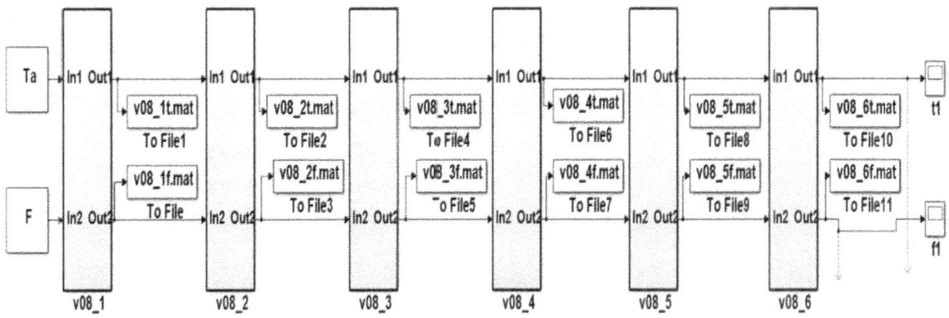

Drying agent parameters *F* and *T*, in the output of each precedent elementary layer, obtained as a result of computation are taken as input parameters for calculations related to the next layer. Calculated values of temperature and moisture content of grain are saved in memory and are used to plot graphs.

Input parameters of drying agent are its temperature T_a and relative humidity *F*. Output (in exit from elementary layer) parameters of drying agent are recorded in files. For instance, file v08_1t.mat contains drying agent temperature change in the output from the first elementary layer. Besides, grain layer parameters, i.e. change of grain moisture content *W* and temperature θ are also saved. Such model structure makes it possible to save results of modeling for each elementary layer and to present results in graphical form. Oscilloscope viewer functions of *Simulink* software can be applied for real-time monitoring results of modeling. In Figure 5, output values of *F* and *T* for subsystem containing six elementary layers are designated as *t1* and *f1*.

In the first stage, modeling of the process was carried out using constant values of drying agent parameters, in the input of grain layer. It was assumed that ambient air relative humidity and temperature are equal to *F*=65% and $T_a = T_0 = 25^o C$, respectively while temperature and moisture content of grain are equal to $\theta_0 = 20$°C and W_0 =18%, respectively. Grain drying process modeling was continued unless the final value of moisture content in grain did not become lower than 14%. Results of modeling are presented in Figure 6.

Results of modeling show that grain drying within layers is not a homogeneous process. On its way through a sequence of elementary layers, air gets wet and then starts to transfer water to the layers located downstream. While the value of grain moisture content of 14% is attained by layers located upstream within the period of three hours from the drying process launch, the 50th elementary layer attains this value of grain moisture content only one and a half hour later. Results of modeling adequately represent the factual situation for drying process in dense grain layer.

Though ambient air parameters feature changeability, we assume that their change within a period of three hours can be neglected. Results of modeling (see Figure 6) show that the first six grain layers attain standard grain moisture content (14%) in three hours after aeration process starting. One hour more is required before the next 19 grain layers attain standard moisture content value. It corresponds to 0.10 m grain layer thickness from the drying agent input point. By installing drying agent and grain parameter monitoring sensors at the distance of 0.10 to 0.11 m from the central air duct of the forced-aerated drying bin we can control the motion of drying zone, in conditions of constant drying agent's input

Figure 6. Drying curves obtained by modeling for 50 elementary grain layers

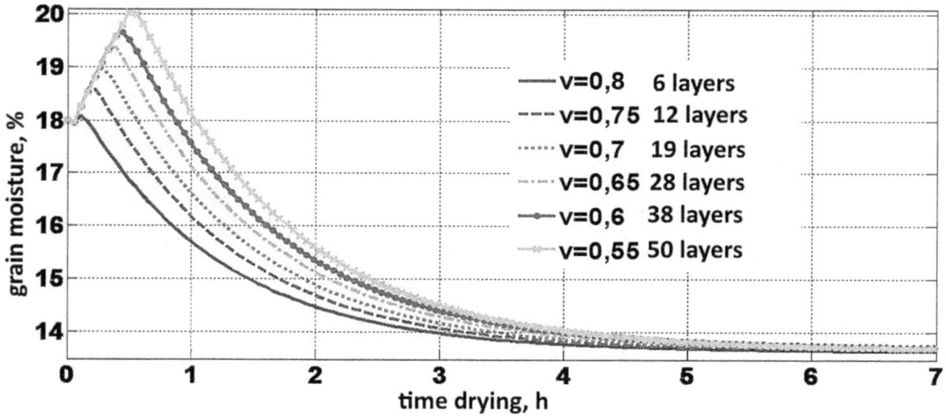

parameter values. Ambient air temperature and relative humidity change will define moisture content of grain input layers. Placing sensors at the distance of 0.10 m far from air duct offers the possibility to make corrections in drying process computation with the use of mathematical model and to have a real picture of drying zone propagation within the grain layer.

In the work Vasilyev A.N., Budnikov D.A., Gracheva N. N. & Severinov O.V. (2016) presents a comparison of simulation results and experimental data on grain drying. The maximum deviation of simulation results did not exceed 10%. This confirmed the quality of the model.

In works Vasilyev & Rudenko (2009) an attempt has been made to reduce the inhomogeneity of grain moisture content owing to the application of microwave heating to grain layer. Authors have developed a drying method based on exploration of this moisture content inhomogeneity.

However, implementing this method causes technical problems since the design of forced-aerated drying bins does not allow separate grain downloading from different drying conditions zones. Intermixing of grain from adjacent zones inevitably occurs, in the course of downloading which reduces the effect of microwave-based method application (recirculation mode drying). In consideration of this problem, a new principle (Severinov, Rudenko, Budnikov & Macynin, 2012) for sectional forced-aerated drying bin architecture has been patented.

Grain Drying Process Control in Sectional Forced-Aerated Bins

Aeration bin with radial air distribution comprises three sections divided by concentric cylindrical surfaces. The sections have equal volumes. While carrying out grain drying by forced aeration in such installations, there is a possibility to transfer grain from one section to another depending on strategy and purposes of process control. It is shown schematically in Figure 7.

Just after the grain has been loaded down into the sections of bin its moisture content is uniform over the entire volume (see Figure 7a). Within the time period of 2.5 to 5 hours after the moment of aeration process launching, moisture content of grain in the first section will attain its standard value (see Figure 7b). In this case, the question arises whether it is expedient to load off dry grain from the first section and to distribute the rest of grain among all sections.

Figure 7. Schematic representation of moisture content distribution in grain over bin sections (left – before drying process launch; right – after 2.5 to 5h of drying

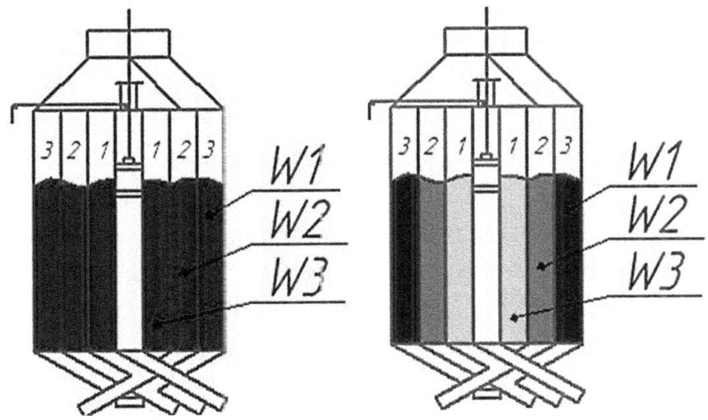

Process modeling was carried out to calculate factual moisture content distribution in grain over the sections of forced-aerated drying bin. In order to perform modeling of grain drying process in three-sectional bin elementary layers were distributed in proportion with grain layer thickness in each section.

In real conditions, ambient air temperature and relative humidity do not remain constant since they change during the day period which may substantially influence the process of drying. That is why model of drying was improved so that daily variations of air parameters could be taken account of and control operations of electric fan heater used to reduce air relative humidity could be modeled (see Figure 8).

Figure 8. Simulink model for grain drying process in sectional aeration bin with changing ambient air parameters

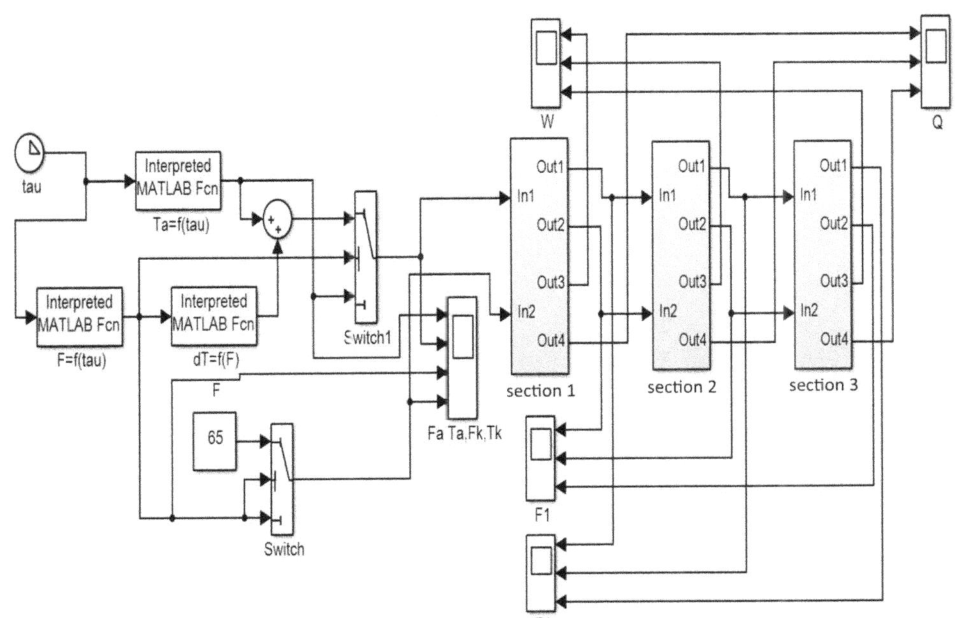

Based on the sampling over aira temperature and relative humidity variations, approximation has been made. Approximation results were included into units *Ta=f(tau)* and *F=f(tau)* of *Simulink* model, in form of functions. In presented model, operation of electric fan heater with the function of air relative humidity control was simulated. In day hours when ambient air relative humidity exceeded 65%, air having humidity of 65% was supplied to the input of grain layer.

In the first stage, modeling was performed with constant values of ambient air parameters. Results of grain drying process modeling for different zones of three-sectional drying bin are presented in Figures 9, 10, 11.

Results of modeling show that both the maximal drying rate and the maximal inhomogeneity of drying within the layer take place in the first zone of drying bin. Inhomogeneity of grain drying in the first zone can be reduced by increasing drying agent velocity. In this case, relative humidity of air blowing through the grain layer will have lower values.

Figure 9. Grain drying curves for the first zone, in direction of air flow, of three-sectional drying bin

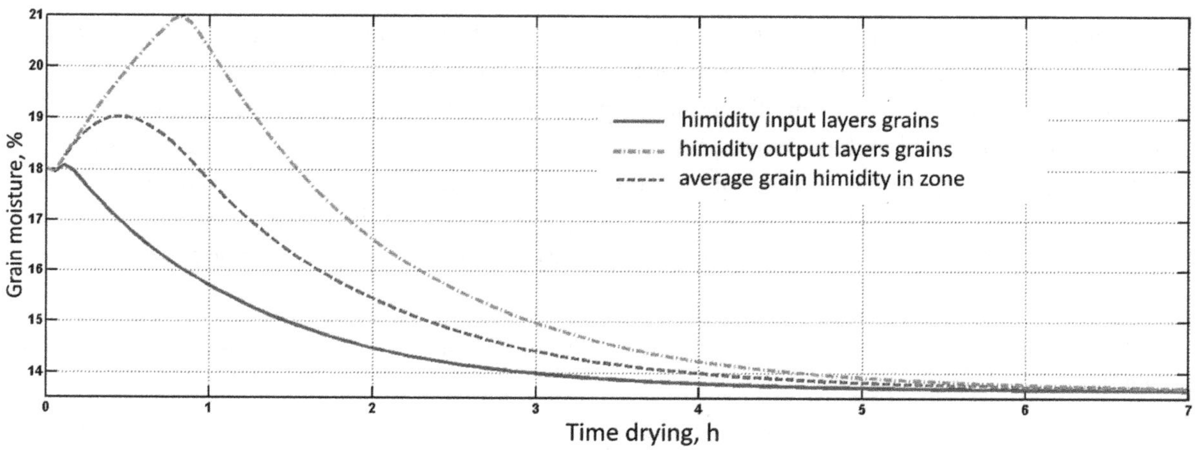

Figure 10. Grain drying curves for the second zone, in direction of air flow, of three-sectional drying bin

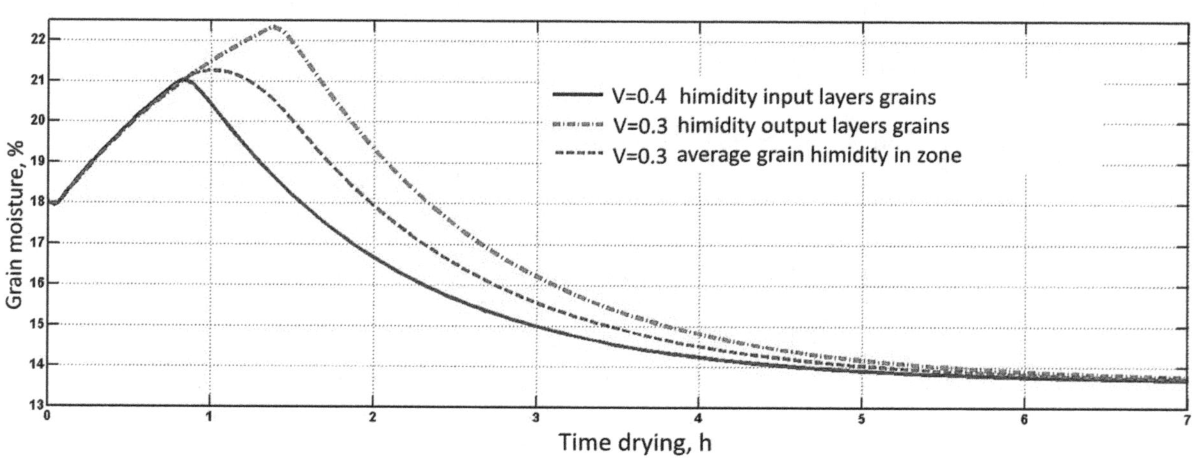

In the second stage, grain drying process was modeled with changing parameters of ambient air. Drying agent parameters change in the input of grain layer, in the process of modeling is shown in Figure 12.

Ta is ambient air temperature, Tk is temperature in the output of electric fan heater, i.e. in the input of grain layer, Fa is ambient air relative humidity, Fk is relative humidity of air in the output of electric fan heater, i.e. in the input of grain layer.

Aeration process can be started in any hour of the day depending on operational need which can cause a change in drying process behavior in dense gran layer.

In the first variant, modeling was carried out for drying process launch moment corresponding to the condition when air relative humidity exceeded 80% and continued its growth. In these conditions, electric fan heater is active and it heats ambient air so that its relative humidity value is within 65%. Results of modeling show that humidification of air passing through the grain layer is a rather fast process. In the output from the first section, air relative humidity is much higher than 65% which makes drying the next grain layers impossible.

Analysis of grain drying graphs for different sections (see Figure 13) makes it evident that, in five hours after starting the drying process, in the first section, grain attains its standard moisture content of 14%. All the time after this moment, grain is getting overdried while in the other sections drying pro-

Figure 11. Grain drying curves for the third zone, in direction of air flow, of three-sectional drying bin

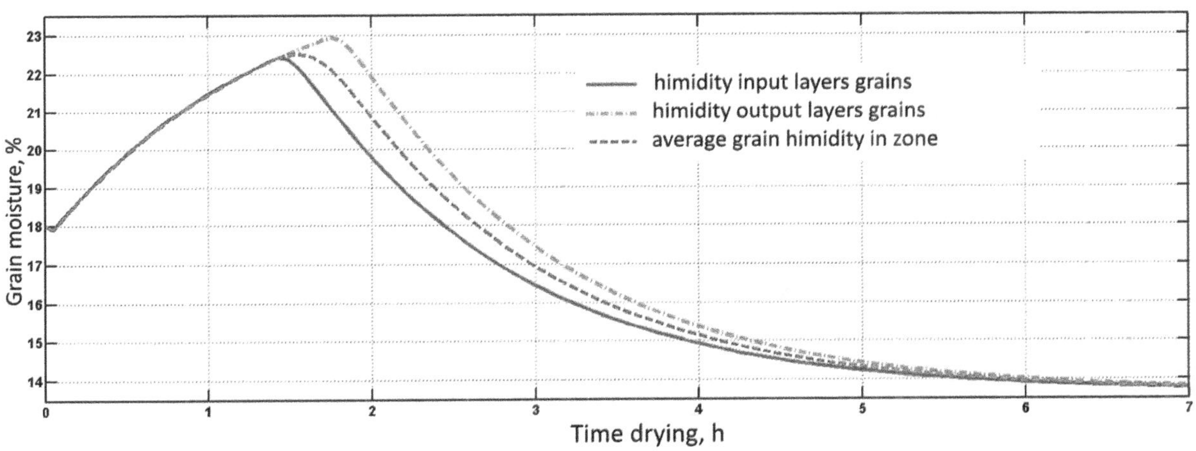

Figure 12. Change of drying agent parameters in the process of modeling

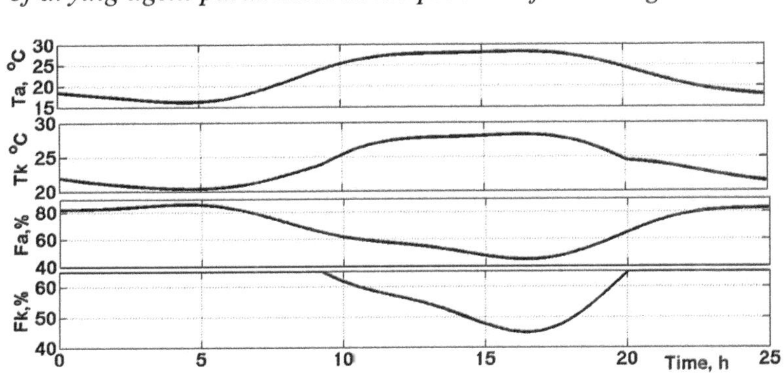

Figure 13. Change of grain average moisture content within a section in conditions of varying ambient air parameters (variant 1)

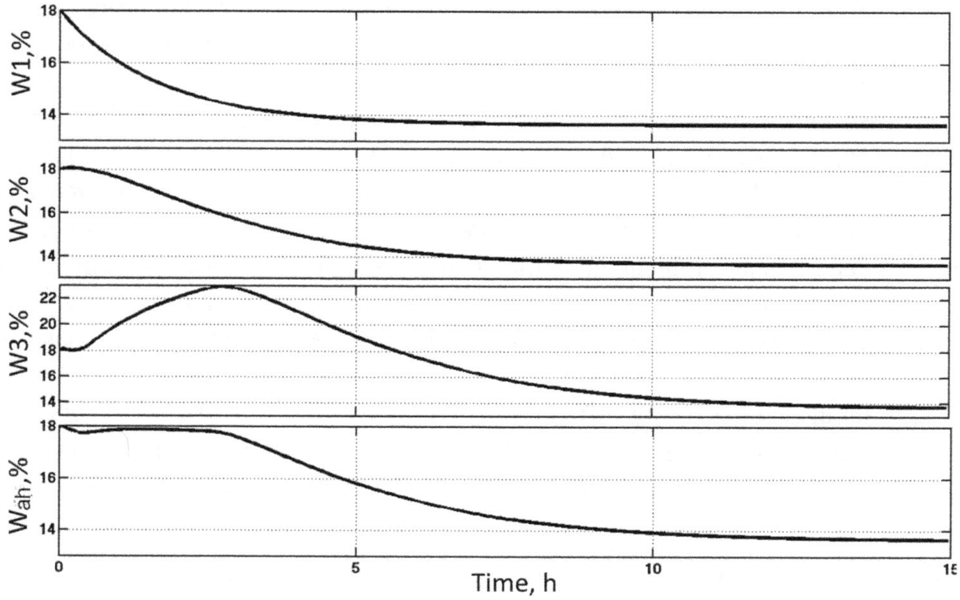

cess has not yet started. Moreover, humidification of grain layer takes place because of high moisture content of drying agent.

W1 is average moisture content of grain in the first section of drying bin, in the direction of air flow, *W*2 is that in the second section, *W*3 is that in the third one, *W*ah is average grain humidity in the hopper.

In the second variant, modeling process is started in the moment when ambient air relative humidity drops below 65%. Thereafter, during the entire drying period, ambient air natural relative humidity values are acceptable for forced aeration drying without the need to switch on electric fan heater. Results of modeling of grain drying process in different sections are shown in Figure 14.

It is clear from graphs that, in the first zone, grain become dry within less than 2.5 h. In the second zone, grain first becomes wet and then attains its standard moisture content during a period of 5.5 h. In the third zone, drying process has a more complicated behavior, and grain becomes dry attaining its standard moisture content value 14% only after 10 hours of drying. Average value of grain moisture content over the entire forced-aerated drying bin attains its standard level of 14% within 7 h period. Results of modeling give reason to conclude that the behavior of drying process, time of drying and, consequently, energy consumption of drying process depend essentially on the choice of aeration starting moment. Therefore, performance of forced-aerated drying bin can be increased owing to improvement of its design.

Results of modeling enable to define certain variants of grain transfer from one section of sectional forced-aerated drying bin to another. In the first variant of transfer, dry grain is transferred from the first section to the storage technological line or to a post-drying treatment line. Grain from the second section is transferred into the first section while grain from the third section is transferred into the second one (see Figure 15).

Figure 14. Results of modeling grain drying process in conditions of changing drying agent parameters (variant 2)

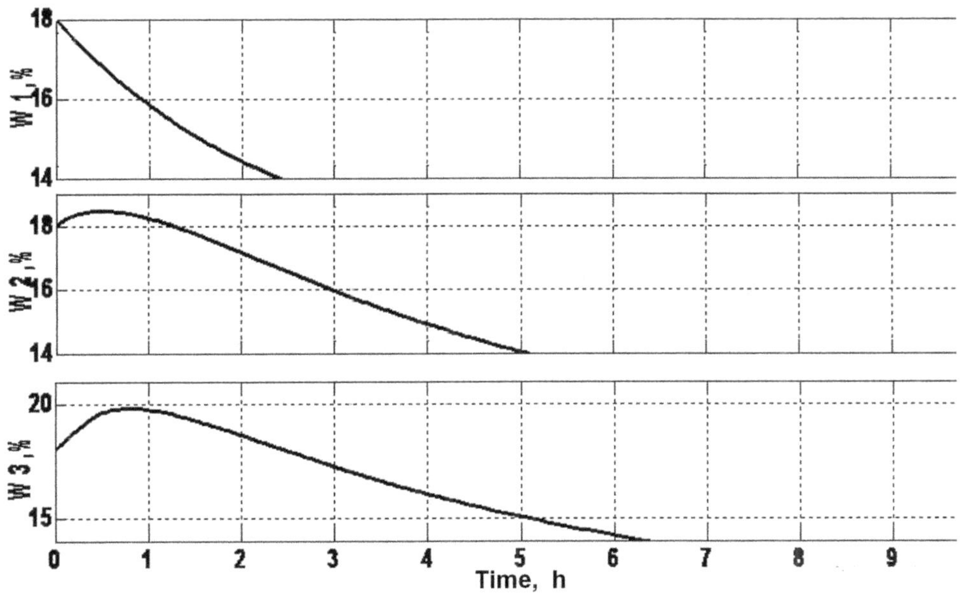

This variant of grain transfer from one section to another does not increase significantly performance of forced-aerated drying bin but it helps to avoid grain overdrying, in the first zone. Besides, grain transfer also contributes to moisture content reduction. At the same time, it has to be considered that frequent transfer of grain may result in its additional damage. That is why this variant can not be recommended for seed-dadicated grain drying.

Another variant of grain transfer from one section of drying bin to another includes grain aeration during 3 hours more, until average moisture content of grain in the first and second sections attains its value of 14%. After that, grain is loaded off from these two sections and grain from section 3 is transferred into section 1 while sections 2 and 3 are charged with fresh portions of grain. As a result, the aggregate volume of grain that attains its standard moisture content of 14% within 15 h will be increased by 33% compared to conventional technology of forced-aerated drying. This method of grain loading and unloading between aeration bin sections can be applied for various aeration launch moments including conditions of both decreasing and increasing relative humidity of ambient air.

It is advisable, for tree-sectional forced-aerated drying bins, to install sensors of drying agent and grain parameters monitoring close to the walls of the first and the second concentric cylinders. It makes it possible to implement various algorithms of drying process control while controlling drying agent parameters.

Figure 15. Variant of grain transfer from one section of drying bin to another

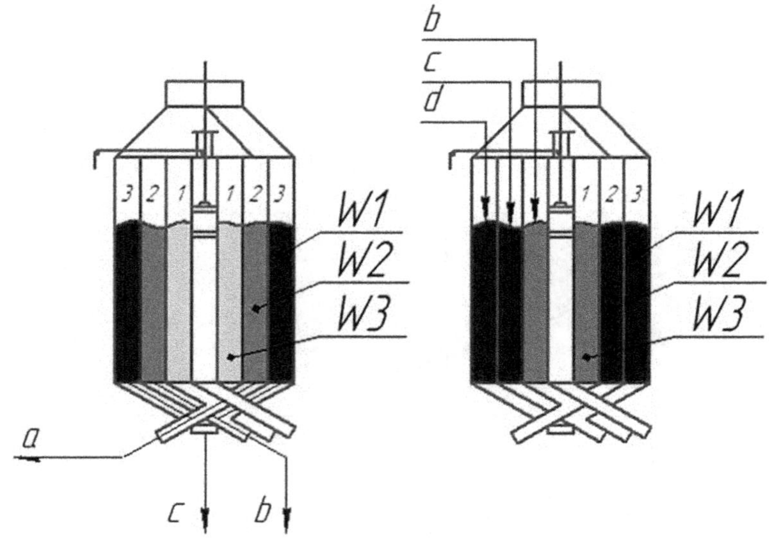

FURTHER RESEARCH DIRECTIONS

Using the developed mathematical model of grain drying allows to optimize the process. As a criterion, the drying speed, or energy consumption, can be adopted. The combination of modeling and sectional bins of active ventilation allows to reduce energy consumption of the process and increase its productivity. The development of optimal control systems using the results presented in this chapter is the main direction of further research.

CONCLUSION

Informational and system-based approaches to the analysis of grain drying process as an interaction of «drying agent» and «grain layer» systems enabled to evaluate informational content of the process. The obtained numeric values of amount of available information make it possible to calculate the number of required channels to transfer data related to the grain drying process and ambient air parameters.

The obtained analytic expression for the change of informational entropy of grain layers made it possible to conclude that parameters of both ambient air and drying agent in its output from the grain layer, as well as those of drying agent and grain in one point of grain layer have to be controlled, for an adequate representation of grain drying process. It requires application of relevant mathematical model of heat-and-moisture exchange in dense grain layer.

Mathematical model for optimal control of grain drying by forced aeration has been developed based on heat balance principal that provides an effective tool for process control upon minimal time of drying criterion. With the help of this model, it was found out that sensors for monitoring drying agent and grain parameters have to be located at a distance of 0.10 to 0.11 m from the central air distribution duct of forced-aerated drying bin.

Application of step-by-step method of heat-and-moisture exchange calculations for thick grain layer makes it possible to build a model of drying that can describe processes inside the grain layer. By choosing the value of elementary grain layer equal to that of single seed it is possible to perform process modeling for changing aira velocity in any point of grain layer.

Representation of drying process mathematical model in form of transfer functions provides an opportunity to simplify developing computational model of drying. The developed mathematical and computation models of grain drying in dense layer adequately describe the process of heat-and-moisture exchange. As a result of modeling, it has been proved that sensors of drying agent and grain parameters have to be installed at a distance of 0.10 to 0.11 m from the central cylinder of forced-aerated drying bin.

Modeling of grain drying process in sectional forced-aerated drying bin made it possible to develop a number of process control algorithms. It has been shown that, for a three-sectional forced-aerated drying bin, it is advisable to install sensors of drying agent and grain parameters monitoring close to the walls of the first and the second concentric cylinders. It will provide a possibility to apply various drying process control algorithms by controlling parameters of drying agent.

REFERENCES

Anisimova, L. V. (1998). *The Russian patent N° 2121782*. Moscow, Russia: Federal Institute of Industrial Property.

Aniskin, V. I. (1972). *Teoriya i tekhnologiya sushki i vremennoj konservacii zerna aktivnym ventilirovaniem* [Theory and technology of drying and temporary preservation of grain by active ventilation]. Moscow, Russia: Kolos.

Aniskin, V. I., Okun', G. S., & Vercman, I. I. (1988). Modelirovanie processa nizkotemperaturnoj sushki zerna v sloe s pomoshch'yu EHVM [Modeling the process of low-temperature drying of grain in a layer with the help of a computer]. Collection of scientific papers VIM, 115, 25-29.

Avdeev, A. V., Nachinov, D. S., & Nachinova, A. A. (2009). Modelirovanie processa i rezhimov konvektivnoj sushki zerna [Modeling of the process and regimes of convective grain drying]. Vestnik Orlovskogo gosudarstvennogo agrarnogo universiteta, 16(1), 39-43.

Chang, C. S., Converse, H. H., & Steele, J. L. (1993). Modeling of temperature of grain during storage with aeration Trans. ASAE. -St.Joseph(Mich.), 36(2), 509-519.

Dimitrov, N. (2014). *Effect of the storage duration and moisture on the temperature pattern of respiration of soft wheat grain*. University of Food Technology.

Dyakonov, V. P. (2008). *SIMULINK 5/6/7. Samouchitel* [SIMULINK 5/6/7. Tutorial]. Moscow, Russia: DMK-Press.

Egorov, G. A. (1973). *Vliyanie tepla i vlagi na processy hraneniya i pererabotki zerna* [Influence of heat and moisture on the processes of storage and processing of grain]. Moscow, Russia: Kolos.

Filatov, A. S., & Paulys, V. Y. U. (2017) Ustrojstva aktivnogo ventilirovaniya zernovyh mass pri hranenii [Devices of active ventilation of cereal masses during storage]. In The *collection of topical issues of science and economy: new challenges and solutions Collection of materials of the LI International Student Scientific and Practical Conference* (pp. 204-206). Academic Press.

Ginzburg, A.C. (1973). *Osnovy teorii i tekhniki sushki pishchevyh produktov* [Fundamentals of the theory and technology of food drying]. Moscow, Russia: Pishchevaya promyshlennost.

Gracheva, N.N., Rudenko, N.B., Kononenko, A.F., & Litvinov, V.N. (2017). *Primenenie EHVM v agronomii. Laboratornyj praktikum* [The use of computers in agronomy. Laboratory Practice]. Zernograd, Russia: Donskoj gosudarstvennyj agrarnyj universitet.

Gulyaev, G. A. (1990). *Avtomatizaciya processov posleuborochnoj obrabotki i hraneniya zerna* [Automation of post-harvest processing and storage of grain]. Moscow, Russia: Agropromizdat.

Gurov, I. P. (2000). *Osnovy teorii informacii i peredachi signalov* [The fundamentals of information theory and signaling]. St. Petersburg, Russia: BHV-Peterburg.

Hilmi, D. (2010). *The Fundamental Properties of Information-Carrying Relations. In Thinking Machines and the Philosophy of Computer Science: Concepts and Principles.* Hershey, PA: IGI Global. doi:10.4018/978-1-61692-014-2.ch002

Lauva, A., Aboltins, A., Palabinskis, J., & Karpova-Sadigova, N. (2006). *Grain drying experiments with unheated air.* University of Agriculture.

Lykov, A. V. (1968). *Teoriya sushki* [Theory of drying]. Moscow: Ehnergiya.

Machihina, L. I., Ushakov, T. I., Lvova, L. S., & Denisova, E. V. (2012). Nauchnyj analiz hraneniya zerna v metallicheskih silosah [Scientific analysis of grain storage in metal silos]. *Hleboprodukty*, *9*, 54–59.

Makushin, A.N. (2012). Izmenenie himicheskogo sostava zerna i tekhnologicheskih dostoinstv krupy pri hranenii zerna sortov prosa [Change in the chemical composition of grains and technological advantages of cereals during storage of grains of millet varieties]. *Izvestiya Samarskoj gosudarstvennoj sel'skohozyajstvennoj akademii, 4,* 82-86.

Mashukov, A. S., & Safonov, Y. A. (2017) Sushka zerna aktivnym ventilirovaniem s ispol'zovaniem reguliruemogo ehlektroprivoda ventilyatorov [Drying the grain with active ventilation using an adjustable electric fan drive]. In The collection of Achievements of Science - Agriculture Materials of the All-Russian Scientific and Practical Conference (correspondence) (pp. 317-322). Academic Press.

Mehdi, K. (2017). *Encyclopedia of Information Science and Technology* (4th ed.). Hershey, PA: IGI Global. doi:10.4018/978-1-5225-2255-3

Melynik, B. E. (1986). *Aktivnoe ventilirovanie zerna: Spravochnik* [Active grain ventilation: Handbook]. Moscow, Russia: Agropromizdat.

Nejadi, Mohammad, & Nikbakht. (2017). Numerical Simulation of Corn Drying in a Hybrid Fluidized BedâInfrared Dryer. *Journal of Food Process Engineering.*

Onhonova, L. O., & Skryabina, I. Y. U. (2017). Ustrojstvo dlya aktivnogo ventilirovaniya i pnevmovygruzki sypuchego materiala [Device for active venting and pneumatic unloading of bulk material]. *Vestnik VSGUTU*, *3*(66), 60–64.

Ospanov, A.B., Vasilyev, A.N., Budnikov, D.A., Karmanov, D.K. & Vasilyev, A.A. (2017). *Improvement of grain drying and disinfection process in the microwave field. Monography.* Almaty, Kazakhstan: Nur-Print.

Petraq, P., & François, P. (Eds.). (2015). *International Journal of Agricultural and Environmental Information Systems*, *7*(2). doi:10.4018/IJAEIS

Petrik, G. G. (2011). O sistemnom podhode k modelirovaniyu ob"ektov i ih svojstv na molekulyarnom i termodinamicheskom urovnyah [On the system approach to the modeling of objects and their properties at the molecular and thermodynamic levels]. *Monitoring. Science and Technology*, *2*, 86–101.

Sabiev, U. K., & Shonov, A. E. (2017). Obzor i analiz bunkerov aktivnogo ventilirovaniya zerna [Review and analysis of active grain ventilation bins]. In *Collection of materials of the XXIII scientific and technical conference of students of the FGBOU VO OMSK GAU* (pp. 104-107). Academic Press.

Severinov, O. V. (2016). *Sovershenstvovanie processa sushki zerna v bunkerah aktivnogo ventilirovaniya s ispol'zovaniem modelirovaniya* [Improvement of the process of grain drying in active ventilation bins using simulation] (Dissertation of the candidate of technical sciences). VIESH, Moscow, Russia.

Severinov, O.V., Rudenko, N.B., Budnikov, D.A. & Macynin, S.N. (2012). Bunker dlya hraneniya i ventilirovaniya zerna [Bunker for storing and venting grain]. *Patent for utility model RUS 113631 byul., 6.*

Tarasenko, F.P. (1963). *Vvedenie v kurs teorii informacii* [Introduction to the course of information theory]. Tomsk, Russia: Izdatel'stvo Tomskogo universiteta.

Vanckul, A. S., Goryacheva, E. M., Garyaev, A. B., & Sorochinskij, V. F. (2017). Issledovanie vliyaniya razlichnyh faktorov na vremya ohlazhdeniya sloya zerna v metallicheskih silosah [Investigation of the influence of various factors on the cooling time of the grain layer in metal silos]. In The *collection Modern problems of engineering sciences, Collection of scientific works of the VI International Scientific and Technical Symposium, International Scientific and Technical Forum* (pp. 101-103). Academic Press.

Vasiliev, A.N., Ospanov, A.B., Budnikov, D.K , Karmanov, D.K, Salginbayev, D.B. & Vasilyev, A.A. (2016). Controlling reactions of biological objects of agricultural production with the use of electrotechnology. *International Journal of Pharmacy & Technology, 8*(4), 26855-26869.

Vasilyev, A. N. (1986). Statisticheskie harakteristiki parametrov atmosfernogo vozduha dlya aktivnogo ventilirovaniya semyan [Statistical characteristics of atmospheric air parameters for active seed ventilation]. Automation and computer facilities in agricultural production - Collection of scientific papers MIISP, 77-82.

Vasilyev, A. N. (2005). Kosvennyj kontrol' vlazhnosti i temperatury zerna pri sushke v plotnom sloe [Indirect control of moisture and temperature of grain during drying in a dense layer]. Mechanization and Electrification of Agriculture, 3, 5-7.

Vasilyev, A. N. (2006). Kriterij optimal'nosti i granichnye usloviya v modeli sushki plotnogo sloya [The optimality criterion and boundary conditions in the dense layer drying model]. Energy-saving technologies in the agroindustrial complex (All-Russian scientific-practical conference, collection of articles), 14-17.

Vasilyev, A. N. (2008). *Ehlektrotekhnologiya i upravlenie pri intensifikacii sushki zerna aktivnym ventilirovaniem* [Electrotechnology and management in the intensification of grain drying by active ventilation]. Rostov-on-Don, Russia: Terra Print.

Vasilyev, A. N. (2009). *EHnergosberegayushchie ehlektrotekhnologii sushki i predposevnoj obrabotki zerna aktivnym ventilirovaniem* [Energy-saving electrotechnology for drying and presowing grain processing by active ventilation] (Doctoral Dissertation). Moscow, Russia.

Vasilyev, A. N., Budnikov, D. A., Gracheva, N. N., & Severinov, O. V. (2016). *Sovershenstvovanie tehnologii sushki zerna v plotnom sloe s ispolzovaniem elektrotehnologiy, ASU i modelirovaniya protsessa* [Improvement of the technology of drying grain in a dense layer with the use of electrotechnologies, automated control systems and modeling of the process]. Moscow, Russia: FSBSI FSAC VIM.

Vasilyev, A. N., & Rudenko, N. B. (2009). Model' teplovlagoobmena pri recirkulyacii zerna [Model of heat and moisture exchange during grain recycling]. Vestnik agrarnoj nauki Dona, 37-41.

Vasilyev, A.N. & Severinov, O.V. (2014). Matematicheskaya model' teplo- i vlagoobmena v ehlementarnom sloe pri sushke zerna aktivnym ventilirovaniem [Mathematical model of heat and moisture exchange in the elementary layer during grain drying by active ventilation]. *AgroEHkoInfo, 2.*

Vasilyev, A. N., & Severinov, O. V. (2016). Ocenka ehffektivnosti usovershenstvovannogo algoritma upravleniya sushkoj zerna v bunkerah aktivnogo ventilirovaniya [Evaluation of the effectiveness of the improved algorithm for controlling grain drying in active storage bins]. *Theoretical and Scientific and Practical Journal Innovations in Agriculture, 3*(18), 165-171. Retrieved from http: //ej.viesh.ru/wp-content/uploads/2016/03/insel16.pdf

Vasilyev, A. N., Severinov, O. V., & Makarova, Yu. M. (2016). Razrabotka komp'yuternoj modeli teplo- i vlagoobmena v plotnom sloe zerna [Development of a computer model of heat and moisture exchange in a dense layer of grain]. *Vestnik NGEHI, 4*(59), 63–71.

Vasilyev, A.N., & Udincova, N.M. (2000). Analiz parametrov vozdushno-teplovoj obrabotki dlya uluchsheniya posevnyh kachestv semyan [Analysis of the parameters of air-heat treatment to improve the seed quality of seeds]. *Mekhanizaciya i ehlektrifikaciya sel'skogo hozyajstva, 1,* 9-10.

Verbickij, V. V., Maratov, B. K., Timoshenko, A. M., & Novoselov, S. V. (1989) Instrukciya po aktivnomu ventilirovaniyu zerna i maslosemyan (tekhnika i tekhnologiya) [Instruction on active ventilation of grain and oilseeds (machinery and technology)]. Moscow, Russia.

Vershinin, V.S. (2011) Izmenenie harakteristik vlazhnosti vozduha [Change of characteristics of humidity of air]. *Santekhnika, otoplenie, kondicionirovanie, 3*(111), 84-86.

VIESH. (1997). *Metodicheskie rekomendacii po matematicheskomu modelirovaniyu processa sushki i ohlazhdeniya zerna v ustanovkah plotnogo sloya* [Methodical recommendations on mathematical modeling of the process of grain drying and cooling in installations of a dense layer]. Moscow, Russia: VIESH.

Yakovlev, A. A. (2005). O sistemnom podhode k formirovaniyu mnozhestv tekhnicheskih reshenij preobrazovatelej ehnergii [About the system approach to the formation of sets of technical solutions for energy converters]. *News of higher educational institutions. Mechanical Engineering (New York, N.Y.)*, *7*, 44–50.

Zelenko, V.N. (1988). *Konvektivnaya sushka sel'skohozyajstvennyh materialov v plotnom sloe. Osnovy teorii* [Convective drying of agricultural materials in a dense layer. Fundamentals of the theory]. Tver', Russia: Tverskoe oblastnoe knizhnoe izdatel'stvo.

Zhidko, V. I., Rezchikov, V. A., & Ukolov, V. S. (1982). *Zernosushenie i zernosushilki* [Drying of grain and grain dryer]. Moscow, Russia: Kolos.

KEY TERMS AND DEFINITIONS

Active Ventilation: Blowing atmospheric air through a grain layer.

Active Ventilation Bunker: A vertically located cylindrical container, which is filled with grain. In the center of the tank there is an air duct through which air is fed into the grain layer.

Mathematical Model: Mathematical description of the relationship between process parameters.

Process Informativeness: The amount of information that control means receive when measuring process parameters.

Chapter 9
Analysis of Experimental Research Results Obtained for Grain Drying With Electrically Activated Air

Alexey Nicolaevich Vasiliev
https://orcid.org/0000-0002-7988-2338
Federal Scientific Agroengineering Center VIM, Russia

Alexey Alexeevich Vasiliev
Federal Scientific Agroengineering Center VIM, Russia

Dmitry Budnikov
https://orcid.org/0000-0002-8729-1300
Federal Scientific Agroengineering Center VIM, Russia

Gennady Samarin
https://orcid.org/0000-0002-4972-8647
Federal Scientific Agroengineering Center VIM, Russia

ABSTRACT

This chapter presents an analysis of factors influencing the heat and moisture exchange for their further use in experimental research to study the process of grain drying with active ventilation. The temperature and velocity of the drying agent and grain moisture content are determined. One centimeter (1 cm) thick layer is considered to be a thin layer. This size appears to be very manageable to transfer the regularities of the drying process to a thicker layer when modeling the drying process in a thick layer. The experiment was carried out for three drying modes: classical mode (i.e., drying of grains with natural and heated air), with a constant concentration of air ions in the drying agent, with cyclic (periodic) presence of air ions in the drying agent. Using the regression equations produced a so-called electro-activation criterion can be described. The dependencies obtained would help to optimize the drying process with respect to drying time criterion.

DOI: 10.4018/978-1-5225-7573-3.ch009

INTRODUCTION

In terms of energy intensity grain drying process holds about 30% of all energy consumption in grain crop production. Energy consumed by the process can be reduced due to the improved design of grain drying equipment, use of various sources of thermal energy, improving process control algorithms and application of electrical technologies. The material proposed presents an application when the electrically activated air is used to increase the grain drying efficiency. Electrically activated air (EAA) means the air treated with the high voltage electrical discharge, and as such, the air enriching with air ions occurs. Using this air as a drying agent will allow increasing the moisture release rate. EAA can be continuously or periodically supplied into the grain layer. If periodically, an alternation of EAA and a conventional drying agent will take place. It is necessary to determine the efficiency of these drying modes and the conditions for their usage.

In order to decide on the best grain drying modes using EAA, it is necessary to find the functional dependencies between grains drying time and drying agent parameters when using electrically activated air. Presently, there are no theoretical studies to identify such dependencies. Therefore, experimental data are used as a basis to find such relationship.

Today, there is no sufficiently clear answer about the grounds for the positive effect of EAA on the grain drying rate. Therefore, another task of the study is to find such explanation.

ANALYSIS OF GRAIN DRYING METHODS USING THE AIR ENRICHED WITH AIR IONS

There are several ways of drying the grain materials using electrically activated air of different composition (Solodova (2002); Tihen'kij (1993); Tkachev (2000); Madden (1987)). The drying methods using EAA include those using ozone-air mixes (OAM) and air enriched with air ions. The proposed analysis aims at identifying the methods, which provide the greatest intensification of the release of internal tightly bound moisture from the grain. At the same time, low energy consumption, preservation of the dried material quality and simple implementation in production plants should be achievable.

Drying involving air ions helps to reduce the duration and energy intensity of the process. To obtain the air containing air ions, corona discharge (CD) ionizers are used in active ventilation bins in the drying agent preparation lines and floor and chamber dryers (Aniskin & Chizhikov, 2002).

The scientists in Chelyabinsk (Krejmeris, Tryukas & Tryukene, 1991) have developed a method for drying the grain seeds in corona discharge (CD) field. The seeds are placed in the produced corona discharge field $E = \left(3.5 - 5.0\right) \cdot 10^5$ V/m and blown with a natural air. As a result, 3 to 6% of free moisture can be released from the seeds.

The disadvantages are low plant performance and extra high electrical hazard for personnel servicing the integrated high-voltage equipment.

There is a combined method known for drying the grain materials (Ilyuhin, Noskov & Babakin, 1983), when a grain layer is fed with a constant electrical potential (of different sign) and blown with an ionized drying agent. Appropriate experiments have been conducted. As a result, it has been found that ionizing the drying agent and imparting it an electrical potential of the opposite or the same sign as the material would allow varying the material drying intensity.

The disadvantage of the above method is its complexity for implementation in the process equipment and low efficiency when drying the low moisture grains.

To produce EAA, various types of ozone generators are used Gorskij (2004). The corona discharge-type and barrier discharge-type ozone generators are the most popular today. Corona discharge sources are simple in design, small-size, unlimited in the discharge range, and effective in producing the air ions. Moreover, they can produce small ozone concentrations (Gorskij, 2004). Air ions do not create hazards for servicing personnel due to their low concentrations and a short life period.

Chizhikov & Golubkovich (2005) was the first who observed the effect of air ions passing through the electric discharge on living organisms. At present, many scientists are concerned with the effect of ozone, ozone-air mixes and air ions on the grain drying process (Table 1).

According to the table above, grain materials are dried using ozone and air ions. The ozone concentrations used for intensification of the process exceed the maximum allowable concentration (MAC), (GOST; Abhijit Sarkar, 2017), which is a significant disadvantage. When drying with air containing air ions, the air ion concentration stays within the MAC range, or slightly exceeds the established limits. Thus, the air enriched with air ions can be used for grain drying with EAA, the concentration of air ions being $3.5 \cdot 10^{10} m^{-3}$.

All these drying methods are suitable for drying grain materials containing moisture content to 30%. With respect to intensification of grain drying process the methods using the air containing air ions are competitive with the effect of ozonized air.

The methods described demonstrate that grain drying with air ions is on the same efficiency level as the grain drying methods using ozone and ozone-air mix. In terms of effect on a biological object, the use of electrically activated air is considered to be equivalent to low-energy effects (Vasilyev, 2008a). Following the maximum information principle of Vasilyev (2008b); Vasiliev et al., (2016); Ospanov, Vasilyev, Budnikov, Karmanov & Vasilyev (2017), it can be assumed that cyclic use of EAA as a drying agent will lead to an intensification of the process. To answer the questions posed, a factor experiment was designed and conducted.

FACTORS AND VARIABILITY LEVELS

To perform experimental research at the first stage it is necessary to determine the main factors. It is also necessary to establish their variation range and levels. Grain drying depends on such process parameters as: initial moisture content of the grain layer W_n; velocity of air passing through the grain layer V_a; drying agent temperature T_a; drying agent relative humidity F; and grain heating temperature Θ.

Depending on the weather conditions and time of harvesting grain crops, grains often enter the field floor with moisture content exceeding the conditioning level. The moisture content of the grains during harvesting varies from 16% to 35% in different years. Grains containing the initial moisture content to 24% are dried in active ventilation plants. Therefore, for the experiment, it is advisable to use grains containing initial moisture content W_n of 16% to 24%.

Active grain ventilation means drying the grains with natural air or air heated by 4 ... 7°C. Active ventilation of fresh harvest grains is carried out in the summer months, when natural air temperature can be above 30 °C. Therefore, we set the drying agent temperature variation within the range of 20°C to 34°C.

Table 1. Grain drying using EAA

Author	EAA Source	Type of Action	Air Ion Concentration	Treatment Mode	Effect Achieved
1	2	3	4	5	6
Kreimeris I.B. Litovsky NIIMESH	barrier discharge-type ozone generator	ozone-air mix	15–30 mg/m³	Periodic T=1–2 h W_z=25%; H_z=3 m Air flow – 850–1500 m³/h	Reduced *energy consumption* by 1.86 times
Gluschenko L.F., Gluschenko N.A. NovGU	corona discharge-type ozone generator	ozone-air mix	2–12 mg/m³	W_z=29–30.3%; H_z=0.2 m	Reduces drying duration by 1.2–1.5 times
Trotskaya T.P. Grodnensky SHI	barrier discharge-type ozone generator	ozone-air mix	0.23–10,00 mg/m³	W_z=23.0–30.5%; Air flow – 1000–1500 m³/h · t t_a=20–24 °C	Reduces drying duration by 1.5–2.0 times
Ksenz N.V., Shtanko R.I.	corona discharge-type ozone generator	ozone-air mix	2.4–10.1 mg/m³	W_z=29–35%; H_z=0.21–0.43 m	Reduced *energy consumption* by 1.7–2.2 times
Golubkovich A.V., Shampanova L.V. VIM	corona discharge-type ozone generator (1986)	air ions	30·10³–500·10³ m⁻³, 3–7 mg/m³	W_z=20–30%; Air flow – 200—300 m³/h; High temperature drying, fluid grain bed	Reduces drying duration by 8–15%; reduced *energy consumption* by 18–20%
Golubkovich A.V., Chizhikov A.G. VIM	barrier discharge-type ozone generator	ozone-air mix	3–7 mg/m³	W_z=20–25%	Reduced *energy consumption* to 25%
Rudobashta S.P., Nuriev N.N. MGAU	barrier discharge-type ozone generator	ozone-air mix	4–60 mg/m³	H_z=0.013; 0.023 m; t=20–45 °C; W_z=15–22%; Air flow – 200–700 m3/h	Reduced *energy consumption* by 20–30%
Borodin I.F. Tkachev R.V. MGAU	corona discharge-type ozone generator	ozone-air mix	to 10 mg/m³	W_z=18, 24%; t_z=24, 40 °C; E=10 50 100,V/m	Reduced *energy consumption* by 1.2-1.8 times
Pavlyk V.A., Tihenky V.I. NPO Nechernozemagromash	corona discharge-type ozone generator «Ion-2»	air ions	6,54·10²⁰–5·10²¹ m⁻³	W_z=20–30%; t_z=20–40 °C; air flow – 10000 m³/h	Reduces drying duration by over 1.3 times

The drying agent temperature T_a is a value closely connected with its humidity. Coupling equation of these factors is written as (Kireev, Grigor'ev & Koval'chuk, 1981):

$$F = a_0 + a_1 \cdot T_a, \tag{1}$$

where F – drying agent humidity, %; T_a – drying agent temperature, °C; a_0, a_1 – slope of the linear effect.

According to Sakun (1974), heating the air by 1°C leads to drop in its humidity by 5%, so the drying agent relative humidity is not an independent factor, and will not be considered in the experimental studies.

The velocity of air passing through the grain layer V_a has a significant effect on the heat and moisture exchange in the drying process. The drying time depends on it. In active ventilation installations, depending on their design, the drying agent velocity varies between 0.1 and 1.5 m/s. So, in the bin-type installations, due to radial air distribution, its velocity varies between 0.8 m/s at the central air duct, to 0.1-0.2 m/s beside the outer cylinder wall. In the floor-type installations using active ventilation, the drying agent velocity depends on the thickness of the ventilated layer and can reach 1.5 m/s at a low height of the grain layer. In view of the above, in our experimental studies we set a variation of the drying agent velocity within the range of 0.6-1 m/s.

Active ventilation refers to soft and long methods of drying. The grain layer is long ventilated by natural air. In this case the grain temperature Θ finally becomes equal to the temperature of natural air, so the grain temperature Θ is not an independent factor, and its variation range will not be set in the experiment.

The moisture that the grain layer will acquire during the drying process not only depends on the air parameters, but also on the time of air effect on the grain, i.e. on the drying time τ. This factor is absolutely independent and is determined by the strategic target of the drying process. If the target is to dry the grains to the conditioning moisture content, then the drying time will be such that this target is achieved. If the target is to reduce the grain moisture content to a certain percentage in order to increase the safe storage period, then the drying time will be set accordingly. As it has been assumed that the grains containing a moisture content of 16 to 24% will be used for experimental studies, the drying time should also be such that the grains are dried to the conditioning moisture. However, it should be taken into account that the grain drying time for a dense fixed layer depends on its thickness. Therefore, before determining the drying time, the grain layer thickness should be determined during the experimental studies.

One of the research tasks is to describe the processes occurring in the grain layer during its drying with EAA. As the drying is performed in a dense layer, the drying zone moves along the layer as the layers located closer to the drying agent entry point are dried. The drying process passes successively from one layer within the drying zone to the next one. The drying process pattern remains unchanged in all consecutive layers being dried. Therefore, there is no need to perform experimental research on a thick grain layer, as it happens in the existing installations. It would be sufficient to determine the regularities of the drying process on a thin layer. A layer 1 cm thick is considered to be such a thin layer. This size appears to be very manageable to transfer the regularities of the drying process to a thicker layer when modeling the drying process for a thick layer. However, it should be borne in mind that during the experiment sampling of grains is required to control moisture content, so the grain volume should be sufficient to ensure the control accuracy. Based on the design of the research facility and the specified conditions, the grain layer thickness is assumed to be 8 cm.

Preliminary experimental studies have demonstrated that the drying time of such a grain layer, depending on the parameters of the drying agent and the initial grain moisture content, will fluctuate within the range of 1 to 6 hours. This is a very wide range of factor variation, and when fully used it may prevent from obtaining adequate models. Therefore, we preferred to split the drying time into two intervals of 1 to 2 hours, when the grains containing low initial moisture content should dry out at a high drying agent velocity, and 2 to 6 hours.

To obtain adequate regression models, the ranges of all factors should be split into three levels: min (-1); medium (0); max (+1).

Thus, the following factors were established as independent significant factors for experimental research (Table 2).

One of the targets of experimental research is to plot drying curves $W = f(\tau)$ for three modes:

- Classical drying, when the drying layer is blown with natural or heated air;
- Grain drying with electrically activated air, when the drying agent is enriched with negative ions, and the drying agent concentrations do not change during the entire drying process;
- Grain drying in the mode of cyclic variation of the ion concentration in the drying agent. In this case, the drying agent is enriched with air ions not continuously, but periodically, at regular intervals. Thus, the grains are dried with air without air ions for five minutes, then with air ions for another five minutes and then again without air ions, etc. The process continues until the grain layer completely dries up.

Air ion concentration in the drying agent was provided with an air ion source at $3.5 \cdot 10^{10}$ m^{-3} (Fig. 1).

Such concentration was set based on the allowable air ion concentration in the air (Borodin, 2005). For grain drying using the air enriched with air ions, a corona discharge field of negative sign was used. The effectiveness of negative air ions used for drying of agricultural materials has been proven by many researchers - Vasant, Alparslan-Gok and Weber (2018); Tanimura, Hirotsuji, Tanaka & Asakura, (2003); Kharchenko & Vasant (2018).

When setting the cyclic mode for enrichment of the drying agent with air ions, the half-cycle time was set to five minutes.

Thus, all independent factors and variability levels were determined.

GRAIN DRYING REGRESSION MODELS FOR DIFFERENT MODES

In accordance with the experimental research design a multi-factorial experiment was conducted to determine how the drying agent parameters (T_a; V_a) influence on changes in grain moisture content, the initial grain moisture content W_n, and the drying time τ. The experiment was carried out for three dry-

Table 2. Independent significant factors for experimental research variability levels

Factors	Variability Levels		
	-1	0	1
Duration of drying, τ, h.	1	1.5	2
	2	4	6
Grain moisture content, W_n, %	16	20	24
Drying agent temperature, T_a, °C	20	27	34
Drying agent velocity, V_a, m/s	0.6	0.8	1

Figure 1. Appearance of the source of aeroions
1 - electrodes in the form of needles; 2, 3 - electrodes in the form of a grid

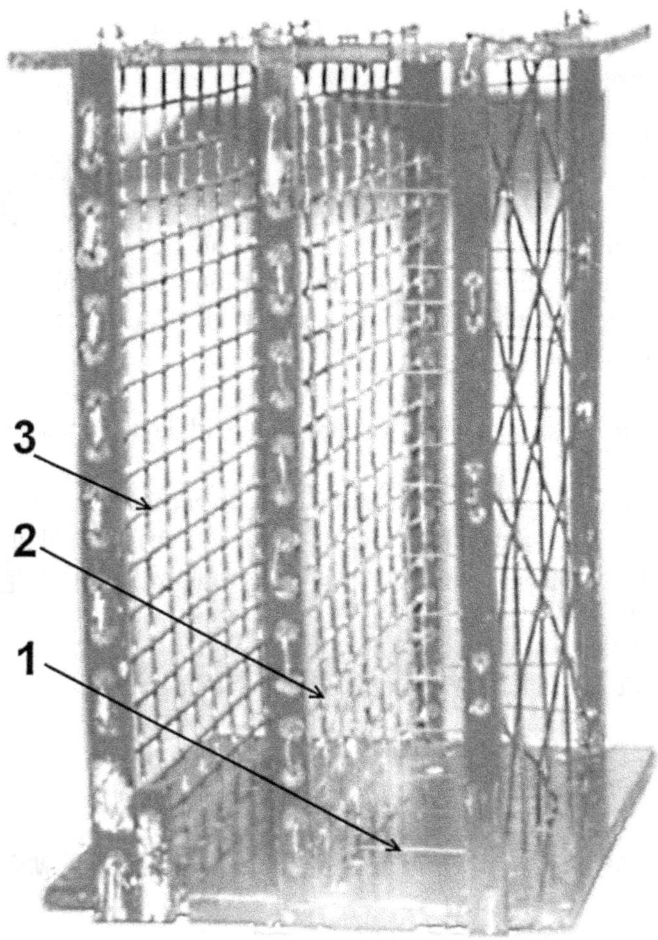

ing modes: classical mode, i.e. drying of grains with natural and heated air; with a constant concentration of air ions in the drying agent; with cyclic (periodic) presence of air ions in the drying agent. As a response, we used the grain moisture content data. Some experimental data are provided in the Table 3.

To obtain the regression models we used a standard function MATLAB *regstats*, having the following syntax:

```
regstats(responses,DATA,'model').
```

As a consequence of multifactorial experiment data processing and after deleting non-significant regression coefficients we obtained the following equation:

$$W = -7.086 + 1.120\tau + 1.503W_n + 0.339T_a - 0.157\tau W_n$$
$$-0.026\tau T_a - 0.268\tau V_a - 0.023W_n T_a + 0.283\tau^2. \quad (2)$$

Table 3. Experimental data on the air parameters and air ions influence on the grain moisture content during drying

N°	τ,h	W_n %	T_a, °C	V_a, m/s	W, %		
					w/o Air Ions	Air Ions Continuously	Air ions Cyclically
1.	1	16	27	0.8	14.15	13.53	13.25
2.	1	24	27	0.8	20.1	19.93	19.1
3.	2	16	27	0.8	12.74	12.78	12.3
4.	2	24	27	0.8	17.15	16.85	15.95
5.	1,5	20	20	1	16.4	16.1	16.3
6.	1	20	27	0.6	17.27	17.23	16.58
7.	2	20	27	1	14.8	14.3	13.63
8.	1,5	16	20	0.6	14.09	14.2	13.8
9.	1	20	20	0.6	18.32	18.1	17.3
10.	2	20	34	0.6	14.14	14.05	13.35
11.	1,5	16	27	1	13.1	12.85	12.05
12.	6	24	27	0.8	12.28	12.13	11.7
13.	4	20	20	0.6	14.2	15.05	14.22
14.	4	24	20	0.6	16.8	16.6	15

Student's t-test table value for number of degrees of freedom: f=32, t^m=2.04. Student's t-test calculated values for regression coefficients respectively equal to t^p=(-5.8139, 5.5985, 23.573, 8.126, -16.034, -6.9234,-6.331, -10.486, 20.022). As a condition $t^p>t^m$ is met for all regression coefficients, then they all are considered to be significant.

The adequacy of the mathematical model was determined from the Fisher test for degrees of freedom f1 = 32, f2=8, with a confidence level of 95%. The calculated Fisher test "bottom" value equals to F^p= 0.0007. The table value, respectively, is F^m=2.93. A condition $F^p<F^m$ is met, then, the model adequately reflects the required dependence.

For evaluation of the quality of the approximation completed the *SSE* and *R-squared* criteria were also used. The *SSE* (Sum of squares due to error) criterion represents the sum of error squares. The fact that *SSE* is close to zero is indicative of the good quality of the data approximation to the parameter-oriented model. *R-square* criterion represents the coefficient of determination.

The *R-square* criterion can only take values from zero to one. As a rule, the closer it is to one, the better the parametric model describes correlations in the experimental data presented. The Root Mean Squared Error (*RMSE*) is also used for evaluation of the quality of approximation made. *RMSE* values close to zero mean a good approximation of the initial data with the parametric model.

The calculations made result in the following criteria values: *R-square* = 0.996; *RMSE* = 0.132, which confirm the high quality of description of the object with the regression equation produced.

Processing of experimental data obtained during multifactorial experiment conducted with continuous effect of air ions, and after deleting non-significant regression coefficients result in the following equation:

$$Wq = 2.845\tau + 1.117W_n - 0.222\tau W_n - 0.0443\tau T_a$$
$$+0.373\tau V_a - 0.154W_n V_a + 0.265\tau^2 - 0.001T_a^2 \quad (3)$$

Student's t-test table value for the number of degrees of freedom: f=33, t^m=2.04. The calculated values of Student's t-test for regression coefficients are, respectively, t^p=(21.485, 80.726, -34.236, -13.47, 3.8192, -10.737, 17.175, -8.8428). Since the condition t^p>t^m is met for all regression coefficients, all of them are considered to be significant.

The adequacy of the mathematical model was determined from the Fisher test for degrees of freedom f1 = 33, f2=7, with a confidence level of 95%. The calculated Fisher test "bottom" value equals to F^p=0.0009. The table value, respectively, is F^m=2.93. A condition F^p<F^m is met, then, the model adequately reflects the required dependence.

The calculations made resulted in the following criteria values: *R-square*=0.995; *RMSE* = 0.154, which just prove high quality description of the object with the regression equation produced.

An analysis of the equations produced demonstrated that the air enriched with air ions and used as a drying agent did not result in a fundamental change in the form of dependences. Equation (3) differs from the equation (2), however, even having these equations it is still difficult to make a conclusion why the grain drying rate changes when electrically activated air is used. When using EAA, the effect of the drying agent temperature is more exhibited. This is expressed with an additional subtrahend in the equation (3) including the agent temperature.

In accordance with experimental research design, data were processed for the mode with cyclic action of air ions. After removing the minor regression coefficients, the following equation is produced:

$$Wqc = -3.845 + 2.003\tau + 1.277W_n + 0.245T_a - 0.197\tau W_n$$
$$-0.032\tau T_a + 0.348\tau V_a - 0.014W_n T_a - 0.121T_a V_a + 0.268\tau^2 \quad (4)$$

Student's t-test table value for the number of degrees of freedom: f=31, t^m=2.04. The Student's t-test calculated values for regression coefficients are, respectively, t^p=(-3.4387, 10.731, 21.824, 6.3409, -22.024, -9.1924, 5.4486, -6.9489, -16.248, 20.665).

Since the condition t^p>t^m is met for all regression coefficients, all of them are considered to be significant.

The adequacy of the mathematical model was determined from the Fisher test for degrees of freedom f1 = 31, f2=9 with a confidence level of 95%. The calculated Fisher test "bottom" value equals to F^p= 0.0008. The table value, respectively, is F^m=2.90. A condition F^p<F^m is met, then, the model adequately reflects the required dependence.

The calculations made resulted in the following criteria values: *R-square* = 0.996; *RMSE* = 0.121, which just prove high quality description of the object with the regression equation produced.

Equation (4) differs from the equations (2 and 3). It contains an additional subtrahend $0.121T_a V_a$, which adds to an effect of the drying agent temperature and rate variation. However, based on this it is still difficult to make a conclusion why the grain drying rate changes when the air cyclically enriched with air ions is used.

Thus, the regression equations produced (2, 3, and 4) did not help to clarify the reasons for intensification of the grain drying process using the electrically activated air. However, these equations are of an applied nature and can be practically applied to calculate the required duration of the drying process using EAA. So, Ukolov equation (V.S. Ukolov) is known for the temperature range of 15 to 45°C, which is recommended for determining the drying time:

$$\Delta W / \tau = 14.8 \cdot 10^{-8} \cdot Q \cdot \left(T - 13\right), \tag{5}$$

where Q – actual specific air supplied, $m^3 / (h \cdot t)$; ΔW – difference between mean moisture content of the bulk grain before and after drying, %; T – drying agent temperature, °C.

As we can see from the equations (2), (3), (4) the grain drying rate depends on the initial grain moisture content and the drying agent velocity, therefore Ukolov equation does not represent all process specifics.

For more precise presentation of the grain drying rate with the parameters τ, W_n, T_a, V_a we use the regression equations (2), (3), (4). Then, we perform the differentiation by time and get the following expressions:

$$dW / d\tau = 1.12 - 0.157W_n - 0.026T_a - 0.268V_a + 0.566\tau \tag{6}$$

or

$$w'_\tau = 1.12 - 0.157W_n - 0.026T_a - 0.268T_a + 0.566\tau, \tag{7}$$

where w'_τ – grain drying rate, %/*h*.

Using this equation we can determine moisture content in the dried grains in a specified time period.

We should not forget that the experimental research has been performed for an elementary layer. As for a thick layer, the variation of the drying agent parameters should be considered when passing through the grain. Thick layer drying can be modeled based on this equation.

Using the same method, there were drying rate equations produced for the mode when the drying agent with a constant concentration of air ions and cyclic enrichment with air ions is used:

$$\left(w_q\right)'_\tau = 2.845 - 0.222W_n - 0.044T_a + 0.373V_a + 0.530\tau, \tag{8}$$

$$\left(w_{qc}\right)'_\tau = 2.003 - 0.197W_n - 0.032T_a + 0.348V_a + 0.536\tau, \tag{9}$$

where $\left(w_q\right)'_\tau$ – grain drying rate at a constant concentration of air ions in the drying agent, %/*h*; $\left(w_{qc}\right)'_\tau$ – grain drying rate at a variable concentration of air ions in the drying agent, %/*h*.

Analysis of the equations (8) and (9) does not either answer the question about the reasons why the grain drying rate varies when the drying agent is enriched with air ions.

It happens very often that application of thermodynamic criteria for presentation of heat and moisture exchange helps to evaluate the driving force of the process. Therefore, we then examined the models produced using thermodynamic criteria.

Table 4. Data for regression analysis of the grain drying process without air ions

Nº	τ	W_n	T_a	V_a	$Lu \cdot 10^{-20}$	$Gu \cdot 10^{-3}$	Ko	$Re \cdot 10$
1.	1	16	27	0.8	1.588041	46.6667	3.2727	18.68456
2.	2	16	27	0.8	1.588041	46.6667	5.4173	18.68456
3.	1,5	20	34	0.6	19.19015	61.8893	4.1107	14.01342
4.	1	20	27	0.6	0.538634	46.6667	3.8685	14.01342
5.	2	20	27	0.6	0.538634	46.6667	6.9019	14.01342
6.	1,5	16	20	0.6	0.01516	37.5427	13.608	14.01342
7.	1	20	20	0.6	0.005142	37.5427	10.336	14.01342
8.	2	20	34	0.6	19.19015	61.8893	5.2096	14.01342
9.	1,5	16	27	0.6	1.588041	46.6667	3.9544	14.01342
10.	6	27	27	0.8	0.457773	46.6667	19.803	18.68456
11.	4	20	20	0.6	0.005142	37.5427	41.162	14.01342
12.	4	27	27	0.6	0.457773	46.6667	16.426	14.01342
13.	6	24	20	0.8	0.004687	37.5427	66.709	18.68456
14.	4	20	27	0.8	0.538634	46.6667	12.571	18.68456

INITIAL DATA FOR REGRESSION ANALYSIS OF THE DRYING USING THERMODYNAMIC CRITERIA

In accordance with the experimental research design the experimental data (Table 3) were represented as variation of thermodynamic criteria Lu, Gu, Ko, Re. Drying time is taken as a response criterion so that it would be more convenient to analyze the effect of the drying mode on the process duration. Some regression analysis data are provided in Tables 4 to 6.

Prior to the regression analysis, it is advisable to perform factor analysis of the data. It should help to determine changes that occur during the drying process when using EAA. Application of thermodynamic criteria for factor analysis should make it possible to interpret the changes occurred in terms of classical theory of drying.

FACTOR ANALYSIS OF EXPERIMENTAL DATA OBTAINED DURING GRAIN DRYING WITH ELECTRICALLY ACTIVATED AIR

Factor analysis of the criteria (Lu, Ko, Gu, Re, Q_e) and the response functions W_n and τ, obtained during the experiment for grain drying with electrically activated air was carried out. Therefore, it was necessary to process five to six parameters, which could be described with two factors *F1* and *F2*, or with some third factor *F3*.

In our case, it is necessary to determine the number of factors, between which the criteria that specify the grain drying process are distributed. It is also important to know the combination of these criteria, i.e. which ones will refer to the same factor. Since we intend to explain the regularities in grain drying

Table 5. Data for regression analysis of the grain drying process with a constant concentration of air ions in the air

N°	τ	W_n	T_a	V_a	$Lu \cdot 10^{-20}$	$Gu \cdot 10^{-3}$	Ko	$\mathrm{Re} \cdot 10$
1.	1	16	27	0.8	1.58804	46.6667	4.4316	18.6845
2.	2	24	27	0.8	0.49102	46.6667	11.9406	18.6845
3.	1.5	20	20	0.6	0.00514	37.5427	22.4769	14.0134
4.	1	20	27	0.6	0.53863	46.6667	4.5907	14.0134
5.	2	20	27	0.6	0.53863	46.6667	8.0501	14.0134
6.	1.5	16	20	0.6	0.01516	37.5427	14.3132	14.0134
7.	1	20	20	0.6	0.00514	37.5427	13.9297	14.0134
8.	2	20	34	0.6	19.1901	61.8893	5.9341	14.0134
9.	1.5	16	27	0.6	1.58804	46.6667	5.1588	14.0134
10.	1.5	20	27	0.8	0.53863	46.6667	7.5785	18.6845
11.	6	27	27	0.8	0.45777	46.6667	22.0164	18.6845
12.	4	20	20	0.6	0.00514	37.5427	38.6299	14.0134
13.	6	24	27	1	0.49102	46.6667	22.5701	23.3557
14.	4	20	27	0.8	0.53863	46.6667	12.4622	18.6845

Table 6. Data for regression analysis of the grain drying process with a cyclic concentration of air ions in the air

N°	τ	W_n	T_a	V_a	$Lu \cdot 10^{-20}$	$Gu \cdot 10^{-3}$	Ko	$\mathrm{Re} \cdot 10$	Q_e
1.	1	16	27	0.8	1.58804	46.666	4.9662	18.684	2111.6100
2.	2	24	27	0.8	0.49102	46.666	13.695	18.684	4223.2200
3.	1.5	20	20	0.6	0.00514	37.542	26.085	14.013	7375.5610
4.	1	20	27	0.6	0.53863	46.666	5.7429	14.013	1583.7080
5.	2	20	27	0.6	0.53863	46.666	9.3988	14.013	3167.4150
6.	2	20	27	1	0.53863	46.666	11.402	23.355	5279.0250
7.	1.5	16	20	0.6	0.01516	37.542	17.652	14.013	7375.5610
8.	1.5	24	34	0.6	17.4940	61.889	6.8784	14.013	7375.5610
9.	1	20	20	0.6	0.00514	37.542	20.108	14.013	1583.7080
10.	2	20	34	0.6	19.1901	61.889	6.7394	14.013	3167.4150
11.	1.5	16	27	0.6	1.58804	46.666	5.2554	14.013	7375.5610
12.	1.5	24	27	1	0.49102	46.666	11.845	23.355	8959.2690
13.	6	27	27	0.8	0.45777	46.666	23.057	18.684	12669.660
14.	4	20	20	0.6	0.00514	37.542	45.940	14.013	6334.8300

with the electrically activated air using the thermodynamic criteria known, then changes in the position of the criteria vectors when performing factor analysis can help.

For factor analysis, we used the experimental data obtained for variation of the criteria (Lu, Ko, Gu, Re, Q_e), and the response functions W_n and τ. Sometimes, when performing a functional analysis, it is possible to obtain larger weights for several parameters, which makes it difficult to evaluate their significance. If this happens, the so-called orthogonal or oblique rotation of the coordinates is used, which allows to effectively evaluate the significance of the parameters.

Processing of experimental data for classical grain drying mode, when describing the process using the criteria Lu, Ko, Gu, Re and the response function W_n, result in obtaining the numerical values of *Loadings1* argument, which allow to determine to what extent the state of the grain layer can be described with two factors:

```
>> [Loadings1,specVar2,T,stats] = factoran(grades,2,'rotate','none')
Loadings1 =
    0.5767    0.5043
    0.8844    0.4614
   -0.8813    0.4672
   -0.0537   -0.0660
   -0.0900   -0.3601
```

In the columns for *Loadings1* argument the weighting coefficient concordance of each parameter to the factor is provided. Thus, in the first column the parameters 1, 2, and 3 have the highest weights. These are the criteria Lu, Ko, Gu, but the criterion vector Ko is of the opposite direction, as indicated by the "minus" sign put beside the coefficient. The slight difference between the first criterion coefficients in the first and the second columns can be considered as equal probability for this criterion to be referred to two factors, i.e. to the first and the second ones. Low fourth and fifth coefficient values in the first and second columns indicate that these parameters refer to another, the third factor.

Resulting from the factor analysis function, the variance estimate values are given in the argument *specVar1*:

```
    0.4131
    0.0050
    0.0050
    0.9928
    0.8623
```

The closer the variance value is to 1, the lesser the examined parameter can be attributed to the selected factor, and vice versa. The obtained variance values substantiate that the second and third criteria (Gu, Ko), used to describe the classical drying method can be referred to the first factor, and the probability of referring Lu criterion to the first factor and the criterion Re to the second factor are very low. They should be attributed to other factors that are not represented in the plane.

The capabilities of MATLAB (D'yakonov, 2004) allow us to draw the vectors of the examined parameters on the factor space. This allows us to visually evaluate the location of vectors against the factors and interpret the picture in terms of heat and moisture exchange processes. For classical drying, the factor space with vectors after oblique rotation is shown in Fig. 2.

It is seen from the figure that the closest to the shift of the response factor vector W_n is Re criterion vector. But it is also more vertically oriented. The angle between the vector Re and the vector W_n is quite high. The vectors Lu and Gu are close to each other, and this would allow us to state that they describe with some approximation the similar processes occurring in the grain layer. The vector of criterion Ko has its own direction and can be referred to the third factor.

Then, factor analysis of the same experimental data but with response function τ was performed.

The following *Loadings1* argument values were obtained:

```
Loadings1 =
   -0.5755     0.5042
   -0.8827     0.4647
    0.8831     0.4639
    0.0540    -0.0654
    0.5782     0.5306
```

The signs of coefficients 1 and 2 in the first column changed for opposite, and the coefficient values actually remained the same. Change of sign means change of vector orientation on the factor plane. This also refers to the vectors of criteria Ko, Re and to the vector of response factor τ.

Resulting from the factor analysis function, the variance estimate values are given in the argument *specVar1*:

Figure 2. Factor analysis results for classical drying.
1 – vector for criterion Lu; 2 – vector for criterion Gu; 3 – vector for criterion Ko; 4 – vector for criterion Re; 5 – vector for response function W_n

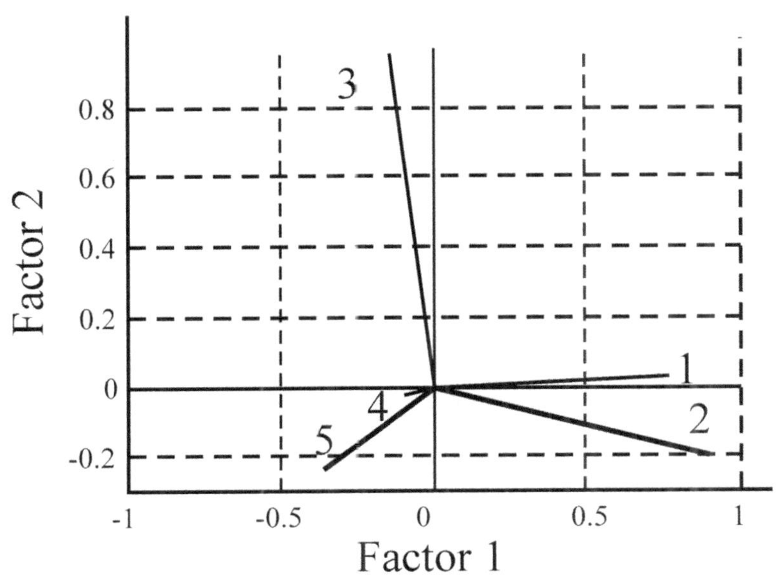

```
specVar1=
    0.4145
    0.0050
    0.0050
    0.9928
    0.3842
```

Variance values also indicate that the second and third criteria may refer to the first factor. The first criterion and the response vector W_n with low probability may refer to the second factor, and the fourth criterion should refer to another, third factor.

The factor space with vectors after oblique rotation is shown in Fig 3.

Factor analysis of the drying mode with a continuous action of air ions was performed based on the results of experimental studies for the criteria Lu, Gu, Ko, Re, and the response functions τ and W_n.

The following *Loadings1* argument values were obtained for the specified criteria and the response factor W_n:

```
Loadings1 =
    0.5751    0.5122
    0.8896    0.4513
   -0.8866    0.4572
   -0.0603   -0.0542
   -0.1212   -0.3056
```

Figure 3. Factor analysis results for classical drying.
1 – vector for criterion Lu; 2 – vector for criterion Gu; 3 – vector for criterion Ko; 4 – vector for criterion Re; 5 – vector for response function τ

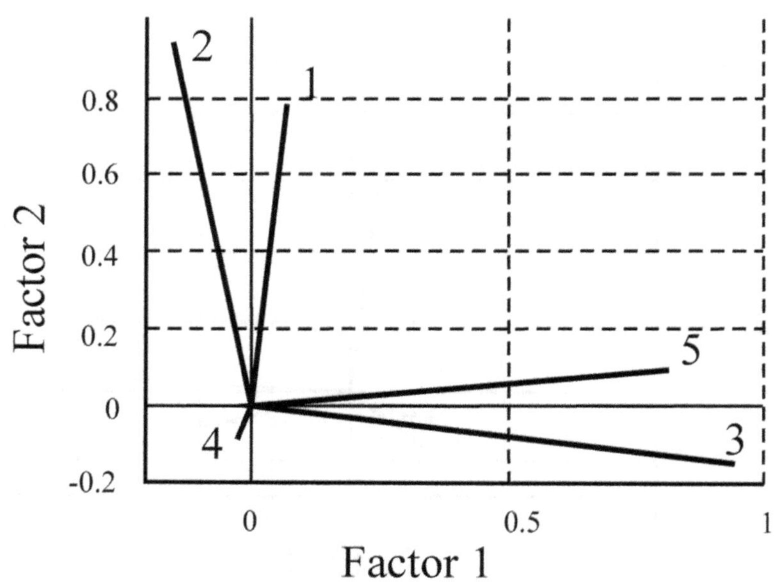

Analysis of the coefficients demonstrates that compared to the classical drying major changes in heat and moisture exchange processes have not occurred. The Lu criterion started to refer more to the first factor, its coefficient in the first column became greater by 0.0016, but for Gu criterion the coefficient value slightly dropped in the first column and increased in the second column. The variance estimate values obtained

```
0.0050
0.0050
0.6021
0.9266
0.9445
```

demonstrate, that Re criteria vector may not refer to either of the factors shown in the plane.

Analysis of the graphs shows that, despite the fact that the drying rate at a continuous use of electrically activated air has changed, the processes occurring in the grains and the grain layer do not fundamentally differ from the processes occurred in classical grain drying modes. The inclination of the criteria vectors has slightly changed, but their mutual arrangement remains, as in the case of classical drying.

Vector arrangement graphs within factor space after oblique rotation are shown in Fig. 4.

For response function τ vector arrangement on the factor space will be as below (Fig. 5).

Analysis of the graphs shows that, despite the fact that the drying rate at a continuous use of electrically activated air has changed, the processes occurring in the grains and the grain layer do not fundamentally differ from the processes occurred in classical grain drying modes. The inclination of the criteria vectors has slightly changed, but their mutual arrangement remains, as in the case of classical drying.

Further, a factor analysis of the experimental data obtained for the cyclic action of air ions on the grain layer was performed. The criteria Lu, Ko, Gu, Re, Q_e were used. In the analysis for three possible factors the following values of the argument *Loadings1*:

```
Loadings1 =
   -0.3004    0.9048    0.0574
   -0.4610    0.6423    0.3985
    0.9022    0.0240   -0.4247
    0.2200   -0.2102    0.4024
    0.9009    0.0001    0.4283
   -0.2856   -0.2623   -0.6317
```

An analysis of the coefficient values demonstrates that the criteria vectors Lu, Ko, Gu no longer refer to the same factor, as it has been in case of classical drying and drying with electrically activated air of constant concentration. The criterion vector Ko together with the criterion vector Q_e belongs to the same factor. The criterion Ko reflects the specific form of the phase transformation criterion and represents the relationship between the heat consumed for evaporation and the heat consumed for grain heating. Referring it to the same factor as the criterion Q_e allows us to state that the cyclic action of air ions is equivalent to heightening the influence of Ko criterion. Since this drying mode demonstrates an increase in the process rate (the rate of moisture release increases), we can state an increase in the

Figure 4. Factor analysis results for drying with continuous feeding of air ions.
1 – vector for criterion Lu *; 2 – vector for criterion* Gu *; 3 – vector for criterion* Ko *; 4 – vector for criterion* Re *; 5 – vector for response function* W_n

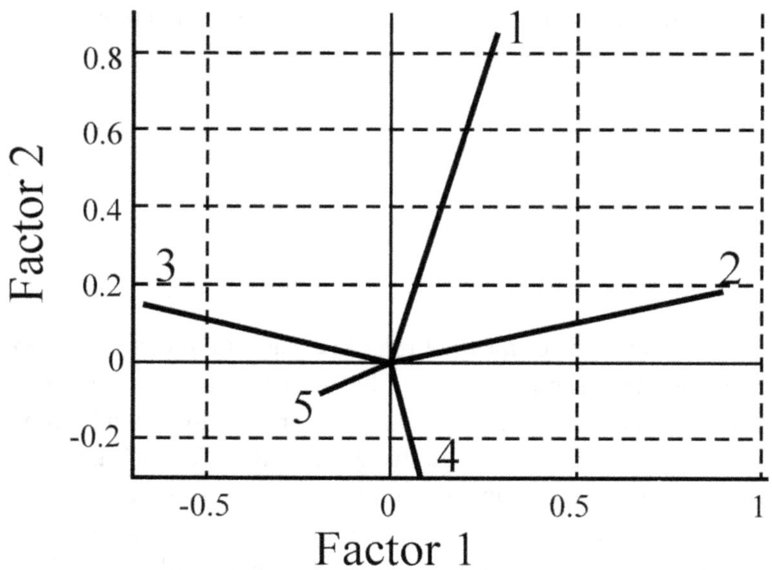

Figure 5. Factor analysis results for drying with continuous feeding of air ions.
1 – vector for criterion Lu *; 2 – vector for criterion* Gu *; 3 – vector for criterion* Ko *; 4 – vector for criterion* Re *; 5 – vector for response function* τ

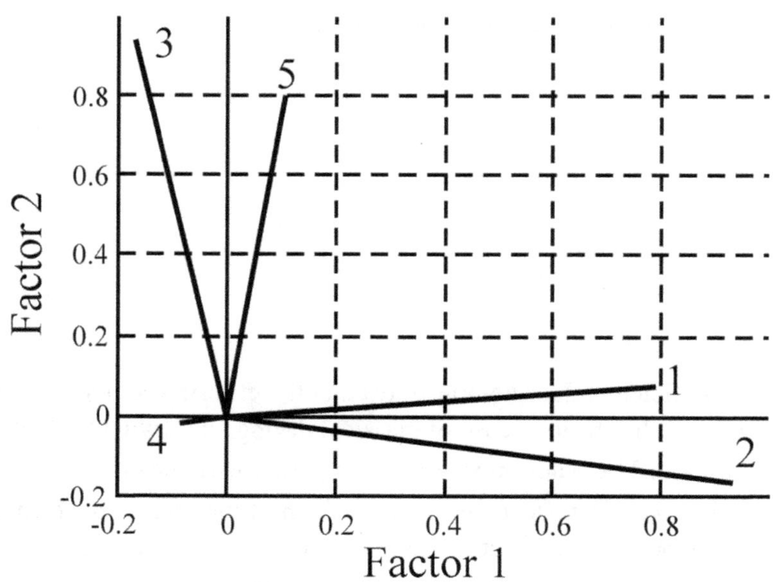

heat consumed for evaporation. In the first column the criterion Gu is the third by the coefficient value. It represents the process of evaporation from the grain surface and specifies the potentiality of the air as a drying agent. Next are the vector coefficients for the criteria Lu and Re. A significant increase of the coefficient of the criterion Re, compared with its value should be mentioned, both for the classical drying mode and drying using electrically activated air with a constant concentration of air ions. At the same time, the value of the coefficient of the Lu criterion vector in the second column sensibly increased. This can also be interpreted as a change in the intensity of change in the moisture content pattern in comparison with the intensity of the development of the temperature pattern..

An image of three dimensional factor space including vector arrangement is shown in Fig 6.

Experimental data factor analysis graph for the response function τ is shown in Fig. 7.

The graphs show that the criteria Ko and Q_e have a significant effect on the drying time (vector 6). The criteria Lu and Gu refer to the second factor having an effect on the drying process, and the criterion Re refers to the third factor.

The factor analysis performed allowed to reveal changes in the heat and moisture exchange pattern applied to the grain layer when using cyclic modes for variation of the air ions concentration in the drying agent, and describe these changes using thermodynamic criteria.

For analytical description of the process, we perform a regression analysis of the data.

Figure 6. Factor analysis results for drying with cyclic feeding of air ions.
1 – vector for criterion Lu; 2 – vector for criterion Gu; 3 – vector for criterion Ko; 4 – vector for criterion Re; 5 – vector for criterion Q_e; 6 – vector for response function W_n

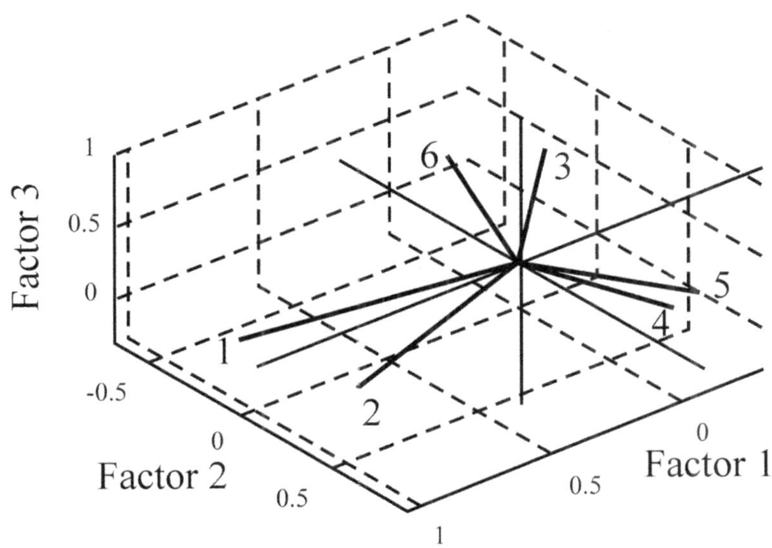

Figure 7. Factor analysis results for drying with cyclic feeding of air ions.
1 – vector for criterion Lu *; 2 – vector for criterion* Gu *; 3 – vector for criterion* Ko *; 4 – vector for criterion* Re *; 5 – vector for criterion* Q_e *; 6 – vector for response function* τ

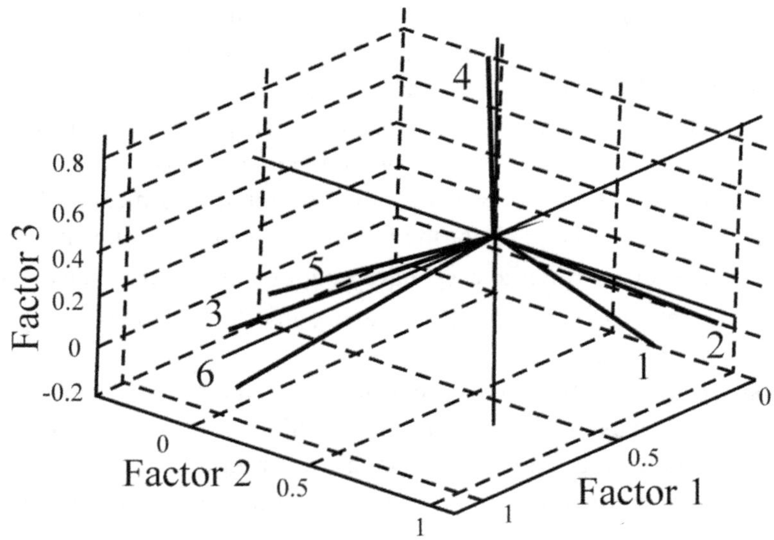

REGRESSION ANALYSIS OF THE DRYING PROCESS USING THERMODYNAMIC CRITERIA

Regression analysis was performed using a mathematical package of MATLAB application programs. For these purposes, the standard *nlintool*, function was used, which allowed performing non-linear regression analysis. The form of the regression equation was as used in the theoretical chapter.

Processing of data provided in Table 4 resulted in the following regression equation:

$$\tau = 1277.7185 \cdot Lu^{0.5759} \cdot Gu^{-6.1123} \cdot Ko^{1.1898} \cdot Re^{-0.1539}. \tag{10}$$

Student's t-test table value for number of degrees of freedom: $dfe = 36, t^m = 2.0281$. Student's t-test calculated values for regression coefficients respectively equal to:

$$t^p = (8.3676; 178.5358; -43.1402; 377.0182; -2.2524).$$

As a condition $t^p > t^m$ is met for all regression coefficients, then they all are considered to be significant.

The adequacy of the mathematical model was determined from the Fisher test. For degrees of freedom $f1 = 4, f2 = 36$ with a confidence level of 95% the table value of the criteria equals to $F^m = 2.63$. The calculated value, respectively, equals to $F^p = 0.000257$. A condition $F^p < F^m$ is met, then, we make a conclusion about statistical significance of the equation.

For the regression equation produced RMSE = 0.0839, then the approximation satisfies the experimental data. R-sguare = 0.9974, i.e. the obtained model explains 99% of the experiment results with a confidence level of 0.95, which proves the high quality of the model.

Regression analysis of data provided in Table 5 (for drying mode with constant concentration of air ions in the air) the following equation is produced:

$$\tau 1 = 9190.482 \cdot Lu^{0.6119} \cdot Gu^{-6.1527} \cdot Ko^{1.2820} \cdot \mathrm{Re}^{-0.3115} . \qquad (11)$$

Student's t-test table value for the number of degrees of freedom: $dfe = 36, t^m = 2.0281$. The Student's t-test calculated values for regression coefficients are, respectively:

$$t^p = (8.8616; 178.4017; -40.2340; 318.1257; -4.3194) .$$

Since the condition $t^p > t^m$ is met for all regression equation coefficients, then these coefficients are considered to be significant.

The adequacy of the mathematical model was determined from the Fisher test. For degrees of freedom $f1 = 4, f2 = 36$ with a confidence level of 95% the table value of the criteria equals to $F^m = 2.63$. The calculated value, respectively, equals to $F^p = 0.000298$. A condition $F^p < F^m$ is met, then, we make a conclusion about statistical significance of the equation.

For the regression equation produced RSME = 0.0885, then the approximation satisfies the experimental data. R-sguare = 0.9970, i.e. the obtained model explains 99% of the experiment results with a confidence level of 0.95, which proves the high quality of the model.

For drying mode with cyclic variation of air ion concentrations the following regression equation is produced:

$$\tau 2 = 13603.68 \cdot Lu^{0.6751} \cdot Gu^{-6.7517} \cdot Ko^{1.4336} \cdot \mathrm{Re}^{-0.2765} . \qquad (12)$$

Student's t-test table value for the number of degrees of freedom: $dfe = 36, t^m = 2.0281$. The Student's t-test calculated values for regression coefficients are, respectively:

$$t^p = (8.2709; 205.9838; -45.4422; 336.8389; -4.0191) .$$

Since the condition $t^p > t^m$ is met for all regression equation coefficients, then these coefficients are considered to be significant.

The adequacy of the mathematical model was determined from the Fisher test. For degrees of freedom $f1 = 4, f2 = 36$ with a confidence level of 95% the table value of the criteria equals to $F^m = 2.63$. The calculated value, respectively, equals to $F^p = 0.000268$. A condition $F^p < F^m$ is met, then, we make a conclusion about statistical significance of the equation.

For the regression equation produced RSME = 0.0845, then the approximation satisfies the experimental data. R-sguare = 0.9973, i.e. the obtained model explains 99% of the experiment results with a confidence level of 0.95, which proves the high quality of the model.

For visual representation of factor influence on the time we plotted the dependencies as shown in Fig. 8 to 10.

We can see on the graphs that air ions used for grain drying significantly change heat and moisture exchange in the grain layer. Thus, electrically activated air used is equivalent to raise of Lykov criterion. This becomes possible due to increasing the moisture diffusion coefficient, or by decreasing the temperature conductivity coefficient. The temperature conductivity coefficient is rather dependent on the grain heating temperature than on its moisture content, and the grain temperature was the same for all drying modes. Therefore, raise of Lykov criterion can be attributed to raise of moisture diffusion coefficient. We can see on the graph in Fig. 9 that change in drying mode do not sensibly affect the value of criterion Gu that specifies the potential ability of the air.

Air enrichment with ions does not influence on its moisture absorption ability. This conclusion is in line with the results obtained with the factor analysis of the grain heating experimental data.

More effect from using the electrically activated air can be observed on the graphs showing the dependence of the drying rate on the criterion Re (Fig.10).

Running the mode using the air ions at a constant concentration will be similar to if the drying agent velocity under classical conditions is increased by 33% for grains with a moisture content of 24% and by 57% when drying the grains with a moisture content of 16%. When cyclic enrichment of drying agent with air ions is used, increasing the criterion Re is the same as to double the drying agent velocity! Since the concentration of air ions in the drying agent for such a mode is not constant, it is not correct to definitely associate the resulting effect only with a variation of the thermo-physical properties of the grains.

Based on the data obtained we can state that the cyclic effect of air ions on the grain layer includes biological reactions in the grains that help more intensive release of moisture.

Figure 8. Dependence of drying time on the criterion Lu.
1 – classical drying (active ventilation), 2 – drying at a constant concentration of air ions, 3 – drying with a cyclic feed of air ions

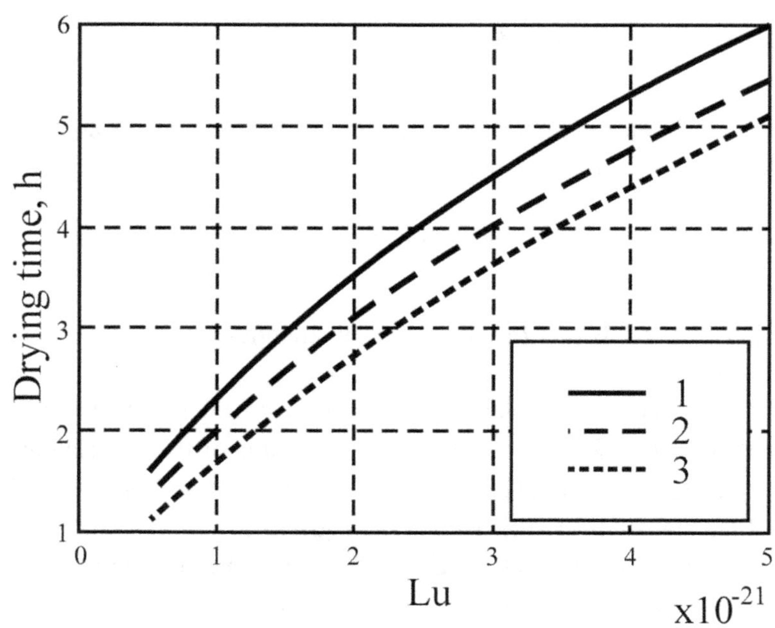

Figure 9. Dependence of drying time on the criterion Gu.
1 – classical drying (active ventilation), 2 – drying at a constant concentration of air ions, 3 – drying with a cyclic feed of air ions

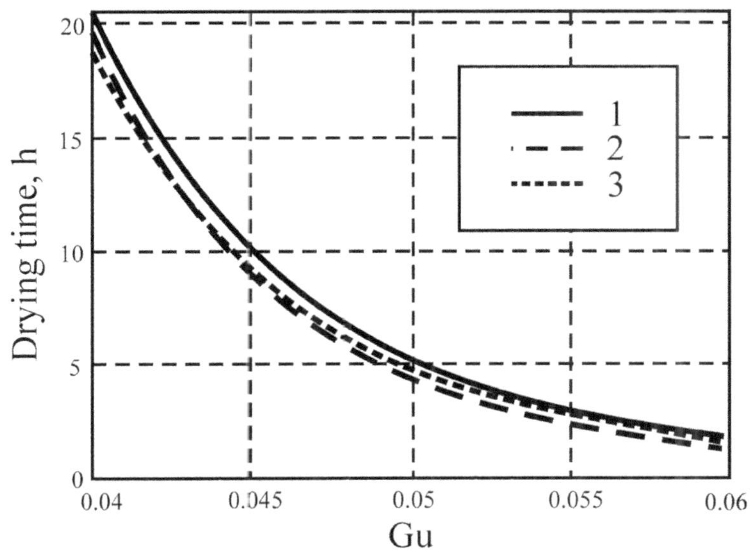

Figure 10. Dependence of drying time on the criterion Re.
1 – classical drying (active ventilation), 2 – drying at a constant concentration of air ions, 3 – drying with a cyclic feed of air ions

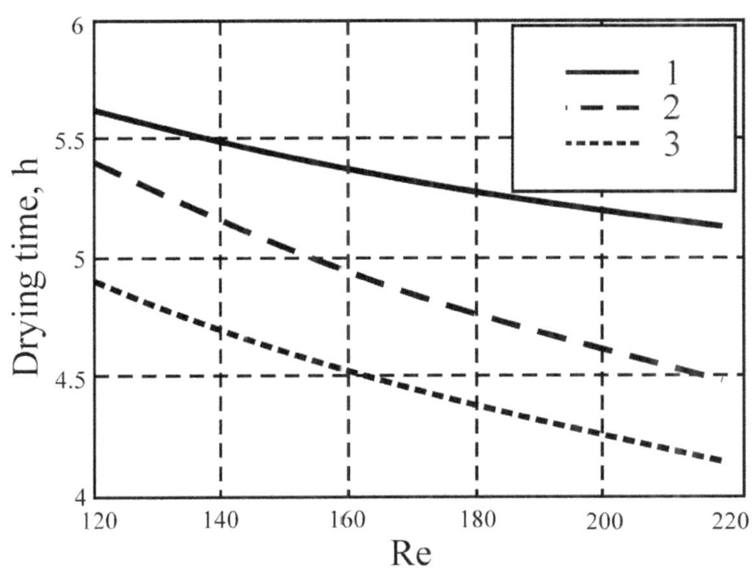

The regression equations produced describe classical drying (10) and the drying using the electrically activated air (11), (12). For the experiments the uniform method was used. The same crops and grain varieties were taken. Experimental studies were performed at the same time. All this allows us to compare these equations with each other and perform algebraic manipulations. Using the obtained mathematical models (9) to (11), we will try to produce an experimental and analytical description of the electro-activation criterion Q_e.

EXPERIMENTAL AND ANALYTICAL DESCRIPTION OF ELECTRO-ACTIVATION CRITERION

For convenience and clarity, we provide the regression equations produced for the grain drying time again:

Equation for grain drying duration without air ions

$$\tau = 1277.7185 \cdot Lu^{0.5759} \cdot Gu^{-6.1123} \cdot Ko^{1.1898} \cdot \mathrm{Re}^{-0.1539}. \tag{9}$$

Equation for grain drying at a constant concentration of air ions

$$\tau 1 = 9190.482 \cdot Lu^{0.6119} \cdot Gu^{-6.1527} \cdot Ko^{1.2820} \cdot \mathrm{Re}^{-0.3115}. \tag{10}$$

Equation for grain drying with a cyclic feed of ions

$$\tau 2 = 13603.68 \cdot Lu^{0.6751} \cdot Gu^{-6.7517} \cdot Ko^{1.4336} \cdot \mathrm{Re}^{-0.2765}. \tag{11}$$

The difference between the equation (10) and the equation (9) in terms of thermal and physical qualities is that the equation (10) represents an additional effect produced by the electrically activated air used. Therefore, the equation (10) can be represented as:

$$\tau 1 = \tau \cdot Q_e, \tag{12}$$

We obtain the dependence of electro-activation criterion $Q_e = \tau 1 / \tau$.

Since we deal with two regression equations for the modes using the electrically activated air, then we will use two criteria Q_e : Q_c – for constant concentration of air ions; Q_v – for variable concentration of air ions.

Therefore, $Q_c = \tau 1 / \tau$, and $Q_v = \tau 2 / \tau$. After performing appropriate mathematical manipulations we obtain:

$$Q_c = 7.1929 \cdot Lu^{0.0360} \cdot Gu^{-0.0404} \cdot Ko^{0.0922} \cdot \mathrm{Re}^{-0.1576}, \tag{13}$$

$$Q_v = 10.6469 \cdot Lu^{0.0992} \cdot Gu^{-0.6394} \cdot Ko^{0.2438} \cdot \mathrm{Re}^{-0.1226}. \tag{14}$$

The above equations demonstrate that electro-activation criterion can be described with the thermo-technical properties.

As we have already mentioned before, running the cyclic mode using enrichment of the drying agent with air ions will allow involving biological reactions occurred in the grain layer. This means that the equation (14) can be represented as:

$$Q_v = Q_c \cdot Q_{cikl}, \tag{15}$$

where Q_{cikl} – electro-activation criterion that reflects biological reaction of grains to the cyclic effect of the electrically activated air, in thermal and physical values.

The manipulations completed result in

$$Q_{cilk} = \frac{Q_v}{Q_v} = 1.402 \cdot Lu^{0.0632} \cdot Gu^{-0.599} \cdot Ko^{0.1516} \cdot \mathrm{Re}^{0.1516} . \quad (16)$$

Substituting the obtained dependencies in the equations (10) and (11) we obtain the following expressions:

$$\tau 1 = 1277.7185 \cdot Lu^{0.5759} \cdot Gu^{-6.1123} \cdot Ko^{1.1898} \cdot \mathrm{Re}^{-0.1539} \cdot Q_c . \quad (17)$$

$$\tau 2 = 1277.7185 \cdot Lu^{0.5759} \cdot Gu^{-6.1123} \cdot Ko^{1.1898} \cdot \mathrm{Re}^{-0.1539} \cdot Q_c \cdot Q_{cikl} . \quad (18)$$

Equation (18) helps to arrange the best drying process control. Manipulations with three drying modes (classical, with continuous effect of electrically activated air, with cyclic effect of electrically activated air) will allow such their combinations that the minimal drying time is provided.

FUTURE DIRECTIONS OF SCIENTIFIC RESEARCH

The carried out researches have allowed to define modes of grain drying. The implementation of the regimes obtained depends on the design features of the drying equipment. It is necessary to take into account the recombination rate of aeroions in the grain layer, the recombination of air ions as they pass through the air ducts. These features impose requirements on the construction of dryers. Therefore, the definition of design parameters of such grain dryers is one of the main directions of further research.

CONCLUSION

The comparative experimental studies on grain drying modes using the cyclic effects of the electrically activated air and factor analysis of the results allow us to state that the intensification of grain drying using EAA depends on the initial and current grain moisture content, and the drying agent temperature and velocity.

The mathematical models developed adequately represent the process of the grain elementary layer drying with a confidence level of 95% in the various modes using EAA, and demonstrate that:

- cyclic modes for changing air ions concentrations in the drying agent with a fluctuation period of 10 minutes and air ions concentration within the range of 0 to $3.5*10^{10}m^3$ is equivalent to an increase in the drying agent velocity by 2 in comparison with the classical method, and in comparison with the mode using a constant concentration of air ions, by 1.35;

- for cyclic mode the influence of grain biological reactions on reducing to 30% the grain drying time is confirmed. Such an influence is equivalent to increase in the drying agent velocity by 1.3 times.

REFERENCES

Aniskin, V. I., & Chizhikov, A.G. (2002). Tekhnologicheskie priemy i tekhnicheskoe obespechenie sushki semyan i zerna v Rossii [Technological receptions and technical support of drying of seeds and grain in Russia]. *1st International Scientific and Practical Conference "Modern Energy-Saving Thermal Technologies (Drying and Thermal Handling of Materials)" (SETT-2002): Proceedings of the Conference, 4*, 7-12.

Borodin, I. F. (2005). Nanoehlektrotekhnologii v sel'skom hozyajstve [Nanoelectrotechnologies in agriculture]. Mekhanizaciya i ehlektrifikaciya sel'skogo hozyajstva, 10, 2–5. (in Russian)

Chizhikov, A. G., & Golubkovich, A. V. (2005). *Tekhnologicheskie i tekhnicheskie priemy ehnergosberezheniya pri sushke zerna v sel'skom hozyajstve* [Technological and technical methods of energy saving during grain drying in agriculture]. The second international scientific and practical conference "Modern energy-saving thermal technologies (drying and thermal processes) SETT-2005", Moscow, Russia.

D'yakonov, V. P. (2004). *Matlab 6.0/6.1/6.5/6.5 + SP1 + Simulink 4/5. Obrabotka signalov i izobrazhenij*. Moscow: Solon-Press.

Gorskij, I. V. (2004). *Obrabotka semyan pshenicy ozonirovannym vozduhom [*Processing of wheat seeds with ozonized air]. Dissertation of the candidate of technical sciences, Russia, Moscow.

GOST 12.1.005-88. (1988). *Obshchie sanitarno-gigienicheskie trebovaniya k vozduhu rabochej zony* [General sanitary and hygienic requirements for the air of the working area]. Moscow: Izd-vo standartov.

Ilyuhin V.V., Noskov B.E. & Babakin B.S. (1983). *Sposob sushki materialov* [Method of drying materials]. Author's certificate 1016643 USSR, MKI5 F 26 V 3/34.

Kharchenko, V., & Vasant, P. (2018). Handbook of Research on Renewable Energy and Electric Resources for Sustainable Rural Development. Academic Press. doi:10.4018/978-1-5225-3867-7

Kireev, M. V., Grigor'ev, S. M., & Koval'chuk, Y. U. K. (1981). *Posleuborochnaya obrabotka zerna v hozyajstvah* [After-harvesting of grain in farms]. Leningrad: Kolos.

Krejmeris I.B., Tryukas K.C. & Tryukene V.A. (1991). Sposob sushki semyan zernovyh kul'tur [Method for drying seeds of cereals]. *Author's certificate N°1630661 SSSR, MKI A01 F25/08.*

Madden, C. (1987). *Electrical process to improve manufacturing efficiency* (Vol. 5). Technology Ireland.

Ospanov, A. B., Vasilyev, A. N., Budnikov, D. A., Karmanov, D. K., & Vasilyev, A. A. (2017). *Improvement of grain drying and disinfection process in the microwave field. Monography*. Almaty: Nur-Print.

Sakun, V. A. (1974). *Sushka i aktivnoe ventilirovanie zerna i zelenyh kormov* [Drying and active ventilation of grain and green forages]. Moscow: Kolos.

Sarkar. (2017). Ozone Pollution, Agriculture, and Food Security. In *Environmental Issues Surrounding Human Overpopulation.* IGI Global. doi:10.4018/978-1-5225-1683-5.ch01

Solodova, E. A. (2002). *Vliyanie aehroionnoj obrabotki na mikroorganizmy, pishchevuyu i biologicheskuyu cennost' produkcii iz morskih gidrobiontov* [Influence of aeroion treatment on microorganisms, food and biological value of products from marine hydrobionts]. Dissertation of the candidate of technical sciences, Russia, Vladivostok.

Tanimura Y., Hirotsuji J., Tanaka K. & Asakura T. (2003). Preserving fruits from putrefying by using negative air ions and ozone J.Japan. *Soc.Agr.Mach., 65*(1), 115-129.

Tihen'kij, V. I. (1993). *Metody i sredstva povysheniya ehffektivnosti processa sushki semyan pri ehlektrorazryadnom vozdejstvii na potok vozduha* [Methods and means of increasing the efficiency of the process of drying seeds in the discharge of electric discharge]. Dissertation of the candidate of technical sciences, Russia, St. Petersburg-Pushkin: NPO «Nechernozemagromash». (in Russian)

Tkachev, R. V. (2000). *Ehlektroaktivirovanie processa sushki semyan* [Electroactivation of the process of drying the seeds]. Dissertation of the candidate of technical sciences, Russia, Moscow.

Vasant, Zeynep, & Gerhard-Wilhelm. (2018). Handbook of Research on Emergent Applications of Optimization Algorithms. Academic Press. Doi:10.4018/978-1-5225-2990-3

Vasiliev, A.N., Ospanov, A.B., Budnikov, D.K., Karmanov, D.K, Salginbayev, D.B., & Vasilyev, A.A. (2016). Controlling reactions of biological objects of agricultural production with the use of electrotechnology. *International Journal of Pharmacy & Technology, 8*(4), 26855-26869.

Vasilyev, A. N. (2008). *Ehlektrotekhnologiya i upravlenie pri intensifikacii sushki zerna aktivnym ventilirovaniem* [Electrotechnology and management in the intensification of grain drying by active ventilation]. Rostov-on-Don: Terra Print.

Vasilyev, A. N. (2008). Informacionnyj podhod k opisaniyu povedeniya zernovogo sloya pri sushke [Information approach to the description of grain layer behavior during drying]. *Vestnik FGOU VPO MGAU im. V.P. Goryachkina. Agroinzheneriya, 3*, 28–31.

Chapter 10
The Magnetic Treatment of Water Solutions and Seeds of Agricultural Crops

Volodymyr Kozyrskyi
National University of Life and Environmental Sciences of Ukraine, Ukraine

Oleksandr Sinyavsky
National University of Life and Environmental Sciences of Ukraine, Ukraine

Mykola Zablodskiy
National University of Life and Environmental Sciences of Ukraine, Ukraine

Rauf Yuldashev
Ministry of Energy and Water Resources of the Republic of Tajikistan, Tajikistan

Vitaliy Savchenko
National University of Life and Environmental Sciences of Ukraine, Ukraine

Svitlana Kalenska
National University of Life and Environmental Sciences of Ukraine, Ukraine

Sławomir Zbigniew Podlaski
Warsaw University of Life Sciences, Poland

ABSTRACT

The reclamation of new areas and distant lands with limited energy and material resources demands using resource- and energy-saving technologies. One of such technologies can be magnetic treatment of irrigating water and artificial manure solutions, and presowing treatment of seeds of agricultural crops in magnetic field. The authors found that magnetic field accelerates the velocity of chemical reactions, enhances salts and fertilizers solvability, and increases oxygen concentration in a solution. Magnetic field improves membrane cells permeability which accelerates molecules and ions diffusion through it. This process improves the ability of a seed to absorb water and increases the concentration of oxygen in a cell. Presowing treatment of seeds increases their ability of germination by 26–50%, sprouting by 20–30%. The best regime of water solutions and seeds of agricultural crops treatment in magnetic field is when magnetic induction is 0.065 Tl and the velocity of a solution is 0.4 m/s.

DOI: 10.4018/978-1-5225-7573-3.ch010

INTRODUCTION

The reclamation of new territories and distant lands with limited energy and mineral resources needs energy and resource-saving technologies which will increase the yield of agricultural crops without using chemicals.

The treatment of water solutions and biological objects of plant origin in magnetic field is a promising technology which enables the farmers to increase crop yield, to ruin pathogenic microflora, to enhance the effect of artificial manure, to impact on biosynthesis, and to intensify processes of anaerobic processing of biomass waste and turning it into biofuel and fertilizers.

Presowing treatment of seeds in magnetic field improves sowing properties of seeds, decreases the rate of plant diseases, and enhances agricultural crop yield and product quality.

But the process of the implementation of magnetic treatment of water solutions and seeds faces a number of obstacles. It can be explained by the fact that not all mechanisms and principles of how magnetic field affects water solutions and biological objects are clear.

The absence of the explanation of the impact of magnetic field on the processes which take place in a water solution provides no possibility to reveal all acting factors and detect their efficiency.

Especially difficult it is to find the most efficient regimes of the treatment of agricultural crops, because seeds have specific biological and electro physical properties. That is why it is important to use such regimes which would suit specific biological properties and achieve the maximum effect.

The aim of this research is to reveal the mechanism of how magnetic field affects water solutions and seeds of agricultural crops, to find the most effective regimes of the treatment and design values of the corresponding equipment.

Estimating relationships between characteristic features of magnetic field and the change of properties of water solutions, sowing features of seeds, and the yield of agricultural crops is highly significant for the development of electrical and physical methods of processing seeds of agricultural crops.

BACKGROUND

Water plays an important role in biological objects of plant origin. Many researchers agree that water treated in electromagnetic, electrical, vibrational or magnetic fields preserves its biological activity for a long period of time (Nakasono, & Saiki, 2000; Lobyshev, 2005; Chaplin, 2007).

Current theory of magnetic treatment of water systems is at the stage of hypothesizing and substantiating, however, experiments confirm that magnetic treatment of water changes its physical and chemical properties, that is such water accelerates coagulation and absorption, changes salts solvability, gases concentration, crystallization and soaking, magnetic susceptibility, viscosity, ion hydration, kinetics of chemical reactions (Malkin, Zhuravska, & Kovalenko, 2015; Klassen, 1982).

The following empirical dependences are revealed when water systems are treated magnetically:

1. The effect of magnetic treatment gradually disappears after a short-term increase.
2. Many cases show extreme dependence of the effect on characteristics of magnetic field such as intensity, field gradient, and frequency.

3. The reverse during magnetic treatment enhances the effect of a solution treatment. Its function is to make ions accumulate at the walls of the pipeline while moving in magnetic field. It causes them to interact further.
4. Almost all research confirms the presence of the best speed of the flow.
5. The effect of magnetic treatment depends on the composition of water system.

There are known cases of the successful use of magnetoactive water to soak seeds, water plants, and remove salts from saline soil (Malkin, Furtat, Zhuravska, & Usachov, 2014; Amaya, et al., 1996)

Watering plants with magnetoactive water affects the absorption of nutrients by plants and accelerates their growth, increases productivity and the content of mineral salts and dry matter.

Moreover, watering plants with magnetoactive water contributes to the transformation of nitrogen, phosphorus, potassium in the form which can be consumed by plants. The amount of these substances in plants increases by 10-15%, and the yield increases by 15-20%. This water has fungicidal properties, ruining the process of sporulation of phytopathogenic fungus (Klassen, 1982).

The existence of various physical and chemical, and biological effects after magnetic treatment do not only make water a passive solvent which fills in the gap in biological objects but also provides it with properties the meaning of which are not fully revealed yes (Hyland, Beloussov, 2003).

All biochemical processes in plant cells take place in water environment. Water volume takes in average 70% of live plant mass, from 5 to 15% in seeds. The volume of potassium ions (K^+), sodium ions (Na^+), calcium ions (Ca^{2+}) is nearly 1%. (Yagodin, Zhukov, & Kobzarenko, 2002).

The transport of plant food compounds into a cell is performed by two autonomous mechanisms such as a passive flow of substance along the electrical and chemical gradient and their active transfer against electrical and chemical gradient (Yagodin, Zhukov, & Kobzarenko, 2002).

Ions pass through the membrane as a result of diffusion or together with solvent. As a result of the diffusion there occurs a concentration gradient. Ions and molecules move through pores in the membrane along this concentration gradient and due to neutral molecules which dissolve well in fats, which dissolve in the membrane. Substance molecules here try to to equalize their concentration in the system. Carbon dioxide, oxygen and water are the main substances which move into cells by simple diffusion.

Magnetic field affects the movement of ions and substances molecules in the cell. It enhances sowing properties of seeds and increases crops yield. Permanent magnetic field is used to treat seeds. The treatment of seeds in alternating magnetic field turned out to be inefficient, that is why it was enhanced by other ways of the treatment such as ultraviolet light, a corona discharge. As a result, the cost of stations with electric magnets is higher than the cost of those with permanent magnets; exploiting expenses increase as well as energy intensity.

The simplest stations for the treatment of seeds consists of two permanent magnets which provide a periodic field (Flóreza, Carbonella, & Martínez, 2007). Some stations are equipped with magnetic module with variable polarity under the conveyer belt (Tkachenko, 1998; Sidortsov, 2007). Other stations treat seeds in a magnetic module which is a cylinder in which cylindrical magnets are set with the help of bolts and dielectric inserts.

While moving through magnetic field, seeds experience a stimulating effect, which is manifested in increasing the permeability of the membrane, rearrangement of ions in cells, the appearance of an electric field as a result of the movement of liquid and particles in magnetic field, etc. (Kolin, Sergeyev, & Gorbatsevich, 2008).

There are known cases of successful presowing treatment of seeds in magnetic field when magnetic induction is 5-30 mTl. Magnetic field was proved to impact positively on sowing properties of seeds (Beno, Dyka, & Skvarko, 2013; Kiva, & Khodurskyi, 2010; Martinez, Pacheco, Aguilar, Pardo, & Ortiz, 2014; Grigor'yeva, 2014, Holubowicz, Kubisz, Gauza, Tong, & Hojan-Jezierska, 2014), as well as biometrical indicators, the yield (Zepeda-Bautista et al., 2014; El-Gizawy, Ragab, Helal, El-Satar, & Osman, 2016), expiration date of agricultural crops (Lysakov, & Ivanov, 2014)) and the decrease of plants diseases (Men'shova, & Nizharadz, 2012).

Many years of field experiments confirm that presowing treatment of barley, wheat, corn, soybean, tomato, pepper, cucumber enhances the productivity by 10–20%.

Pregrowing treatment of potatoes in magnetic field enhances the productivity by 18–20%, the mass of marketable potato tubers by 3.5–6.0%, and improves plants resistance to diseases. Plants are injured by potato blight 5–7 days later, when the disease dispersion is already 8–15%, the resistance to Rhizoctonia blight by 15–20%. The content of starch, protein, and vitamin C increases in potato tubers. The protective skin of tubers gets thicker during the season of picking potatoes and it reduces storage loss from 25–30% to 4–5% (Kolin, Sergeyev, & Gorbatsevich, 2008).

However, the absence of the explanation of how magnetic field affects water solutions and seeds of agricultural crops does not enable us to figure out all regime parameters of the treatment.

The implementation of the technique of magnetic treatment of water solutions and seeds of agricultural crops in the process of yielding demands from researchers to find the best parameters of the treatment and to substantiate design value of the corresponding equipment.

MAIN FOCUS OF THE CHAPTER

The Impact of Magnetic Field on PH and Oxidation-Reproductive Potential of Water and Artificial Manure Solutions

Magnetic field accelerates the process of chemical reactions which results in the increase of the concentration of reaction components in a water solution:

$$dC_i = \omega dt, \tag{1}$$

C_i – the concentration of substance, mol/L; ω – the speed of a chemical reaction, mol/L·s; t – time, s.

The following formula calculates the speed of chemical reactions during the process of magnetic treatment (Kozyrskyi, Savchenko, & Sinyavsky, 2018):

$$\omega_B = \omega \exp m(K^2B^2 + 2KBv)N_a / 2RT, \tag{2}$$

ω– the speed of a chemical reaction without being affected by magnetic field, mol/(L·s); m – total mass of ions, kg; B – magnetic induction, Tl; v – ions velocity, m/s; K – coefficient which depends on the concentration and the type of ions and on the number of back magnetization, m/(s·TL); N_a– Avogadro's number, molecule/mol; R – universal gas constant, J/mol·K; T – temperature, K.

The interaction of electrolytes and water is a chemical reaction which destroys ionic or molecule crystals and creates hydrated ions. Magnetic field accelerates the process of creating new ions in the chemical reaction.

The process of dissolving electrolytes is characterised by the degree of electrolytic dissociation which can be presented in the following way:

$$\alpha = \frac{n}{N} = \frac{\omega t}{C}, \tag{3}$$

n – the number of molecules which has split into ions; N – the total number of ions; ω – the speed of the reaction of electrolytic dissociation, mol/L·s; t – the time of the reaction, s; C – the total concentration of the solution, mol/L.

If to take (2) into account, then we get:

$$\alpha_m = \frac{\omega t e^{\frac{m(K_i^2B^2+2K_iBv)}{2RT}}}{C}, \tag{4}$$

or

$$\alpha_m = \alpha e^{\frac{m(K_i^2B^2+2K_iBv)}{2RT}}, \tag{5}$$

α_m and α – the degree of an electrolytic dissociation before and after the treatment in magnetic field.

The change of the speed of chemical reactions and salts solvability affects pH and oxidation-reproductive potential of water environment.

The change of pH can be calculated using the formula:

$$\Delta pH = \lg fC_{H_1^+} - \lg fC_{H_2^+} = \lg C_{H_1^+} - \lg C_{H_2^+}, \tag{6}$$

f – the activity coefficient; C_H – the concentration of hydrogen's ions, mol/L.

If to take (1) into account, then we get:

$$\Delta pH = \lg \omega_{H_1^+} - \lg \omega_{H_2^+}. \tag{7}$$

If we substitute the equation of calculating the speed of a chemical reaction (2) into the formula (7), then the formula of the change of pH is as follows when a water solution is treated magnetically:

$$\Delta pH = \frac{mN_aK}{2,3RT}\left(\frac{KB^2}{2} + v_nB\right), \tag{8}$$

or

$$\Delta pH = A_1 B^2 + A_2 Bv, \tag{9}$$

A_1 and A_2 – coefficients.

The change of oxidation-reproductive potential (ORP) of a solution is calculated using Nernst equation:

$$\Delta ORP = 2{,}3\frac{RT}{zF}\left(\lg fC_2 - \lg fC_1\right), \tag{10}$$

z – an ion valence; F – Faraday number, C/mol; C_1 – the concentration of ions before magnetic treatment, mol/L; C_2 – the concentration of ions after magnetic treatment, mol/L, or

$$\Delta ORP = 2{,}3\frac{RT}{zF}\left(\lg \omega_2 - \lg \omega_1\right). \tag{11}$$

If to take (2) into account then we get:

$$\Delta ORP = -\frac{mN_a K}{zF}\left(\frac{KB^2}{2} + v\Box\right). \tag{12}$$

or

$$\Delta ORP = A_3 B^2 + A_4 Bv, \tag{13}$$

A_3, A_4 – coefficients.

Coefficients, which are included in (9) and (13), cannot be calculated analytically. They are calculated on the basis of the data of experiments.

The experimental study of the change of parameters of water solutions of salts were conducted on the laboratory station. Solutions were passed through magnetic field in Petri dishes. The magnetic field was provided by permanent magnets made from intermetallic composite NdFeB, these magnets were set side by side above and under the belt of the conveyer with variable polarity. The temperature of the water was 20 ° C.

Magnetic induction was regulated by changing the distance between magnets and was measured using a teslameter 43205/1. The velocity of the conveyer belt on which there stood a Petri dish with a solution was regulated by changing the speed of a driving engine of the conveyer using a frequency changer.

ORP and pH of irrigation water and solutions of artificial manure were measured before magnetic treatment and after it using ionometer И-160M.

We used the method of experiment planning (Adler, Markova, & Granovskiy, 1976). The values of top, bottom, and main levels for magnetic induction were 0, 0.065 and 0.13 Tl correspondingly, the values of the velocity of water and artificial manure solutions were 0.4, 0.6 and 0.8 m/s.

We found out that when magnetic induction changes from 0 to 0.065 Tl, pH of water increases, with further increase of magnetic induction it begins to decrease (Figure 1, *A*). The higher the velocity of solutions is, the worse effect of magnetic treatment.

This multi-factor experiment resulted in an equation of regression which in physical values is as follows:

$$\Delta pH = 3.1B - 0.009Bv - 21.17B^2. \tag{14}$$

Magnetic field changes ORP of water. ORP of water decreases when magnetic induction is from 0 to 0.065 Tl, and when magnetic induction increases, it also starts to increase (Figure 1, *B*). The increase of water velocity changes ORP less.

Regression equation which describes the change of ORP of water after magnetic treatment can be presented like this:

$$\Delta ROP = -276.923B + 38.462Bv - 153.8B^2. \tag{15}$$

The bigger number of back magnetization gives the better effect of magnetic treatment of water. The most efficient number of back magnetization is four-time back magnetization; because its increase does not change water parameters significantly (the change of pH does not exceed 0.01, and ОВП – 1 mV).

The experimental study of the change of pH and ORP of artificial manure solutions were done using the method of an experiment planning, that is the method of orthogonal central composition planning was used. Top, bottom, and main values for magnetic induction were 0, 0.065 and 0.13 Tl respectively, the values of solution velocity were 0.4, 0.6 and 0.8 m/s. the number of back magnetization was four and the temperature of solutions was 20 ° C. ORP and pH of a solution were measured before and after magnetic treatment with the help of ionometer И-160М.

If magnetic induction is from 0 to 0.065 Tl, pH of solutions of artificial manures increases, but with further increase of magnetic induction it starts to decrease (Figure 2).

Figure 1. The change of pH and ORP of water affected by magnetic field

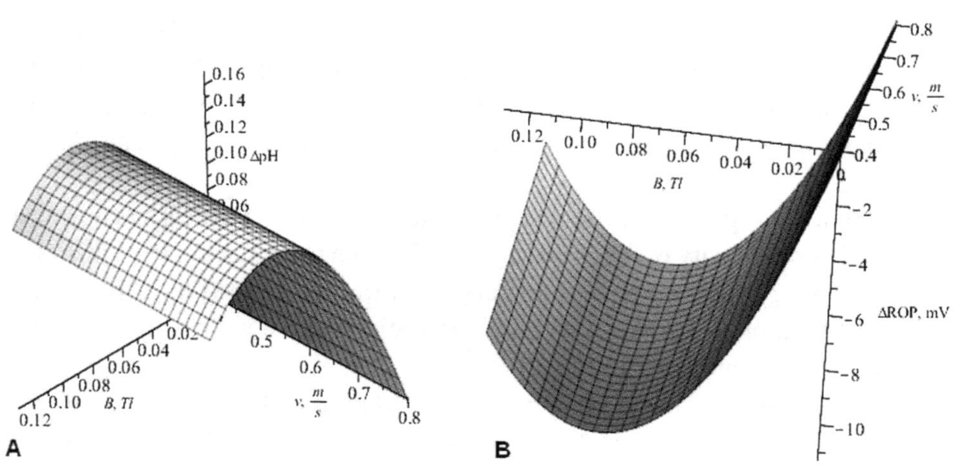

Figure 2. The dependence of the change of pH of solutions of artificial manures under the effect of magnetic induction and the velocity of artificial manure solutions: A – ordinary superphosphate; B – potash chloride; C – potassium saltpeter; D – lime saltpeter; E – ammonium saltpeter; F– magnesium sulphate heptahydrate

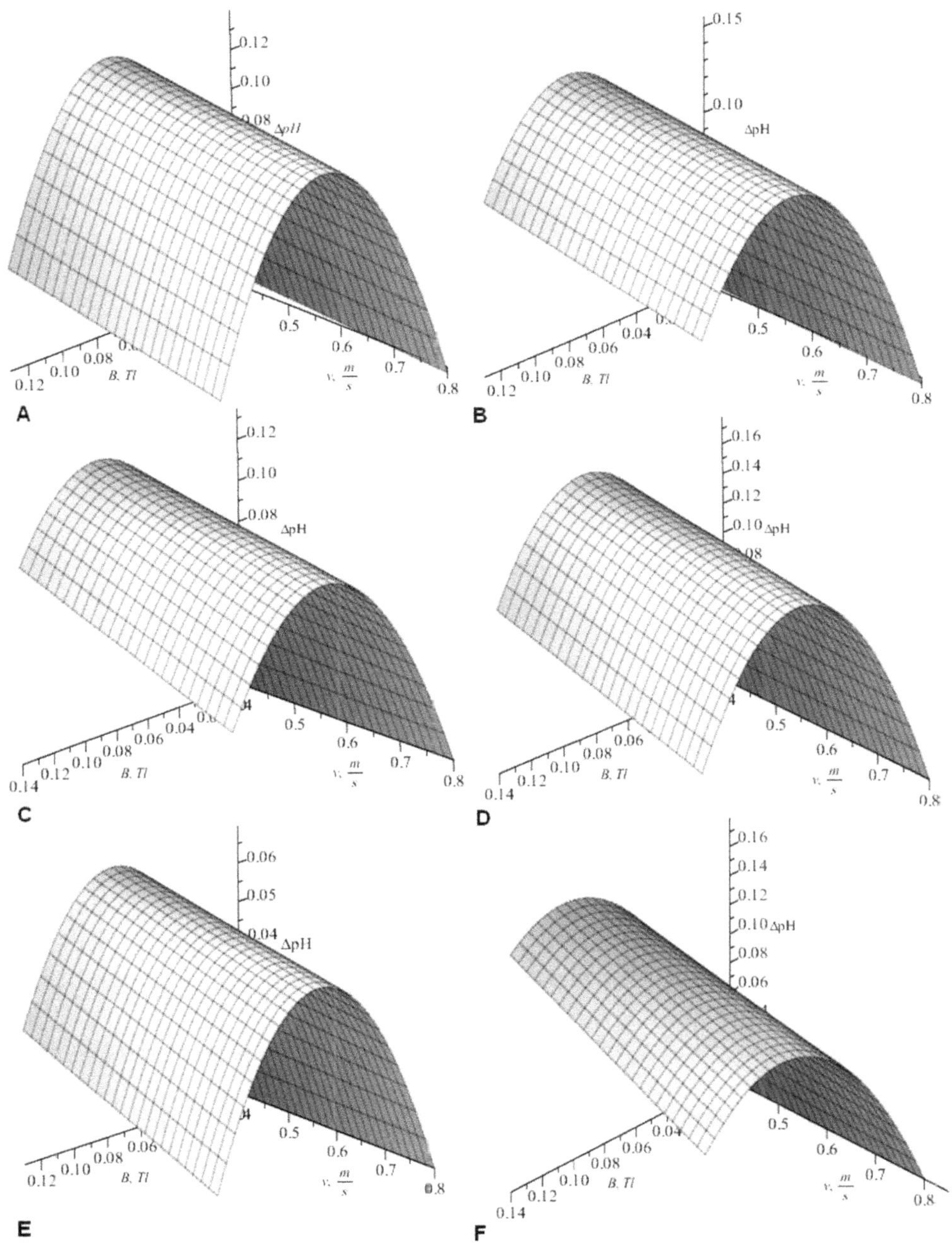

Many multifactor experiments resulted in a regression equation, which can be presented in physical values for pH salts solutions in the following way:

for ordinary superphosphate

$$\Delta pH = 3.859B - 0.321Bv - 25.904B^2; \quad (16)$$

for potash chloride

$$\Delta pH = 3.915B - 0.385Bv - 23.011B^2; \quad (17)$$

for potassium saltpeter

$$\Delta pH = 3.256B - 0.769Bv - 16.174B^2; \quad (18)$$

for lime saltpeter

$$\Delta pH = 4.41B - 0.769Bv - 23.669B^2; \quad (19)$$

for ammonium saltpeter

$$\Delta pH = 2.004B - 0.449Bv - 11.966B^2; \quad (20)$$

for magnesium sulphate heptahydrate

$$\Delta pH = 3.825B - 1.218Bv - 15.516B^2; \quad (21)$$

The ch ange of magnetic induction from 0 to 0.065 Tl decreases ORP of a solution, the further increase of magnetic induction increases ORP (Figure 3). The bigger change of ORP occurred when the velocity of a solution was lower. However, the velocity of a solution is a less significant factor than magnetic induction.

Regression equations for ORP of artificial manure solutions can be presented using physical values like this:

for ordinary superphosphate

$$\Delta ROP = -297.436B + 38.462Bv + 1854B^2; \quad (22)$$

Figure 3. The dependence of the change of oxidation-reproductive potential of solutions of artificial manures on magnetic induction and the velocity:A – ordinary superphosphate; B – potash chloride; C – potassium saltpeter; D – magnesium sulphate heptahydrate

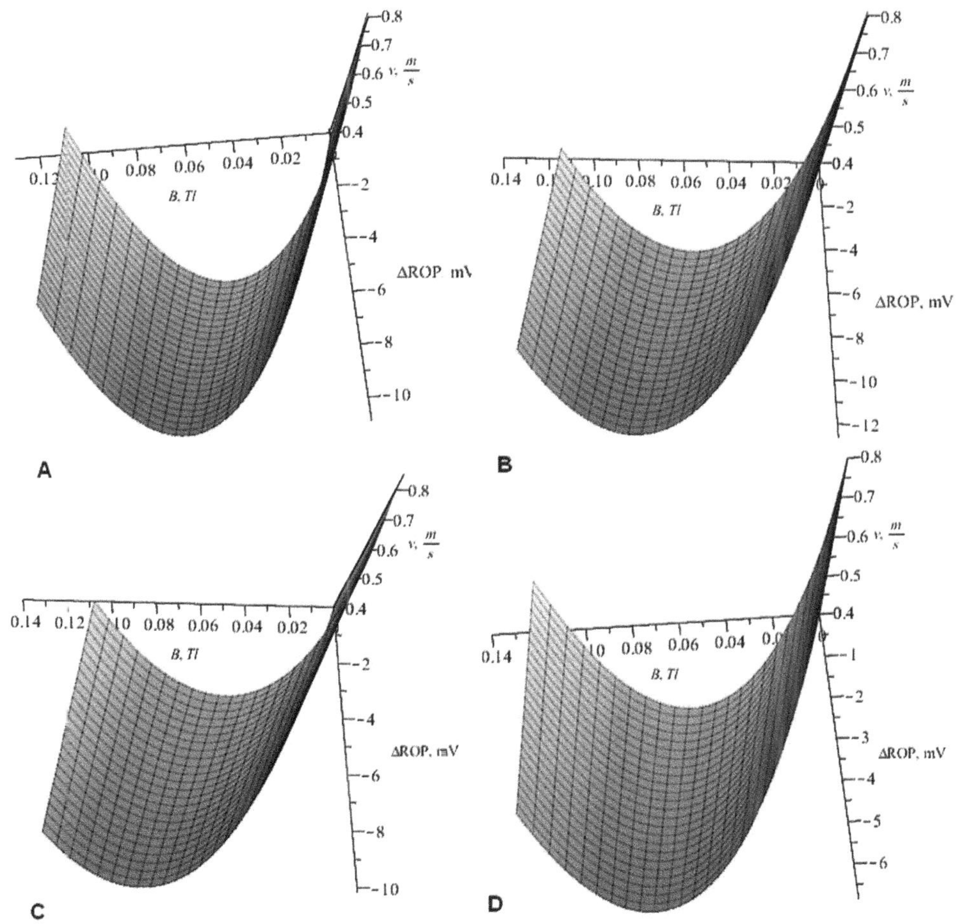

for potash chloride

$$\Delta ROP = -292.308B + 38.462Bv + 1538B^2; \tag{23}$$

for potassium saltpeter

$$\Delta ROP = -236.325B + 70.513Bv + 1065B^2; \tag{24}$$

for magnesium sulphate heptahydrate

$$\Delta ROP = -170.513B + 32.051Bv + 907B^2. \tag{25}$$

These dependences show that the best value of magnetic induction to treat water solutions is 0.065 Tl. Successful magnetic treatment also depends on the velocity, the number of back magnetization and chemical composition of a solution that is the concentration and the composition of ions. Solutions which contain ions-stabilizers of water structure such as Ca^{2+}, Mq^{2+}, SO_4^{2} are influenced more than solutions containing ions such as K^+, NO_3^- which, let us say loosen the structure of water.

The Impact of Magnetic Field on the Concentration of Oxygen in a Water Solution

Magnetic field affects oxygen solvability in a water solution. Magnetic field changes the concentration of oxygen in a solution:

$$C_{CO_2m} = C_{CO_2} \exp(K_1^2 B^2 + K_2 Bv), \qquad (26)$$

C_{CO_2m} and C_{CO_2} – the concentration of oxygen in a water solution before and after magnetic treatment, mg/L; K_1 and K_2 – coefficients.

We used the method of experimental planning to study the impact of magnetic field on the change of the concentration of oxygen in water solutions, particularly the method of orthogonal central and composition planning.

We passed the water and the solution of potassium nitrate through magnetic field. The concentration was 0.5 g/l. The values of top, bottom, and main level for magnetic induction were 0, 0.065 and 0.13 Tl, for the velocity of a solution it was 0.4, 0.6 and 0.8 m/s. The number of back magnetization was four and the temperature of solutions was 20 ° C.

The concentration of oxygen was measured before and after a water solution was treated in magnetic field using the oxygenmeter АЖА-101.1М.

When magnetic induction changes from 0 to 0.065 Tl, the concentration of oxygen in water (Figure 4, *A*) and in the potassium nitrate solution increases (Figure 4, *B*) increases. When magnetic induction continues to increase, they start to decrease. The concentration of oxygen was bigger when a solution velocity was lower. The velocity of a solution is a less significant factor than magnetic induction.

Multifactor experiments have resulted in regression equations which are as follows in physical values for the concentration of dissolved oxygen:

Figure 4. The dependence of the concentration of dissolved oxygen on magnetic induction and velocity (A) and potassium nitrate (B) in magnetic field

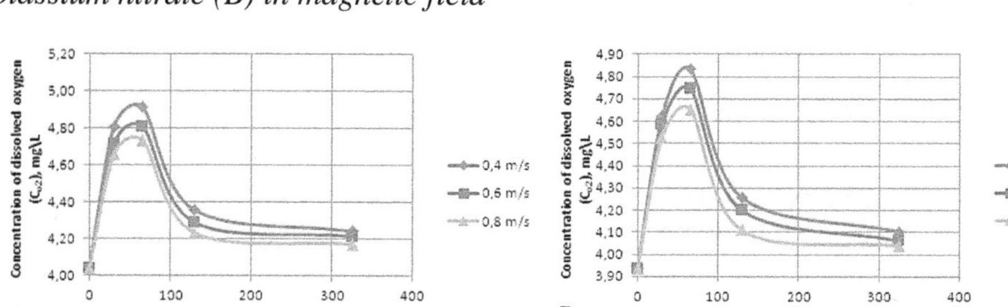

for water

$$C_{0_2} = 4,095 + 23,4B - 0,99v - 2,372Bv - 154,72B^2; \tag{27}$$

for a solution of potassium nitrate

$$C_{0_2} = 3,992 + 24,62B - 0,9v - 2,759Bv - 161,604B^2. \tag{28}$$

We found out that when magnetic induction is 0.065 Tl and the velocity of a solution in magnetic field is 0.4 m/s, the concentration of oxygen in water increases by 24%, in a potassium nitrate solution by 25%.

The Impact of Magnetic Field on Specific Conductivity of a Water Solution

Ions are formed during the process of the dissociation of salts, acids and the base as a result of a chemical reaction:

$$K_{\varsigma_+} A_{\varsigma_-} = \varsigma_+ K^{z_+} + \varsigma_- A^{z_-}.$$

Thus, a dissociation process forms ς_+ cations with the valence of z_+ and ς_- anions with the valence of z_- from one molecule. We will use β to present the product of these two values:

$$\beta = \varsigma_+ z_+ = \xi_- z_-. \tag{29}$$

Correspondingly, the number of cations and anions will be

$$n_{i_+} = \varsigma_{i_+} n_i; \tag{30}$$

$$n_{i_-} = \varsigma_{i_-} n_i. \tag{31}$$

The amount of molecules which split into ions can be calculated with the degree of an electrolytic dissociation (Antropov, 1984).):

$$\alpha = \frac{n}{N} = \frac{n}{N_a C}, \tag{32}$$

n – the amount of molecules which split into ions; N – the total amount of molecules; N_a – Avogadro's number, molecule/mol; C – molar concentration of the substance, mol/l.

Ions provide electric current in electrolytes. Let us hypothesize that there is a solution of electrolyte with sectional area Ω between two electrodes which are at l, between each other. An external electric field with intensity E is applied to these electrodes. This field makes ions move orderly, creating electric current I. If a solution consists of k electrodes, then the amperage is

$$I = \sum_{i}^{k} (I_{i+} + I_{i-}) \tag{33}$$

I_+ and I_- – components of general current which are formed by cations and anions.

The amperage will be equal to the total charge of positively and negatively charged ions which cross sectional area Ω per time unit that is ions which are in the volume $v_i\Omega$. Then current components which are formed by ions of i-type will be

$$I_{i+} = n_{i_+} \Omega v_{i_+} z_{i_+} e, \tag{34}$$

$$I_{i-} = n_{i\_} \Omega v_{i\_} z_{i\_} 5, \tag{35}$$

e – elementary charge, C.

An ion starts to move with the velocity v_i under the impact of exterior electric field:

$$F_{e_i} = z_i eE = z_i e \frac{U}{l}, \tag{36}$$

U – the voltage which is applied to electrodes, V; l – the distance between electrodes, m.

A friction force will occur since the motion takes place in the environment with finite viscosity

$$F_{m_i} = k_{m_i} v_i; \tag{37}$$

k_m – the coefficient of friction, N·s/m.

Forces of interaction $F_{в}$ also occur between electrolyte ions in a solution as well as between electrolyte ions and solvent molecules. As well as the force of fraction, this force is proportional to the velocity of ions. We make a correction on the forces of interaction with the help of the coefficient of electric conductivity:

$$f_i = \frac{F_m}{F_m + F_b}. \tag{38}$$

It will accelerate the movement of an ion to some extent. According to Newton's second law:

$$m_i \frac{dv_i}{dt} = F_{e_i} - F_{m_i} - F_{b_i}, \tag{39}$$

m_i – the mass of an ion, kg.

When we substitute (36), (37) into equation (39), we get:

$$m_i \frac{dv_i}{dt} = z_i e \frac{U}{l} - \frac{1}{f_i} k_{m_i} v_i, \tag{40}$$

or

$$m_i \frac{dv_i}{dt} + \frac{1}{f_i} k_{m_i} v_i = z_i e \frac{U}{l}. \tag{41}$$

Having solved (41), we received the equation of the movement of an ion:

$$v_i = \frac{f_i z_i e U}{k_{m_i} l} \left(1 - e^{-\frac{k_{m_i}}{f_i m_i} t} \right). \tag{42}$$

As the coefficient of friction and the one of ions interaction considerably exceed the mass of an ion so the value $e^{-\frac{k_{m_i}}{f_i m_i} t}$ can be neglected (Antropov, 1984).) So we can believe that an ion moves with uniform velocity

$$v_i = \frac{f_i z_i e U}{k_{m_i} l} = v_i^0 f_i \frac{U}{l}, \tag{43}$$

v_i^0 – absolute velocity of an ion.

When we substitute the equation of the velocity of an ion (43) into equation (34) and (35), we will get the equation of current components:

$$I = n_{i_+} \Omega v_{i_+}^0 f_{i_+} z_{i_+} e \frac{U}{l} \tag{44}$$

$$I = n_{i_-} \Omega v_{i_-}^0 f_{i_-} z_{i_-} e \frac{U}{l}. \tag{45}$$

So

$$I = \sum_{i=1}^{k} \left(n_{i_+} \Omega v_{i_+}^0 f_{i_+} z_{i_+} e \frac{U}{l} + n_{i_-} \Omega v_{i_-}^0 f_{i_-} z_{i_-} e \frac{U}{l} \right). \tag{46}$$

As Faraday's number is

$$F = eN_a. \tag{47}$$

Then formula (46) can be presented in the following way

$$I = \frac{\Omega F U}{N_a l} \sum_{i=1}^{k} \left(n_{i_+} v_{i_+}^0 f_{i_+} z_{i_+} + n_{i_-} v_{i_-}^0 f_{i_-} z_{i_-} \right). \tag{48}$$

To take (29) – (32) into consideration, we will get:

$$I = \frac{\Omega F U}{l} \sum_{i=1}^{k} f_i \alpha_i \beta_i C_i \left(v_{i+}^0 + v_{i-}^0 \right), \tag{49}$$

or

$$I = \frac{\Omega U}{l} \sum_{i=1}^{k} f_i \alpha_i \beta_i C_i \left(\lambda_{i_+}^0 + \lambda_{i_-}^0 \right) = \frac{\Omega U}{l} \sum_{i=1}^{k} f_i \alpha_i \beta_i \lambda_i^0 C_i, \tag{50}$$

λ_i^0 – ion mobility, Sm·m²/mol.
According to Ohm's law

$$I = \sigma U, \tag{51}$$

σ – conductivity of a substance.
Then

$$\sigma = \frac{\Omega}{l} \sum_{i=1}^{k} f_i \alpha_i \beta_i \lambda_i^0 C_i. \tag{52}$$

As

$$\sigma = \kappa \frac{\Omega}{l}, \tag{53}$$

κ – specific conductivity, S/m, then

$$\kappa = \sum_{i=1}^{k} f_i \alpha_i \beta_i \lambda_i^0 C_i. \tag{54}$$

The degree of electrolytic dissociation and the coefficient of interaction depend on the concentration of an electrolyte. That is why the dependence (54) for a wide range of the change of the concentration has a complicated character.

Magnetic field increases the degree of electrolytic dissociation (5), which results in a solution specific electric conductivity changing:

$$\gamma = \sum_{i=1}^{k} f_i \alpha_i \beta_i \lambda_i^0 C_i e^{\frac{\mu(K_i^2 B^2 + 2K_i Bv)}{2RT}} . \tag{55}$$

We conducted an experimental study of the change of specific electric conductivity during the process of treating a water solution in magnetic field with the following composition of a water solution: ammonium nitrate – 11.36 g/l; sodium nitrate – 0.12 g/l; potassium sulfate – 1.56 g/l; calcium phosphate – 6.6 g/l; magnesium sulphate heptahydrate – 1.05 g/l. By its composition and the concentration of mineral elements, this solution is similar to a cell solution of plants.

We passed the solution through magnetic field which was provided by permanent magnets with the velocity of 0.4 m/s. The temperature of the solution was 20 ° C. Specific electric conductivity was measured before and after magnetic treatment using a laboratory conductormeter КЛБ-1М.

We found out that when magnetic induction changes from 0 to 0.065 Tl, specific electric conductivity increases in water solutions. It decreases with further increase of magnetic induction (Figure5).

As magnetic treatment increases specific electric conductivity of a water solution, it also improves salts solvability.

Figure 5. The dependence of the change of specific electric conductivity of salts solutions on magnetic induction and the velocity of solution of 0.4 m/s

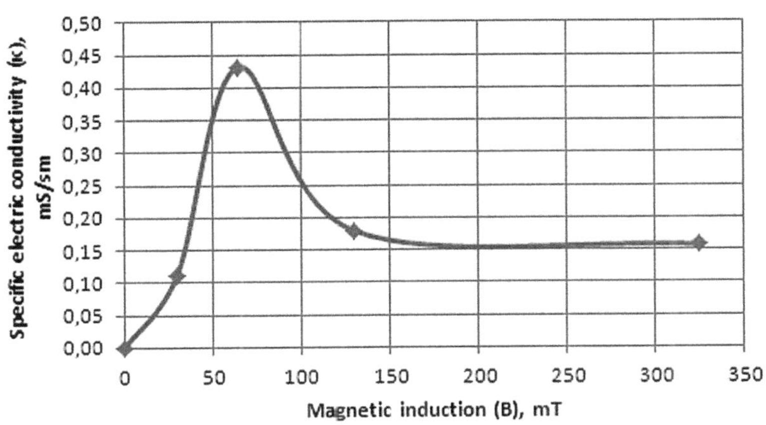

The Impact of Magnetic Treatment of Irrigation Water on the Growth and Development of Plants and Productivity of Agricultural Crops

The research of the impact of magnetic field on the change of pH, ORP, electric conductivity and the concentration of oxygen of a water solution shows that the best regime of magnetic treatment of a water solution occurs when magnetic induction is 0.065 Tl, four-time back magnetization and the velocity of a solution of 0.4 m/s.

We did the research of the impact of magnetic treatment of irrigation water on the growth and the development of plants using the method of field trial (Dospekhov, 1985). We did trials according to the scheme: the 1st variant (control): seeds were soaked in tap-water and this water was also used to water plants, the 2nd variant: seeds were soaked in magnetoactive water and this water was used to water plants.

Irrigation water was treated in magnetic field with induction of 0.065 Tl, four-time back magnetization and the velocity of a solution of 0.4 m/s.

We did experiments using cucumbers of hybrid "Topoliok" in a ground greenhouse which is equipped with a drip irrigation system.

Each pilot plot area was 8 m^2 and they were located using the method of usual repetition.

The cultivation of soil, the application of mineral and organic fertilizers, the preparation of seeds, sowing, and the care of plants were done to meet the requirements or agricultural technology of growing plants under cover.

The results of the study show that magnetic treatment of irrigation water impact positively on the growth and the development of plants. Watering cucumbers with magnetoactive water accelerates their growth. Moreover, the growth was more noticeable and statistically more significant at the beginning of blossoming and fruiting and was accordingly 14.75 and 15.75 sm.

Magnetic treatment of irrigation water stimulated cucumber early blossoming and fruiting 2-3 days earlier and increased their productivity by 14.7%.

It also improved the quality of the product. The amount of nitrates in cucumbers reduced by 6% after cucumber had been watered with magnetoactive water.

Due to the better consumption of mineral fertilizers by plants their consumption decreased by 10-15%.

Soaking seeds in magnetoactive water improves their sowing properties. The experiments were conducted using oat seeds which had been treated in magnetic field. Seeds were soaked in tap water and in the water which was treated in magnetic field with magnetic induction of 0.065 Tl and the velocity of 0.4 m/s. It was revealed that the germination of both oat seeds which had been treated in magnetic field and soaked in tap water increased by 24% in comparison with untreated seeds and their germination energy increased by 20%. Oat seeds which had been treated in magnetic field and soaked in magnetoactive water there values were better that is germination increased by 34% and germination energy increased by 30% in comparison with the control group.

The Impact of Magnetic Field on Water Substrates of Bio Gas Stations

Experimental studies of the change of pH and ORP of water substrates of bio gas stations were conducted using a laboratory station. Substrates in Petri's dishes were passed through magnetic field which was provided by four pairs of permanent magnets made from intermetallic composite NdFeB.

We used the method of experimental planning that is the method of orthogonal central and composite planning. The values of top, bottom and main levels were 0, 0.065 and 0.13 Tl respectively for magnetic induction, 0.4, 0.6 and 0.8 m/s cor the velocity of a solution. The number of back magnetization was four and the temperature was 20° C.

ORP and pH of a solution were measured before and after magnetic treatment using ioncmeter И-160M. The difference in values was the evidence of the efficiency of magnetic treatment.

Multifactor trials resulted in regression equation of pH of water substrates which can be presented in physical values as follows (Figure 6):

for a water substrate made from maize

$$\Delta pH = 0.024 + 7.94B - 0.039v - 0.641Bv - 40.763B^2; \quad (56)$$

for a water substrate made from buckwheat

$$\Delta pH = 0.034 + 6.376B - 0.058v - 1.41Bv - 30.638B^2; \quad (57)$$

for a water substrate made from soy beans

$$\Delta pH = 0.017 + 5.389B - 0.029v - 1.218Bv - 25.1153B^2; \quad (58)$$

for a water substrate made from sunflower seeds

$$\Delta pH = 0.018 + 5.748B - 0.029v - 0.833Bv - 27.876B^2. \quad (59)$$

These dependences show that when magnetic induction changes from 0 to 0.065 Tl, the value of pH increases. With further increase of magnetic induction it starts to decrease. The effect of magnetic treatment is worse when the velocity of solutions increases.

Regression equation for ORP of water substrates is as following when to use physical values (Figure 7):

for a water substrate made from maize

$$\Delta ORP = -0,972 - 1163B + 1,25v + 211,538Bv + 6075B^2; \quad (60)$$

for a water substrate made from buckwheat

$$\Delta ORP = -3,287 - 836,325B + 5,417v + 134,615Bv + 4392B^2; \quad (61)$$

for a water substrate made from soy beans

$$\Delta ORP = -1,25 - 602,137B + 2,639v + 108,974Bv + 3314B^2; \quad (62)$$

Figure 6. The dependence of the change of pH of water substrates on magnetic induction and the velocity of a substrate: A – from maize; B – from buckwheat; C – from soy beans; D – from sunflower seeds

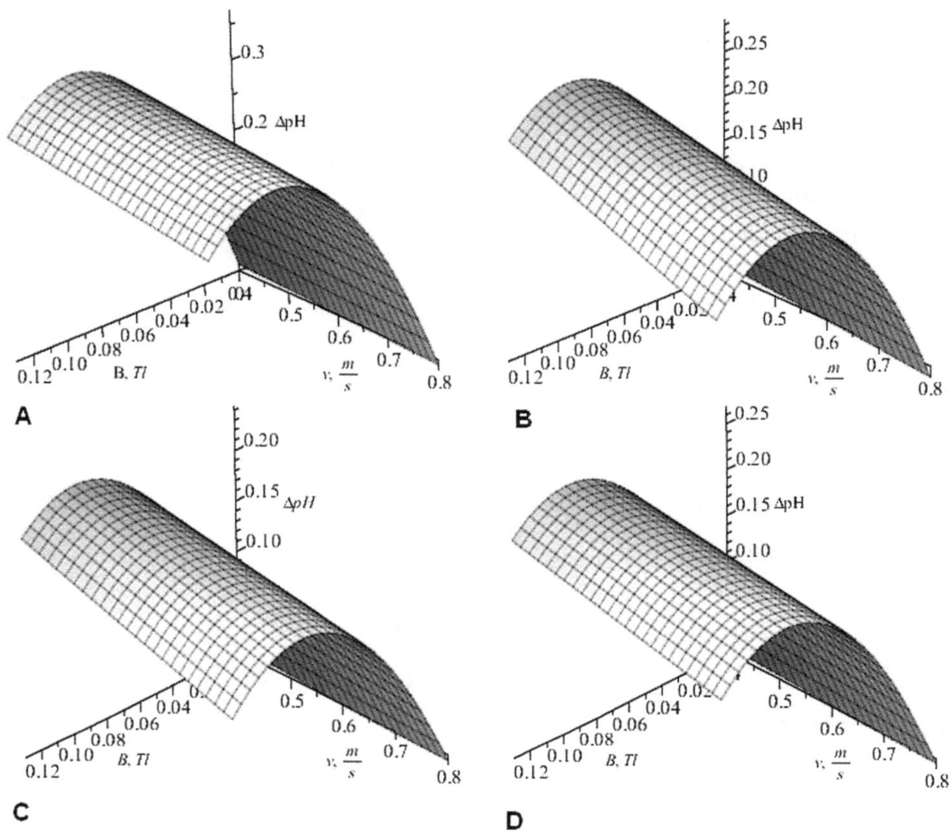

for a water substrate made from sunflower beans

$$\Delta ORP = -4,213 - 1230B + 7,083v + 185,897Bv + 6141B^2 \,. \tag{63}$$

When magnetic induction changes from 0 to 0.065 Tl, ORP of a solution increases. ORP increases when magnetic induction exceeds 0.065 Tl. The change of ORP was more significant when the velocity of a solution was lower.

Anaerobic microflora (bacteria) based on poultry manure changes the initial values of ORP and pH significantly. However, the effect of magnetic field remains the same as in previous experiments (Figure 8).

The conducted experiments show that the best value of magnetic induction when to treat water substrates of bio gas stations is 0.065 Tl, the number of back magnetization is four and the velocity of a solution is 0.4 m/s.

Figure 7. The dependence of the change of oxidation-reproductive potential of water substrates on magnetic induction and the velocity of a substrate: A – from maize; B – from buckwheat; C – from soy beans; D– from sunflower seeds

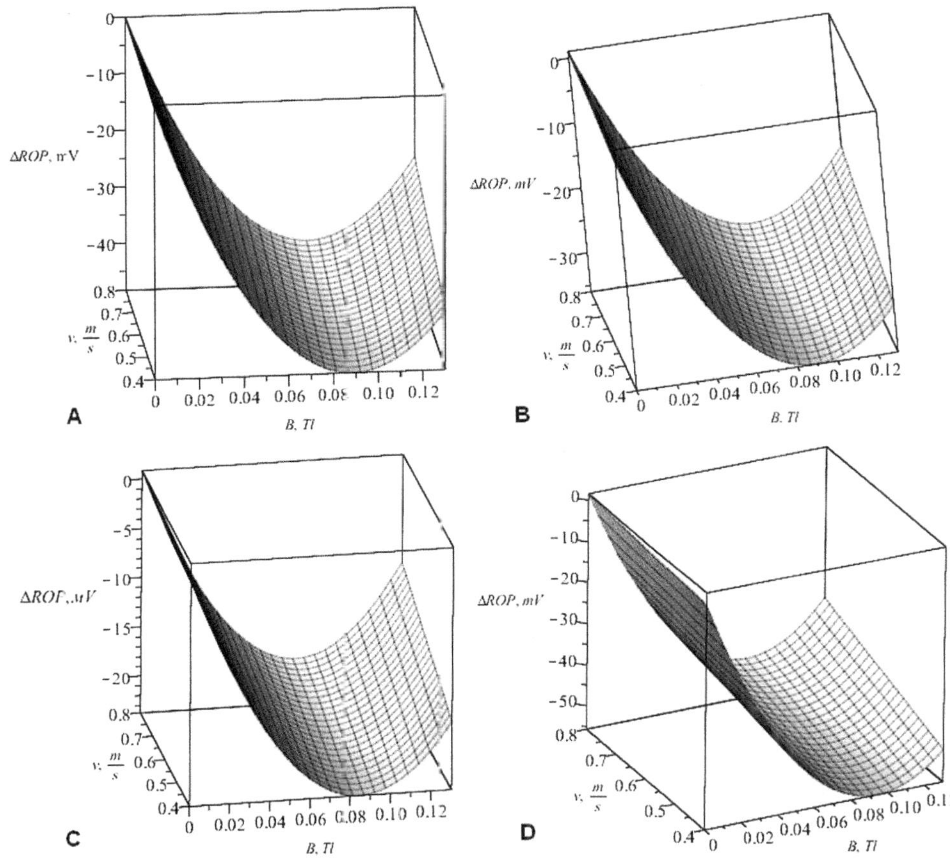

Figure 8. The change of pH (a) and oxidation-reproductive potential (b) of water substrates under magnetic treatment: 1 – maize; 2 – maize and bacteria; 3 – buckwheat; 4 – buckwheat and bacteria

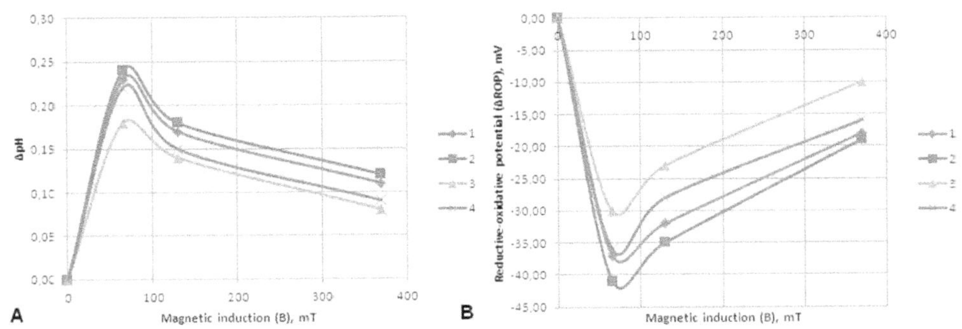

The Impact of Magnetic Field on Seeds of Agricultural Crops

All biochemical processes in plant cells occur in water environment. Water is used in all chemical reactions, it transfers nutrients and supports the elasticity of bodies.

Plant cells contain between 20 and 25 chemical elements, the part of macro elements is 98%, the part of microelements is nearly 1%. They react easily, actively interact with hydrogen, are able to create common electronic pairs with atoms of other elements, thus forming a variety of chemical organic compounds, soluble in water.

Magnetic field accelerates the speed of chemical and biochemical reactions (Kozyrskyi, Savchenko, & Sinyavsky, 2018), which are mostly oxidation-reproductive. As a result, the concentration of reaction components increases and seed stimulation occurs.

Hereby, it enhances the solvability of salts and acids which are in a plant cell.

The change of the speed of chemical and biochemical reactions which occur in a plant cell as well as the change of salts solvability influences on the bio potential and pH of the environment.

Ions move better through a cell membrane affected by magnetic field and by Lorentz force, thus increase the concentration of mineral substances in a cell which take part in chemical reactions (Kozyrskyi, Savchenko, & Sinyavsky, 2018):

$$\Delta C_{i_2} = C_{i_1} f l_i \frac{\phi}{L}(a + 2K_m B / \tau)(a + 2K_m B / \tau + K_K K_b Bv)e^{\frac{m(K_i^2 B^2 + 2K_i Bv)}{2RT}}. \tag{64}$$

C_{i1}, C_{i2} – the concentration of substances in plant cells which are separated by the membrane, mol/L; l_i – the path of an ion movement in magnetic field, m; φ – diffusion potential, B; L – the thickness of the membrane, m; a – the size of a pore in the membrane, m; K_m, K_b, K_i – coefficients; τ – a polar division, m.

Magnetic field accelerates the diffusion of molecules through the membrane. The change of the concentration of a substance which moves through the membrane by means of diffusion can be calculated by the following equation (Kozyrskyi, Savchenko, & Sinyavsky, 2018):

$$\Delta C = \frac{C_1 - C_2}{2}\left(1 - e^{-\frac{2k_d (a + K_m B / \tau)^2 e^{-\frac{E_a}{kT}}}{L^2} t}\right), \tag{65}$$

k_d – coefficient, s^{-1}, E_a – the energy of diffusion activation, J; k – Boltzmann constant, J/K.

The increase of the membrane permeability capacity leads to the increase of its concentration in the cell:

$$\Delta C = \frac{\left(C_{1O_2} - C_{2O_2}\right) e^{K_1^2 B^2 + K_2 Bv}}{2}(1 - e^{-\frac{2k_d (a + K_m B / \tau)^2 e^{-\frac{E_a}{kT}}}{L^2} t}). \tag{66}$$

The increase of the concentration of oxygen in cells ruins the process of sporulation of fungi and increases the productivity of agricultural crops.

Water penetrates faster into a cell under the influence of magnetic field. As magnetic treatment improves membranes permeability capacity, so water absorption also improves (Kozyrskyi, Savchenko, & Sinyavsky, 2018):

$$\Delta m = \frac{C_1 - C_2}{C_1 + C_2} \rho V \left(1 - e^{-\frac{k_d (a + K_m B/\tau)^2 e^{-\frac{E_a}{kT}}}{L^2} t} \right), \tag{67}$$

ρ – the density of water, kg/m^3; V – the amount of water which moved through the membrane, m^3.

By these analytical equations (64) – (67) we can conclude that main acting factors of seeds treatment in magnetic field are magnetic induction and its gradient, and the velocity of seeds.

That is why seeds presowing treatment in magnetic field should be done in inhomogeneous magnetic field. The use of period magnetic field enhances the effect of the treatment. The change of physical and chemical properties of seeds depends on the square of magnetic induction and its velocity in magnetic field.

The Impact of Magnetic Field on PH and Bio Potential of Seeds

The increase of the speed of chemical and bio chemical reaction and salts solvability lead to the change of pH of water environment in a seed. The change of pH for any water solutions can be calculated using the following equation:

$$\Delta pH = A_1 B^2 + A_2 Bv. \tag{68}$$

Magnetic field also changes ORP:

$$\Delta ORP = A_3 B^2 + A_4 Bv. \tag{69}$$

Albert Szent-Györgyi introduced the term bio potential to study biological objects. It is connected with ORP by the following:

$$BP = 820 - ORP. \tag{70}$$

Then the change of bio potential during the process of seeds treatment in magnetic field will be as follows

$$\Delta BP = A_5 B^2 + A_6 Bv, \tag{71}$$

A_5 and A_6 – coefficients which are calculated experimentally.

Figure 9. The dependence of the change of pH of seed onion (A) and garlic (B) on magnetic induction and their velocity in magnetic field

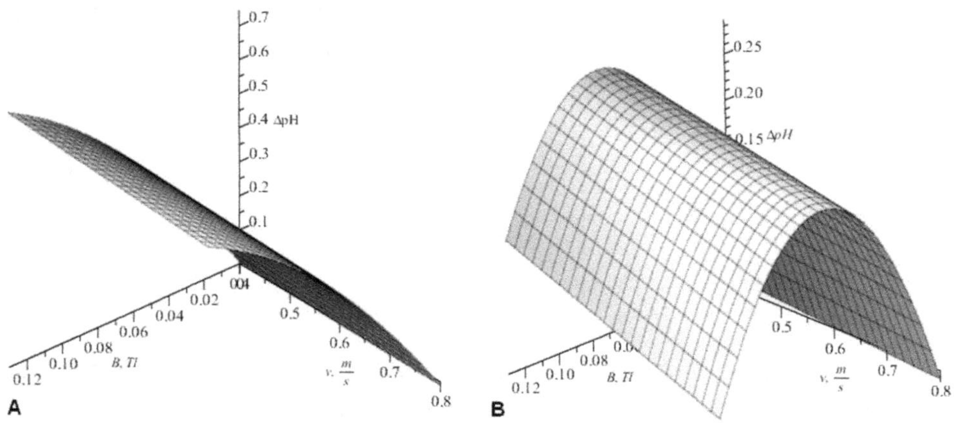

The experimental study of the impact of magnetic field on the change of pH was conducted using seed onion «Luhanskyi» and garlic «Parus». Onions moved on the conveyer through magnetic field which was provided by four pairs of permanent magnets which were set with variable polarity.

Bio potential and pH of onion and garlic was measured using ionometer И-160 before and after the treatment in magnetic field directly in an onion and a clove of garlic. Bio potential was measured using the developed measuring electrode in the form of platinum plate with a sharp end; pH was measured using a glass electrode. In addition, we used a standard silver chloride electrode.

To study the effect of magnetic field on the change of bio potential of seeds of grains, grain legumes, and green crops were first treated in magnetic field, then they germinated. And then the bio potential of young shoots was measured with the help of И-160 completed with a platinum plate and a silver chloride additional electrode.

We used the method of experimental planning that is the method of orthogonal central and composition planning.

Factors were magnetic induction and the velocity of seeds. The basic values were pH and bio potentail. The values of the top, bottom and main levels of a factor were 0.015; 0.065 and 0.115 Tl respectively for magnetic induction; they were 0.4; 0.6 and 0.8 m/s for the velocity of seeds.

Experimental dependences of the change of pH of onion on magnetic induction and the velocity of onions in magnetic field are presented in Figure 9, *a*, garllic – in Figure 9, *b*, and bio potential – in Figure 10, *a* and Figureture 10, *b* correspondingly. These dependences show that when magnetic induction changes from 0 to 0.065 Tl, the values of pH and bio potential increase; with the further increase of magnetic induction they start to decrease. The increase of the velocity of an onion worsens the effect of magnetic induction. However, this factor is not as important as magnetic induction when the range of the velocity is between 0.4 and 0.8 m/s.

Magnetic field impacts on the change of bio potential of grains, grain legumes and green crops. The dependence of the change of bio potential of young shoots of rye «Kharkivskyi», barley «Solntsedar», cucumber «Skvyrskyi», peas «Adahumskyi» on magnetic induction and the velocity during the process of presowing treatment in magnetic field are shown in Figure 11.

Figure 10. The dependences of the change of bio potential of seed onion (A), garlic (B), rye young shoots (C), barley young shoots (D), cucumber (E), peas (F) on magnetic induction and the velocity of seeds in magnetic field

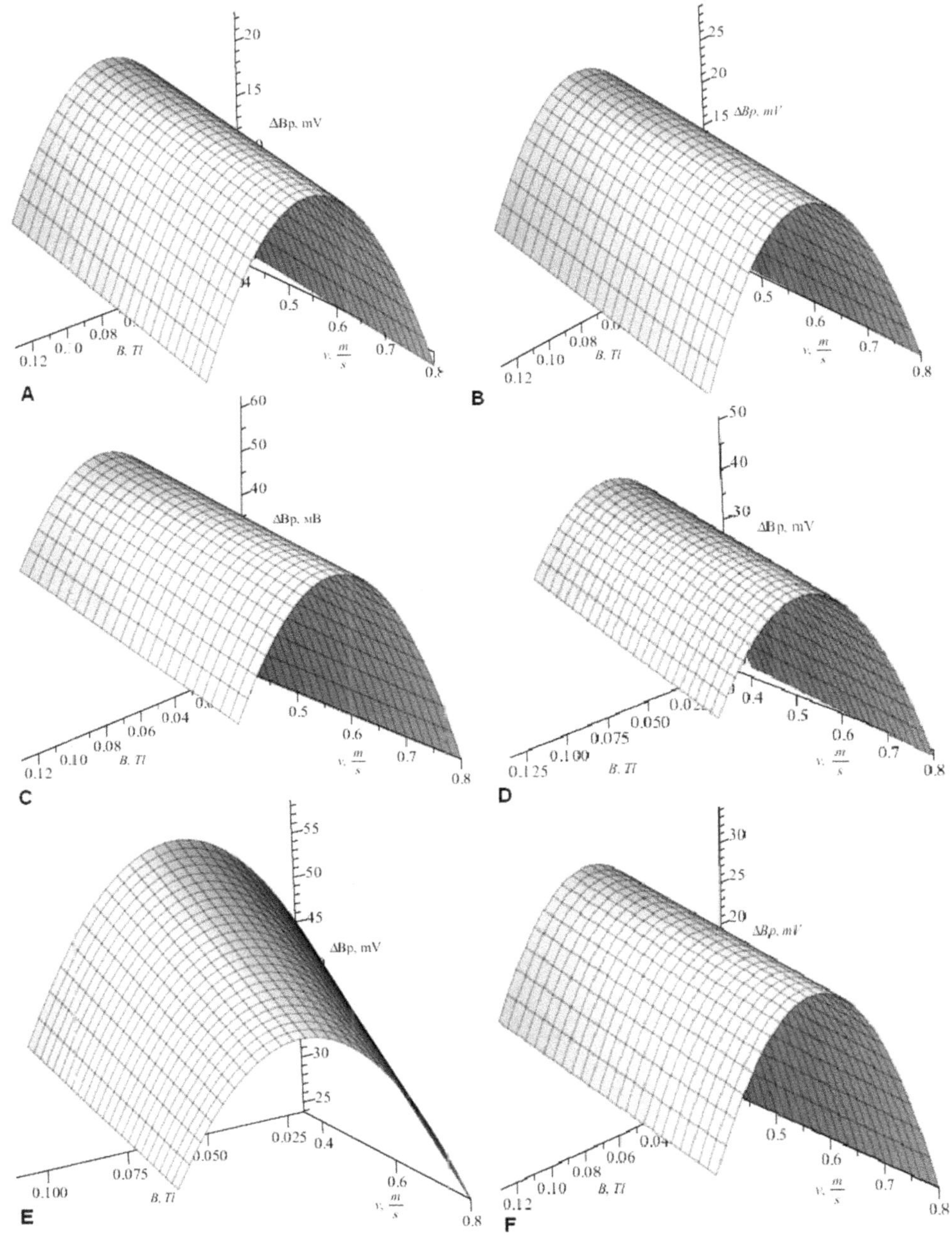

When magnetic induction changes from 0 to 0.065 Tl, bio potential increases; with the further increase of magnetic field, it starts to decrease. When magnetic induction exceeded 0.13 Tl, bio potential hardly changed. However, the value was higher than the value of seeds which had not been treated in magnetic field.

Figure 11. The dependence of germination energy on magnetic induction and the velocity of seeds in magnetic field: A – rye; B – peas; C – beans; D – oat

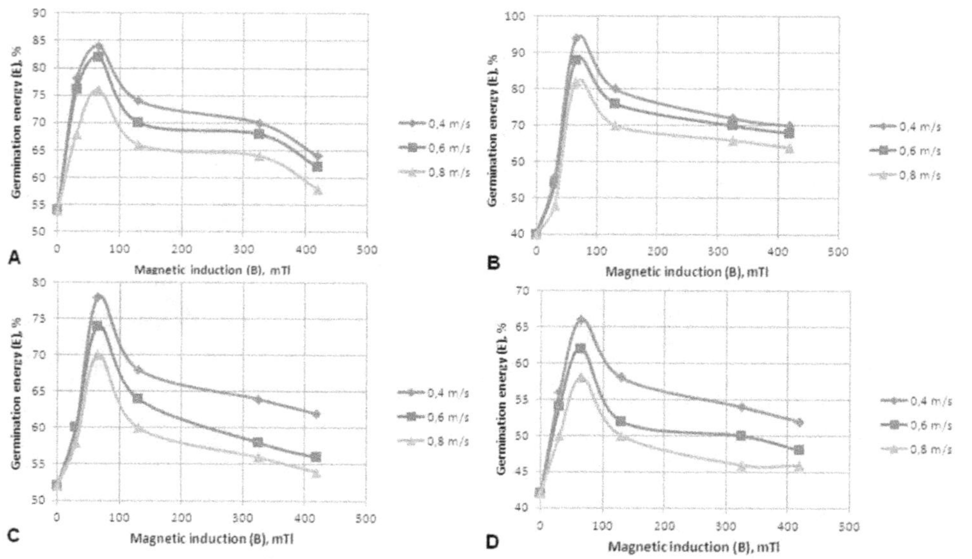

Regression equation which links seed bio potential and regime parameters of the treatment is as follows: for seed onion

$$\Delta Bp = 3.787 + 598.718B - 6.25v - 70.513Bv - 3800B^2; \quad (72)$$

for garlic

$$\Delta Bp = 1.537 + 787.189B - 2.5v - 128.205Bv - 4747B^2; \quad (73)$$

for rye

$$\Delta Bp = -0.907 + 1621B + 0.833v - 282.051Bv - 9007B^2; \quad (74)$$

for barley

$$\Delta Bp = 5.528 + 1218B - 8.472v - 185.897Bv - 6746B^2; \quad (75)$$

for cucumber

$$\Delta Bp = 45.891 + 955.075B - 53.565v + 324.074Bv - 8615B^2; \quad (76)$$

for peas

$$\Delta BP = 0.472 + 906.941B - 0.417v - 121.795Bv - 5404\ B^2. \tag{77}$$

Bio potential of seeds changed greatly when magnetic induction was 0.065 Tl. The velocity of seeds in magnetic field during their treatment also influences on their bio potential. However, when the velocity is between 0.4 and 0.8 m/s, it is not as important as magnetic induction. The best results turned out to be when the velocity was 0.4 m/s.

The Impact of Magnetic Field on Sowing Properties of Seeds

Germination capacities improve under the effect of magnetic field.

Experimental studies of the impact of magnetic field on germination capacities of seeds were conducted using rye «Kharkivskyi 98», peas «Adahumskyi», beans «Hrybovskyi», oat «».

Seeds moved on the conveyer through magnetic field which was provided by four pairs of permanent magnets which were orderly set above and under the conveyer belt with variable polarity.

Magnetic induction was regulated by changing the distance between magnets within 0 – 0.5 Tl and was measured by teslameter 43205/1. The velocity of seeds through magnetic field was regulated by the frequency of the driving engine of the conveyer belt by means of a frequency changer.

Seeds which were treated in magentic field germinated. Then its germination energy was measured (Kozyrskyi, Savchenko, & Sinyavsky, 2018).

Experimental dependences of germination energy of seeds on magnetic induction and the velocity in magentic field are shown in Figure 11. When magnetic induction changes from 0 to 0.065 Tl, germination energy increases; with further increase of magnetic induction it starts to decrease. When magnetic induction is over 0,13 Tl, the energy of germination hardly changes, though it is higher than germination energy of the control group.

Germination energy of rye increased by 16%, of peas – by 54%, of beans – by 16%, and of oat – by 24% after the seeds had been treated in magnetic field with magnetic induction of 0.065 Tl and the velocity of seeds of 0.4 m/s.

We used the method of experimental planning to find out regression equations which link the energy of germination and parameters of seeds treatment in magnetic field. Single-factor experiments show that the values of bottom, main and top levels of the factor were 0.015; 0.065 and 0.115 Tl for magnetic induction, and for the velocity of seeds they were 0.4; 0.6 and 0.8 m/s.

The multi-factor experiment helped us to receive regression equations of germination capacities of seeds which can be presented as follows:

for rye

$$E = 55.5 + 821B - 3.611v - 153.846Bv - 4536B^2 \tag{78}$$

for peas

$$E = 43.315 + 1327B - 5.278v - 192.308Bv - 7232B^2; \tag{79}$$

for beans

$$E = 69.491 + 504.701B - 2.917v\text{-}96.154Bv - 2827B^2\,; \tag{80}$$

for oat

$$E = 43.556 + 620.513B - 3.333v - 153.846Bv - 3393B^2 \tag{81}$$

The dependence of germination of seeds on magnetic induction and the velocity in magnetic field are presented in Figure 12. When magnetic induction changes from 0 to 0.065 Tl, the germination energy of seeds increases; with further increase of magnetic induction it starts to decrease. When magnetic induction is over 0.130 Tl, germination energy of seeds hardly changed but it was better than of those in the control group.

A multi-factor experiment resulted in a regression equation of germination energy of seeds which can be presented in the following way:

for rye

$$G = 72.611 + 712.821B - 3.611v - 153.846Bv - 4339B^2 \tag{82}$$

for peas

$$G = 68.111 + 730.769B - 4.444v - 115.385Bv - 4103B^2\,; \tag{83}$$

Figure 12. The dependence of germination energy of seeds on magnetic induction and the velocity of seeds in magnetic field: A – rye; B – peas; C – beans; D – oat

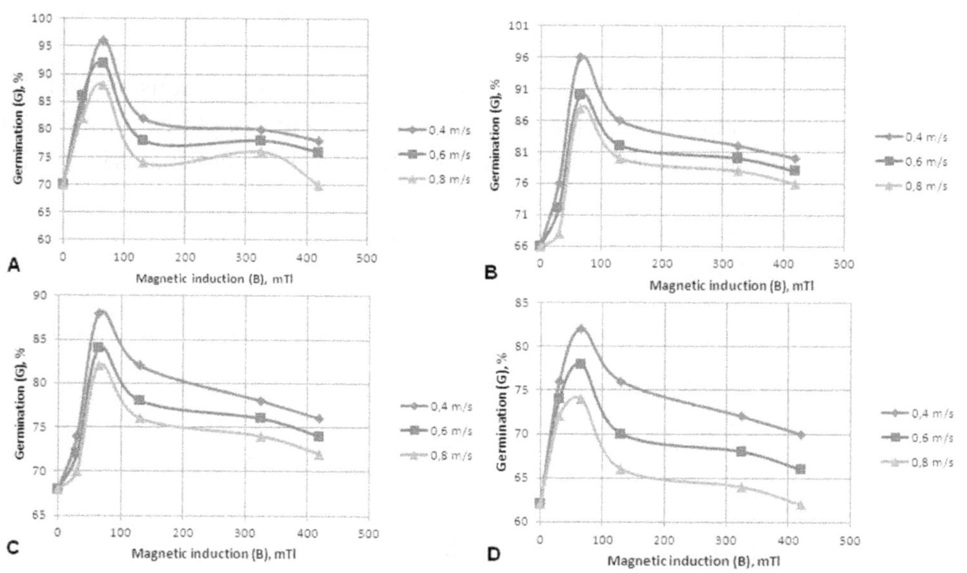

for beans

$$G = 54.75 + 626.923B - 4.583v - 96.154Bv - 3629B^2\,; \tag{84}$$

for oat

$$G = 63.407 + 544.444B - 2.778v - 192.308Bv - 2788B^2 \tag{85}$$

Germination of rye increased by 26%, of peas - by 30%, of beans - by 10%, and of oat - by 20% after the seeds had been treated in magnetic field with magnetic induction of 0.065 Tl and the velocity of seeds of 0.4 m/s.

We figured out that the energy of germination of seeds have the best value when magnetic induction is 0.065 Tl. The effect of magnetic treatment depends on the velocity of seeds. The best value of germination energy of seeds was received when the velocity was 0.4 m/s.

Magnetic field changes the way the seeds absorb water (Figure 13). The best value of water absorption by seeds was received when magnetic induction was 0.065 Tl and the velocity of seeds was 0.4 m/s. The relative water absorption by wheat seeds increased by 6%, by barley seeds – by 11%.

We conducted the study of how germination energy of seeds changed during a month after being treated in magnetic field in order to calculate the duration of the effect of magnetic field. To achieve this aim we treated rye seeds in magnetic field and every five days we measured the energy for germination of seeds.

We have found out that the effect of magnetic treatment of seeds is valid for fifteen days after the treatment.

Figure 13. The dependence of specific water absorption by wheat seeds (A) and by barley seeds (B) on magnetic induction and the velocity of seeds

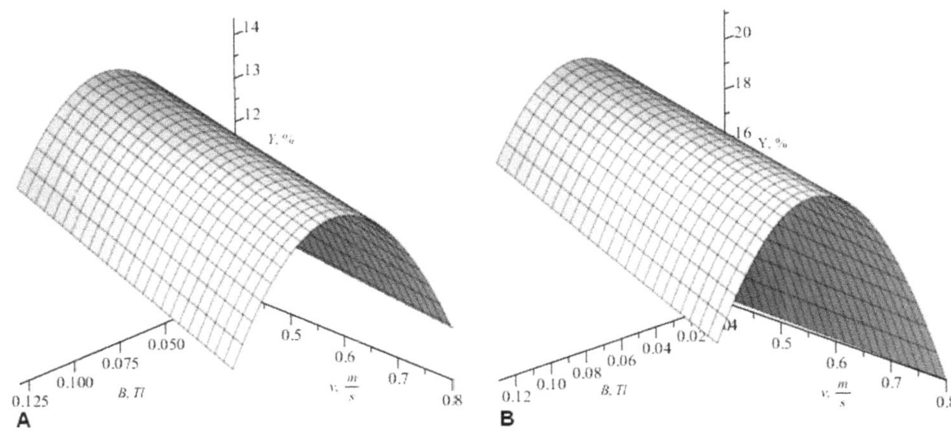

The Change of Activation Energy of Seeds Treated in Magnetic Field Before Sowing

The analytical equation of the speed of a chemical reaction during magnetic treatment of seeds (2) can be presented in the following way:

$$\omega_m = \omega \exp\left(-\frac{E^* + \Delta E^*}{kT}\right), \tag{86}$$

ω_m– the speed of a chemical reaction during magnetic treatment of seeds of a water solution; ω – the speed of a chemical reaction without the effect of magnetic field; ΔE – the change of activation energy, J/mol.

The change of oxidation-reproductive potential is calculated by means of Nernst formula (11):

$$\Delta ROP = 2,3\frac{RT}{zF}\left(\lg \omega_m - \lg \omega\right), \tag{87}$$

R – a universal gas constant, J/mol·K; T – temperature of a solution, K; z – valence of an ion; F – Faraday number, C/mol; C_1 – the concentration of ions before magnetic treatment, mol/L; C_2 – the concentration of ions after magnetic treatment, mol/l.

When we take (86) into account, ORP is:

$$\Delta ROP = -\frac{\Delta E}{zF}, \tag{88}$$

and the change of bio potential can be calculated by the following equation:

$$\Delta BP = -\Delta ROP = \frac{\Delta E}{zF}. \tag{89}$$

Then

$$\Delta E = zF\Delta BP. \tag{90}$$

That is why the change of activation energy during the process of treating seeds in magnetic field, kJ/g-equal, can be calculated by means of the following formula:

$$\Delta E = 96,48\Delta BP. \tag{91}$$

The formula (91) calculates the change of activation energy during the process of treating seeds in magnetic field. Activation energy of seeds is 2.4-5.7 kJ/g-equal when magnetic induction is 0.065 Tl and the velocity is 0.4 m/s.

The Impact of Magnetic Field on Bio Metrical Parameters and the Productivity of Agricultural Crops

Presowing treatment of seeds in magnetic field has a positive impact on the growth and the development of plants.

Hereby, bio metrical parameters of plants also improve. Figure 14 shows pea shoots after peas seeds had been treated in magnetic field with different magnetic induction. We received the best results when magnetic induction was 0.065 Tl.

Bio metrical parameters of peas were the best when presowing treatment of pea seeds had been done in magnetic field with magnetic induction of 0.065 Tl and the velocity of seeds of 0.4 m/s (Figure15).

Presowing treatment of seeds in magnetic field enhances the productivity of agricultural crops. Experimental dependences of the productivity of peas and onions on magnetic induction and the velocity of 0.4 m/s in magnetic field are presented in Figure 16. The productivity of agricultural crops increases when magnetic induction changes from 0 to 0.065 Tl; it starts to decrease with further increase of magnetic induction.

Figure 14. Pea shoots after seeds had been treated in magnetic field with the following magnetic induction: 1 – control; 2 – 0.03 Tl; 3 – 0.065 Tl; 4 – 0.13 Tl; 5 – 0.325 Tl; 6 – 0.42 Tl

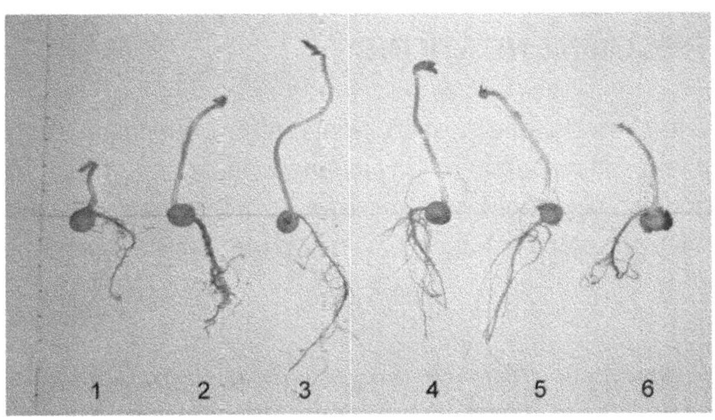

Figure 15. Bio metrical parameters of peas after pea seeds had been treated before sowing in magnetic field: A –4 weeks after sowing; B –7 days after sowing

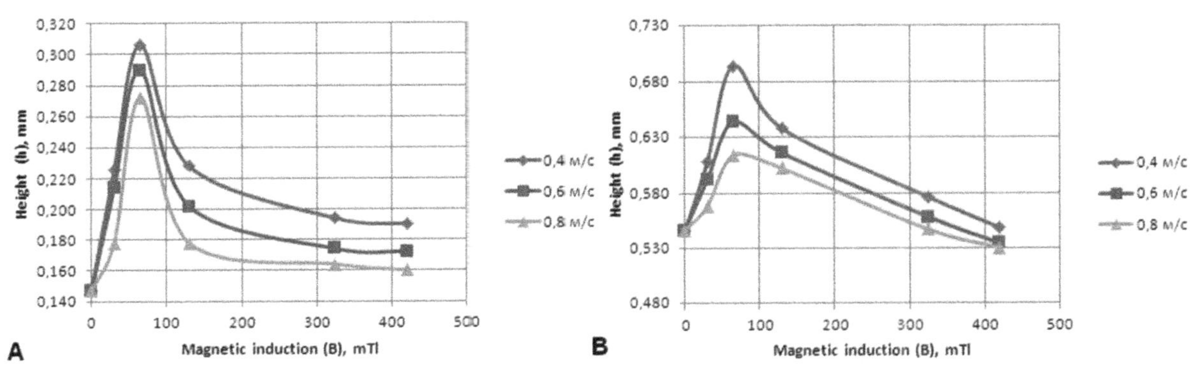

Figure 16. The dependence of the productivity of peas (A) and the mass of onions (B) on magnetic induction

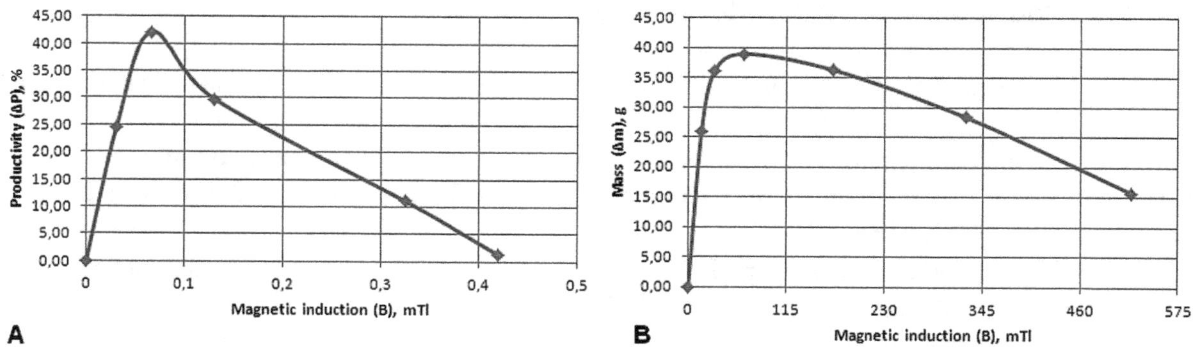

The productivity of peas on experimental plots increased by 42%, of onions by 80% after seeds had been treated before sowing in magnetic field with magnetic induction of 0.065 Tl and the velocity of seeds of 0.4 Tl.

Production check confirmed that presowing treatment of grains in magnetic field increases their productivity by 20 – 25%.

SOLUTIONS AND RECOMMENDATIONS

The conducted experiments show that the most efficient regime of magnetic treatment of water solutions and seeds of agricultural crops in magnetic field is magnetic induction of 0.065 Tl in the centre of an air gap, the treatment in period magnetic field with four-time back magnetization and the velocity of 0.4 m/s.

The scheme of the device presented in Figure 17 can provide with such a regime of magnetic treatment.

Figure 17. The device for magnetic treatment of agricultural crops:1 – the conveyer; 2 – permanent magnets; 3 – the steel plate; 5 – the electric engine; 6 – the frequency changer

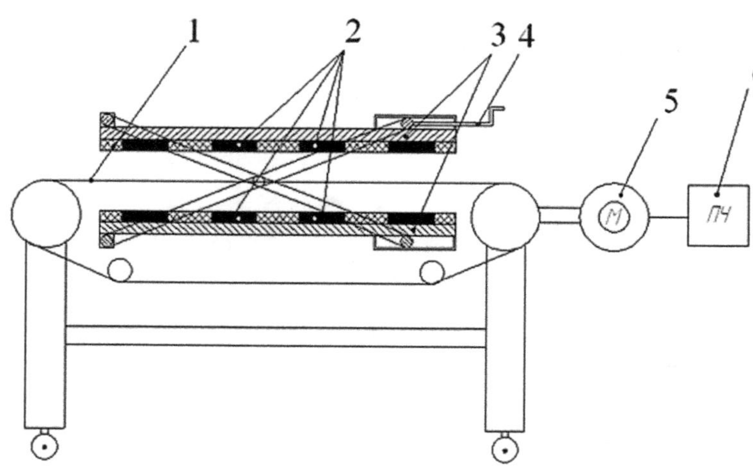

The device consists of four pairs of magnets with variable polarity 2 made from intermetallic composite NdFeB, which are set parallel to each other above and under the conveyer belt. Magnets are on plate *3*, made from steel. Gaps between magnets are filled with textolite. The needed value of magnetic inducted is regulated by the distance between magnetic cores with the help of magnetic induction regulator *4*.

The design of the conveyer provides four rollers to set up a magnetic plate under the conveyer belt.

A three-phase asynchronous electric engine with the power of 0.23 kW starts the conveyer *1* by means of a reduction gear. The engine is powered with the help of a frequency changer *6*, which provides the needed velocity of seeds in magnetic field.

This design enables us to make a simple and reliable device which provides with the most efficient regime of magnetic treatment of agricultural crops.

We used Ansys Maxwell program to experiment the device in order to treat agricultural crops in magnetic field. We found out that in the centre of an air gap of the area of permanent magnets there occurs magnetic induction of 0.065 Tl. When the experimental material moves along the conveyer belt from one magnet to another one, magnetic induction reduces, approaching 0 in the centre of magnets and after that it starts to increase change its sign (Figure 18).

The application of presowing treatment of grains in magnetic field can provide with present net profit of 71.700 $ on the area of 100 ha. The payback of the device for the magnetic treatment of seeds is one year. The profitable index during the first year of this device exploitation is 2.6. It confirms the efficiency of this development.

FUTURE RESEARCH DIRECTIONS

Magnetic treatment of irrigation water, artificial manure solution, and seeds of agricultural crops proves its efficiency.

One of the promising directions of further research is the treatment of flower seeds and bulbs in magnetic field. In addition, it is necessary to further investigate how magnetic field can be applied to fight against diseases of agricultural crops.

Magnetic treatment of water solution can be applicable in industry and construction. The research should focus at revealing the most successful regime of magnetic treatment of different products.

Figure 18. The distribution of vector magnetic potential (force lines), Wb/m

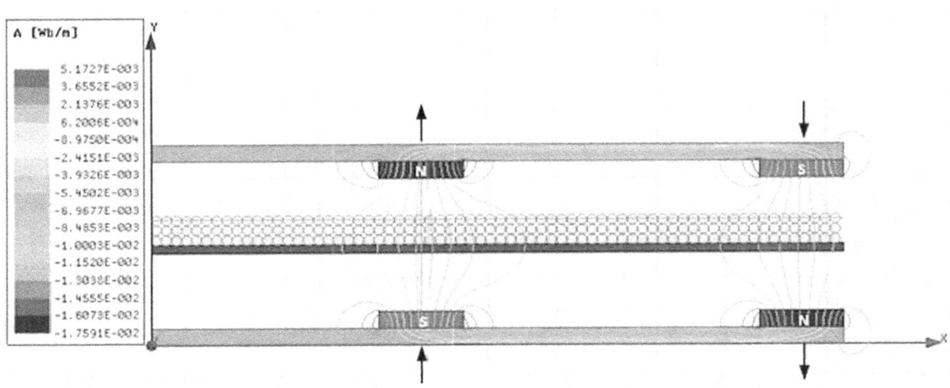

It is necessary to optimize characteristic features of magnetic field which is provided by permanent magnets while designing devices for magnetic treatment of water solutions and seeds in magnetic field. It will help to reduce the cost of such a device.

CONCLUSION

1. Magnetic field changes the speed of a chemical reaction, increases the degree of electrolytic dissociation when water solutions are treated in magnetic field. It enhances salts and acids solvability.
2. Main acting factor of treating water solutions in magnetic field are magnetic induction and the velocity of a water solutions. The treatment of water solution should be done in inhomogeneous magnetic field. Periodic magnetic field enhances the treatment. Successful treatment depends on the square of magnetic induction, the velocity of solution in magnetic field and its chemical composition.
3. Salts solvability improves and the concentration of oxygen increase when water and artificial manure are treated in magnetic field. The increase of the concentration of oxygen ruins the process of sporulation of fungi and decreases disease rate of agricultural crops.
4. The concentration of oxygen and specific electric conductivity increases when magnetic induction changes from 0 to 0.065 Тл pH; it starts to decrease after. Oxidation-reproductive potential of a solution first decreases and then starts to increase.
5. Field trials show that the treatment of irrigation water and artificial manure solutions impacts positively on the growth and the development of plants. Herewith, the values of bio metrical parameters improve, the period of blossoming and fruiting is earlier, the mass of plants increases, and the productivity increases by 20 – 25%. The amount of nitrates reduces by 6 – 10%. The better consumption of mineral fertilisers by plants leads to the reduction of their use by 10 – 15%.
6. The treatment of seeds in magnetic field accelerates the speed of chemical and biochemical reactions which occur in plant cells. It stimulated the germination of seeds and the development of plants.
7. Magnetic field affects membrane cells, their permeability increases; it leads to the increase of the diffusion of molecules and ions through a cell membrane. The water absorption by plants improves, the oxygen solvability increases, and the diffusion of its molecules through a cell membrane improves. It ruins the process of sporulation of fungi and reduces plants disease rate.
8. Lorentz force accelerates the movement of ions and increases the concentration of mineral elements which feed a cell when seeds are treated in magnetic field.
9. Main factors of the treatment of seeds in magnetic field are magnetic induction and its gradient, and the velocity of seeds in magnetic field. The success of the treatment of seeds in magnetic field depends on the square of magnetic induction and the velocity of seeds in magnetic field.
10. Magnetic field improves the quality of seeds. The best values of the energy for germination and sprouting are when magnetic induction is 0.065 Tl and the velocity of seeds is 0.4 m/s. This regime of the treatment increases the energy for germination by 26 – 50%, sprouting by 20 – 30% in comparison with the control seeds. Magnetic induction which exceeds 0,13 Tl hardly change the energy for germination and sprouting, though it is better than that of the control seeds.
11. Presowing treatment of seeds in magentic field changes bio potential. It increases when magnetic induction changes from 0 to 0,065 Tl and it decreases with further increase of magnetic field. The change of bio potential of seeds can be the factor of the efficiency of its presowing treatment. It

can also help to calculate the change of activation energy. Activation energy changes by 3.0– 6.0 kJ/ g equal when magnetic induction is 0.065 Tl and the velocity is 0.4 m/s.

12. Experimentations show that the change of bio potential, activation energy and seeds sprouting depend on the energy dose of seeds treatment in magnetic field. The most successful regime of presowing treatment of seeds in magnetic field occurs when energy was of the treatment is 1.5 – 2.5 J•s/kg.
13. The results of the studies of the change of sowing properties of seeds, their water absorption and bio potential show that the most successful regime of the treatment occurs when magnetic induction is 0.065 Tl, with four-time back magnetization and the velocity of seeds is 0.4 m/s. This regime of the treatment of seeds in magnetic field increases the productivity of grains by 20 – 25%, onions by80%, peas by 40%.
14. The increase of the productivity of agricultural crops after magnetic treatment can be explained by the increase of salts and acids solvability, better consumption of mineral fertilizers by plants, water degassing, and the enrichment of water with oxygen, better permeability of biological membranes.
15. The best results of the treatment in magnetic field as for the water solutions so for seeds are when magnetic induction is 0.065 Tl, with four-time back magnetization and when the velocity is 0.4 m/s. We can hypothesize that the stimulation of seeds can be explained by the impact of magnetic field on water solutions which are in a biological object.
16. The presowing treatment of grains in magnetic field can provide with present net profit of 71.700 $ on the area of 100 ha. The payback of the device for the magnetic treatment of seeds is one year of its exploitation. The profitable index is 2.6.

ACKNOWLEDGMENT

The authors are deeply grateful to the editors of the book for the Prof. Valeriy Kharchenko and Pandian Vasant and the reviewers for valuable comments that were taken into account in the preparation and finalization of the chapter.

REFERENCES

Adler, Y. P., Markova, E. V., & Granovskiy, Y. V. (1976). Planirovaniye eksperimenta pri poiske optimal'nykh usloviy [Planning an experiment when searching for optimal conditions]. Moskow: Nauka.

Amaya, J. (1996). Effects of stationary magnetic fields on germination and growth of seeds. *Hortic. Abst*, *68*, 1363.

Antropov, L. I. (1984). Teoreticheskaya elektrokhimiya [Theoretical Electrochemistry]. Moskow: Vysshaya shkola.

Basiuk, O.I. (2006). Pro efektyvnist zastosuvannia mahnitnoi obrobky vody [On the effectiveness of the application of magnetic water treatment]. *Nova tema, 2*, 36.

Beno, Iu. I., Dyka, M. V., Skvarko, K. O. (2013). Doslidzhennia vplyvu postiinoho mahnitnoho polia na prorostannia nasinnia Physostegia virginiana Alba (L.) benth [The influence of constant magnetic field on seed germination of Physostegia virginiana Alba (L.) benth]. *Biolohichni Studii / Studia Biologica, 7*(3), 107-114.

Chaplin, M. F. (2007). The memory of water: An overview. *Homeopathy, 96*(3), 143–150. doi:10.1016/j.homp.2007.05.006 PMID:17678809

Dospekhov, B. A. (1985). Metodika polevogo opyta [Technique of field experience]. Moskow: Agropromizdat.

Flóreza, M., Carbonella, V., & Martínez, E. (2007). *Exposure of maize seeds to stationary magnetic fields: Effects on germination and early growth.* Academic Press.

Grigor'yeva, O. (2014). Sposoby podgotovki semyan k posevu [Methods for preparing of seeds for sowing]. *LesProm, 6*(104), 176-177.

Hyland, G. J. (2003). Bio-Electromagnetism. In Integrative Biophysics-Biophotonics. Kluwer Academic.

Kiva, O. V., & Khodurskyi, V. Y. (2010). Doslidzhennia ta rozrobka prystroiu dlia peredposivnoi obrobky nasinnia tsukrovoho buriaku [Research and development of the device for pre-sowing processing of seeds of sugar beet]. *Visnyk Poltavskoi derzhavnoi ahrarnoi akademii, 4*, 176 – 178.

Klassen, V. I. (1982). Omagnichivaniye vodnykh system [Magnetization of water systems]. Moskow: Khimiya.

Kolin, A. R., Sergeyev, V. V., & Gorbatsevich, N. A. (2008). *Vozdeystviye gradiyentnym magnitnym polem na posadochnyy material i vegetiruyushchiye kartofel'nyye rasteniya* [Effect of a gradient magnetic field on planting material and vegetative potato plants]. Russkiye vysokiye tekhnologii. Available at: http://skutis.ucoz.ru/publ/26-1-0-13

Kozyrskyi, V., Savchenko, V., & Sinyavsky, O. (2018). *Presowing Processing of Seeds in Magnetic Field. In Handbook of Research on Renewable Energy and Electric Resources for Sustainable Rural Development* (pp. 576–620). IGI Global. doi:10.4018/978-1-5225-3867-7.ch024

Lobyshev, V. I. (2005). Water is a sensor to weak forces including electromagnetic fields of low intensity. *Electromagnetic Biology and Medicine, 24*(3), 449–461. doi:10.1080/15368370500382248

Lysakov, A. A., & Ivanov, R. V. (2014). Vliyaniye magnitnogo polya na sokhrannost' kartofelya [The influence of the magnetic field on the conservation of potatoes]. *Uspekhi sovremennogo estestvoznaniya, 8*, 103-106.

Mahnitna obrobka vody pered posadkoiu. (2018). [Magnetic water treatment before planting]. Available at: http://budivnik.in.ua/magnitna-obrobka-nasinnya-pered-posadkoyu.html

Malkin, Y.S., Furtat, I.Y., Zhuravska, N.Y., & Usachov, V.P. (2014). Perspektyvy stvorennia resursozberihaiuchykh tekhnolohii shliakhom mahnitnoi obrobky vody ta vodnykh rozchyniv [Prospects for resource-saving technologies through magnetic treatment of water and water solutions]. *Ventyliatsiia, osvitlennia ta vodopostachannia, 17*, 120 – 127.

Malkin, Y.S., Zhuravska, N.Y., & Kovalenko, N.O. (2015). Protses obrobky vody v mahnitnykh poliakh [The process of water treatment in magnetic fields]. *Ventyliatsiia, osvitlennia ta vodopostachannia, 18*, 70 – 74.

Martinez, F. R., Pacheco, A. D., Aguilar, C. H., Pardo, G. P., & Ortiz, E. M. (2014). Effects of magnetic field irradiation on broccoli seed with accelerated aging. *Acta Agrophysica*, *21*(1), 63–73.

Men'shova, E. A., & Nizharadz, T. S. (2012). Vliyaniye predposevnoy obrabotki semyan yachmenya na ego ustoychivost' k boleznyam i urozhaynost' [Effect of presowing treatment of barley seeds on its resistance to disease and yield]. *Izvestiya Samarskogo nauchnogo tsentra Rossiyskoy akademii nauk*, *14*(5), 241-244.

Nakasono, S., & Saiki, H. (2000). Effect of ELF magnetic fields on protein synthesis in Escherichia coli K12. *Radiation Research*, *154*, 208–216.

Podleoeny, J., Pietruszewski, V., & Podleoena, A. (2004). Efficiency of the magnetic treatment of broad bean seeds cultivated under experimental plot conditions. *International Agrophysics*, *18*, 65–71.

Sidortsov, I. G. (2007). Ustanovka dlya predposevnoy obrabotki semyan [Plant for presowing seed treatment]. *Tekhnika v sel'skom khozyaystve, 3,* 61-62.

Tkachenko, Y. (1998). *Method of crops growing*. Patent of Russian Federation A01G7/04, A01C1/00. N°2110910 declared. 25.10.96; published 20.05.98, N° 10.

Vasant, P., Weber, G. V., & Dieu, V. N. (2016). *Handbook of research on modern optimization algorithms and applications in engineering and economics*. IGI Global. doi:10.4018/978-1-4666-9644-0

Yagodin, B. A., Zhukov, Y. P., & Kobzarenko, V. I. (2002). Agrokhimiya [Agrochemistry]. Moskow: Kolos.

Zepeda-Bautista, R., Hernandez-Aguilar, C., Suazo-Lopez, F., Dominguez-Pacheco, A. F., Virgen-Vargas, J., Perez-Reyes, C., & Peon-Escalante, I. (2014). *Electromagnetic field in corn grain production and health*. Academic Press.

KEY TERMS AND DEFINITIONS

Biopotential: The difference of potentials between two points of tissue which reflects its bioelectric activity.

Aqueous Solution: A kind of solution in which water is the solvent.

Diffusion: The process of mutual penetration of molecules or atoms of one substance among other atoms or molecules, which usually leads to equilibrium of their concentrations in the occupied volume.

Germination: The ability to form well developed seed sprouts.

Germination Energy: Seed capacity to germinate fast and together.

Magnetic Field: The component of electromagnetic field, due to which the interaction occurs between moving electrically charged particles.

Magnetic Induction: A vector physical value, the main characteristic feature of the magnitude and direction of the magnetic field.

Oxidation-Reduction Potential (Redox): Is a measure of the chemical activity of elements or their compounds in reversible chemical processes associated with a change in the charge of ions in solutions.

pH (Hydrogen): The value which shows the extent of the activity of hydrogen ions ($H+$) in a solution.

Rate of Chemical Reactions: The change of the concentration of reactants or a reaction product per unit time.

Chapter 11
Monitoring the Condition of Mineral Nutrition of Crops Using UAV for Rational Use of Fertilizers

Dmytro Sergiyovich Komarchuk
National University of Life and Environmental Sciences of Ukraine, Ukraine

Vitalii Pylypovych Lysenko
National University of Life and Environmental Sciences of Ukraine, Ukraine

Oleksii Oleksandrovych Opryshko
National University of Life and Environmental Sciences of Ukraine, Ukraine

Nataliia Anatoliivna Pasichnyk
National University of Life and Environmental Sciences of Ukraine, Ukraine

ABSTRACT

The reviewed problem deals with the metrological support of sensory equipment using cameras with standard and infrared lenses. The elaborated method to use UAVs for remote shooting of agricultural crops and obtaining images is to be used to determine the level of nitrogen provision for grain crops has been developed. For radiofrequency correction of lighting changes while mineral nutrition state monitoring, it is recommended to use a method with the application of official data to be calculated on the basis of camera's regular camera exposure meter measurements. It is advisable to use an infrared lens for monitoring the state of individual crops using digital non-specialized cameras instead of optical one. Exampling GoPro Hero 4 digital camera, it has been found that there are three channels to use IR lens which have the characteristics of fixing radiation either in the infrared range or in the red area of the optical range.

DOI: 10.4018/978-1-5225-7573-3.ch011

INTRODUCTION

The use of unmanned aerial vehicles (UAVs) for monitoring the state of mineral nutrition (elements providence) of grain crops deals with the need in rational use of fertilizers. This is especially important in the context of a significant fluctuation of prices for mineral fertilizers and final products while operational information about the state of mineral nutrition may be used to program the yield using maximum profit as optimality criterion.

The use of the UAV to create an electronic field map to detect the state of mineral nutrition would allow to use differential fertilization by equipment to be equipped only with the GPS system without complicated and expensive onboard optical sensors for leaf planting

diagnostics. However, in industrial practice while planning the fertilization of crops, agrarian enterprises limitely use UAVs for operational monitoring of the supply of plants with nutrition elements.

As experience shows that successful use of these promising technologies is possible only with the correct methodological approach, first of all, due to operational receipt and further processing of information.

The methodologically correct organization of work to be dealt with the monitoring of plant nutrition using UAVs is the result of scientific research, planning and practical work out of actions, forces and means.

Therefore, this section is devoted to the further development of methods for remote sensing of the state of agricultural crops mineral nutrition using UAVs to be based upon the analysis of existing practical experience and conducted research.

IMPORTANCE OF NITROGEN NUTRITION FOR PLANTS

In the context of the increasing level of chemicalization of agriculture and the cost of fertilizers, one of the pressing problems is to increase the efficiency of their application. Successful resolution of this issue involves the need in a comprehensive techno-agrochemical approach to optimize the plant's mineral nutrition, colon and balance of nutrients in the plant-soil-fertilizer system.

Fertilizer technologies should not ensure only the requirements of agriculture but the environment too (Horodnii, 2006).

Nitrogen is the most widely used macro element, one of the most important building elements to increase the green (vegetative) mass of plants and, consequently, yield. They do not mean the nitrogen to be the «element of life» in vain.

In plant organism it participates in the synthesis of proteins. As an important component it is found in nucleoproteins and nucleic acids of the RNA and DNA, it is part of the molecule of chlorophyll, vitamins, alkaloids. The deficit of nitrogen in a plant nutrition suppresses the pace of intragranular opagans, which, in its turn, influences on the formation of phytosanitary leaflet growth and causes the reducing of yield and its qualitative features (Horodnii, 2006; Horodnii and Melnychuk, 2006; Coruzzi and Bush, 2001).

Features of winter wheat nitrogen nutrition. For winter wheat nitrogen is a priority element to be responsible for the growth of vegetative mass, the formation of productive stems, and the pouring of grain. While sowing, it is necessary to provide the plants with nitrogen for sufficient autumn bait but the amount of this element should not cause plant overgrowth. Autumn is a season to be important to create a sufficient phosphoric-potassium background of the soil and to give «start» nitrogen to the plants.

All further methods of establishing fertilizers are nutritious and are carried out during the spring-summer period. First of all, in spring plants need to leave the winter rest and quickly restore the vegetation. At this time, it is important to give nitrogen for stimulation. During this period plants need readily available nitrogen compounds but the dose of these elements should be moderate because the root system is still weakly developed and a sharp increase of soil solution concentration may cause toxic poisoning of plants and inhibition of their growth processes (Horodnii, 2006; Horodnii and Melnychuk, 2006; Hospodarenko, 2016).

Weather conditions are one of the key factors in spring nutrition. The release of the field from the snow is extremely important in the nitrogen nutrition of winter wheat. The possibility to use machinery for field works should be combined with the presence of a large amount of moisture in the soil which increases the efficiency of nitrogen fertilizers and stimulates the restoration of plant growth. It is no need in too early nitrogen injection during so-called «February's windows» because future snow may cause large losses of nitrogen by washing out.

Manufacturers know that the 3-4th stage of vegetation is extremely important for winter crops because the future harvest is being laid during this period. Nitrogen fertilization during this period is aimed on strengthening plants and laying productive stems. Nitrogen deficiency at this stage of development causes significant crop losses. During the first nutrition under winter wheat it is in need to inject mostly 30% of nitrogen from the full required dose. Good results are noted with the use of ammonium nitrate, CAS or ammonium sulfate.

The second spring nitrogen nutrition is carried out at the end of the bushing or at the beginning of the phase of tubing. While adding nitrogen during this period it is also necessary to take into account the reserves of productive moisture in the soil. The amount of nitrogen to be injected during this period is 40-50% from the total dose of the element. Nitrogenous fertilizers are used in the form of ammonium nitrate and if the temperature allows they carry out the foliar application of urea. If it is adequate machinery CASs are injected in the interlinear way to avoid burns of plants.

Due to the lack of moisture that has been observed in recent years, preference is given to foliar feeding (second to third) with urea solution. In order not to cause burns of plants, the concentration of the solution is precisely chosen.

During this period due to the rapid growth of the wheat vegetative mass, excess nitrogen reduces the resistance of plants to pathogens of fungal diseases, therefore, it is necessary to avoid excessive amounts of it, it is need to apply fungicides and other plant protection products together with carbamide.

The third fertaliization of wheat, in the phase of the tubing – starting of earing is to be aimed on the proteinness of the grain and its quality itself. The yield of wheat is almost not increased because the number of productive stems and the number of ears in the ear have been already formed. Becides, late nitrogen fertilization prolongs vegetation, and therefore 0 work of the synthetic apparatus and, consequently, the productivity (Hospodarenko, 2016; Marchuk, Makarenko, and Kravchenko, 2002; Lisoval, Makarenko, and Kravchenko, 2002).

Features of maize nitrogen nutrition. For maize nitrogen is also an element that determines either the magnitude or quality of yield. The yield of maize and its quality are caused by the biochemical processes of the plant. The direction of these processes depends either upon the biological properties of the plant itself or upon the environment.

For plant growth and development energy is needed. The plant receives it from the sun. An important role in increasing the productivity of plants belongs to fertilization. For the application of fertilizers on a visual basis it is necessary to know what value in the life of plants nutrients have, when, in which quantities and ratios with other elements of nutrition they are digested.

According to the International Institute of Plant Nutrition, maize hybrids for 1 ton of grain and the corresponding amount of by-products spend 18-23 kg of nitrogen (N). However, over 112 kg of N / ha they consume during the first 50 days of vegetation. Maximum consumption of nitrogen per day may reach 3.9 kg / ha (Shpaar et. al, 2009; Kuperman and Andryenko, 1959).

The assimilation of nitrogen significantly depends upon the level of plant providence with available forms of phosphorus and potassium. Viewing high levels of phosphorus and potassium in the soil as well as the consistent injection of nitrogen, phosphorus and potassium fertilizers, the use of nitrogen increases due to the improvement of the synthesis of organic compounds in plants.

While using phosphorus or phosphorus and potassium fertilizers, the use of nitrogen from soil is also being increased in comparison with non-fertilized areas (Lisoval, Makarenko, Kravchenko, 2002).

Nitrogen forms affect the carbohydrate and protein metabolism of maize. In plants to be grown on a nitrate background, carbohydrate content in vegetative organs is higher and total or protein nitrogen is lower than for plants to be grown on the ammonia background.

Synthetic activity of roots in ammonia nutrition is higher than in nitrate one. It is known that ammonia acts on the living matter of plant cells like poison. When it enters the cells of the roots in small quantities the plant can quickly incorporate it into the composition of amino acids, and then of proteins where the ammonium ion to be converted to an amino group loses its toxicity.

At a high concentration of ammonium ions in the substrate, they enter the root in large quantities and the plant does not have time to include them into exchange and free ammonium ions may poison the plant body (Lisoval, Makarenko, Kravchenko, 2002). .

For maize (which has large maternal carbohydrate reserves) the first leaves can first easily deal with ammonia form of nitrogen but then, when the stock of "antidotes" is exhausted and its own carious carbohydrate assimilation device is still poorly developed, sensitivity to ammonia in young plants increases, and then, with increasing leaf surface it decreases again (Kuperman and Andryenko, 1959; Kroyzer et. al, 2014).

The high potential productivity of maize is caused by ability to absorb a large number of elements of mineral nutrition, primarily nitrogen, phosphorus and potassium.

Their assimilation correlates with the accumulation of organic matter. The process does not occur evenly, increasing from the beginning of autotrophic feeding to the flowering of plants and lasts until the wax and physical maturity of the grain. If in the beginning of vegetation, the absorption of nutrients is insignificant, then before volatiling per one day the plants are assimilated at the rate of 1 hectare: potassium 3,44 kg, nitrogen 3,16 and phosphorus 0,87 kg. In the first half of the vegetation, maize uses 70% of potassium, 40% of nitrogen and 28% of phosphorus from the general doze (Lisoval, Makarenko, Kravchenko, 2002). .

In the phase volatiling the absorption of mineral elements is somewhat ahead of the accumulation of dry matter and in future these processes occur almost in parallel.

However, the dynamics of consumption of each element is different. Starting from seedlings and during the entire phase of stinging, potassium is most actively absorbed by reaching a maximum of 10-12 days before the volatiling, then the intensity decreases, although the dry mass intensively increases.

After flowering finish, the absorption of potassium is stabilized. Its lack weakens the outflow of photosynthes products from leaflets, main and supporting roots are weakly developing, spear-like ears with a badly lined upper part are being formed, the grain reaching grain is delaying (Prystash, 2003).

Nitrogen is intensively absorbed by plants from the beginning of the vegetation, reaching a maximum in the phase of volatiling and flowering. At the beginning of lactic ripeness, the absorption of nitrogen is weakened. Its lack sharply reduces grain yield of maize, in comparison with other grain crops.

Increasing the dose of nitrogen gives a greater increase in grain yield than the vegetative mass. Optimum harvest levels are possible with a content of nitrogen in the leaves from 2,8%, and its decrease up to 2,2%, although it does not significantly affect on the coloration of leaves and reduces the yield of grain. Yields may be significantly reduced with insufficient supply of plants by nitrogen during the volatiling - blooming (Kuperman and Andryenko, 1959; Prystash, 2003).

Enough nitrogen nutrition increases the intensity of photosynthesis, improves the amino acid composition and water regime of crops, promotes the accelerated synthesis of proteins and increases their content in the grain, improves pollination and fertilization processes, increases resistance to depletion, especially related to damages of the root system by nematodes contributing to rapid growth of new roots (Lisoval, Makarenko, Kravchenko, 2002; Prystash, 2003).

The specific effect of nitrogen fertilizers is on reaching the grain. The greater the growth of grain yield from nitrogen fertilizers, the faster the grain moisture is attained when it reaches.

Nitrogen is necessary for plants in the phases of differentiation of ears, flower formation and fertilization. To obtain high maize yields of good quality it is important to determine the required rate of nitrogen input accurately.

The critical period of nitrogen assimilation is flowering and seed formation. The excess of this element inhibits the ears formation and increases the accumulation of nitrates in the leaves and stems. Its shortage accelerates the aging of the leaves, weakens the assimilation of other nutritious elements (Coruzzi and Bush, 2001; Kroyzer et. al, 2014).

Thus, the nitrogen providence must be consistent with the need of this element in a particular phase of plants. This is also important from an environmental point of view because while injecting nitrogen fertilizers there is a danger of soil leakage, if they are injected inappropriately and not in accordance with the needs of plants.

The need in fertilizers while maize cultivating is determined by:

- calculation of the injection of nutrients according to the planned harvest;
- monitoring of the nutrient content in the soil;
- operational monitoring of the content of plants elements to have high dynamics in the soil.

Nitrogen is necessary for plants in the phases of ears differentiation, the flowers formation and fertilization.

To obtain high maize yields of good quality it is important to determine the required rate of nitrogen injection accurately.

Due to the fact that maize with stem height of about 20 cm has the greatest need in nitrogen and absorbs at this time more than 50% of the required standard during recent years retail nitrogen injection is to be recommended.

The second dose should not be taken too late to ensure the full use of nitrogen especially during the formation of reproductive organs and during fertilization.

It is recommended to add no more than 50% of nitrogen fertilizers before sowing and 50% - until the plants reach 20 cm height (Kuperman and Andryenko, 1959; Krozyer et. al, 2014). Many authors note that increasing plant nutrition with nitrogen causes increasing the duration of the growing season.

However, with a balanced ratio of nitrogen, phosphorus and potassium fertilizers plants are ripen even earlier than untreated ones. It is generally noted that in such cases when nitrogen significantly increases the yield, i.e. it is used effectively, the ripening of plants does not slow down. It is known that nitrogen takes the first place according to the amount of nutrients to be consumed by maize. Lack of phosphorus and potassium in the soil does not lead to such a decrease of the maize yield which even a slight lack of nitrogen causes.

However, this fact does not diminish the value of other elements. Numerous experiments (exampling to be carried out by the Department of Agrochemistry and the Quality of Plant Products of the NULES of Ukraine) have proved that only a balanced full fertilization provides a high yield of maize zebras (Prystash, 2003; Ermokyn and Krasnytskyi, 2009).

General recommendations to apply nitrogen fertilizers are compiled according to the average data of field experiments, are not able to take into account the features of the hybrid, soil, agrotechnical and weather conditions.

Timely diagnostics and corrections into the application of nitrogen fertilizers is a way of highly efficient use of soil nitrogen and fertilizers nitrogen during the formation of the high quality field crops yield.

NEW INSTRUMENTAL METHODS TO DIAGNOSE PLANT NUTRITION

The aim of the plant diagnostic methods to be included into the complex nutrition diagnostics is to provide continuous control over the conditions of cultivation and plant nutrition correction during the process of vegetation, which makes it possible to utilize the nutrients of soil and fertilizers more fully.

The rapid pace of development of production, the need in operational analysis and decision making about plant nutrition regulation, as well as monitoring the state of crops during their vegetation, the introduction of elements of precise agriculture - all of this results the vital interest to rapid analysis eiother by scientists or by producers.

Nowadays, individual methods are becoming popular. The use of electrophysical soil research methods attracts the attention of scientists as an operational and accurate way of monitoring modern crop production.

One such method is determining the electrical conductivity of the soil, which depends on the concentration and mobility of ions in the soil (Lohinova and Smyk, 2012).

Experienced agronomists can visually establish a tentative level of nitrogen supply to plants but these definitions are not highly accurate. The accuracy of visual diagnosis is influenced by a number of subjective and objective factors: illumination, vision, period of day, etc.

Traditionally, in order to establish the provision of plants with nitrogen, a layer (tissue) chemical diagnostics is used. For its conducting in the appropriate phase of growth and development, whole aboveground part of plants is selected. In this case, the determination of the total nitrogen content in plants is possible only in the conditions of the agrochemical laboratory, which requires considerable time, equipment, chemical reagents, strict observance of the methodology of conducting analyzes and trained specialists.

Chemical express-diagnostics of nitrogen nutrition of plants with the help of portable mini-laboratories does not allow to characterize the supply of plants with nitrogen accurately and in the shortest possible time. The express diagnosis with the N-tester device (Fig. 1) or other biosensors is enough fast and convenient. However, sensors have particular application: for detailed analysis of the field one should t

Figure 1. N-tester device

1. Measuring head; 2. Finger rest; 3. Power switch; 4. ESC: Cancel / return to the previous display; 5. Enter button: Confirm selection; 6. USB connection cap; 7. Hand strap; 8. Battery chamber cover; 9. Strap eyelet; 10. DOWN: Moves the cursor / line marking down or to the right; 11. UP: Moves the cursor / line marking up or to the left; 12. Calibration disc; 13. LCD panel: Displays data and other information; 14. Sliding depth stop: In order to allow the leaf to be measured with the same distance to the leaf edge each time. Can be adjusted or removed when required; 15. Sample slot; 16. Center line: Indicates the center of the measuring area.

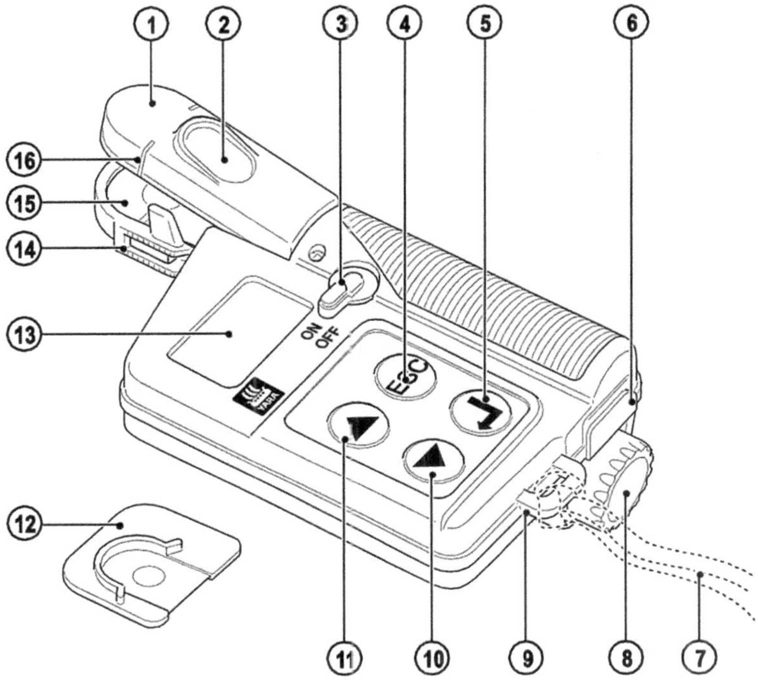

pass the entire area, they require accurate data correlation, because measurements in different parts of one leaf of the plant often give different results.

It is established that during the formation of grain crop yields, as a rule, the largest part of efficient action - up to 73% belongs to nitrogen fertilizers themselves. Bing injected in accordance with the physiological needs of grain crops, 1 kg / ha of nutrients of nitrogen fertilizers provides an average increase in their yields from 10 to 25 kg / ha of grain. However, the constant rise in prices of energy resources and nitrogen fertilizers (N) prompts agrarians to apply them reasonably not only from an agronomic point of view but also from an economic one (Rachkulyk and Sytnykova, 1981).

Timely and direct impact of nitrogen fertilizer on the formation of high yield of maize on grain is possible only on the basis of accurate diagnosis. Establishment of the need to apply nitrogen fertilizers using the N-tester device in critical periods of growth and development of crops. The N-tester is manufactured by the Japanese company «Minolta».

The essence of the method of rapid diagnosis by N-tester device: conversion of solar energy into organic matter is caused by the process of plants photosynthesis.

The direct participant of this process is the unique plant organelles - planet solar energy accumulator – chlorophyll to determine the green color of the leaf surface and plant stems.

It has been established that in the process of photosynthesis of plants a number of elements of mineral nutrition are involved but the greatest correlation is observed between their nitrogen provision and the content of chlorophyll.

Relying on the strong correlation between chlorophyll content and nitrogen content of plants, N-tester device allows to determine their need for nitrogen in critical phases of growth and development according the intensity of leaf coloring with chlorophyll.

While conducting field nitrogen express diagnosis with the device N-tester one must remember that the result should reflect the actual state of the study area. In order to obtain objective and precise results it is necessary to follow the proposed method of conducting crop surveys with the help of the "N-tester" device.

The measurements are carried out on the diagonal of the plot at exactly the relative points, retreating from the edge of the field. To conduct the experiment, it is necessary to choose fully developed leaves of plants. During growth, when visually it is possible to differentiate productive stems and shoots, measurements are carried out on the upper leaves of different stems in proportion to their number, giving priority to the main shoots. In this case it is not necessary to look for the most vivid areas and plants.

The middle of the leaf (in length) is inserted into the measuring head of the device, and the "meter assistant" on the device is exposed in such a way that the measuring sensor is in the middle of the leaf (in width).

If there are spots, yellow spots with external symptoms of disorder of plant mineral nutrition, their growth lag because of the decrease in relief, measurements in these areas do not spend, if the area does not exceed 20% of the field surface. If the variability of the field at the visual estimate is approximately 50% the measurement is performed throughout whole field area.

The measurement principle: light transmission along two wavelengths (650 and 960 nm), measuring area– 2-3 mm, light source 2 LEDs (light diades), SPD receptor (silicon photodiode).

REFLECTION OF LIGHT ON LEAVES OF PLANTS AS A BACKGROUND OF REMOTE SENSING METHODS

The optical properties of plants are determined by the values of reflection, transmission and absorption of electromagnetic radiation by tissue of the plant and may be quantified by using the appropriate coefficients (Rvachëv, 1987; Rozenberh, 1959; Moldau, 1965; Hizhnyi et. al, 2014).

While interacting solar radiation with plants, the parameters of the reflected light flux in the visible and near infrared bands are mostly depending upon the peculiarities of transmission and reflection of light elements from the plant cap.

In the IR range, starting from 3 microns, the value of the heat radiation of plants becomes more significant. This is due to the fact that the size of the angle of incidence of solar radiation to be visible from Earth is 0.5 ° and solar energy is scattered by the earth's surface at a 2π angle.

Within the area of 0.1-1 microns, actual plant radiation is practically absent, and therefore the reflection of solar radiation is almost the only source of secondary radiation because plant fluorescence and other types of redundancy include only few percent of the absorbed light (Brandt and Taheeva, 1967).

According to the optics laws to describe the behavior of electromagnetic waves on the boundary between two media, a part of the incident radiation passes into the middle of another medium, and part of it is reflected from the boundary of the section and remains in the same environment where the incident beam propagates.

The intensity of the reflected beam, as well as the phase and polarizing characteristics of the reflected wave, depend upon the ratio of the optical density of the phases at the border of which reflection occurs and upon the magnitude of the incidence angle.

For the mirror image, the relationship between these values is described by Fresnel's formulas.

The nature of the spatial distribution of the intensity of light to be reflected from the boundary of the section depends upon the quality of the reflecting surface.

There are cases of mirror and diffuse reflection.

The first one is realized if the surface irregularities are much smaller than the wavelength of the incident light. With the size of the inequalities to be compared with. the reflection has diffuse nature.

In this case, the spatial distribution of the intensity of the reflected light depends upon the size and location of the inequalities (Kondratev and Fedchenko, 1982; Rvachëv, 1987).

Let's observe some patterns of mirror reflection. The surface of the leaves is the boundary of the distribution of air-internal leaf tissue to separate the optical medium from less dense to the more dense. Relation of refractive indices is $n_{;ucm} / n_{no2.} > 1$ In such system the reflection coefficient is more when the ratio of the refractive index of the two media is more.

At very small angles of the light incidence $(\phi \to 0)$, that is in cases of normal fall, the reflection coefficient does not depend on the polarization of the incident light and equals $(n_2 - n_1)^2 (n_2 + n_1)^2$ where n_1 and n_2 - the coefficients of refraction of the first and second phases. At $\phi \to \pi / 2$ (sliding fall) the reflection is complete, i.e, its value goes to 1. For $\phi = \pi / 2 - x$, where x is the refractive angle, the reflected beam is completely polarized. The angle $\pi / 2 - x$ is called as the Brewster angle, and its value depends on the ratio of two media refractive index because $\pi / 2 - x = arctg \cdot n_2 / n_1$.

Note that these regularities are performed for the case of reflection from the boundary of non-absorbing surfaces. Absorption by one or two media introduces specific changes depending on the intensity of the reflected beam from the parameters of the incident one (Brandt and Taheeva, 1967; Kondratev and Fedchenko, 1982; Rachkulyk and Sytnykova, 1981; Rvachëv, 1987).

Viewing the reflection from the boundary of the two media with changes not only for intensityand the incident wave amplitude but also its other parameters exampling polarization, we will observe changes in this characteristic.

Recently, there have been papers to study polarized radiation characteristics to be reflected from plant leaves and plant cap in order to clarify the information capabilities of this parameter for testing the state of plants.

Light radiation is the electromagnetic waves that belong to the corresponding frequency range. Their distribution may be detected by periodic changes in the intensity of electric and magnetic fields at any point in space.

The simplest case is the so-called flat waves, when fluctuate of the electric $(\vec{E})$ and magnetic $(\vec{H})$ fields are on mutually perpendicular areas.

The direction of wave passage is perpendicular to the direction of the vector $\vec{E}$ and $\vec{H}$ (Fig. 2). The light wave is linearly polarized if the vectors $\vec{E}$ and $\vec{E}$ have a well-defined orientation to be relative to the vector.

Figure 2. Direction of vibrations of vectors of electric $\left(\vec{E}\right)$ and magnetic field $\left(\vec{H}\right)$ relatively to direction of propagation of an electromagnetic wave $\vec{V}$

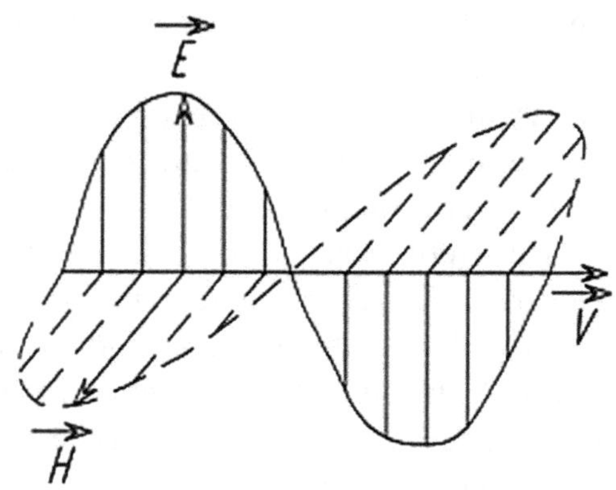

Let's review the case of mirror image from the boundaries of two media (Fig. 3) (Rozenberh, 1959). As the relative position of the vectors $\vec{E}$ and $\vec{H}$ is tightly bound, let's observe changes only for one of them. The vector $\vec{E}$ is decomposed into two components, one of which $\left(E_p\right)$ lies in a area with falling beam and a normal N to the boundary of the distribution of two media with refractive index n_1 and n_2. The second component is perpendicular to the fall plane.

Similar components of the electric vector are also detected in the reflected wave R_p to be polarized in the incidence area R_s and to be perpendicular to it. Intensities R_p and R_s depend upon the magnitude of the ratio n_2 / n_1 and the incidence angle.

Fig. 3 depicts the dependence of reflection coefficients $r_s = (R_s / E_s)^2$ and $r_p = (R_p / E_p)^2$ on ϕ for $n_2 / n_1 = 1,52$, which approximates the system of air-internal tissue of the leaf.

Figure 3. Location of vectors components $\vec{E}$ and $\vec{H}$ in incident and reflected waves with mirror reflection from the boundaries of two media

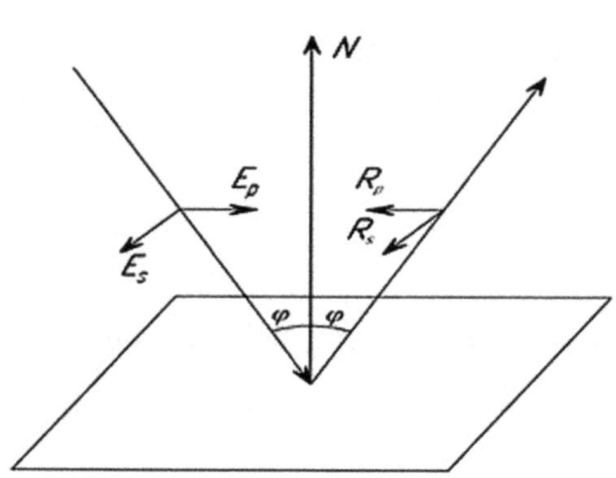

Figure 4. Dependences of reflection coefficients for perpendicular (r_s) and parallel polarized waves (r_p) from the incidence angle o ϕ

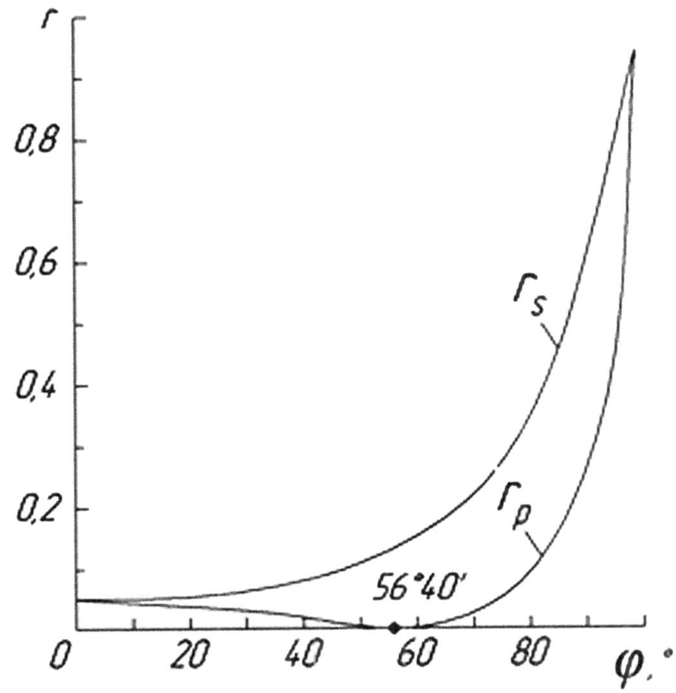

From the graph (Fig. 4) it is seen that when ϕ is near to 0, $r_s = r_p$; when ϕ approximates to $\pi / 2$, the value of the general components goes to 1.

At $\phi = 56°40'$ a corresponding Brewster angle $(arctg(n_2 / n_1))$, for this ratio of refractive indices $r_p = 0$ and the reflected beam is only r_s, thus it is completely polarized.

Such a dependence is possible only if the distribution boundary is an ideal mirrorless plane in the space. But even well-polished metal or glass surfaces differ in their properties from the ideal surface, so the light to be reflected from them at Brewster angle and even the light to be reflected from the surface of the leaf is partially polarized. But the degree of polarization in this case is higher comparing with other incidence angles.

The degree of polarization is characterized by $x = (I_{II} - I_I / I_{II} + I_I)$ indicators where I_{II} and I_I correspond to the intensity of light to be transmitted by the polarization device at polarization, parallel or perpendicular to the area of observation.

Let's observe the laws of diffusion reflection which is in fact the scattering of light in all directions when it hits the distribution boundary that has roughness whose dimensions may be compared with the wavelength or in the case of a heterogeneous internal structure of the second medium.

Both of these moments are essential while looking at the reflected light from the plant leaves. If the irregularities of the distribution boundaries are a set of definitely oriented areas, the size of which is larger than the wavelength, the surface is smooth but uneven.

The angular dependence of the intensity and polarization of the reflected light is caused by the light to be reflected from these areas and is found according to the Fresnel formulas, taking into account the statistical areas distribution to be relative to the orientation.

The highest brightness corresponds to the reflection from the most numerous group of areas with a certain orientation. If the location of areas is chaotic, the scattering indicatrix corresponds to a more or less spatial distribution of the brightness of the reflected light. If the size of the areas on the rough surface corresponds to the order of the wavelength, the regular role of the planes diffraction on the distribution between adjacent planes and the interference of the diffracted waves will play a major role. In this case as well as in the case of large incidence angles, an intense component of mirror reflection may be formed.

Exampling angular distribution of the intensity of the light to be scattered from the leaves surface, one can note the data to be obtained by H. Moldau (1965). Fig. 5 shows the indications of light scattering from maize leaves. The nature of the scattering strongly depends upon the angle of incidence to be projected from the above laws of optics.

Viewing the chaotic arrangement of inequalities, the spatial distribution of the intensity of scattered light is confirmed by the " M theory ". The scattering indicatrix is partially elongated in the direction of propagation of the light wave. The larger is scattering indicatrix the larger is the particle size.

The scattering radiation is partially polarized.

In the case when the surface inequalities are smaller than the wavelength, light scattering takes place according to Rayleigh law, it is polarized at an angle of 90 ° to the surface of the fall and has a strong spectral dependence: the intensity of the scattering light is proportional $1/\lambda^4$ and the short-wave part of the spectrum dissipates much more strongly.

While lighting such surfaces with white light, they are appearing to be blue- lightly blue.

An ideal scattering surface to give uniform scattering along all directions is called as «Lambertian» or orthotropic but the most of real surfaces are non-rigid.

At small angles of falling light on the leaf surface much of it comes inward and is reflected from the inhomogeneous internal structure.

Such situations are also considered in the theory of diffusion reflection. In this case the phenomenon has a volumetric nature and its laws are determined by the effects of multiple light scattering to fall into the middle of the leaf.

If the heterogeneous medium is absorbing, the intensity of the absorbed light drops sharply in the spectral bands corresponding to the absorption bands.

The angular distribution of the reflected light is determined by the type of scattering matrix and varies with the changes in the ratio of the volumetric scattering γ and absorption γ coefficients.

The polarization of the reflected light also greatly depends on the ratio γ.

As a rule, the degree of polarization in this case is low. Thus, depending on the angle of incidence of light, you can obtain information about the condition of the leaf surface or its interior fabrics. If the leaf is illuminated by scattered light the separation of information of two kinds may be obtained by means of the separation of polarized and unpolarized reflected light.

Summarizing the above, we note that information about the condition of plantings may be obtained from the following parameters of reflected light plant leaves: integral intensity, spectral and angular intensity distribution.

Figure 5. Angular dependence of the intensity of radiation to be diffused from the surface of the maize leaf in blue (solid curves) and near infrared (dashed curves) at angles of incidence 0 ° (curve 1) and 60° (curve 2)

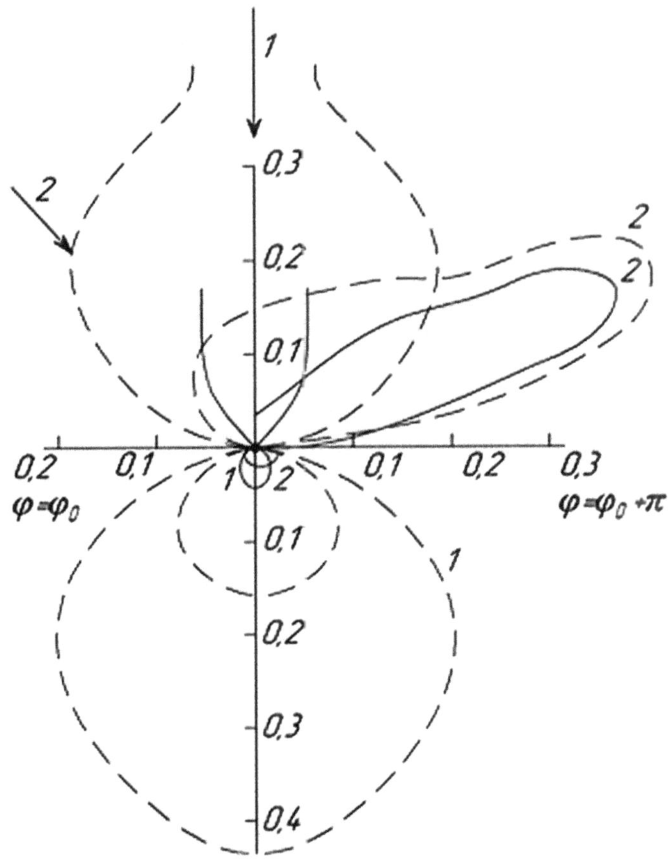

The state of the reflected object leaf is caused by genetic deterministic features: anatomical structure, regularity of time development, physiological functional activity and also depends on the environment in which the plant grows.

Therefore, changing light parameters of light to be determined by changes in reflected light flux carry information about all components.

Thus, the interconnections between these or other parameters of the reflected light and the physiological characteristics of the leaves make it possible to implement some methods to monitor plant cap conditions.

THE METHODS TO ESTIMATE SPECTRAL CHANNELS WITH A REFERENCE SPECTRAL SENSOR

To assess the probability of the results of the above approach, we conducted a comparative study using the Slantrange 3 sensor equipment as well as GoPro HERO4 Black digital cameras with standard IR and IR lens.

The pictures were taken in jpeg format. Determination of color constituents intensity was carried out using the software of its own development - Land Damage Expert (Fig. 6).

LDE program allowed to use the manipulator (mouse) in photo to allocate 2 experimental areas for which statistical processing of graphic data was carried out. Noting the standard Jpeg format not to define RGB values for each specific pixel but for areas with same parameters the LDE program converts Jpeg data into BMP format in which the brightness values of the additive color model components are determined for each pixel.

The program predicted the possibility of data statistical processing. So, to remove pixels to be associated with contamination of the prototype with additional substances, it is possible to ignore the calculations of pixels of magnitude, RGB of which differ from the average for the corresponding percentage, which is determined when the analyzer is set up.

To be separate research parameters different plants were selected in the botanical garden of the National University of Life and Environmental Sciences of Ukraine. The photo shooting took place one-hour round in cloudy weather without artificial lighting inside and outside greenhouse under polycarbonate. For measurements, the same object, which was recorded by all three cameras, was visually selected. The results are presented in Table 1. Basing upon the given data about measurements of color constituents

Figure 6. Land damage expert interface and block setup

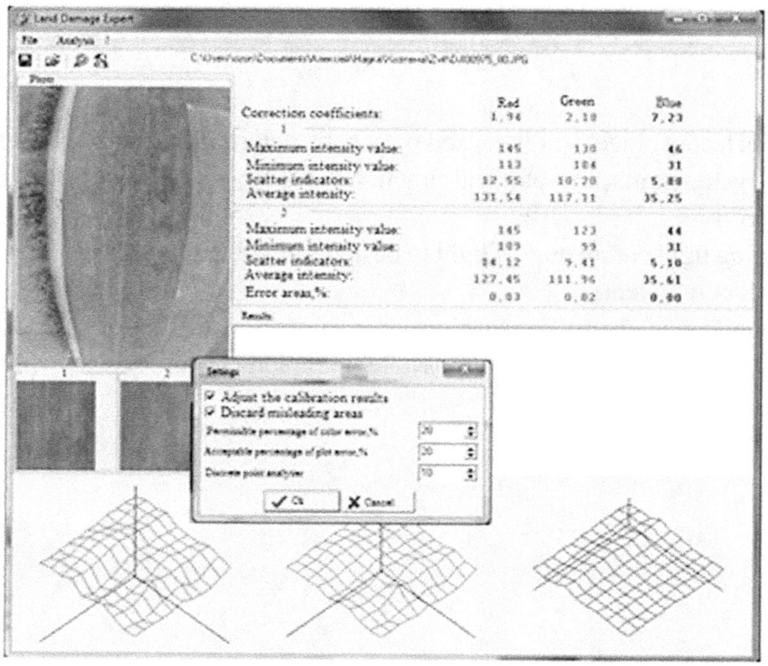

Table 1. Results of a comparative study of the intensity of color constituents by spectral channels for plant objects

Im\Ch		R	G	B	NIR	GP - Sl			GP nir – Sl nir		
						R	G	B	R	G	B
	GP	118	129	115	-	33	40	40	-	-	-
	GP nir	178	57	147	-	-	-	-	113	-8	82
	Sl	85	89	75	65	-	-	-	-	-	-
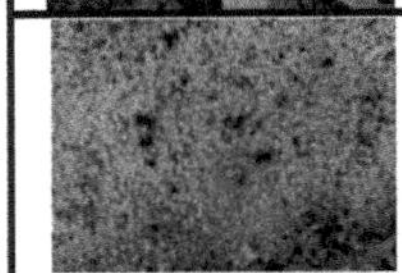	GP	141	134	104	-	61	55	36	-	-	-
	GP nir	194	78	168	-	-	-	-	134	18	108
	Sl	80	79	68	60	-	-	-	-	-	-
	GP	104	145	43	-	4	42	-14	-	-	-
	GP nir	199	80	176	-	-	-	-	125	6	102
	Sl	100	103	57	74	-	-	-	-	-	-
	GP	105	128	114	-	32	16	54	-	-	-
	GP nir	193	69	165	-	-	-	-	132	8	104
	Sl	73	112	60	61	-	-	-	-	-	-

intensity by different cameras, it has been found that accuracy is unacceptably low, therefore, the red channel's difference was 57 units, equaling almost 20% of the total scale. Observing IR range, it was found that the nearest value to the results of the measurements by Slantrange infrared channel there was 2d channel (green) but the results fluctuated in significant ranges - up to 26 units.

Thus, the approach to comparative study by different cameras to derive corrective relationships did not yield acceptable results.

METHOD OF INSTRUMENTAL MEASUREMENT OF SPECTRAL CHANNELS

As a source of continuous light spectrum a CFL-type incandescent lamp was used. To obtain an image of an object in the light of a certain wavelength, a modified double monochromator DFS-12 was used, which under certain experimental conditions allowed to emit light with a wavelength λi with an accuracy of $\Delta\lambda < 0.8$ nm.

The input slot of the monochromator played the role of the subject. The image of the object was projected on the output slot of the monochromator where the camera was located. The size of the output and the input slit was selected to obtain a bright image without the «illumination» of the camera. In detail, such laboratory equipment is described in paper by Yu.A. Hizhnyi et. al (2014). The picture of the laboratory equipment is shown on Fig. 7.

In the studies the step change in wavelength was 10 nm. For a camera with a standard lens, a range of wavelengths ranging from 400 to 850 nm was investigated.

Figure 7. Laboratory equipment for calibration of digital cameras spectral channels (Kyiv National Taras Shevchenko University, Faculty of Physics)

While using IR filter, a range of 500-1000 nm was observed. Below it is an image where the intensity of any of the channels did not exceed 5 units. The hole, through which the photomatrix illumination was carried out, had the shape of a square. Noting different saturation within the frame to minimize the possible effect of illumination three options were separately observed:

1. A sample of square shape, in which the dimensions are determined by length because in all frames the height was of constant value and the image width with clear boundaries of the colored image varied.
2. A sample of a rectangular shape in which the height is equal to the previous sample and the width is reduced to three (IR) and four (optical range) times, with the centering was performed relatively to the areas with the maximum saturation of the channels.

Figure 8. The spectra to be obtained in the optical range by GoPro HERO4 Black camera with regular and infra-red lenses, respectively

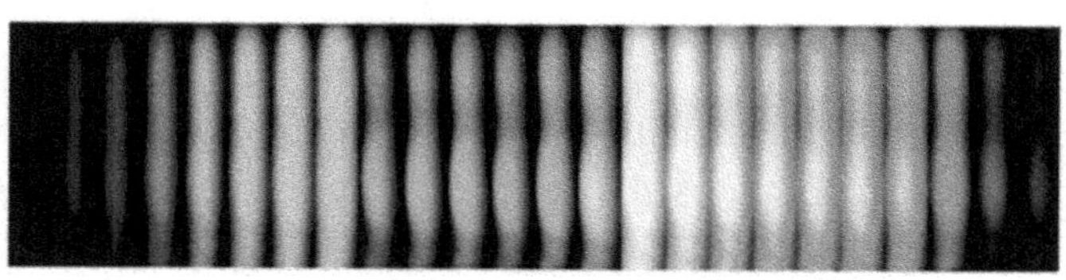

Figure 9. The spectra to be recorded by GoPro HERO4 with IR lens MP

3. The sample was taken from the upper part of 2 type sample and had a square shape which side width equaled the width of the original sample.

The given results are presented on Fig. 10.

While using IR lens the radiation was fixed at two channels – 640–690 nm, that is actually part of the "red" section was of the optical spectrum and 820-880 nm was of the infrared range.

In this case three available channels have their own specifics, namely:

Figure 10. Intensity of spectral channels depending upon the wavelength for the GoPro HERO4 Black Camera

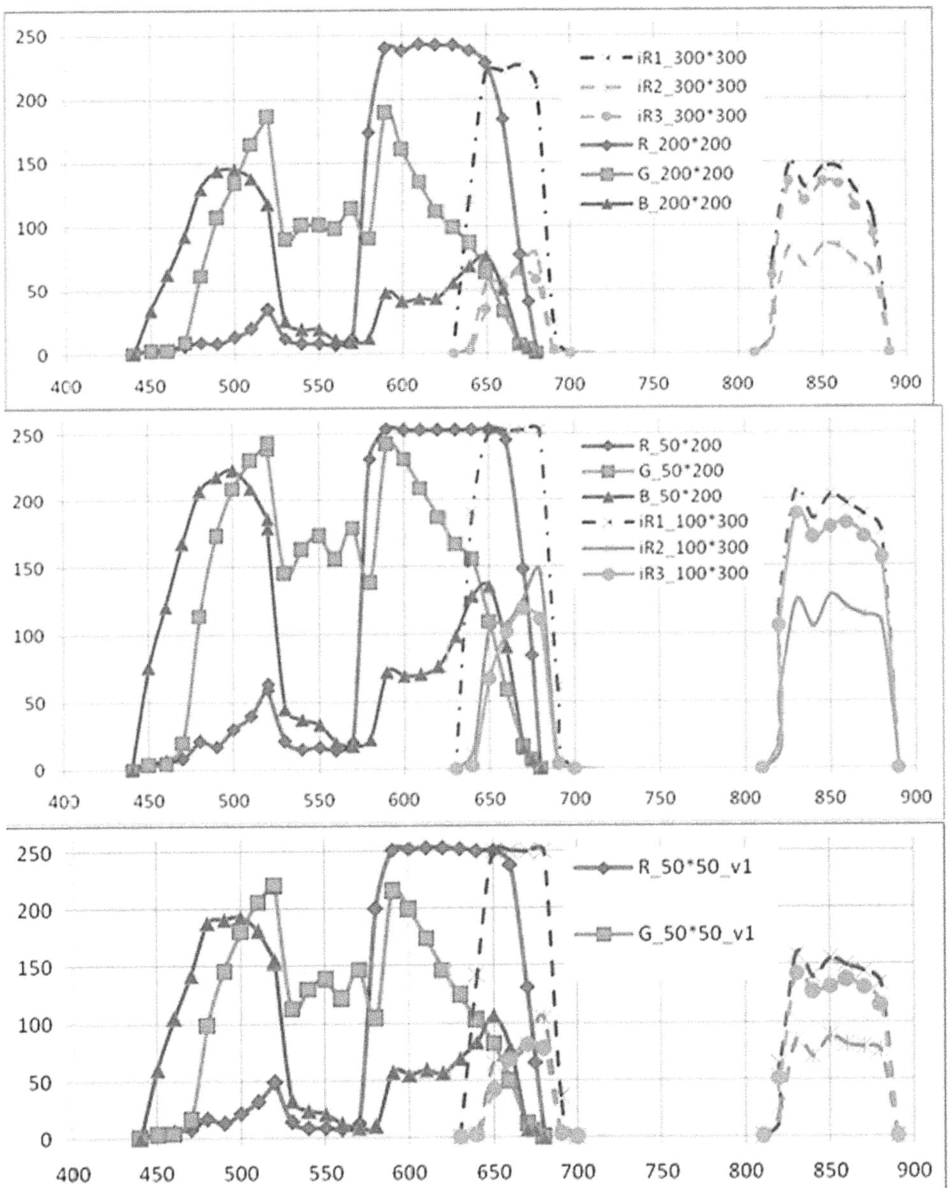

- iR1 to be corresponded to the red channel in jpeg format has a maximum limit in the optical range and a maximum (130-160 units) in the IR spectrum;
- iR2 to be corresponded to the green channel in jpeg format has approximately the same intensity values either in optical or in infrared spectra;
- iR3 to be corresponded to the blue channel in jpeg format has maximum values in the IR spectrum and the lowest of all channels in "red" range.

Thus, we can conclude that it is possible to observe the corresponding components of the additive model of color formation as separate informative channels while using IR lens digital cameras for monitoring.

THE GROWTH INDICES TO BE ADAPTED FOR UAV

Currently, the vegetation indices to be created for satellite platforms are the most widespread. Previously dealt with the peculiarities of the organization of vegetation indices, basing upon specific parameters exampling first of all, the atmosphere but there are additional agronomic determinants. Two basic assumptions were made during the development of vegetation indices for satellite platforms:

1. the spectral indices of plants are informative to determine their condition;
2. open ground in the picture will form in the spectral space so-called «ground line» (GL) to be a straight line.

For serial UAVs used to survey industrial fields, standard solutions for changes in light value are the use of an additional anti-aircraft sensor, such as the Light Sensor for the MicaSense RedEdge camera complex, represented by (T. Duan et al., 2017) and (Jyun-Ping Jhan et al., 2018). Hardware lighting fixation will not require ground patterns used in satellite technology, but will increase the cost of UAV, moreover, this approach will be relevant only in conditions of even lighting, but in the presence of clouds may be get large errors at analyses of image. An alternative to a specialized sensor is the possibility of using calibration of the data from the built-in meter of the main touch-sensitive equipment proposed by (Saberioona M. et al., 2014). The technique for using the standard digital camera exposure meter was developed by (Korobiichuk I. et al., 2018) for the Canon A460 camera with regard to the white balance adjustment parameters for field photography. Mathematical equations for taking into account the changes in lighting were presented in (Lysenko V. et al., 2017) where the calculation of the intensity of the color component was based on the value of Light Value for the DJI PHANTOM 2+ VISION FC200 DPL camera. The authors have experimentally established that in a certain range the dependence of the intensity of the components of color on the value of Light Value is linear in nature. It was used for lighting correction.

Spectral indices for spring crops were considered exampling the grade Xiayvo barley, the phase of «tubing». vegetation. The research was conducted on 07.06.2017 in clear weather (fig. 11).

Camera debugging options were based on metadata: Exposure Time while shooting in 0.00252 s (FC200) optical range -; Exposure Time while shooting in 0.00325 s (GoPro); infrared range. The number of diaphragms and «white balance» are identical to the previous study for winter wheat. In order to avoid the possibility of other objects falling into the frame except the plants leaves it was «dropping off »the results to be carried out to differ from the average value more than 10%. Due to the non-consideration

Figure 11. Experimental barley plot in optical and infrared ranges (filming date -07.06.2017)

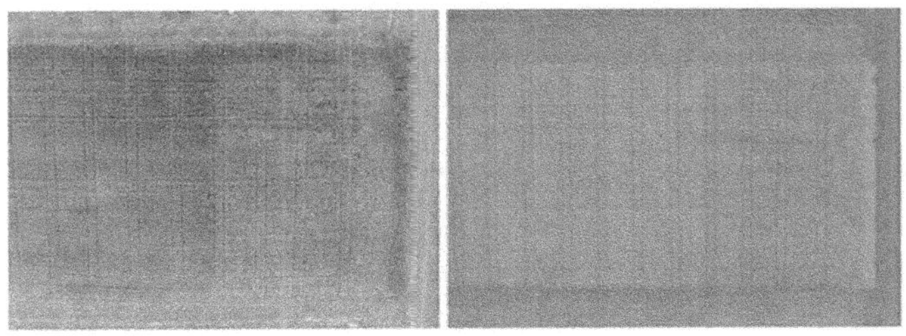

of plots to be caused by the soil and other random objects, the intensity of the components of the color did not change by more than 3 units. As for the previous crop the amount of nitrogen was calculated separately in the terrestrial part of the plant and in three upper leaves. The given data are presented graphically (see Fig. 12). Perspective vegetation indices are given in Table 2.

Spectral indices for cultivated crops were observed exampling maize, a hybrid Pioneer P 8816, a stage of 4-6 leaves development. Experiments were carried out on 07.06.2017, as with barley, using optical and infra-red lenses (Fig. 13). In the study of spectral indices for a field with maize at the first stage, a preliminary statistical treatment approach was worked out exampling wheat and barley. The LD expert software to be used on previous crops was created basing upon the assumption that the number of pixels in the image corresponding to the soil and other objects other than plants does not exceed 10%. For crops of continuous sowing, a larger percentage of pixels to be caused by soil, is possible at the initial stages of vegetation, when the plants are small in size. Particularly clearly this is noted for cultivated crops. It should be noted that while probing winter crops in the early stages of vegetation, it is estimated, first of all, not the state of mineral nutrition but the loss of plants after winter.

For cultivated crops exampling maize, a high percentage of pixels with non-growthal objects is also observed in the vegetative stages to be recommended for nutrition.

Either for vertical or for lateral sounding practically the whole leaf part of the plant falls into the frame, in contrast to solid crops. Therefore, the amount of nitrogen was determined throughout whole terrestrial part without upper leaves separating. The results of maize crops studies using LD expert program, are presented in Table 3. The obtained data showed that the method of «dropping off » the soil was based on extraction of parts of the image the intensity to differ from the average more than a certain percentage for the experimental area was ineffective for the cultivated crops. For the optical range mentioning difference between the components color intensity values of the soil and of the field with 6-16 maize units, the «dropping off » the plots according to the described principle managed to adjust the result throughout whole field by less than 1 unit.

While using an infrared lens the difference between the spectral indices of the soil and the maize fields was lower than for the optical range and therefore the analysis method to be described above showed even worse results. While approximating the experimental data between the intensity of the color components and the nitrogen content in the form of linear dependence the highest 0.88 determination coefficient was found for red component and it was 0.66 for green component while the infrared lens had the largest 0.62 index in iR1 channel.

The obtained results indicate that the optical range is more informative. It is expedient to be used in further data processing to be related to the recognition of the soil image which will allow to create

Figure 12. Dependence of color components intensity upon the amount of nitrogen in barley dry matter, the vegetative phase - the tubing: for the above-ground part of the plants (top graph) and three upper leaves (below graph)

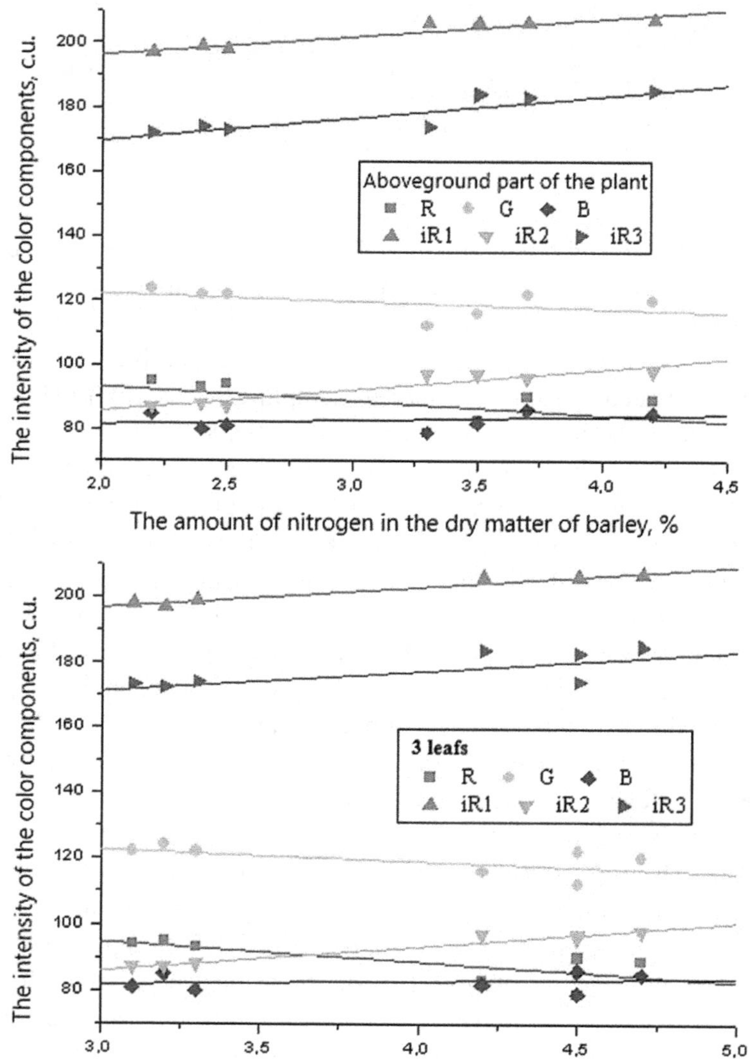

spectral indices for cultivated crops more accurately and probably. Dropping off the areas with the soil image was carried out basing on its spectral parameters. In the industrial use of the system, it is expedient to determine the soil parameters basing upon the ground roads to be available on the fields, necessarily taking into account the road side because there soil itself is not pressed with wheels and has the nearest optical indicators to the field ground.

While analyzing a jpeg image, the bmp format photos were converted into bmp format, for which each pixel corresponded to all three components of the color: red, green and blue for the optical range and, accordingly, iR1 ... iR3 while using IR lens. If all three indicators of the pixel spectrum of the experimental plot were within the range of values to be corresponded to the soil such pixel was removed

Table 2. Perspective spectral indices for barley «tubing») phase

№	Range	Name	Equation	R2		hmax/hmin
				above-groundpart	3 leafs	
1	RGB	R	R	0,17	0,41	1.2
2	RGB	G	G	0,02	0,21	1.11
3	RGB	B	B	-0,1	-0,1	1.09
4	RGB	GR	GR = (G - R)/(G + R)	0,28	0,54	1.305
5	RGB	GB	GB = (G - B)/(G + B)	0,61	0,74	1.218
6	NIR	iR1	iR1	0,88	0,95	1.05
7	NIR	iR2	iR2	0,86	0,95	1.13
8	NIR	iR3	iR3	0,75	0,5	1.08
9	NIR	iRa	iRa = (iR1 - iR3)/(iR1 + iR3)	0,011	-0,16	1.49
10	NIR	iRb	iRb = (iR1 - iR2)/(iR1 + iR2)	0,845	0,945	1.09
11	NIR	iRc	iRc = (iR3 – iR2)/(iR3 + iR2)	0,31	0,63	1.164
12	RGB+NIR	RiR3	RiR3 = (iR3 - R)/(iR3 + R)	0,54	0,7	1.31
13	RGB+NIR	GiR1	GiR3 = (iR1 - G)/(iR1 + G)	0,357	0,57	1.17

Figure 13. Maize plot (hybrid Pioneer P 8816) in optical (a) and infrared (b) bands

Figure 14. Experimental stationary plot, 4-6 leaves stage maize field

Table 3. Spectral indices of maize crops without the use of soil recognition technologies

Nitrogen, %	% Error	R		G		B		iR1		iR2		iR3	
		I	False, %	I	False, %	I	False, %	I	False, %	I	False, %	I	False, %
Soil	10	152,7	2	138,3	2	149,8	1	185	0	56	0	137,1	0
2,7	10	136,1	12	132,3	12	136,3	12	181,1	0	58,6	0	137,9	0
2,7	6	136,2	26	132,2	28	136,4	26	180,9	2	58	3	137,4	3
2,7	4	136,4	42	132,6	46	136,5	42	180,5	11	57,3	11	136,8	13
2,7	2	136,6	68	133,5	72	137,4	66	180,6	43	57,6	38	137,4	44
2,9	10	134,9	14	133,1	15	132,5	14	183,4	0	62,3	1	141,7	1
3,1	10	137,8	16	140,8	16	133,4	16	186,9	0	66	0	146,1	1
3,7	10	139,6	14	140,9	14	135,5	14	187,4	0	65,2	1	146,1	1
3,5	10	139,9	13	142,3	13	135,4	13	187,7	0	66,6	0	147,2	1
4,2	10	141,4	12	143,5	13	136,5	13	188	0	65,4	0	146,5	1
4,2	8	141,5	19	143,3	20	136,5	19	188	0	65,2	2	146,2	2
4,2	6	141,6	29	143,5	30	136,5	29	187,7	3	64,6	5	145,7	6
4,2	4	141,7	45	144,3	47	136,5	45	187,3	13	64,1	15	145,5	19
4,2	2	141,7	70	144,6	71	136,7	71	187,6	43	64,6	48	146,5	56

for further calculations. By the pixels that were not recognized as soil, the average intensity of each channel was deduced.

Besides, it was the percentage of pixels in the experimental plot to be determined exactly for the plants. The obtained results are presented in Table 4. In assessing the results to be obtained after soil recognition and filtration (method 2), it was found that they fundamentally differed from the values to be obtained while "dropping off" data to differ from the average more than the established percentage (method 1). Thus, using IR lens according to method 2, regardless of the channel, with the increase in the amount of nitrogen, the intensity of the color constituents decreased, in contrast to the results to be obtained in the studies according to method 1. To our mind this fact is explained by the following: most pixels in the experimental plot corresponded to the soil itself but not to the plants. Using method 1 increasing plants area with a better state of nitrogen nutrition, the magnitude of the average intensity increased.

In the optical range the number of pixels to be recognized as plants is significantly less than the red-infrared range due to the more refined filter settings. IR lens is the best due to the greater difference between the soil and plants comparing with the purely optical range. Proceeding from this it was assumed that it is expedient to use the IR lens itself. For field studies in optical range irrespective of the filtration technique with the increase of nitrogen amount the magnitude of the color intensity constituents were increasing while they were decreasing in the phytotron. It may be explained by the following factors:

- in the phytotron the methods of making samples were worked out and such limit values were studied as stress to be associated with excess nitrogen (up to 7%), whereas in the stationary conditions the maximum nitrogen value equalled optimal one. (≈4.2%).

Table 4. Spectral indices of plants with soil filtration

N, %	R	G	B	Plants, %	iR1	iR2	iR3	Plants, %
2,7	129	131	128	5,8	206	109	172	16,2
2,9	133	138	128	18,9	207	113	175	23,6
3,1	133	145	128	20,6	201	94	165	37,2
3,7	135	148	126	24,5	200	92	164	38,7
3,5	137	150	128	16,3	201	95	166	41,3
4,2	138	150	129	23,9	201	92	164	43,8

- taking photographs in laboratory conditions was carried from close distance, thus there were fixed such elements as veins on leaves not to be fixed in field conditions from 100 meters distance.

While creating samples in a phytotron for spectral studies, these factors should be taken into account, as well as during preparation of samples and during measurements. While approximating the results to be obtained in the optical spectrum in the form of a linear dependence, the maximum determination coefficient was obtained for the red and infrared components – of 0.76 and 0.67 in accordance. The dependence for blue component to be obtained in field conditions, as well as in studies in the phytotron, is lower. While observing the amount of nitrogen in the range 2.7-4.2%, the maximum difference for the maize color intensity component was observed for green channel to vary in the range of 131-150 units, that is, the sensitivity is low. It is interesting fact that with increasing nitrogen content in plants their vegetative mass was increasing too, and hence, the percentage of the area under plants in the picture. We have fixed such changes from 6 to 24%. It is expedient to be used while assessing the state of maize mineral nutrition either separately or in combination with the assessment of color constituents intensity.

Fig. 15 shows the dependence of the index involving either the intensity of the color component or the percentage of the plants on the experimental plot. The use of integral indices, that is, those that take

Figure 15. Dependence of the integral index IR2_PR upon the amount of nitrogen in maize ground part

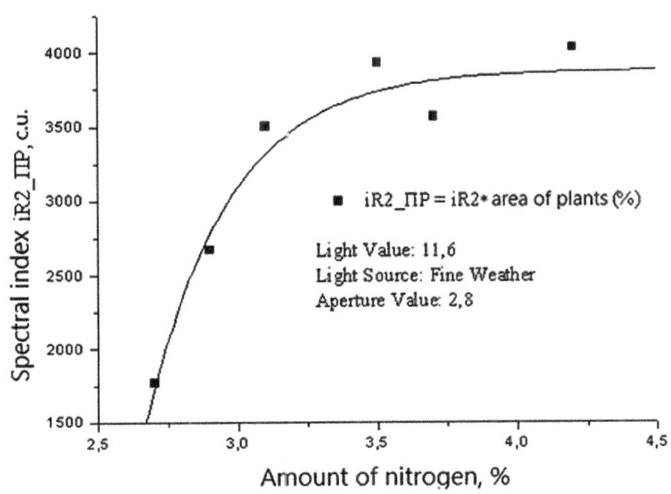

into account either the intensity of radiation to be reflected by plants or the area underneath them, may be promising for cultivated crops to detect sensitivity and selectivity.

It has been established that the spectral channels iR (RED + NIR) have a good sensitivity in the range up to 3% nitrogen and the geometric component is within the range from 3 to 4%. The combination of these two indicators in the integral index, namely the color component intensity (spectral component) and the dimensions of plants (geometric parameter) allowed to achieve good sensitivity in a critical range of maize nitrogen nutrition up to 3.5-4%. We consider this stress index to be more acceptable for the automation of differential fertilizer processes. In addition, we carried out research of crops by thermal range sensors (Fig. 16). The use of the thermal imager was caused by the fact that the deviation of the power increased comparing with the optimal state of temperature influence on the plants state

We could not find a reproducible relationship between the state of nitrogen supply and the thermal portrait of plants. However, such a range may be used to assess the moisture state and plant density state.

Probably, to obtain the reproducible data while using a thermal imager, it is necessary to develop methods of calibration viewing reference objects separately because metadata to be analogous to digital photographs were not stored by the used device.

Analysis of literary data shows that at the current level of development of sensory equipment and technologies for tweed sensing information processing is possible and may be realized on a production scale for specific crops, in certain phases of vegetation.

CONCLUSION

In laboratory conditions the elaborated method of spring crops physical modeling is developed to allow operative creation of gauge dependencies for new varieties and hybrids.

Own elaborated method to use UAVs for remote shooting of agricultural crops and obtaining images is to be used to determine the level of nitrogen provision for grain crops has been developed.

For radiofrequency correction of lighting changes while mineral nutrition state monitoring, it is recommended to use a method with the application of official data to be calculated on the basis of camera's regular camera exposure meter measurements.

Figure 16. Photo from the experimental stationary plot using FLIR thermal imager

It is advisable to use infra-red lens for monitoring the state of individual crops using digital non-specialized cameras instead of optical one.

Exampling GoPro hero 4 digital camera it has been found that there are three channels to use IR lens which have the characteristics of fixing radiation either in the infrared range or in the red area of the optical range.

The special features of channels to be obtained with the use of an IR filter, allow them to be used either independently or when constructing spectral indices for monitoring the condition of plantings.

There are components of the additive model of color formation, the most informative for the spectral diagnosis of nitrogen nutrition to be established, namely:

- for wheat, the most informative channels are R and G where the determination coefficient for linear approximation is greater than 0.89.
- for barley, for the GoPro hero 4, the most informative channels are iR1 and iR2 to be obtained using IR lens for which the determination coefficient for linear approximation would be 0.95.
- for cultivated crops (for example, maize) it is advisable to use technologies that use algorithms of the preliminary recognition of plants images or additional objects.

The most optimal are integral spectral indices that take into account either the spectral indices of plants or their geometric parameters.

REFERENCES

Brandt, A. B., & Taheeva, S. V. (1967). *Optical parameters of plant organisms*. Moscow: Science.

Coruzzi, G., & Bush, D. (2001). Nitrogen and Carbon Nutrient and Metabolite Signaling in Plants. *Plant Physiology*, *125*(1), 61–64. doi:10.1104/pp.125.1.61 PMID:11154297

Duan, T., Chapman, S. C., Guo, Y., & Zheng, B. (2017). Dynamic monitoring of NDVI in wheat agronomy and breeding trials using an unmanned aerial vehicle. *Field Crops Research*, *210*, 71–80. doi:10.1016/j.fcr.2017.05.025

Ermokhyn, Yu. Y., & Krasnytskyi, V. M. (2009). *Modern approaches to the diagnosis of nitrogen nutrition and the use of nitrogen fertilizers: methodical recommendations*. Omsk: Variant-Omsk.

Hizhnyi Yu, A., Nedilko, S. G., Chornii, V. P., Slobodyanik, M. S., Zatovsky, I. V., & Terebilenko, K. V. (2014). Electronic structures and origin of intrinsic luminescence in Bi-containing oxide crystals BiPO4, K3Bi5(PO4)6, K2Bi(PO4)(MoO4), K2Bi(PO4)(WO4) and K5Bi(MoO4)4. *Journal of Alloys and Compounds*, *614*, 420–435. doi:10.1016/j.jallcom.2014.06.111

Horodnii, M. M. (Ed.). (2006). Agrochemicals (4th ed.). View. LLC "Ariste".

Horodnii, M. M., Melnychuk, S. D., & Honchar, O. M. (2006). Applied biochemistry and quality management of crop production. Ariste.

Hospodarenko, H. M. (2016) Fertilizer of agricultural crops. Kiev: LLC "SIC GRUP UKRAINE".

Jyun-Ping, J., Jiann-Yeou, R., & Haala, N. (2018). Robust and adaptive band-to-band image transform of UAS miniature multi-lens multispectral camera. *ISPRS Journal of Photogrammetry and Remote Sensing*, *137*, 47–60. doi:10.1016/j.isprsjprs.2017.12.009

Kondratev, K. Ia., & Fedchenko, P. P. (1982). *Spectral reflectivity and vegetation recognition*. Gidrometeoizdat.

Korobiichuk, I., Lysenko, V., Opryshko, O., Komarchyk, D., Pasichnyk, N., & Juś, A. (2018). Crop Monitoring for Nitrogen Nutrition Level by Digital Camera. *Automation 2018, AISC, 743*, 595–603. Retrieved from https://link.springer.com/chapter/10.1007/978-3-319-77179-3_56

Krozyer, Dzh. Hekhl, Khardy, & Kheinyher. (2014). Nutrition management of maize at high plant density in broad-leaved and narrow-row crops. *Plant Nutrition, 2*, 17-22.

Kuperman, F. M., & Andryenko, S. S. (1959). *Physiology of corn*. Moscow: Publishing House of the Moscow Unitary.

Lisoval, A. P., Makarenko, V. M., & Kravchenko, S. M. (Eds.). (2002). Fertilizer application system. Vyshcha shkola.

Lohinova, I. V., & Smyk, S. I. (2012). Prognosis of fertilizer efficiency for corn on grain according to soil diagnostics. Scientific reports of NUBiP, 3(32).

Lysenko, V., Opryshko, O., Komarchuk, D., Pasichnyk, N., Zaets, N., & Dudnyk, A. (2017). Usage of Flying Robots for Monitoring Nitrogen in Wheat Crops. *The 9th IEEE International Conference on Intelligent Data Acquisition and Advanced Computing Systems: Technology and Applications*, 30–34. 10.1109/IDAACS.2017.8095044

Marchuk, I., Makarenko, V., Rozstalnyi, V., & Savchuk, A. (2002). *Fertilizers and their use*. Kiev: Directory.

Moldau, X. (1965). *Vegetative course of the brightness coefficients of leaves of some plants. Issues of the radiation regime of vegetation cover*. Tar-tu. Institute of Physics and Astronomy of the Academy of Sciences of the Estonian SSR.

Prystash, I. V. (2003). Yield and quality of corn grain depending on fertilizer system on meadow-chernozem ground. *Collection of scientific works of the Institute of Agriculture of the Ukrainian Academy of Agrarian Sciences*, *3*, 58–63.

Rachkulyk, V. Y., & Sytnykova, M. V. (1981). *Reflective properties and the state of vegetation cover*. Gidrometeoizdat.

Rozenberh, H. V. (1959). Absorption spectroscopy of dispersed substances. Uspekhy fyz. *Nauk.*, *69*(1), 57–104.

Rvachëv, V. L. (1987). *Methods of optics of light scattering media in physics and biology*. Minsk: Belarusian Publishing House.

Saberioona, M., Amina, M., Anuarb, A., Gholizadehc, A., & Wayayokd, A. (2014). Assessment of rice leaf chlorophyll content using visible bands atdifferent growth stages at both the leaf and canopy scale. *International Journal of Applied Earth Observation and Geoinformation*, *32*, 35–45. doi:10.1016/j.jag.2014.03.018

Shpaar, D., Hinapp, K., Dreher, D., Zakharenko, A., & Kalenska, S. (2009). *Growing, harvesting, preserving and using*. Alfa-Steeve LTD.

Yara N-Tester™. (n.d.). Retrieved from http://www.yara.ua/crop-nutrition/tools-and-services/n-tester

Chapter 12
Application of Solutions of the Electrochemical Processed Mineral "Bishofit" in Plant Production

Valerij Drevin
Volgograd State Agrarian University, Russia

Valerij Fomichev
Volgograd State Agrarian University, Russia

Igor Yudaev
Don State Agrarian University, Russia

Lyubov Minchenko
Volgograd State Agrarian University, Russia

Gulnara Gizzatova
Volgograd State Agrarian University, Russia

Inna Kucherova
Volgograd State Agrarian University, Russia

Tatyana Shipaeva
Volgograd State Agrarian University, Russia

Valeriya Komarova
https://orcid.org/0000-0002-9541-3435
Volgograd State Agrarian University, Russia

Ivan Rybintsev
State Agrarian University of Moldova, Moldova

ABSTRACT

Hexaqua magnesium chlorine, bishofite, almost never occurred in the fossil salt deposits until recently, while industry has used artificially produced magnesium chloride. The discovery of bishofite deposits in the Lower Volga region (both for raw materials purity of MgCl2 up to 95-98%, and for reserves about 500 billion tons) was a unique event. The use of bischofite consists in the salt compositions of sand to protect roads from icing yet. The chapter represents the results summary of the bishofite solutions electrochemical processing with the purpose of obtaining solutions with the antibacterial properties complex the use of which can be recommended for pre-plant seeds treatment, plant protection from fungal effects, which allowed, as shown by experiments on the crops cultivation, to increase yields up to two-three times, and also to use for disinfection of domestic and industrial wastewater in agricultural enterprises. The technological characteristics of the using electrochemical device are considered in the chapter.

DOI: 10.4018/978-1-5225-7573-3.ch012

INTRODUCTION

The working out and use of new technologies for the production of plant protection products and for their development and growth stimulation that have an effective integrated effect on the crops yield increase, as well as have environmental safety for users in contrast to currently used synthetic preparations, are the urgent tasks and their solution will improve the food security level of our country. The solutions of the regional mineral «Bischofite» – magnesium chloride – were used as a basis for the preparation of this kind.

The presented research is devoted to the study and investigation of the preparations obtain based on the mineral «Bischofite» mined in the Lower Volga region, which, being environmentally safe, can be used in agricultural production to improve the positive effect on the subjected to electrical processing seeds and plants. The processes regularities of the bischofite solutions diaphragmless electrolysis, the electrode materials choice ground, the electrochemical reactions possible mechanisms, the obtained products properties are considered in this research.

BACKGROUND

General characteristics of bischofite. Hexaqua magnesium chloride, bischofite, until recently did not occur in fossil salt-bearing scales, and in the industry was used artificially obtained magnesium chloride.

For the first time the mineral bischofite was discovered in the form of a component in the famous Shtatsfurtsky salt-bearing scales of Germany by the German geologist and chemist Karl-Gustav Bischof at the end of the 19th century. The year of the official bischofite opening was the 1877. It is commonly believed that this mineral is a product of the dissolved salts crystallization of the ancient sea, which precipitated during the water evaporation while reaching the limiting ion concentrations, more than two hundred million years ago, in the Permian period. Bischofit is an ancient crystalline salt. In the first decades after its discovery, bischofite was considered as a relatively rare mineral and was of interest only as a source of magnesium and its compounds (Derevyagin, Sedlechkiy & Ermakov, 1989; Shand, 2006).

Bischofite is easily disolved and therefore is extracted by the artesian water dissolution method (leaching) of the dry underground mineral deposits at the depth of occurrence. The resulting brine is pumped upward

In the seventies of the twentieth century, the Lower Volga region was the discovery place of the bischofite deposits, magnesium chloride ($MgCl_2 \times 6H_2O$) with a high content of up to 95-98%, and reserves – over 500 billion tons, which was a unique event. It should be noted the presence of numerous microelements in the composition of natural bischofite. The occurrence depth is 800-2000 m. The total occurrence area is 10,500 km^2 (Batalin & Svidzinsky, 2001; Bondarenko, 2001). The chemical composition of "Bischofit" is given in Table 1.

For the first time, the assumptions about the presence of chloromagnesium salts in the Volgograd region have appeared as far back as 1958-1959 during oil reconnaissance. By the beginning of the sixties there was a sufficient number of indirect factors for the possible existence of chloromagnesium salts deposits in the interior of the salt-bearing Volgograd region. There were not any similar analogues in the world (Valyashko, 1976; Grebennikov & Ermakov, 1982)

The Volgograd bischofite occurrence discoverers in the mid-1970s by right are considered Volgograd specialists V.A. Ermakov and N.P. Grebennikov. They made an assumption about the presence of

Table 1. The bischofite solution chemical composition in the Volgograd occurrence

Indicator's Name	Value
Specific gravity, g/l	1.320-1.330
PH, unit	5.5-6.8
Mineralization, g/l	400-450
Salt Content to Dry Residue,%	
$MgCl_2 \times 6H_2O$	90-96
$MgSO_4 \times H_2O$	0.1-2.5
$MgBr_2$	0.4-0.95
NaCl	0.1-0.4
$CaSO_4$	0.1-0.7
$KCl \times MgCl_2 \times 6H_2O$	0.1-5.5
Microelements, %	
Iron	0.003-0.03
Bismuth	0.0005-0.001
Molybdenum	0.0005-0.001
Bor	0.002-0.08
Aluminum	0.001-0.02
Titanium	0.0005-0.001
Cuprum	0.0001-0.0006
Silicon	0.02-0.2
Barium	0.0001-0.0006
Strontium	0.001-0.02
Cobalt	0.003-0.005
Rubidium	0.0001-0.002
Cesium	0.0001-0.001
Lithium	0.0001-0.0003

bischofite occurrences, and in the late 80s S.A. Svidzinsky continued research and engineering surveys activities devoted to the bischofite occurrences and there were expanded geological studies of the bischofit content of the Privolzhsky monocline (Grebennikov & Ermakov, 1982; Svidzinsky & Moscow, 2008).

The bischofite of the Volgograd occurrence is the compacted layers, which are pressed under high pressure, lying in the Kungur salt-bearing stratum. Depth of occurrence of their top varies from 900 to 1900 m, thickness from 10 to 60 m.

The Volga bischofit basin, with a width of 10 to 50 km, runs along the river bed Volga from the south of the Saratov region to the northern part of Kalmykia. The available geological material confirms that the bulk of the bischofit-bearing basin lies in the Volgograd region (Ermakov & Grebennikov, 1985; Derevyagin, Sedlechkiy & Ermakov, 1989).

In the Volgograd region the bischofite occurrence stretched along the river Volga at a distance of more than 400 km, with a width of 10-50 km, administratively located from north to south in Staropoltavsky, Nikolayevsky, Bykovsky, Kamyshinsky, Olkhovsky, Dubovsky, Gorodishchensky, Svetloyarsky, Kalachevsky and Kotelnikovsky districts.

To date, the total undiscovered resources of bischofite beds raw salts in the Volgograd region are 182.2 billion tons, including administrative regions (in billion tons) (Table 2)

Composition of bischofite beds. The method of bischofite salt determining allowed to distinguish it in the context of other types of salts, such as carnallite and sylvinite. The revealed regularities were also confirmed by rock material raised from the boreholes.

Bischofite, as a rock-forming material, – hexaqua magnesium chloride $MgCl_2 \times 6H_2O$ - is basically 88-98%, in some cases up to 99%. This monomineralty defines a monotonous monolithic structure with a weak layering feature. Bischofite has a banded texture, due to structural and texture features of the rock, coloring and the presence of other salts. Inside the bischofite packs, there are halite and carnallite inclusions, insignificant admixtures of gypsum, kieserite and anhydrite (Grebennikov & Ermakov, 1982; Derevyagin, Sedlechkiy & Ermakov, 1989).

Bischofite, depending on the impurities content in it, is colored differently: more often water-transparent and colorless, sometimes milky-white, smoky gray, pink or cream. Gray tones are characteristic for the lower parts of the bischofite beds. Colorless and transparent differences, usually coarse-grained, are confined to the uppermost part of the formation, contain a very small amount of insoluble residue, which imparts a gray shade to the rock, or does not contain it at all. Pink coloration is often caused by an admixture of carnallite, evenly dispersed along the bed or encountered in the form of «eyes», partial nests and thin interlayers. Such areas have unique features and are distinguished by a certain increase in gamma activity (Grebennikov & Ermakov, 1982; Derevyagin, Sedlechkiy, & Ermakov, 1989).

Studies of cores taken on the territory of the Volgograd region show that the bischofite rock has a structure from small to coarse-grained. Colorless monoclinic bischofite crystals have the following crystallographic parameters: a = 0.99 nm, b = 0.715 nm, c = 0.61 nm, $\beta = 94°$. While heating in a dry state at 120 °C, the bischofite melts with the recoil of four water molecules and boils at 150 °C with the recoil of six water molecules. The physical-chemical properties of magnesium chlorides are given in Table 3 and averaged indicators of the mineralogical composition of the Volgograd bischofite in Table 4 (Derevyagin, Makarov & Kovalsky, 1982; Derevyagin, Sedlechkiy & Ermakov, 1989).

The abnormally high bromine content in the Volgograd bischofite rock has no analogues and exceeds the world standards. The chemical composition of bischofite deposits of the Lower Volga region is described in details by B.C. Derevyagin and co-authors (Derevyagin, Makarov & Kovalsky, 1982; Derevyagin, Sedlechkiy & Ermakov, 1989).

Table 2. Total undiscovered resources of bischofite beds raw salts

Administrative District	Blns of Tons
Svetloyarsky	11.2
Gorodishchensky	41.8
Dubovsky	47.0
Bykovsky	48.6
Nikolayevsky	17.0
Kamyshinsky	16.6

Table 3. Physical and chemical properties of magnesium chlorides

Properties	$MgCl_2$	$MgCl_2 \times 6H_2O$
Molecularmass	95.22	203.31
Color	colorless	colorless
Crystallineform	hexagonal	monoclinic
Refractive index	1.378; 1.39	1.495; 1.507; 1.528
The true density, g /cm3	2.316; 2.41	1.56
Melting temperature, °C	708; 714	-4 H_2O;120
Boiling temperature, °C	1412; 1417	-6 H_2O;150
Solubility in g per 100 g: - Cold water (20°C) - Hot water (100°C)	54.6 73.4	306 rather soluble

The composition of bischofite mineral when extracted by underground dissolution depends on many factors: the age and location of the salt pool, the depth of extraction and technogenic impurities (impurities introduced by water to dissolve the bischofite and emerging from the metal structures of the installation and the brine storage tanks). Despite the fact that the bischofite mineral in the deposits of the Lower Volga region is contained from 67 to 99.7%, there are impurities of carnallite and kieserite in its rocks. These magnesium salts are carriers of a number of chemical elements that may be of some interest in terms of their pharmacological and toxicological properties. To date, it has been established that bromine, rubidium, cesium, boron and strontium accumulate in hydrochloric salt bischofite rocks (Valyashko, 1976; Derevyagin, Makarov & Kovalsky, 1982; Derevyagin, Sedlechkiy & Ermakov, 1989).

The average chemical content of the Volgograd bishofite is presented in Table 5 (Derevyagin, Makarov, & Kovalsky, 1982; Derevyagin, Sedlechkiy & Ermakov, 1989).

The use of bischofite mineral. The spectrum of bischofite application is huge: from construction (tile, stone) to agriculture (pre-sowing treatment of seeds, plants in the growing season), in oil production for the preparation of grouting and hardening mixtures, in the chemical industry for obtaining magnesium compounds of high purity (Shand, 2006; Fomichev et al., 2016).

The bischofite brine obtained directly from the borehole has a whole set of useful properties and is both a raw material and a multi-purpose product that has been used in various branches of the national economy.

Table 4. The Volgograd bischofite mineralogical composition average rates

Component Name	Chemical Symbol	Mass Content, %
Bischofite	$MgCl_2 \times 6H_2O$	88 – 98
Carnallite	$KCl \times MgCl_2 \times 6H_2O$	0.1 – 5.5
Kiserite	$MgSO_4 \times H_2O$	0.1 – 2.8
Magnesium bromide	$MgBr_2$	0.45 – 0.98
Anhydrite	$CaSO_4$	0.1 – 0.7
Halite	NaCl	0.1 – 0.4

Table 5. The average chemical content of the bishofite of the Volgograd occurrence

N° p/p	Salts and Elements	Mass Content, %
1	Magnesiumchloride	90-96
2	Calciumsulphate	0.2-0.7
3	Sodiumchloride	0.1-0.4
4	Potassium and Magnesium Chloride	0.5-5.55
5	Magnesiumsulfate	0.2-2.5
6	Magnesiumbromide	0.4-0.9
7	Silicon	0.02-0.2
8	Bor	0.002-0.08
9	Cadmium	0.003-0.005
10	Strontium	0.001-0.02
11	Iron	0.003-0.03
12	Aluminum	0.001-0.02
13	Cuprum	0.0001-0.003
14	Barium	0.0002-0.0006
15	Rubidium	0.0001-0.0002
16	Lithium	0.0005-0.001
17	Bismuth	0.0005-0.001
18	Titanium	0.0005-0.001
19	Molybdenum	0.0005-0.001
20	Cesium	0.0001-0.001

Let us consider, for example, the use of bischofite for medical purposes and in agricultural production.

Medicine. In recent years, the composition and physicochemical properties of bischofite, possible ways of its use in medicine have been studied in sufficient detail. The unique combination of macro and microelements in it gives the bischofite unusual properties, which are deprived of its individual constituent elements. Attempts to use it in the treatment of joint diseases go back to the distant past. Employees of the Volgograd State Medical University have studied many aspects of the bischofit influence on the animal and human organism. The absence of carcinogenic, allergenic properties, absence of pathogenic microflora, lack of toxicity, anti-inflammatory effect, stimulating effect, normalizing influence on immune reactions are proved (Spasov, 2000; Zborovsky, 2001).

For example, bischofite is used as a balneological agent in the treatment of diseases of the musculoskeletal system, nervous and cardiovascular system. It is successfully used in dentistry as a disinfectant. Bischofite is approved for use by the Ministry of Health of the Russian Federation.

Studies carried out by Volgograd scientists confirm the presence in bischofite, a natural environmentally friendly mineral, medicinal properties that are not inferior to expensive medicinal products in efficiency. In the studies course, a number of medications were developed on the basis of the Volgograd bischofite, namely: paste «Bischal», cream «Bischolin», solution «Polycatan», plaster, etc. (Spasov, 2000; Zborovsky, 2001).

Agriculture. Bischofite is used as an extra nutrition for animals and birds, fertilizer for trees, shrubs and plants, a preservative of green forages, anti-radiation protection of cotton with seeds processing, etc.

In the process of testing carried out in the Volgograd department of the Scientific-Research and Technological Institute of Chemicalization of Agriculture (SRTICA), a high efficiency of bischofite in the daily seeds processing, vegetative fertilizing of all agricultural plants, late crops desiccation was obtained. The research has allowed to develop new technologies that confirm the effectiveness of bischofite in dry land conditions. Crop yield increases from 2 to 7 centner/ha (25-30%) are noted, the quality of winter wheat is improved, practically all grain is grocery. The resistance of crops to pests and diseases increases by 25-30%.

The Volgograd Institute of VNIALMI has identified the effectiveness of bischofite as a means to increase the productivity of fruit and berry crops and resistance to pests and diseases (Belitskaya & Gribus, 2009; Belitskaya & Kryukova, 2010).

The saturation of the natural bischofite mineral with chemical elements makes it promising for the desiccation of late-sowing crops and for biological stimulation of plant development and productivity. Also crops sowing studies, carried out 1-1.5 months before harvesting and allowing to shorten the vegetation period of crops are advanced (Okonov, Khalgaeva & Unkanzhinov, 2012).

Due to bischofite processing of fruit and berry crops in the early spring period before bud blossoming, pests and pathogens are destroyed, and yields, sugar content, taste qualities are also improved. From the treated trees obtained product does not contain residual toxic substances (Fomichev, Yankovskaya & Abyzov, 2014).

Bishofite is a reliable means of pre-winter processing of the earth's surface in the process of irrigation and drainage works. Smolensk VNIIG and MSM bischofite was used to thaw the surface of the earth during the laying of the meliorative pipes network in the winter time, which made it possible to shorten the payback period of the development in 1.5-2 times.

The results of long-term researches conducted by scientists and specialists of the Department of Feeding and Breeding of Domestic Animals of the Volgograd Agricultural Academy testify to the positive effect of bischofite's natural salt on the productivity of cattle, sheep, pigs and poultry (Yudaev & Kostychev, 2013; Yudaev et al., 2016).

In addition, bischofite is used in cosmetology, wood processing, power engineering, construction, oil industry, fire extinguishing, to prevent icing, for anti-dust purposes, as frost protection and disinfectant, etc. (Gritsenko & Ovchinnikov, 1986; Kulikova, 2003; Kulikova, 2004; Lavrikova & Fomichev, 2009; Fomichev & Filimonova, 2010)

MAIN FOCUS OF THE CHAPTER

To increase the potential productivity of cultivated crops and the ecological sustainability of agrocoenoses and agroecosystems, today inventions based on flexible scientific intensive technologies that will solve the problem of crop production phytosanitary optimization and whole yield increase, for example, of cereals, are required. Very promising in this regard is the use of growth regulators and microelements, which can solve the problem of plants production process regulating that determines the maximum realization of their genetic potential.

From the viewpoint of plants providing with vitally important nutritional elements, the use of various raw materials of natural origin is worth noticing, among them the most natural mineral is bischofite. The basis of bischofite is magnesium, and namely it is the part of enzymes that intensify the synthesis of proteins, carbohydrates, lipids and a number of other substances, regulates the supply of nutrients to plants, and besides this bischofite contains more than 70 macro- and microelements.

The exploration work results in different soil and climatic conditions made it possible to conclude that the use of this natural mineral salt as a source of microelements is promising. Consequently, bischofite in small doses stimulates the plants development, improves their productivity and resistance to harmful organisms, is characterized by low toxicity and high processability.

A new, very progressive trend has been the use of an aqueous solution of electrolytic bischofite (or electrochemically nanostructured aqueous bischofite solution) in plant growing, which is obtained by electrochemical treatment of an aqueous solution of a natural mineral using a copper anode.

As a treatment result colloidal (nano-sized) structures containing magnesium ions and copper hypochlorite and hypobromite ions are formed, bound into micellar forms of 20-90 microns in size with hydroxide dispersions based on magnesium and copper, which have excess activity, which ensures close interaction with the plant. The ions of magnesium and copper presented in the electrolytic bischofite participate in the regulation of the transfer process through the cell membrane of nutrients and products of vital activity.

In order to identify the most effective indexes of the researching solutions used for presowing seeds treatment, the experience was laid on light brown soils in the Volgograd region (Belitskaya & Gribust, 2009, Belitskaya & Gribust, 2012; Fomichev et al., 2016). Objects of researches were selected, they were winter and spring wheat, cultivated in field crop rotation. Control was an option sown with seeds treated with water. The area of the experimental variants was 10 hectares. Seeds incrustation was performed a day before sowing.

Seeds treated with the solution were wet to the touch, but after 30-40 minutes, the solution was completely absorbed. The use of an aqueous solution of electrolytic bischofite, as well as natural bischofite, eliminates the need to introduce sticking agents into the working fluid. This agent has good adhesive properties. The solution completely envelops the seeds, penetrates inside and protects the fissures and microdamages of the embryo from penetration of pathogenic microflora, soil mites and decay reliably. However, since this composition is characterized by high hygroscopicity, it passes moisture to the seeds well and does not interfere with their germination.

Observations testified that the use of the working solution application high rate (15 l/ton) is inadvisable, since in this case the grains stick together, which subsequently makes seed sowing difficult. The most effective was the use of a solution in a dose of 8-10 liters per ton of seeds.

The solution application to the seeds ensures the growth of harmonious and uniformed seedlings, increases seed germination in 1.3-5.6%. The use of these solutions for the treatment of poor quality seed grain leads to an increase in germination in 8.0-13.7%. A high result of the stimulating effect of electrolytic bischofite aqueous solution is explained by an increase in the course of biological processes in the seeds by additional energy and nutrients obtaining.

The researches results showed that a high effect was noted due to solution use for the presowing treatment of winter wheat seeds. Seedlings of most varieties appear 1-5 days earlier than on a chemical standard (Table 6). The tillering phase also occurs with a break in 1-2 days. At later stages, determining the crop productivity (earing and yellowing), the effect of the studied solution was less pronounced. It should be noted that the winter wheat growing period under the influence of electrolytic bischofite

Table 6. Effect of presowing seed treatment with electrolytic bischofite aqueous solution on phenology and winter wheat yield

Variety of Winter Wheat	Variant	Phenology				Yield Structure		
		Seadlings October	Tillering September	Earinge May	Yellowing June	Crop Productivity t/ha	Grain Quality	
							Grain Unit,g/l	Mass of 1000 Grains, g
Don Surprise	1	12	25	11	21	0.398	802	39.2
	2	10	24	11	21	0.393	804	39.2
	3	8	22	10	20	0.367	801	38.8
Tanais	1	16	25	11	21	0.392	921	37.2
	2	14	25	12	21	0.397	822	37.3
	3	9	25	12	20	0.398	819	37.1
Don-95	1	9	24	12	21	0.340	809	41.6
	2	10	24	13	21	0.372	812	43.0
	3	9	23	12	20	0.387	813	42.6
Don-93	1	10	23	15	22	0.350	818	41.3
	2	8	23	15	22	0.366	817	41.7
	3	9	23	16	21	0.361	819	43.1
Guarantee	1	12	26	18	27	0.399	786	36.4
	2	11	26	18	27	0.389	788	36.9
	3	10	25	17	26	0.391	790	36.2
Don unbearded	1	17	27	22	23	0.352	798	43.8
	2	16	27	22	27	0.358	804	41.3
	3	16	26	22	27	0.348	803	41.6
NCP 095						1.6		
Accuracy Experience						2.3		

Note: 1-premis, 1.2 kg / ha; 2-10.6 kg / ha of premis + 1/2 ENRB, 5%; 3 electrochemically nanostructured bischofite solution, 10%.

aqueous solution passed more rapidly. This is important for the dry land region as it affects the conjugate development of winter wheat pests, which ensures the weakening of biocenotic connections between them.

Seed treatment with electrolytic bischofite solution promoted the development of a larger number of stems, including productive ones, and the yield increment in individual varieties was 0.6-4.7 c/ha.

The analysis of the biological crop structure showed that the grain increase from the incrustation of the seed grain by the electrolytic bischofite and its mixture with the premise was obtained due to the greater number of remaining productive harvesting and the number of grains in the head. According to the mass of 1000 grains and the grain nature between the experimental variants and the control, no significant differences were observed.

In the research course, the different responsiveness of winter wheat varieties to incrustation with electrolytic bischofite solution was also noted. A more pronounced positive effect of the nanostimulator was recorded on such varieties as Don Surprise, Tanais and Garant. Sorts of Don-95 and Don-93 showed significantly less responsiveness.

In the experiments, the effectiveness of grain crops spraying in the tillering phase with an electrolytic bischofite solution was evaluated in comparison with the fertilizer-stimulating composition (GUSS) containing, along with microelements, biologically active compounds. Experience has shown that the introduction of a nanostimulant into a stalky plant has contributed to a stable decrease in the population of plantings by phytophages (in 1.9-7.89 times) and suppression of the infectious agents activity: septoriosis, ascohitosis, brown rust, dust bunt (2-4.3 times). Only the leafhoppers reacted equally to the use of these preparations. Electrochemically nanosized bischofite solution stimulated attraction of indexed species, parasitic and predatory insects to crops. Their density in this variant was 24.1% and 26.0%, respectively, higher than by the drug ZHUSS using.

In addition, the researches of M.N. Belitskaya and co-authors (Belitskaya & Kryukova, 2010; Belitskaya & Gribust, 2012) showed that the use of dilute bischofite solutions, including electrolytic bischofite, increases the productivity of cereal agrocenoses and protects crops plants (Table 7).

Experimental data show that the use of natural bischofite aqueous solutions allows actively reducing the influence of major pests on plant development.

The structure of the resulting barley crop, as shown in the table, has increased productive characteristics. Productivity increases by 58% in comparison with the control.

At the same time, the observed decrease in the crop shortage (table 10) is 9.3%, and in control 12.3%.

Table 7. Effectiveness of electrolytic copper hypochlorite based on bischofite aqueous solution action on barley disease at various methods of use

Option	Powdery Mildew	Helminthosporium	Septoria	Helminthosporium-fusarial Root Rot
	R,%	R,%	R,%	R,%
Control	30.0	20.0	15.0	10.6
Presowing Seed Treatment				
Bischofite solution	16.5	14.0	10.0	4.6
Electrolytic hypochlorite based on bischofite aqueous solution	16.0	10.0	8.2	4.2
Plants Foliar Top Dressing				
Bischofite	19.0	16.0	12.0	8.8
Electrolytic hypochlorite based on bischofite aqueous solution	18.0	14.0	9.0	8.5
Presowing Treatment + Foliar Top Dressing				
Bischofite +Bischofite	10.5	10.5	8.0	3.6
Electrolytic hypochlorite based on bischofite aqueous solution	10.0	9.3	7.6	4.0

Table 8. Natural bischofite influence on the sprouting energy and germination

Option	Sprouting Energy, %	Germination
Control	79.3	86.0
Natural bishofite	85.7	97.1
Electrolytic cuprum hypochlorite based on bischofite aqueous solution	89.2	98.4

Table 9. Influence on the barley yield structure (seeds treated with a natural bischofite solution)

Option	Number of Productive Stems, pcs.	Length of Head, cm	Number of Grains in Head, pcs.	Weight 1000 Grains, g	Productivity, Centner/ha
Control	78	7.1	16.8	40.9	16.0
Natural bishofite	96	8.1	19.4	49.8	25.3
Electrolytic cuprum hypochlorite based on bischofite aqueous solution	120	7.7	20.9	48.3	36.3

Table 10. Influence on pests and crop shortage

Option	Pest Number, ind./m^2	Development of Diseases,%	Number of Grains in Head, pcs.	Crop Shortage	
				%	Centner/ha
Control	3.2	35.6	16.8	12.3	2.0
Natural bishofite	4.0	14.6	19.4	9.3	2.4
Electrolytic cuprum hypochlorite based on bischofite aqueous solution	1.1	13.5	20.9	4.4	1.6

Research of I.A. Kulikova and co-authors showed that bischofite solutions obtained as a result of electrochemical exposure possess a complex of antibacterial properties, the use of which was recommended for disinfection of domestic and process wastewater, can be used for presowing seed treatment, protecting plants against fungal impact. The effect of an electrochemically treated bischofite solution influence on plants showed the increase in their qualitative characteristics.

The introduction of copper ions into the bischofite solution has shown that the use of this solution in the process of preparing crops seeds allows to increase the yield up to two-three times (Tables 7-10). To explain the effect obtained, it is necessary to consider the roles of ions obtained during the electrochemical treatment of natural bischofite solutions: magnesium, copper and hypochlorite.

The death of microorganisms in water at the concentrations of metals in it, amounting to millions of parts of the amount of solution, i.e. traces observed by many researchers. These properties of copper and silver ions were named oligodynamic, from the Greek words «oligos» and «dynamis». In the literal translation: «action of the trace». Further researches have shown that silver has the greatest oligodynamic effect. The action of copper is 4-5 times weaker. But taken together, these metals greatly enhance the properties of each other. And not only antimicrobial properties. Copper plays a key metabolic role in the metabolism of all living organisms, starting from the simplest cell. It is a part of biological catalysts - enzymes. Without them, life is impossible. That is why biologists called copper «the metal of life». Directly or indirectly, copper participates in most metabolic processes and is their main regulator. The main biochemical function of copper in the body is participation in enzymatic reactions as an activator or in the composition of copper-containing enzymes. The amount of copper in plants varies from 0.0001 to 0.05% (on dry matter) and depends on the type of plant and the content of copper in the soil. In plants, copper is a part of enzymes-oxidases and protein of plastocyanin. In optimal concentrations, copper increases the cold resistance of plants, contributes to their growth and development. For copper MPC_v (for copper ion), 1 mg/dm^3 is established (the limiting indicator of harmfulness is organoleptic)

MPC_{vr} – 0,001 mg/dm^3 (the limiting indicator of toxicity is toxic). By the degree of chemical elements danger (GOST.17.4.1.02-83), copper is a moderately hazardous chemical element.

The copper deficiency in the soil causes copper starvation of plants, which, however, is also due to the inaccessibility of copper compounds for them, despite the considerable magnitude of copper gross content in the soil in a number of cases. The lack of copper in plants affects a number of abnormalities associated with the retardation of development and the violation of many important physiological functions, leading to a decrease and even to the death of the crop. Copper has a stabilizing effect on chlorophyll, protecting the latter from destruction. Without affecting the very process of photosynthesis, copper promotes an increase in the assimilative surface of plants, prolongs photosynthetic activity and thereby enhances the assimilation process, which causes an increase in the productivity of plants, the crop increase. The influence of copper on the water regime of plants is revealed. With copper starvation, the weakening of protein synthesis leads to a change in hydrophilicity - to a decrease in the water-retaining and water-absorbing capacity of the plant, which causes the plants to wither. There is an increase in the overall stability of the plant organism under the influence of copper, in particular, the cold resistance of winter plants is increasing. Copper improves the stage of vernalization.

Green leaves of plants contain chlorophylls, which are magnesium-containing porphyrin complexes involved in photosynthesis. Magnesium is also involved in biochemical processes in organisms. The importance of magnesium, as a macronutrient, in life activity is manifested in the fact that it is a universal regulator of biochemical and physiological processes in organisms. Magnesium, entering into reversible bonds with many organic substances, provides the possibility of metabolism of about 300 enzymes, in particular, creatine kinase, adenylate cyclase, phosphofructokinase, K-Na-ATPhase, Ca-ATPhase, protein synthesis enzymes, glycolysis, transmembrane ion transport, etc. Magnesium is necessary to maintain the structure of ribosomes, nucleic acids and some proteins. It participates in the reactions of oxidative phosphorylation, protein synthesis, the exchange of nucleic acids and lipids, in the formation of energy-rich phosphates. Good supply of the organism with magnesium promotes better tolerance of stressful situation, suppression of depression. A significant amount of magnesium is found in nuts and cereals (wheat bran, coarse flour), apricot, dried apricots, plums (prunes), dates, cocoa (powder). They are rich in fish (especially salmon), soy, nuts, bread with bran, chocolate, fresh fruit (especially bananas), watermelons. Magnesium contains cereals (oatmeal, millet, buckwheat), legumes (beans, peas), sea cabbage, squid, meat, eggs, bread (especially rye from coarse flour), greens (spinach, parsley, lettuce, dill), lemons, grapefruits, almonds, nuts, halva (sunflower and tahini), apples. In lieu of the above facts, which show the large value of copper and magnesium ions for processes occurring in the plant organism and directly affecting the yield, it becomes evident the need for using copper-containing fertilizers. Until now, forms and technologies for the application of bischofite mineral solutions in agriculture have not been found, in which all of its most valuable properties would have been manifested with maximum efficiency.

To improve the bischofite effectiveness in plant growing, methods for natural bischofite solutions modifying were studied, an electrochemical method was used.

In the course of the bischofite solution electrochemical treatment, a number of processes are observed, they can be schematically represented by the stages:

1. **Electrochemical Stage:** The ionization of chloride and bromide ions, as well as metallic copper, takes place at the anode.

$$2\ Cl^- - 2e^- \rightarrow Cl_2 \qquad (1)$$

$$2\ Br^- - 2e^- \rightarrow Br_2 \qquad (2)$$

$$2\ Cu - 2e^- \rightarrow 2Cu^{2+} \qquad (3)$$

In this case, the evolved chlorine dissolves in the electrolyte with the formation of hypochlorous and hydrochloric acids:

$$Cl_2 + H_2O = HClO + HCl \qquad (4)$$

With the accumulation of hypochlorite in the electrolyte, the amount of hydrogen decreases due to the predominance of the electrochemical reduction of the hypochlorite ion on the cathode:

$$C1O^- + H_2O + 2e^- = C1^- + 2OH^- \qquad (5)$$

flowing together with the reaction of water molecules discharge (3):

$$2H_2O + 2e^- = H_2\uparrow + 2OH^- \qquad (6)$$

The hydrogen atoms are recovered from the solution as a gas after reduction on the cathode, the remaining OH^- ions form near the cathode with Mg^{2+} ions magnesium hydroxide particles, and with Cu^+ ions, copper oxide:

$$Mg^{2+} + 2\ OH^- = Mg(OH)_2 \qquad (7)$$

$$Cu^{2+} + 2\ OH^- = Cu(OH)_2 \qquad (8)$$

In this case, due to the mixing of anolyte with catholyte, the hypochlorous and hypobromous acids react with alkali to form hypochlorite and hypobromite of magnesium and copper:

$$2\ HClO + Mg(OH)_2 = Mg(ClO)_2 + 2H_2O \qquad (9)$$

$$4\ HClO + Cu_2O = 2\ Cu(ClO)_2 + 2H_2O \qquad (10)$$

$$2\ HBrO + Mg(OH)_2 = Mg(BrO)_2 + 2H_2O \qquad (11)$$

$$2\ HBrO + Cu_2O = Cu(BrO)_2 + H_2O \qquad (12)$$

Electrolytic oxidation of a solution of natural bischofite produces chlorites, hypochlorites, hypobromites of magnesium and copper, the mutual action of which creates a synergistic effect, thereby enhancing the fungicidal, bactericidal activity of the final disinfectant, which distinguishes the product from the currently used products.

2. **The Chemical Stage:** Dilution of the electrochemically obtained bischofite solution with water is accompanied by a hydrolysis process:

$$Cu(ClO)_2 + HOH \rightarrow CuOHClO + H^+ \rightarrow Cu(OH)_2 + HClO \quad (13)$$

The formation of the ultradisperse suspension (sol) of $Cu(OH)_2$ initiates the micellization stage according to the assumed scheme:

$$[Cu(OH)_2]_n \cdot mCu(OH)^+ + 2ClO^- \rightarrow |\, [Cu(OH)_2]_n \cdot mCu(OH) + (m-x)ClO \,|\, x^+ \cdot xClO^- \quad (14)$$

As the result of mixing the solution, micellar structures containing hypochlorite, copper, magnesium ions are formed in it, which allows using this product for processing agricultural crops seeds. Due to the mixing of anolyte with catholyte, the anions of hypochlorous and hypobromous acids react with magnesium hydroxide ions:

$$ClO^- + MgOH^+ \rightarrow MgOHClO, \quad (15)$$

$$BrO^- + MgOH^- \rightarrow MgOHBrO, \quad (16)$$

with the formation of complex magnesium compounds with the corresponding hypochlorite and hypobromite ions.

When metallic copper is used as an anode material, along with oxidation of chloride ions, oxidation of copper with formation of cuprate- and cuprite-ions occurs:

$$CuO \rightarrow Cu^+ \rightarrow Cu^{2+} \quad (17)$$

When interacting with catholyte, complex formations are formed:

$$Cu^+ + ClO^- \rightarrow CuClO, \quad (18)$$

$$Cu^{2+} + ClO^- \rightarrow Cu(ClO)_2, \quad (19)$$

Along with this, there are also processes:

$$Cu^{2+} + 2OH^- \rightarrow Cu(OH)_2, \quad (20)$$

$$Cu(OH)_2 + ClO^- \rightarrow CuOHClO + OH^-, \quad (21)$$

Taking into account the comparatively low solubility of magnesium and copper hydro-compounds, it is possible to form high-dispersion zeolite structures (with dimensions of ~ 30-150 nm) that stabilize in micellar form:

$$[Mg(OH)_2]_n \cdot m\,Mg(OH)^+ + 2\,Cl^- \rightarrow \{[Mg(OH)_2]_n \cdot m\,Mg(OH)^+ \cdot (m-x)\,Cl^-\}\,x^+ \cdot x\,Cl^-, \quad (22)$$

$$\{[Mg(OH)_2]_n \cdot m\,Mg(OH)^+ \cdot (m-x)\,ClO^-\}\,x^+ \cdot x\,ClO^-, \tag{23}$$

$$\{[Cu\,(OH)_2]_n \cdot m\,Cu\,(OH)^+ \cdot (m-x)\,ClO^-\}\,x^+ \cdot x\,ClO^-. \tag{24}$$

The proof of the presence of such micellar or nano-sized dispersed particles in the electrolytically treated solution under the developed technological regimes are the data in which quantitative characteristics of the presence in the solution of nanoscale micelles are given.

The presence of micellar structures, as the researches showed, are the influence on bischofite solution properties - as a true solution, which is a homogeneous environment. Ionic structures of such a solution are "blocked" during the interaction with the cell membrane of the plant. This "blockage" arises from processes, in particular, coagulation, occurring in the structure of cell membranes due to the presence of magnesium (II) ions in the solution. These processes slow the penetration of active magnesium ions into the cellular space. Thus, all the most valuable properties of bischofite solution appear to be outside the cell, not being able to fully manifest themselves.

On the other hand, the presence of ions in a homogeneous solution leads to an increase in the osmotic pressure in the cell, which leads to a disruption in the processes of its vital activity.

Hence the need for the development of «anti-lock» technologies aroused, i.e. such technologies that would facilitate the penetration of magnesium ions from bischofite solution into the cellular space.

The ions providing the vital activity of the cell are transported to it through a network of conducting capillaries with dimensions of the order of 30-40 nm, which permeate the molecular-protein structure of the cell membrane.

The technology that ensures the ions penetration from the solution into the cell through the net of conductive nano-sized capillaries is a nanoprocess technology that takes into account the nanoscale structure of the cell.

Basic Methods of Experimental Research

The membraneless electrolysis of weakly mineralized waters is not fully investigated, but the factors limiting the efficiency of the technological parameters rational values process and range are not studied in practice. The purpose of this research is to select a rational operating mode for electrolytic cells using bischofite solutions based on the study of the electrochemical factors influence on the efficiency of obtaining a complex disinfectant.

Laboratory researches were carried out with the bischofite solution, which composition is presented in the Table 1. The laboratory installation is a nonflown electrolytic cell with a volume of 0.5 liters. The material of the electrodes was chosen based on the researches (Fomichev & Filimonova, 2010; Fomichev, Naumova & Akhremenko; 2013; Fomichev et al., 2016) anode is copper, cathode is stainless steel. Electrodes are placed vertically in the center of the electrolytic cell, the distance between them was 10 mm.

The scheme for switching on the laboratory installation is shown in Figure 1.

Method to Obtain Electrochemically Treated Bischofite Solutions

A 0.5 liter of working solution is poured into the electrolysis cell. The installation operated at a constant and pulsed current. The direct current was supplied from a constant current source LIPS-35. The electrical circuit includes a voltmeter and an ammeter. The electrolysis process is carried out for 0.5 hours. The

Figure 1. Diagram of a laboratory installation for obtaining disinfectant solutions: 1 - electrolyzer, 2 - source of regulated impulse current, 3 - ammeter, 4 - voltmeter, 5 - constant current source

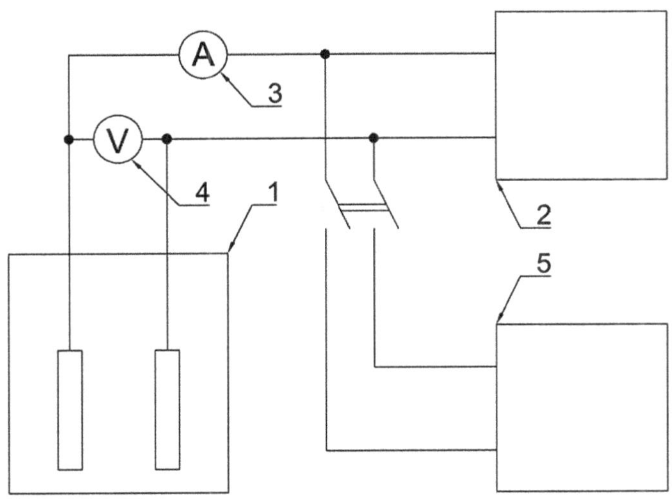

value of the hydrogen index of the environment (pH) of the initial and obtained solutions was determined according the pH meter of the millivoltmeter pH-121.

In this research, we propose a technology for converting a true bischofite solution, changing its qualitative characteristics and transferring it from a homogeneous state to a heterogeneous (micellar), with particle sizes comparable to the nanosized capillaries of a cell membrane. This solution transformation is carried out by electrochemical solution treatment, by means of a specially designed regimen.

It is known that the colloidal (heterogeneous or micellar) form of the solution provides for more intensive delivery and absorption of substances necessary for the development of plants at the cellular level, which allows the body to use each ingredient with maximum efficiency and facilitates their cellular assimilation. On the other hand, in a heterogeneous solution, ions are bound and, therefore, have a lower osmotic pressure on the cell than in an ionic homogeneous solution, which provides a more gentle mode of cell activity.

In a heterogeneous (micellar), electrochemically treated bischofite solution, hypochlorite anions, copper (II) ions, hydroxide ions, which bind copper and magnesium ions to colloidal structures (micelles), which are neutral and do not cause coagulation.

Thus, in the applying process of the developed technology of a homogeneous bischofite solution electrochemical treatment, a colloidal heterogeneous structure is formed, which is a nanostructured micellar medium. The developed technology for converting a homogeneous solution into a heterogeneous nanostructured solution allows a technological method – electrolysis, to control the qualitative and quantitative (particle size) characteristics of a given solution in a given process regime.

Technological Aspects of Electrochemical Treatment of Bischofite Solution

The most important factor of the electrochemical process is the yield of the product of the electrochemical reaction from the current density, i.e. on the speed of the process, which determines the mass of the product, the productivity of the process obtained per unit time.

The experimental data presented in Fig. 2 show the dependence of the products current yield of the copper ions anodic oxidation and chloride ions in terms of active chlorine during an electrochemical process using direct current.

It follows that the current yield of active chlorine and copper depends on the current density and the concentration of the bischofite solution (Fig. 2). At the same time, the maximum current output can be obtained with a bischofite concentration of 10%.

Figure 2. Dependence of current (BT) yield of copper (a) ions and active chlorine ions (b) in bischofite solutions on the current density, solution temperature 25°C, constant current

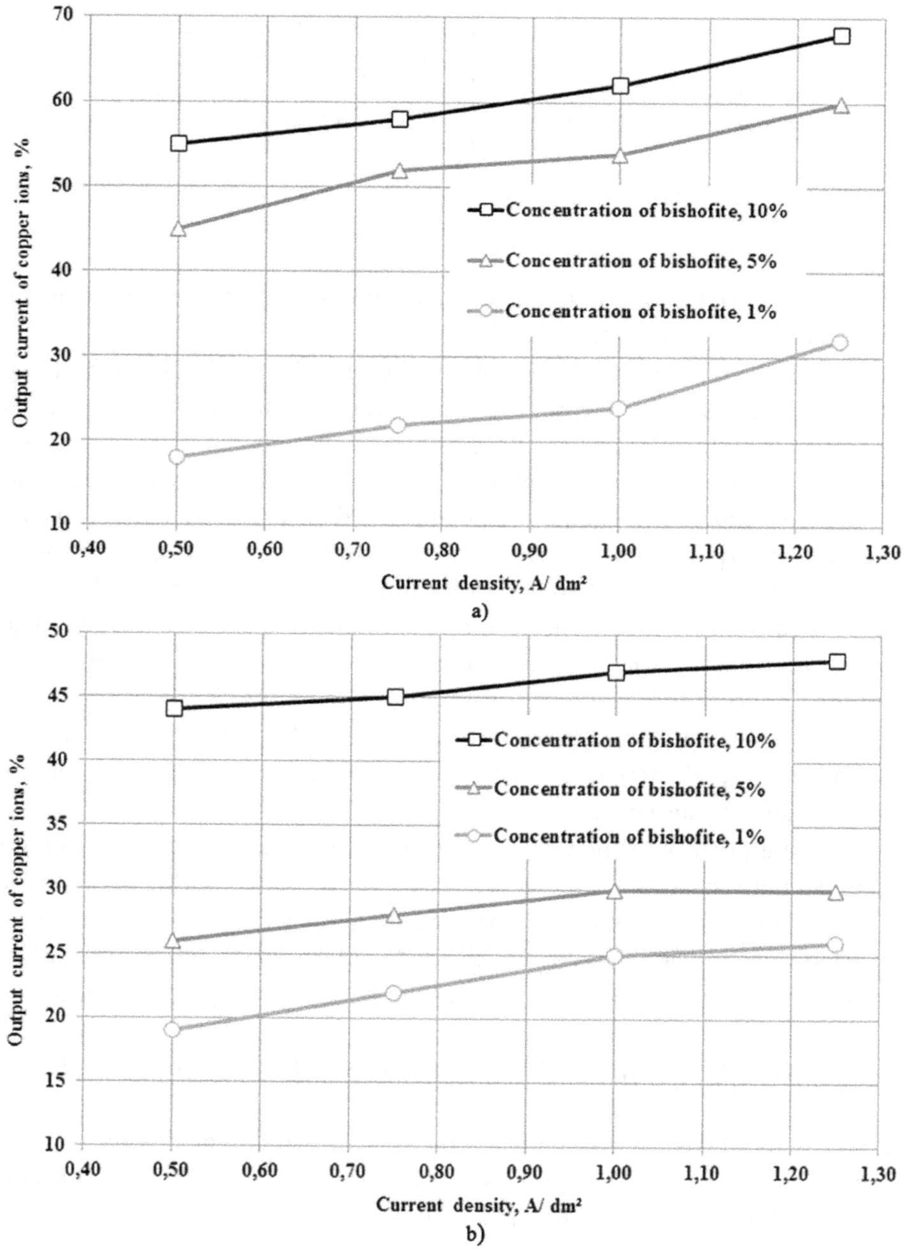

These data allow to determine the specific electricity consumption. From the data presented in Figure 3 it follows that the energy consumption of the process increases with increasing current density. At the same time, the greatest relative costs refer to a minimum concentration of 1%, the minimum costs correspond to a bischofite concentration of 10%. In the range of current densities from 0.1 to 1.5 A/dm^2, energy consumption increases in 2 times.

The concentration of copper ions and active chlorine (Fig. 3, 4) increases with increasing both the concentration of the initial solution and the current density, especially for current densities of 1 and 1.5 A/dm^2.

The final product of the process of electrolysis of chlorine-containing salts is at the anode: gaseous oxygen and hypochlorite, and hydrogen at the cathode.

Unlike the bound chlorine, which is a part of various chloramines, such compounds as hypochlorous acid, chlorine oxide, molecular chlorine and hypochlorite ions are referred to contain free chlorine.

Free chlorine is 25 times more bactericidal and has a higher oxidation potential, which is reflected in the hygienic requirements of drinking water.

It is proved that hypochlorous acid is 50-100 times more bactericidal than virucidal. The bactericidal activity of hypochlorous acid is many times greater than bactericidal hypochlorites, which have approximately the same bactericidal and virucidal effect. The virucidal effect of hypochlorites is 3-5 times higher than their bactericidal activity, which indicates different mechanisms of bactericidal and virucidal activity of free chlorine.

The acids formed in the solution are present in the form of the corresponding ions. Hydrogen ions, recovering at the negative electrode (cathode) to the molecular state, are evolved as a gas. The ions OH- accumulating in this solution interact with magnesium ions, forming hydroxide ions: MgOH+ (catholyte).

Figure 3. Specific energy costs for obtaining disinfectant in bischofite solutions at constant current, solution temperature, 25°C

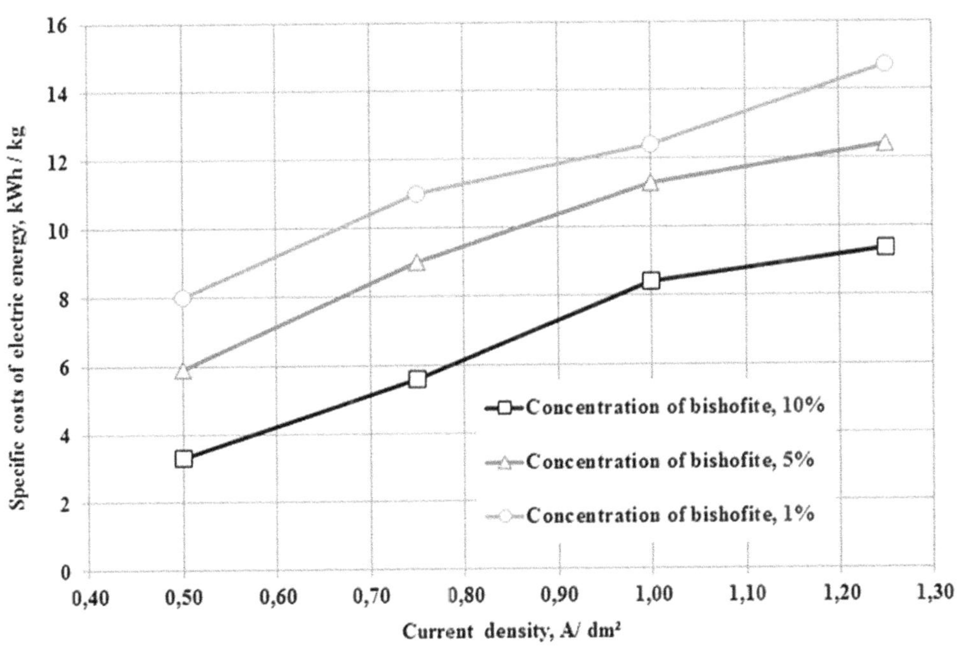

Figure 4. Dependence of the concentration of copper ions (a) and active chlorine (b) in a bischofite solution on the current density during electrolysis at a constant current; the temperature of the solution is 25°C

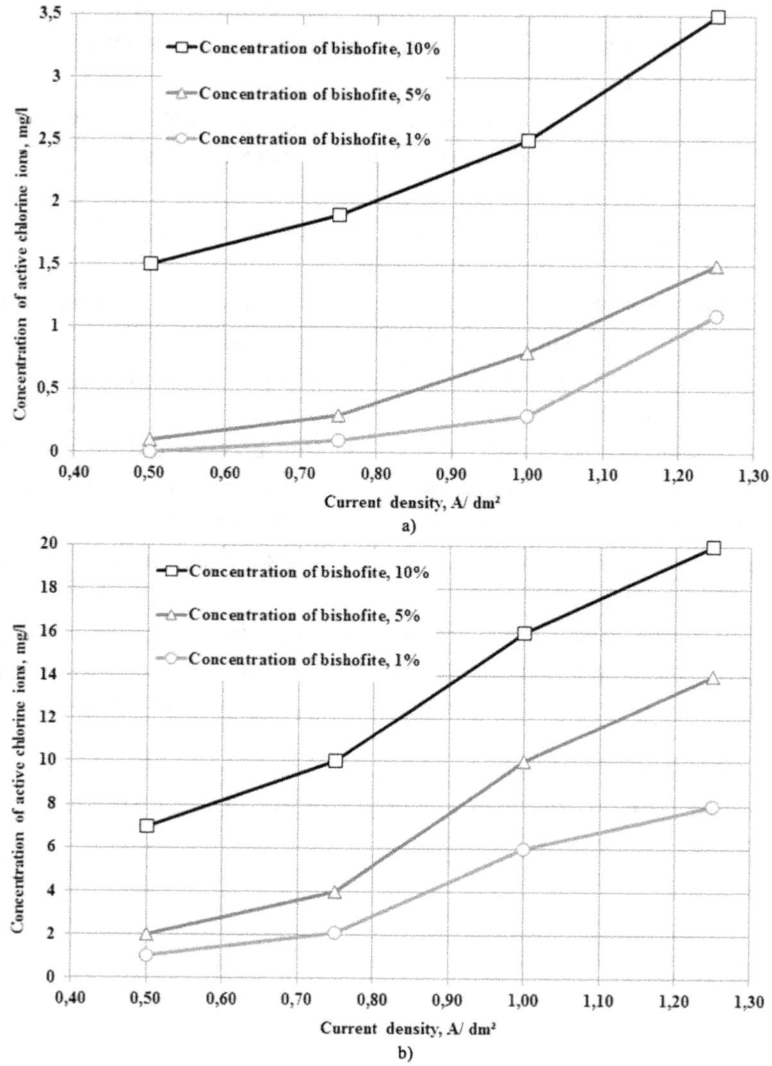

Due to the mixing of anolyte with catholyte, the anions of hypochlorous and hypobromous acids react with magnesium hydroxide ions:

$$ClO^- + MgOH^+ \rightarrow MgOHClO \quad (25)$$

$$BrO^- + MgOH^+ \rightarrow MgOHBrO \quad (26)$$

with the formation of complex magnesium compounds with the corresponding hypochlorite and hypobromite ions.

When metallic copper is used as an anode material, along with oxidation of chloride ions, oxidation of copper with formation of cuprate- and cuprite-ions occurs:

$$CuO \rightarrow Cu^{+} \rightarrow Cu^{2+} \quad (27)$$

When interacting with catholyte, complex formations are formed:

$$Cu^{+} + ClO^{-} \rightarrow CuClO \quad (28)$$

$$Cu^{2+} + ClO^{-} \rightarrow Cu(ClO)_2 \quad (29)$$

Along with this, there are also processes:

$$Cu^{2+} + 2OH^{-} \rightarrow Cu(OH)_2, \quad (30)$$

$$Cu(OH)_2 + ClO^{-} \rightarrow CuOHClO + OH^{-} \quad (31)$$

Taking into account the comparatively low solubility of hydro-compounds of magnesium and copper, it is possible to form high-dispersion zeolite structures (with dimensions of ~ 30-150 nm) that stabilize in micellar form:

$$[Mg(OH)_2\ ]_n \cdot m\ Mg(OH)^{+} + 2\ Cl^{-} \rightarrow \{[Mg(OH)_2]_n \cdot m\ Mg(OH)^{+} \cdot (m - x)\ Cl\ \text{-}\}x^{+} \cdot x \cdot Cl^{-}, \quad (32)$$

$$\{[Mg(OH)_2]_n \cdot m\ Mg(OH)^{+} \cdot (m - x)\ ClO\ \text{-}\}x^{+} \cdot x \cdot ClO^{-}, \quad (33)$$

$$\{[Cu\ (OH)_2]_n \cdot m\ Cu\ (OH)^{+} \cdot \cdot (m - x)\ ClO\ \text{-}\}x^{+} \cdot x \cdot ClO^{-}\ . \quad (34)$$

The proof of the presence of such micellar or nano-sized dispersed particles in the electrolytically treated solution under the developed technological regimes are the data in which quantitative characteristics of the presence in the solution of nanoscale micelles are given.

The presence of micellar structures, as researches showed, influence the properties of bischofite solution – as a true solution, which is a homogeneous environment. Ionic structures of such a solution "are blocked" during the interaction with the cell membrane of the plant. This "blockage" arises from processes, in particular, coagulation, occurring in the structure of cell membranes due to the presence of magnesium (II) ions in the solution. These processes slow the penetration of active magnesium ions into the cellular space. Thus, all the most valuable properties of a bischofite solution appear to be outside the cell, not being able to fully manifest themselves.

On the other hand, the presence of ions in a homogeneous solution leads to an increase in the osmotic pressure in the cell, which leads to a disruption in the processes of its vital activity.

Hence the need for the development of «anti-lock» technologies, i.e. such technologies that would facilitate the penetration of magnesium ions from bischofite solution into the cellular space.

The ions providing the vital activity of the cell are transported to it through a net of conducting capillaries with dimensions of the order of 30-40 nm, which permeate the molecular-protein structure of the cell membrane.

The technology that ensures the penetration of ions from the solution into the cell through a net of conductive nano-sized capillaries is a nanoprocess technology that takes into account the nanoscale structure of the cell.

This paper proposes the technology for converting a true bischofite solution, changing its qualitative characteristics and transferring it from a homogeneous state to a heterogeneous (micellar), with particle sizes comparable to the nanosized capillaries of a cell membrane. This transformation of the solution is carried out by electrochemical solution treatment, by means of a specially designed regimen.

It is known that the colloidal (heterogeneous or micellar) form of the solution provides more intensive delivery and absorption of substances necessary for the development of plants at the cellular level, which allows the organism to use each ingredient with maximum efficiency and facilitates their cellular assimilation. On the other hand, in a heterogeneous solution ions are bound and, therefore, have a lower osmotic pressure on the cell than in an ionic homogeneous solution, which provides a more gentle mode of cell activity.

In a heterogeneous (micellar), electrochemically treated bischofite solution, hypochlorite anions, copper (II) ions, hydroxide ions, which bind copper and magnesium ions to colloidal structures (micelles), are neutral and do not cause coagulation process.

SOLUTIONS AND RECOMMENDATIONS

The use of bischofite and products made from its main advantage is environmental friendliness and safety in agriculture, food industry, medicine, etc. Therefore, as recommendations, the possibility of bischofite wider use in the technological processes of such branches of agriculture as crop production, cattle breeding, poultry farming, fur farming, as well as in medicine, construction and directly where animals and people live, should be noted.

FUTURE RESEARCH DIRECTIONS

To continue the work and future researches, as well as the electrolytic bicshophite solutions subsequent practical application, first of all, the possibility of its application for the various materials and products creation and production that can be used to improve the agricultural products quality, cultivated crops yields and animals weight gain increase should be mentioned. The seeds and seedlings pre-sowing stimulation before sowing and planting, seeds disinfection before storage, use as feed additives in the breeding of poultry and animals, application in veterinary practice, etc. can be referred to such technological operations. To implement these technologies, it is necessary not only to understand the mechanism of the ongoing processes in the electrolytic bischofite solution production, but also to know the parameters and characteristics of its practical use, and namely the results of these researches will allow to formulate technological conditions for agriculture and processing industry various branches.

CONCLUSION

As the conclusion, it should be noted that in the process of applying the homogeneous bischofite solution electrochemical treatment developed technology, a colloidal heterogeneous structure is formed, that is a nanostructured micellar environment. The developed technology for converting a homogeneous solution into a heterogeneous nanostructured solution allows to control the qualitative and quantitative (particles size) characteristics of the given solution in the given process regime by the technological method - electrolysis.

The application of the presented technology is very promising in agricultural production, especially in such an industry as crop production, allowing not only to increase the crops yield, but also to improve the crops phytosanitary condition.

REFERENCES

Batalin, Y. V., & Svidzinsky, S. A. (2001). Perspektivy osvoeniya kalijnyh mestorozhdenij Volgogradskoj oblasti i osnovnye problemy razrabotki brom-magnievogo syr'ya – bishofita [Prospects for the development of potash deposits in the Volgograd region and the main problems of developing bromine-magnesium raw materials – bischofite]. *Osvoenie i ispol'zovanie prirodnyh resursov Volgogradskoj oblasti – put' ustojchivogo razvitiya regiona: materialy nauchno-prakticheskoj konferencii 15-16 dekabrya 2000* [Development and use of natural resources of the Volgograd region - the way of sustainable development of the region: materials of the scientific and practical conference. December 15-16, 2000], 63-64.

Belitskaya, M. N., & Gribust, I. R. (2009). Stabilizaciya i produktivnost' zernovyh agrocenozov pri ispol'zovanii nanostrukturirovannogo rastvora bishofita [Stabilization and Productivity of Grain Agrocenoses with the Use of a Nanostructured Bischofite Solution]. *Materialy 2-oj Vserossijskoj nauchno-tekhnicheskoj konferencii «Nanotekhnologii i nanomaterialy: sovremennoe sostoyanie i perspektivy razvitiya v usloviyah Volgogradskoj oblasti»* [Proceedings of the 2nd All-Russian Scientific and Technical Conference «Nanotechnologies and Nanomaterials: Current State and Prospects of Development in Volgograd Region»], 53-58.

Belitskaya, M.N., & Gribust, I.R. (2012). Razrabotka tekhnologii primeneniya ehlektroliticheski nanostrukturirovannogo rastvora na osnove bishofita [Development of technology for the application of electrolytically nanostructured solution based on bischofite]. *Ehkologiya Rossii: na puti k innovaciyam* [Ecology of Russia: on the way to innovation], *6*, 192-194.

Belitskaya, M. N., & Kryukova, E. A. (2010). Bishofit pomogaet zashchitit' posevy [Bishofit helps protect crops]. *Zemledelie [Agriculture]*, *5*, 41–43.

Bondarenko, Ya. N. (2001). Material'no-syr'evaya baza Volgogradskoj oblasti kak osnova ustojchivogo razvitiya regiona [Material and raw materials base of the Volgograd region as a basis for sustainable development of the region]. Osvoenie i ispol'zovanie prirodnyh resursov Volgogradskoj oblasti – put' ustojchivogo razvitiya regiona [Development and use of natural resources of the Volgograd region - the way of sustainable development of the region], 11-21.

Derevyagin, B.C., Sedlechkiy, V.N., & Ermakov, V.A. (1989). *Bishofity Nizhnego Povolzh'ya* [Bishofits of the Lower Volga Region]. Rostov: Izd-vo Rostovskogo universiteta [Publishing house of the Rostov University].

Derevyagin, V. S., Makarov, A. S., & Kovalsky, F. I. (1982). Geologiya i kalienosnost' nizhnepermskoj galogennoj formacii Severnogo Prikaspiya [Geology and Kalianosity of the Lower Permian Halogen Formation of the Northern Caspian Region]. Stroenie i uslovie obrazovaniya solenosny`x formacij [The structure and conditions of formation of saline formations], 98-109.

Ermakov, V. A., & Grebennikov, N. P. (1985). Zakonomernosti bishofitovyh zalezhej solenosnoj tol [Regularities of bischofite deposits of saline tol]. In Problemy solenakopleniya [Problems of salt accumulation]. Moscow: Nauka [Science].

Fomichev, V.T., Drevin, V.E., Filimonova, N.A., Kulikova, I.A., Kamkova, S.V., & Yudaev I.V. (2016). *Perspektivy osvoeniya i tekhnologiya fiziko-himicheskoj aktivacii mineral'nogo syr'ya Volgogradskih mestorozhdenij* [Perspectives of development and technology of physical and chemical activation of mineral raw materials of Volgograd deposits]. Volgograd: VolGAU.

Fomichev, V.T., & Filimonova, N.A. (2010). Poluchenie fungicidov na osnove minerala bishofita [Preparation of fungicides based on bischofite mineral]. *Internet-Vestnik VolgGASU* [Internet-Bulletin of VolgGASA], *1*, 4-8.

Fomichev, V.T., Naumova, G.A., & Akhremenko, O.Y. (2013). Innovacionnaya tekhnologiya povysheniya biologicheskoj aktivnosti rastvora minerala bishofita [Innovative technology of increasing the biological activity of a solution of bischofite mineral]. *Vestnik VolGU. Seriya 10. Innovacionnaya deyatel'nost'* [Bulletin of the VolSU. Series 10. Innovative activity], *1*(8), 131-138.

Fomichev, V.T., Yankovskaya, M.B., & Abyzov, V.V. (2014). Izuchenie rezul'tativnosti dejstviya kompleksnogo preparata «Bilator-6» na porazhaemost' gribnymi zabolevaniyami rastenij zemlyaniki sadovoj i barbarisa obyknovennogo [The study of the effectiveness of the complex preparation «Bilator-6» on the damage caused by fungal diseases of strawberry and barberry plants of the common]. *Vestnik VolGU. Seriya 11. Estestvennye nauki* [Bulletin of the VolSU. Series 11. Natural Sciences], *2*(8), 36-42.

Grebennikov, N. P., & Ermakov, V. A. (1982). Bishofit: problemy i perspektivy [Bishofit: problems and prospects] *Tekhnika i nauka. Engineering and Science*, *4*, 56–64.

Gritsenko, A. I., & Ovchinnikov, A. D. (1986). Nekotorye voprosy ustojchivosti bishofita v pristvol'noj zone skvazhiny [Some problems of bischofite stability in the near-well zone of a well]. Neftepromyslovaya geologiya i stroitel'stvo skvazhin v Prikaspijskom regione [Oilfield geology and well construction in the Caspian region], 75-84.

Kulikova, I. A. (2003). Poluchenie dezinfekanta na osnove prirodnogo syr'ya dlya processov obezzarazhivaniya prirodnyh vod [Preparation of disinfectant based on natural raw materials for disinfection of natural waters]. In *Ehkologiya. Ohrana sredy. Stroitel'stvo: doklady VIII Regional'noj konferencii molodyh issledovatelej Volgogradskoj oblasti* [Ecology. Protection of the environment. Construction: reports of the VIII Regional Conference of Young Researchers of the Volgograd Region]. Volgograd: VolgGASA.

Kulikova, I.A. (2004). Ehkologicheskaya bezopasnost' processa hlorirovaniya prirodnyh i stochnyh vod [Ecological Safety of the Process of Chlorination of Natural and Wastewater]. *Vestnik VolgGASU. Seriya Estestvennye nauki* [Vestnik VolgGASU. Series of Natural Sciences], *3*(10), 159-160.

Lavrikova, N. A., & Fomichev, V. T. (2009). Ehkologobiologicheskaya bezopasnost' stroitel'nyh materialov i izdelij [Ecological and Biological Safety of Building Materials and Products]. *Kachestvo vnutrennego vozduha i okruzhayushchej sredy: materialy VII Mezhdunar. nauch. konf* [Quality of Internal Air and the Environment: Materials VII Intern. sci. Conf.], 8-12.

Okonov, M. M., Khalgaeva, K. E., & Unkanzhinov, G. D. (2012). Urozhajnost' i kachestvo zerna ozimoj pshenicy pri ispol'zovanii Binorama, Biosila Bishofita na fone mineral'nyh udobrenij [Yield and quality of winter wheat grain using Binorama, Biosila Bishofit against mineral fertilizers]. *Plodorodie [Fertility]*, *1*(64), 15–16.

Shand, M. A. (2006). *The chemistry and technology of magnesia*. Wiley. doi:10.1002/0471980579

Spasov, A. A. (2000). *Magnij v medicinskoj praktike* [Magniy in medical practice]. Volgograd: Otrok.

Svidzinsky, S.A. & Moscow, G.A. (2008). *Metodika poiskovo-razvedochnyh rabot na mineral'nye soli v usloviyah solyanokupol'nyh struktur* [Methods of prospecting for mineral salts in salt-dome structures]. Saratov: Izdatel'stvo Saratovskogo universiteta [Saratov University Press].

Valyashko, M.G. (1976). *Brom v solyanyh otlozheniyah i rassolah kak geohimicheskij indikator ih genezisa, istorii i poiskovyj priznak* [Bromine in salt deposits and brines as a geochemical indicator of their genesis, history and search sign]. Academic Press.

Yudaev, I.V., Rodionov, S.N., Gamaga, V.V., Zotov, V.M., Kostychev, K.V., & Panteleev, A.K. (2016). Analiz harakteristik ehlektrolitov ispol'zuemyh pri ehlektricheskoj obrabotke rastitel'nyh kormov [Analysis of the characteristics of electrolytes used for electrical processing of plant feeds]. *Mekhanizaciya i ehlektrifikaciya sel'skogo hozyajstva* [Mechanization and electrification of agriculture], *11*, 28-30.

Yudaev, I.V., & Kostychev, K.V. (2013). Obrabotka ehlektricheskim tokom zernofurazha, kak process uluchsheniya ego perevarimosti pri otkorme cyplyat-brojlerov [Electricity grain feed processing as a process of improving its digestibility in fattening chickens-broilers]. In *Integraciya nauki i proizvodstva – strategiya ustojchivogo razvitiya APK Rossii v VTO. Materialy mezhdunarodnoj nauchno-prakticheskoj konferencii* [The integration of science and production is the strategy of sustainable development of Russia's agro-industrial complex in the WTO. Materials of the international scientific-practical conference]. Volgograd: VolGAU.

Zborovsky, A. B. (2001). Itogi i perspektivy ispol'zovaniya prirodnyh faktorov Volgogradskoj oblasti v medicine [Results and perspectives of the use of natural factors in the Volgograd region in medicine]. *Osvoenie i ispol'zovanie prirodnyh resursov Volgogradskoj oblasti – put' ustojchivogo razvitiya regiona: materialy nauchno-prakticheskoj konferencii 15-16 dekabrya 2000* [Development and use of natural resources of the Volgograd region - the way of sustainable development of the region: materials of the scientific and practical conference December 15-16, 2000], 94-99.

ADDITIONAL READING

Fomichev, V. T., & Kulikova, I. A. (2004) *Method of obtaining hypochlorite*. RU Patent 2238348 C25B 1/26, 1/18.

Fomichev, V. T., & Naumova, G. A. (2012) *Method of electrochemical preparation of a solution of hypochlorites of magnesium and copper*. RU Patent 2466214 A01C C25B 1/26

Krasovskiy, G. N. (1976) Podhody k obosnovaniyu klassifikacii opasnosti himicheskih zagryaznenij vody [Approaches to the substantiation of the classification of the danger of chemical contamination of water] Gigienicheskie aspekty ohrany okruzhayushchej sredy [Hygienic aspects of environmental protection]. Moscow, 45-52.

Krasovskiy, G.N., & Egorova, N.A. (1990) Principy i kriterii novoj koncepcii kontrolya kachestva vody [Principles and criteria for a new concept of water quality control] *Gigiena okruzhayushchej sredy [Environmental Hygiene]*. Moscow: Vysshaya shkola [Higher school].

Kulikova, I. A. (2004) Bezopasnost' primeneniya hlorsoderzhashchih dezinfekantov v processah vodopodgotovki [Safety of chlorine-containing disinfectants in water treatment processes] Proizvodstvo pishchevyh produktov v sootvetstvii s trebovaniyami koncepcii zdorovogo pitaniya i drugie voprosy: materialy Vserossijskoj nauchno-prakticheskoj konferencii [Production of food products in accordance with the requirements of the concept of healthy nutrition and other issues: materials of the All-Russian Scientific and Practical Conference]. Volgograd, 281-284.

Kulsky, L.A. (1985) *Primenenie ehlektrohimicheskih processov i apparatov dlya obezzarazhivaniya vody* [Application of electrochemical processes and devices for water decontamination]. Kiev: UkrNIINTI.

Kulsky, L.A., Slipchenko, A.V., & Matskevich, E.S. (1988) Osobennosti raboty gipohloritnyh ehlektrizerov pri nizkih koncentraciyah hloridov v vode [Features of the operation of hypochlorite electors at low chloride concentrations in water] *Himiya i tekhnologiya vody [Chemistry and Technology of Water]*. 10(5), 438-441.

Mohnach, V. O. (1968). *Teoreticheskie osnovy biologicheskogo dejstviya galoidnyh soedinenij* [Theoretical Foundations of the Biological Action of Halogen Compounds]. Leningrad: Nauka. [Science]

Samokhotsky, A. I. (1990). *Metallovedenie* [Metal Science]. Moscow: Metallurgiya. [Metallurgy]

Sharma, R. A. (1997) *Electrolytic Magnesium Production Process*, US Patent, 5665220.

Wehner, G. (1953). *Basic Chloride of Magnesium, Z. Anorg. AUg*. Chern.

KEY TERMS AND DEFINITIONS

Bischofite (Hexaqua Magnesium Chlorine): Is a hydrous magnesium chloride mineral with formula $MgCl_2 \cdot 6H_2O$. Bischofite is an ecologically pure natural magnesium poly-mineral with a unique composition. It contains many macro- and micro-elements vital for human health, in much higher concentrations than can be found in sea or ocean salt.

Electrolysis: Is a technique that uses a direct electric current (DC) to drive an otherwise non-spontaneous chemical reaction.

Electrolyte: Is a substance that produces an electrically conducting solution when dissolved in a polar solvent, such as water.

Electrolytic Bischofite: Is an aqueous solution of bischofite subjected to the electrolysis process.

Ion: Is an atom or molecule that has a non-zero net electrical charge (its total number of electrons is not equal to its total number of protons).

Micellar Particle: Is a particle of colloidal dimensions that exists in equilibrium with the molecules or ions in solution from which it is formed.

Chapter 13
Innovative Technologies for E-Learning of Safety of Agro-Industrial Production

Evgeny Vasilevich Khalin
FGBNU FNAC VIM, Russia

Elena Evgenievna Mikhaylova
FGBNU FNAC VIM, Russia

ABSTRACT

An important factor of impact on the decrease in operational injuries is the high-quality and timely training, certification, preparation, and retraining of personnel in safety of production. Procedures of training and control of knowledge of various directions of safety of production are regulated by the existing normative documents developed by authors of the chapter . The systems of e-learning of personnel for safety of agro-industrial production have to be built in digital technologies of management of agro-industrial production and have the modular structure allowing the customization of systems on requirement of concrete production.

INTRODUCTION

Important factor of impact on decrease in operational injuries it is also production the caused incidence of employees of the agro-industrial organizations and enterprises high-quality and timely training, certification, preparation and retraining of personnel in safety of production are. Procedures of training and control of knowledge of various directions of safety of production are regulated by the existing normative documents developed by authors of the present chapter now. The e-learning systems of personnel for safety of agro-industrial production have to be built in digital technologies of management of agro-industrial production, have the modular structure allowing to customize systems on requirement of concrete production.

The e-learning systems of safety of production are executed in the form of easily adaptable user shell programs on the basis of the expert technologies consisting of independently operating modules selected

DOI: 10.4018/978-1-5225-7573-3.ch013

on the functional purpose and type of knowledge and actions localized in each module. Modules have the multi-level correlations excluding accumulation of contradictory knowledge and incorrect data. Each system module of e-learning provides rational execution of the functions with an opportunity for the user to change their set in use of the module by adding by the user of new data items. Systems keep different shape of the organization of e-learning of safety of production (mobile, network, independent, the mixed, co-education) and types of distant educational technologies.

One of the main conditions of formation of the effective and personally focused programs of e-learning of safety of production is use of e-portfolios of employees on safety of agro-industrial production. E-portfolios are necessary also for the successful solution of problems of professional staff recruitment, questions of employment, assessment of quality of professional education, social demand of employees of certain professional groups.

E-portfolios provide a possibility of control of individual achievements, skills and competences on safety of production in each process of preparation and after its end, during all production activity of the employee. The high production functionality of electronic portfolios of employees on safety of production is provided with integration of e-portfolios in a learning management system and personnel resources of the organization, use of filling, components and opportunities of e-portfolios with use of computer network information technologies.

Application of information models of the competences of employees which are built in the e-learning systems of area safety of agro-industrial production is the effective tool for the organization of the personified e-learning of employees guaranteeing acquisition of steady competences on safety of agro-industrial production. Information models of competences of employees on safety of production provide monotonous, same submission of information on competences on safety of agro-industrial production. Information models of competences promote the solution of problems of rational use of personnel resources at exchange of information about competences between various organizational agro-industrial structures and functionally compatible the e-learning systems.

The models of competences used in the e-learning systems of safety of production have to provide the authorized access, extraction and the analysis of competences and the related information for assessment of quality and the continuity of training in a production cycle of the employee and with use of different systems. Models have to provide the coordinated interaction and comparison of the network personified information competency models on safety of production in various subject domains and branch structures.

MOTIVATION

The present chapter contains results of researches in the field of development and application of networked e-learning systems for the safety of agro-industrial production which are an important factor of stable development of modern agriculture, the safe use of technological processes, machines and aggregates, prevention of production injuries and prophylaxis of production the caused incidence. As necessary basic elements of effective e-learning for the safety of agro-industrial production the systems of e-portfolios and information models of competences of employees on safety of agro-industrial production are considered.

BACKGROUND

The author of chapter conducts researches in the field of e-learning of safety of production, including safety of agro-industrial production, more than 30 years. The first control training complexes were developed in the field of electrical safety, then the e-learning systems of safety of thermal power stations, to the general rules on labor protection, safety of the equipment operating under pressure, safety of lifting constructions, gas industry, compressor installations, refrigeration installations, automobile and railway transport, safety of work in construction, safety of agricultural production were developed. The unique knowledge base on safety of industrial installations and objects according to the existing normative documentation is created.

In recent years intensive researches in the field of development of the multifunction intellectual e-learning systems including the systems of e-portfolios and information models of competences of employees on safety of production are conducted. In the last 15 years development of national standards in the field of application of information and communication technologies and information systems at the organization of e-learning of employees of safety of production is conducted. It became possible by the results of many years of industrial application of the developed software systems and due to the active position of users of these software tools.

LITERATURE REVIEW

By preparation of keeping of the chapter the accumulated domestic and foreign experience in the field of intellectual information and communication technologies and knowledge formalization (Kharchenko V., Vasant, P. (2018), Zelinka, I., Vasant, P., Duy, V.H. & Dao, T.T. (2018), Vasant, P., Litvinchev, I. & Marmolejo, J. A. (2018)) and the results of using it to create of intellectual e-learning systems of safety of agro-industrial production (Khalin, Eu.V. (2017), (2018), (2018a), (2018b)) was used. The advanced information and communication technologies which are based on the formalized skills experience and knowledge of highly qualified specialists and professionals in the field of safety of production are considered as a basis of creation of the effective e-learning systems of safety of agro-industrial production including the systems of e-portfolios and information models of competences of employees on safety of agro-industrial production.

E-LEARNING SYSTEMS

E-learning systems for the safety of agro-industrial production include the systems of preparation, instructing and checking the knowledge of personnel on safety of agro-industrial production with use of means of the information technologies and information systems applied directly on production in training centers, professional educational institutions and other educational institutions of all levels providing preparation and retraining of personnel.

The preparation, instructing and certification of personnel for safety of agro-industrial production united in the general concept of e-learning of safety of production should be constant (continuous) multilevel character and be carried out in the organizations both in the course of labor activity, and with a separation from it.

At realization of processes of preparation, instructing and certification for safety of agro-industrial production it is necessary to fulfill requirements of the national standard (GOST, 2016), to use various forms of the organization of e-learning, including mobile, network, autonomous, mixed, joint, and various educational technologies, including to apply remote educational technologies.

The order of preparation, instructing and certification of personnel for safety of agro-industrial production on the basis of e-learning has to be directed to maintenance of timeliness and efficiency of preventive measures for decrease in operational injuries, occupational diseases and to improvement of the working conditions employees and establishes the general obligatory provisions of e-learning, instructing and checking the knowledge on safety of agro-industrial production.

The intellectual e-learning systems of safety of agro-industrial production in the form of the universal open user program covers including both the basic and additional rules and requirements, standard and instructive materials which can be included in processes of training, instructing and certification of personnel at the discretion of the person responsible for e-learning have to be applied to preparation, instructing and test knowledge.

For preparation, instructing and test knowledge on the basis of e-learning it is necessary to use the formalized knowledge on the basis of the existing specifications and technical documentations and rules in the form of the text and graphic descriptions including the system of the interconnected text descriptions in the form of questions, answers and explanations to them and the system of graphic descriptions supplementing them in the form of set of the dynamic pictures changing depending on the performed operations.

The e-learning systems of safety of agro-industrial production should be implemented in the form of easily adaptable user covers on the basis of the expert technologies consisting of independently working modules allocated on functional purpose and type of the knowledge and actions having the multilevel interrelations excluding accumulation of contradictory knowledge and incorrect data localized in each module.

Each module of e-learning system should perform rationally the functions with an opportunity to change the user their set from the original list, without breaking their structure in general, and it is necessary to allow modification in use of the module by adding by the user of new data items (Khalin Eu.V. (2016b)). The approximate set of the main functional capabilities of the e-learning system for the safety of agro-industrial production is given in figure 1.

The module of the interface should contain set of interdependent elements of knowledge shown in the volume necessary for the user, include language knowledge in the natural form and multi-level electronic documentation and to serve for maintenance of interaction of the trainee, instructed, certified and the expert organizing preparation, instructing and certification with modules of the e-learning systems.

Electronic documentation should reflect an analysis result of actions taken by alleged users by operation with the e-learning systems with presentation to the user of data in any requested volume. Necessary simplicity and accessibility of maintenance of the e-learning systems should be reached by setup of systems on a certain professional grade of the examined or trained employee, on execution of the separate task, group of tasks or on operation with their full range.

The e-learning systems should be intended for implementation of the following main functions:

- maintaining of a basis of professional information on personnel having training, instructing and certification and about the personnel organizing these processes;
- maintaining of a basis of information on instructing at the workplace;

Figure 1. Functional and technological structure e-learning systems for the safety of agro-industrial production

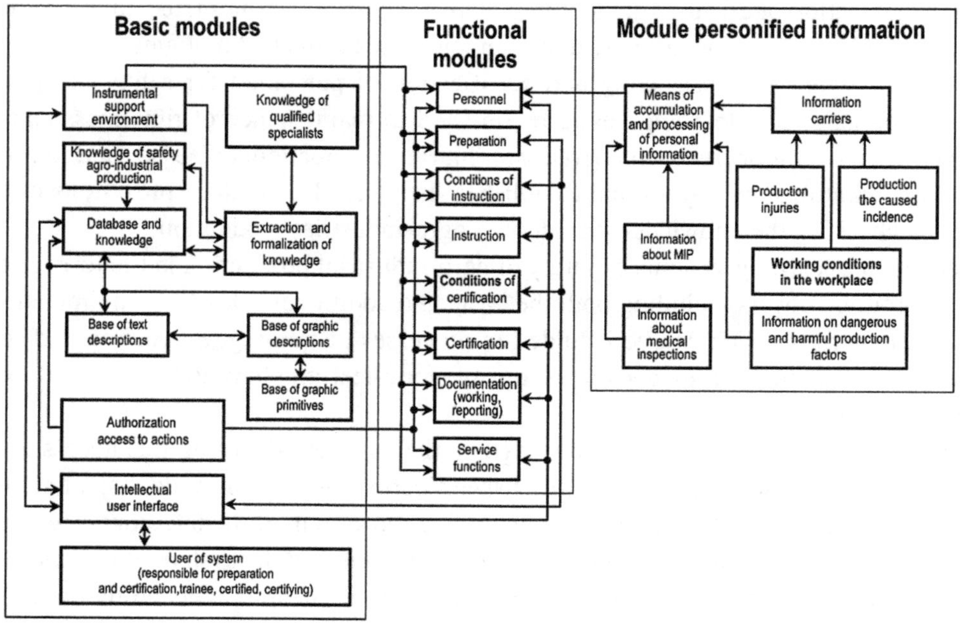

- electronic control and reminders on periods of instructing and knowledge checks;
- formations of additional direct e-learning systems, questions and comments to them, constructions of graphic descriptions according to requirements and rules of any subject area;
- preparation of the document or the created program (on the selected group of textual and graphic descriptions in the required volume) in full accordance with an order of instructing and certification both in the mode of viewing questions and responses, and in the mode knowledge control, at the request of the trainee;
- control of knowledge with assessment only correctness of the answer, with a possibility of obtaining explanations to the incorrect answers, with assessment of total time of certification of the employee;
- formations and printing of the current working documentation – electronic lists and protocols of preparation and certification taking into account the information fields which are in addition entered by the user.

The intellectual systems of e-learning it is necessary to realize the following additional functions:

- maintaining of a basis of information on harmful production factors, medical inspections, means of individual protection, short information about accidents with employees involved to preparation, instructing and certification;
- formation of electronic programs of preparation, instructing and certification with use of system signs (document types, types of installations, characteristics of installations), keywords and word combinations;

- formation and the printing of the report output documentation, including in professional editors, including the journal of registration of instructing at the workplace, the journal of knowledge testing on labor protection, the minutes of the meeting of the commission for the examination of knowledge;
- centralized updating of questions and graphic representations of targeted e-learning systems and adding of the targeted e-learning systems included in the industrial supply;
- authorization of group and personal access to modules, actions and functions of e-learning systems depending on category of the user.

The functional modules of the e-learning system should support:

- adaptation of the e-learning system by the user to change of external knowledge;
- creation of the friendly intellectual interface;
- acquisition of sustainable professional skills trained and certified.

Means of functional modules of the e-learning systems should be supported in addition:

- formalization of knowledge and experience of highly qualified specialists with application of text and graphic procedures;
- development of the e-learning system by addition of functions of blocks and modules with modification of the realized functions by not programming professional.

Networked e-learning systems should support:

- possibility of operating control and intervention in the modes of training and knowledge testing;
- comparison of the formalized knowledge and results obtained in the process of preparation and attestation on network objects;
- use of means of the operational network e-cards;
- confidential character of all the formalized information and knowledge on the safety of agro-industrial production circulating in the networks (Khalin Eu.V. (2016), Khalin Eu.V. (2016a), Khalin Eu.V., Strebkov D. S. & Mikhaylova E. E. (2016), Khalin Eu.V., Pozdneev B. M. & Mikhaylova E. E. (2017)).

The information about personnel applied in e-learning systems for the safety of agro-industrial production are subject to single input and repeated use in various modules, the accompanying, working and reporting documentation and makes the personified knowledge of e-learning systems with the base of text and graphic descriptions corresponding to them.

The module of information about personnel of the e-learning system should include the updated knowledge of employees of a particular organization (enterprise) containing the following as the main information about:

- professional groups of the enterprise and skill levels;
- frequency of certification and terms of preparation;
- type of knowledge testing;

- general work experience and length of service by profession;
- frequency, type and reason of instruction;
- character and duration of the internship at the workplace;
- responsible for carrying out certification and instructing;
- conclusion of the commission.

The module of information about personnel should include the updated knowledge of employees of a certain organization (enterprise) containing in quality additional information about:

- harmful and dangerous production factors;
- means of individual protection (MIP);
- medical inspections;
- place and circumstances the accidents which have happened to trainees;
- type of a traumatic situation and consequences of a injury;
- professional features and working conditions of specific employees.

It should be possible to increase the list of information about employees by the user of the e-learning system.

Knowledge of the safety of agro-industrial production should form the basis of all procedures of training, instructing and control of knowledge, the formation of accompanying documentation, to be applied in filling of all system actions and functions of e-learning systems. The technology of accumulation and replenishment of the knowledge base should allow the user nonprogrammer to develop and adapt easily the e-learning system to concrete production features, to independently reflect changes in subject area (the operating system of rules and requirements), to accumulate knowledge, experience and skills of professionals.

The technology for replenishing professional textual knowledge in the e-learning system should include the simple easily acquired procedures of formalization of textual information with use of multilevel structures of obligatory data input with control of inaccuracy, consistency and repeatability of data. The standard technology of auto formalization of textual knowledge used in the e-learning systems should be designed for no programming professional and to include use of semantic (textual) structures (QUESTION – ANSWERS – JUSTIFICATION of the CORRECT ANSWER) in the form of production (rules) such as "IF...TO", having modularity and allowing to organize their effective search, adjustment and compilation.

Answers to the question (correct and incorrect) should be under construction with semantic distinctions, the having different semantic accents and with the minimum differences in completeness, accuracy and visual configuration, without allowing use trivial (generally known, extremely simple) information as distinctive signs of answers. The finished semantic (textual) representation describing some finished task and beginning with the functional verb indicating the main operation of the performed work should be the main form of the answer.

When using in the e-learning systems of graphic descriptions for formalization of safety knowledge of work these descriptions should be performed in the form of the interconnected drawings reproduced by sets of primitives in which transition from display of current state of the situation to everyone to the subsequent should be performed at the corresponding actions with elements current accompanied with text messages.

The volume of textual and graphic descriptions according to the concrete normative document (subject area) should fully reflect completely the requirements of this document available to formalization by text, text and graphic and graphic procedures, and if necessary can be complemented with audio and video materials.

For preparation and certification of personnel for a concrete profession, specialty, technological process, a type of works, etc., responsible for training and control of knowledge form the corresponding groups of questions and graphic descriptions (in the presence) to which unique names are appropriated.

Appointment procedures of passwords and authorization of access for certain groups of the users allocated on the volume of access to modules, actions and functions have to be provided and at the personified access and also procedures of start of notifications about terms of carrying out the next instructing and knowledge checks. The e-learning systems are developed and delivered as the program covers developed in use accumulating and showing knowledge not only the qualified specialists responsible for preparation and certification, but also knowledge of the instructed and trained employees.

The module of preparation should be available to the trainee without preliminary adjustment by the specialist responsible for preparation, at any convenient time, both before the certification and after it. By e-learning system realization of several techniques of electronic preparation with use of various technologies of training reflecting experience and knowledge of the specialists organizing preparation can be supported.

It is necessary to regulate:

- rate of preparation;
- mode of dialogue;
- types and methods of hints;
- possibility of correction of the made mistakes;
- order of complication of the training procedures and the repetition of the passed.

The results of electronic preparation should be registered with preservation of records of its intermediate results between training sessions with definition if necessary by data about the trainee, where the training process was interrupted and from where it should be continued. It is necessary to provide a possibility of logical summation of results of electronic preparation for the trainee in case of several cycles of preparation in days or other determined time period.

As a basis of the universal processes of preparation it is necessary to use the standard emergency situations structured with use of descriptions of violations of operation modes of the equipment and installations, accidents and injuries and formalized in the form of rules and graphic images as a part of the knowledge base and if necessary to include audio and video representations. It is necessary to maintain by the e-learning systems confidentiality of preparation which excludes the negative emotions by search of the correct answer and a possibility of obtaining information by other trainees about the degree of correctness of answers.

It is necessary to provide with network versions of e-learning systems rational current and professional training of personnel with expeditious finishing the fullest and relevant knowledge for formation of steady knowledge, skills and abilities in process of accumulation of knowledge in the form of the allocated resource of a certain enterprise, the organization with attraction of the available personnel resources of the organization, and resources of network knowledge bases which each network user could use effectively.

Processes of personnel certification in the e-learning systems should to allow to accumulate knowledge about the specialists responsible for certification, and of the certified employees. In the special module of the e-learning system it is necessary to concentrate the necessary relevant knowledge and information on conditions of control of knowledge of employees, formed responsible for certification, including as obligatory:

- the name of the group of the certified employees and the skill level corresponding to it, reflecting degree of complexity of questions;
- order of presentation of justification of the correct answer at certification;
- number of questions for certification;
- quantity of questions in the ticket on a certain type of the document (subject area) from among certified which are earlier selected according to this document for this group;
- conditions of calculation of assessment of certification for the "tough" scheme of the correct and incorrect answer or on the "soft", considering degree accuracy of answers;
- time of the answer to one question;
- condition of presentation the rest of the time for the answer certified;
- general time spent by the certified employee for control of knowledge.

Conditions of electronic certification of personnel should be regulated depending on originating production needs and most fully describe control electronic programs of different professional levels and production spheres of application. Process of certification should be carried out according to earlier created conditions (options) and to show a series of questions and graphic descriptions by separate fragments of knowledge (subject area). When organizing the process of attestation of personnel, the technology of forming assignments and programs of electronic certification should exclude unauthorized access of the attested person to the correct answers to questions, to unreasonable choice of correct decisions.

The output documentation of e-learning systems should have the flexible structure which is adjusted by the user and reflect the established existing requirements to reporting documentation and need of the user for fixation of the information, knowledge and data necessary for the organization of effective preparation, instructing and control knowledge of personnel in production and in training centers.

The set of the electronic documentation formed by the e-learning system of safety of agro-industrial production should allow to open most fully the maintenance of the e-portfolio of employee on safety of the agro-industrial production built in the form of the interconnected information on results of preparation, instructing and certification of the employee for safety of agro-industrial production about harmful and dangerous production factors and other factors of working conditions, about production injuries, is production the caused incidence, the applied means of individual protection for the considered period of work of the employee.

E-PORTFOLIOS OF EMPLOYEES

Maintaining an e-portfolio of employee on safety of agro-industrial production (EPE SAP) with the resources of the e-learning system allows to estimate the professional potential, skills and competences of the employee in the field of safety of agro-industrial production by tracking of types and results of learning (instructing, certifications), places of work and working conditions in workplaces, personal information on the applied means of individual protection, medical inspections, is production the caused incidence and production injuries and also to estimate professional development in a certain workplace.

E-portfolio of employee on safety of agro-industrial production contains necessary information for management of processes of preparation, instructing and certification and also increase in efficiency of educational activities in the field of safety of agro-industrial production.

Application of e-portfolio of employee in the field of safety of agro-industrial production is the effective tool for the organization of personally focused training of employees, the successful solution of tasks of professional selection of employees, questions of employment, assessment of quality of professional education, social demand of employees of certain professional groups.

EPE SAP provide the possibility of control of individual achievements, skills and competences on safety of agro-industrial production in each process of preparation and after its end, during all production activity of the employee. The high production functionality of e-portfolios of employees on safety of agro-industrial production is provided with integration of EPE SAP in a learning management system and personnel resources of the organization, use of filling, components and opportunities of EPE SAP with use of computer network information technologies.

When organizing and conducting e-learning processes for employees in the safety of agro-industrial production, the personal and professional characteristics of the trained employee, the safety of working conditions in the workplace, the indicators of production injuries and production the caused incidence for improvement of quality and efficiency of future training.

In e-learning processes for the safety of agro-industrial production it is necessary to use electronic portfolios of employees in which should be interconnected personal and professional information, knowledge and skills of employees on safety of agro-industrial production protected from unauthorized access and which allow to accumulate information about employees necessary for adoption of effective decisions on their employment taking into account their safe production skills are submitted (GOST, 2017).

EPE SAP should be formed in the form of the flexible information model of professional characteristics focused on application in processes of training for decrease in production injuries and is production the caused incidence for the reasons of low-quality training of personnel.

EPE SAP is considered as set of the electronic information including data on results of preparation, instructing and certification of the employee for safety of agro-industrial production, on harmful and dangerous production factors and other factors of working conditions, about production injuries, production the caused incidence, the applied means of individual protection for all (or any considered) the period of the employee's work activity.

EPE SAP must contain the module of the main personal professional information about the employee, the module of additional personal professional information about the employee, the module of knowledge and skills of the employee creating single flexible information model of professional characteristics of the employee in the form of the information environment saturated by the relevant personified professional information (Khalin Eu.V. & Pozdneev B. M. (2016), Khalin, Eu.V., Pozdneev, B. M. & Mikhaylova, E. E. (2018)). The example of realization of structure of EPE SAP is given in figure 2.

The structure of EPE SAP is intended for import, export and combination of information with support of compatibility, transmission and exchange of information by network means, for use by the trainees and persons who are carrying out training. The structure of EPE SAP should reflect scope of application, the characteristic of content and normative documents, to be followed by reference information about structure and the maintenance of e-portfolios with the description of types of the e-portfolios used in the e-learning systems of safety of agro-industrial production (for their classification).

Figure 2. Structure of the e-portfolio of the employee on safety of agro-industrial production

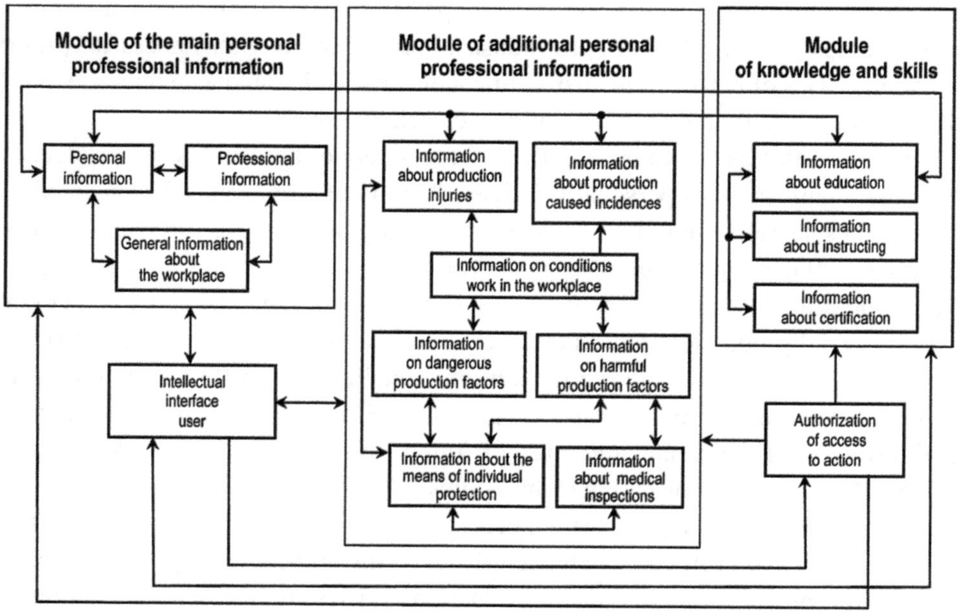

In the module of the main personal professional information about the employee of EPE SAP it is necessary to accumulate the personal information about the employee, including a surname, a name, a middle name, sex of a person, date of birth, number of the insurance policy, total work experience, the professional information about the employee, including names of professions, work experience by each profession, general information about the employee, including the name of the organization for the place of work, the name of division of the organization, date of revenues to work, the personnel number.

In the module of additional personal professional information about the employee of EPE SAP the following information characterizing production activity of the employee should be accumulated and replenished:

- information on production injuries, including date of the accident, the characteristic of consequences of the accident, number of days of disability;
- information on the production the caused incidence, including a type of a disease, date of establishment of a disease, a disability outcome, number of days of disability, data on dispensary account;
- information on working conditions at the workplace, including the name of a workplace, a type of technological process, the name of dangerous and harmful factors of a workplace;
- information on the dangerous production factors that affected the employee during the labor activity including information on dangerous factors on the approved list according to the existing normative documentation;
- information on the harmful production factors influencing the employee during work including information on harmful factors on the approved list according to the existing normative documentation;

- information on the means of individual protection applied by the employee during his labor activity, including, for example, date of issue, norm of issuance, term socks, the basis for extradition;
- information on medical inspections, including, for example, date of survey in treatment-and-prophylactic establishment, date of survey in the center of professional pathology, frequency of surveys, date of the following medical inspections.

In the module of information on knowledge and skills of the employee of safety of production in EPE SAP the following data on knowledge and skills of the employee should collect and be replenished:

- information on the general education, professional first education, professional second education and the subsequent in the presence;
- information on instructing on the last place of work, including, for example, date of the introductory instructing, date of unplanned instructing, date of primary instructing in a workplace, date of repeated instructing, date of target instructing;
- employee on certification of the employee for the entire period of work, including, for example, date of certification, a type of certification, including primary, regular (planned), extraordinary, the characteristic of subject area of certification with a volume and content of the material shown to the employee at certification.

EPE SAP should intend for realization of the following main functions:

- compilation relevant individual electronic programs of preparation, instructing and certification according to information model of professional characteristics of the employee;
- increases in efficiency of programs of preparation, instructing and certification of personnel;
- formations of sustainable skills and knowledge on the safety of agro-industrial production affecting decrease in industrial injuries and production the caused incidence;
- estimates of level of safety of performance by the employee of production targets during all cycle of his production activity with use of indicators of industrial injuries, it is production the caused incidence actually of employee or arisen through his fault;
- adoptions of effective decisions on employment of the employee and when changing a kind of activity of the employee;
- detection of statistical information on not observance of terms of instructing and certification for a fault of the employee, other indicators and characteristics agreed and accepted by the employee.

It is necessary to support by network means of e-portfolios of employees on safety of agro-industrial production:

- possibility of operating control of contents and timeliness of instructing and examinations;
- comparison of the formalized employee of EPE SAP on structural divisions, production objects;
- confidential character of all the formalized information and knowledge on safety of agro-industrial production circulating in networks which are contained in e-portfolios.

The trainee must interact freely with an e-portfolio in the e-learning system, have access to management of process of studying of content, without repeated confirmation of personal authenticity and check of authorization of any changes of a portfolio, starting with provided access The trainee is presented with any new comments, assessments, sections of the general access and informs about users interacting with distributed data and knowledge of the e-portfolio.

At access to EPE SAP for the next training the trainee must download necessary information on the previous training according to structure of a portfolio for identification of the available experiences and knowledge and to choose the strategy of training according to requirements responsible for training (preparation and certification) on safety of agro-industrial production with use of network content of other trainees who were trained on the strategy appointed to the new trainee.

The administrator of network training (system administrator) connects or disconnects e-portfolio from the e-learning system without the need for repeated authentication of the trainee and supports use of the instrumental environment of an e-portfolio in system.

The administrator of network training provides creation and control entered, both group and personal portfolios, ensures safety of information about employees and their confidentiality, limits access (executes access authorization) to procedures of adjustment and deleting information, to actions with structure of a portfolio, excludes access to resources of unauthorized and external users.

The access to EPE SAP is executed automatically on the basis of identification of users according to the directory of users and their access rights, updated in a constant mode.

If operational storage of some groups of information of EPE SAP is not required, constant deleting not used information with saving regularly backup copies and storage of those from them which can recover all history of training of the employee is provided. Temporal accounting entries of users of e-portfolios are created for external users for their tracing and are deleted as required.

After completion of an educational online course on safety of agro-industrial production in a workplace the responsible specialist (administrator) confirms its completion in EPE SAP, along with earlier finished test operation in a workplace of the employee.

The responsible specialist creates the information models of trainees reflecting their advance to the training purposes, stages of a study of content and check of skills in a workplace, results of certification with appropriate level of personification.

The user of EPE SAP has an opportunity to view, comment, edit and evaluate the general contents of an e-portfolio depending on the access rights set by the owner of content during the time specified by the system administrator.

INFORMATION MODELS OF COMPETENCES OF EMPLOYEES

The e-portfolio of employee in the safety of agro-industrial production should ensure the formation of an information model for the competences of the employee in the safety of agro-industrial production (IMC SAP), which makes it possible to determine and promptly adjust the individual professional e-learning program for subsequent procedures of electronic certification of personnel for realization of the operated personnel policy and creation of safe working conditions on production.

IMC SAP provide monotonous, same submission of information on competences on safety of agro-industrial production, promote the solution of problems of rational use of personnel resources at exchange of information about competences between various organizational structures and functionally compatible of e-learning systems. Competence of the employee in the field of safety of agro-industrial production and the related information should be unambiguously formalized for effective management and exchange of information in the e-learning systems on safety of agro-industrial production.

IMC SAP should be formed with application of software in the form of network information model of the formalized interconnected qualitative and quantitative characteristics of competences and increase quality and efficiency of e-learning, reliability and stability of safe professional and production competences, accuracy of assessment of professional suitability for safety of agro-industrial production.

In processes of the e-learning safety of agro-industrial production it is necessary to use the information models of competences of employees containing interconnected the formalized personal and professional information about the employee, about professional competences and assessment of professional competences of the employee on safety of agro-industrial production which allow to personify the training processes (programs), to reach the necessary importance of saturation and the appropriate qualification level.

IMC SAP should be formed in the form of the flexible information model of professional competences focused on application in processes of training for decrease in production injuries and production the caused incidence for the reasons of low-quality training of personnel and professional discrepancy.

Information saturation of model of competences on safety of agro-industrial production is used when forming programs of training for increase in efficiency and quality of training of personnel in safety of agro-industrial production, for the organization of safe and effective work of the specific employee according to his professional level.

IMC SAP should allow to define and to quickly adjust the program of individual vocational electronic preparation of the employee for the subsequent procedures of electronic certification of personnel for formation of steady competences and creation of safe working conditions on production.

Information on competences in the IMC SAP should be provided by semantic unambiguous consecutive structures with a possibility of extension of descriptions and introduction of elements of additional information supporting management of competences, completeness of formalization of information on competences and assisting mutual understanding, unambiguous interpretation and perception of semantic designs and results of formalization in different contexts of human resource management, education and training for safety of agro-industrial production.

IMC SAP must meet the requirements of functional compatibility and mutual consistency at realization of network architecture, the method of generalization, management and exchange of information about knowledge, skills, competences, information related to them and the educational purposes between various organizational structures in the e-learning systems for safety of agro-industrial production and career strategy connected with this process and career development of employees.

IMC SAP must contain an interconnected among themselves the module of personal and professional information about the employee, the module of formation of professional competences of the employee on safety of agro-industrial production, the module of professional competences of the employee on safety of agro-industrial production and the module of assessment of professional competences of the employee on safety of agro-industrial production connected by the intellectual interface of the user to authorization of access to operations of modules (Khalin Eu.V. & Pozdneev B. M. (2018), Khalin, Eu.V. & Mikhaylova, E. E. (2018)). The example of implementation of structure of an information model of competences of the employee on safety of agro-industrial production is given in figure 3.

The structure of IMC SAP is intended for formalization of information and data with support of compatibility, transmission and exchange (import and export) of information by network means, for use by the trainees and persons who are realizing training.

Figure 3. Structure of the information model of competences of the employee on safety of agro-industrial production

1 - module of the personal and professional information about the employee; 2 - module of the information on the processes of forming the professional competences of the employee on the safety of agro-industrial production; 3 - module for monitoring the professional competences of the employee on the safety of agro-industrial production; 4 - module for assessing the professional competences of the employee on the safety of agro-industrial production; 5 - block of the personal information; 6 - block of the professional information; 7 - block of the intellectual interface of the user; 8 - block of the information about education; 9 - block of the information on training in safety of production; 10 - block of the information about the experience of production activities; 11 - block of the authorization of access to actions; 12 - block of the control of competences on safe performance of professional duties; 13 - block of the control of competences on application of means of individual protection; 14 - block of the control of competences on prevention of occurrence and influence of dangerous production factors; 15 - block of the control of competences on prevention of occurrence and influence of harmful production factors; 16 - block of the characteristics of consequences of emergency situations and accidents as a result of actions of the employee; 17 – block of the characteristics of consequences of production injuries of the employee 18 - block of the characteristic of the consequences of production injuries of other persons as a result of the actions of the employee; 19 - block of the characteristic of the consequences of production caused incidence of the employee; 20 - block of characteristic of the consequences of production caused incidence of other persons as a result of actions of the employee.

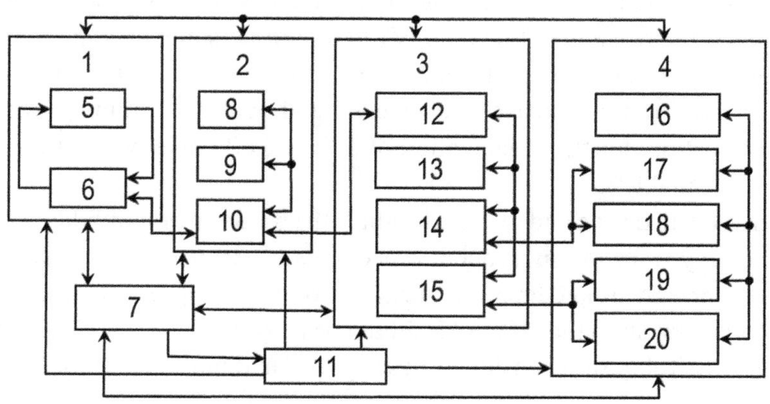

The structure of the IMC SAP should reflect scope of application, the characteristic of content and normative documents, to be followed by the help information about structure and the maintenance of model of competences with the description of types of models of the competences used in the e-learning systems of safety of agro-industrial production.

In the module of personal and professional information about the employee of IMC SAP it is necessary to accumulate using application-oriented software interconnected the formalized (digitized) relevant personal information, including a surname, a name, the employee's middle name, date of birth, number of the insurance policy, total work experience, and professional information, including names of professions, work experience for each profession.

In the module of formation of professional competences of the employee on safety of agro-industrial production of IMC SAP should collect and be replenished permanently using application-oriented software interconnected the formalized data about:

- education, including information on the general education, professional first education, professional second education, information on training of safety of agro-industrial production, including information on instructing on the last place of work with date of introductory instructing, date

of unplanned instructing, date of primary instructing, date of repeated instructing, date of target instructing, with the description of content of repeated and target instructing;
- certifications of the employee for the entire period of work with date of certification, a type of certification, including primary, regular, extraordinary, with the characteristic of subject area of certification with a volume and content of the material shown to the employee in case of certification;
- experience of production activity, including the name of a type of activity and a time frame of execution of a type of activity, the name of technological process and a time frame of an involvement in technological process, the name of a type of the equipment, machine and a time frame of operation with the equipment, machine.

In the module of professional competences of the employee on safety of agro-industrial production of IMC SAP should collect and be replenished permanently using application-oriented software the following interconnected the formalized information on competences of the employee on:

- safe execution of professional duties, including the list of competences on the organizations of safe agro-industrial production, the list of competences for safe execution of production operations, the list of competences on safe preventing of emergency situations on production and the list of competences on safe elimination of consequences of emergency situations on production;
- use of means of individual protection, including the list of means of individual protection which the employee, and the period of work with means of individual protection;
- preventing of influence of dangerous production factors, including the list of dangerous production factors on the approved list according to operating normative documentation under the influence of which the employee can execute production operations, an operation time frame with dangerous factors, the name of devices and the equipment on monitoring and measurements of dangerous production factors which the employee can use;
- preventing of influence of harmful production factors, including the list of harmful production factors on the approved list according to operating normative documentation which influence is possible in case of execution of production operations by the employee, an operation time frame with harmful factors, the name of devices and the equipment on monitoring and measurements of harmful production factors which the employee can use.

In the module of assessment of professional competences of the employee on safety of agro-industrial production of IMC BP should collect and be replenished permanently using application-oriented software the following interconnected the formalized information on consequences:

- emergency situations and accidents as a result of actions of the employee, including information on production consequences of emergency situations over time restoration of technological process after elimination of an emergency situation, time of a complete recovery of a production cycle after elimination of an emergency situation, about the material consequences of emergency situations, information on production consequences of accidents over time restoration of technological process after accident, time of a complete recovery of a production cycle after accident, about the material consequences of accidents,

- production injuries of the employee, including date of accident, number of days of disability, information on production consequences of accident, on the material consequences connected to each accident of the employee,
- production injuries of other persons as a result of actions of the employee, including date of accident, number of days of disability, data on production consequences of accident, on the material consequences connected to each accident of the other person as a result of actions of the employee,
- production the caused incidence of the employee, including the type of disease, date of establishment of the disease, the disability outcome, the number of days of disability, information on dispensary account, about the material consequences, connected to each disease of the employee,
- production the caused incidence of other persons as a result of actions of the employee, including the type of disease, date of establishment of the disease, the disability outcome, number of days of disability, information on dispensary account, about the material consequences, connected to each disease of the other person as a result of actions of the employee.

Each IMC SAP module should be considered as group of the blocks providing implementation of a complete cycle of operations and actions with interdependent information and the personified professional information in the form of qualitative and quantitative characteristics of the competences used for formation of effective personal programs of training of safety of agro-industrial production and in case of assessment of professional suitability.

Each block of the information model of competences for safety of agro-industrial production should be considered as the hardware-software device providing functions of interdependent formalization, accumulation and addition of information (data and knowledge), including mathematical and logical descriptions, a certain group of qualitative and quantitative information characteristics of competences.

The structure of IMC SAP and its filling should provide the completest formalization of information on competences of the employee on safety of agro-industrial production (in the form of a set of semantic elements with qualitative and quantitative attributes) and automation of information exchange about competences for search of personnel with the required competences or for the purpose of bringing competences to the required level with the minimum resource expenses with monitoring and documenting of process of increase in level of competences on safety of agro-industrial production.

IMC SAP are base for appointment educational purposes, the control device of achievement of these is more whole and provide the interactive platform for recommendations about training, including provide the personified planning of training of safety of agro-industrial production with use of accurately formulated strategy for acquisition of new steady competences, increase in level of the available competences, concentration on specific professional features of the trainee, high motivation of employees on acquisition of profound knowledge and with involvement of trainees to formation of programs of training, relevant for them, increasing their professional suitability.

IMC SAP should intend for realization of the following main functions:

- drawing up relevant individual electronic programs of preparation, instructing and certification according to information model of competences of the employee, his professional and production characteristics;
- increase the effectiveness of programs of preparation, instructing and certification of personnel;
- formation of sustainable skills, knowledge and competences of safety of agro-industrial production influencing decrease in production injuries and production the caused incidence;

- estimates of professional competences of the employee on safety of performance of production targets with use of indicators of consequences of emergency situations, accidents as a result of actions actually the employee or arisen through his fault during all cycle of his production activity;
- estimates of professional competences of the employee on safety of performance of production targets with use of indicators of production injuries, production the caused incidence actually of the employee or arisen through his fault during all cycle of his production activity;
- adoptions of effective decisions on increase in professional suitability of the employee;
- identification of statistical information and other quality and quantitative indicators and characteristics on professional competences including agreed and accepted by the employee.

It is necessary to support by network means of IMC SAP:

- operating control of the status of competences on safety of agro-industrial production, contents and timeliness of knowledge checks;
- comparison and change of the formalized information of IMC BAP on structural divisions, production objects;
- confidential character of all formalized information circulating in the networks about competences and knowledge of safety of agro-industrial production which are contained in information models of competences.

The trainee should interact freely with information model of competences on safety of production both in the e-learning system, and independently of the e-learning systems, at authorization of access with the established powers and confirmation of personal authenticity.

IMC SAP should provide a possibility of assessment of dynamics of change of competences on safety of agro-industrial production in time both concerning one employee, and when comparing information on competences of one employee with another with reduction of information to comparability conditions at the uniform system of criteria, methods and processes of assessment.

Constant deleting not used supplied and the information accumulated in blocks IMC SAP for the operational purposes with saving regularly backup copies and storage of those from them which can recover all history of competences of the employee is provided. Temporal accounting entries of users of IMC SAP are created for the external authorized users for their tracing and are deleted as required.

After completion of an educational online course on safety of agro-industrial production in a workplace with use of IMC SAP for confirmation of competence in practice the responsible specialist (administrator) fixes its completion in IMC SAP, along with earlier finished test operation in a workplace of the employee.

The responsible specialist makes changes to information models of competences of the trainees on safety of agro-industrial production reflecting the level of their competences, content study stages, results of check of skills in a workplace and certification according to the set requirements.

The user of IMC SAP has an opportunity to view, comment, edit and evaluate the general contents of IMC SAP depending on the access rights set by the owner of model during the time specified by the system administrator.

EFFICIENCY AND QUALITY INDICATORS

The efficiency of the e-learning system on safety of agro-industrial production, the systems of e-portfolios and the systems of models of competences of employees can be estimated by a complex of technical, economic, social indicators by results of improvement of security status of agro-industrial production and working conditions, to be considered proceeding from interests and opportunities of concrete production and to make a main objective of life cycle of systems.

The generalized parameters of the effectiveness of the e-learning system, including the e-portfolios and models of competences of employees, can be classified as:

- depth and stability of the acquired skills and knowledge promoting adoption of rational safe decisions in real production regardless of location of the trainees and certified (at distance learning and certification) with involvement of the available specialist;
- expeditious bringing to users of system of the new and changed information and knowledge with application of means of information technologies and information systems;
- possibility of improvement of modules and components of system by not programming professional;
- individualization of preparation and certification, reducing the cost of one hour of classes;
- release of teachers from routine work, intellectualization of their activity;
- reduction of number of the involved specialist at the organization of training and certification for safety of agro-industrial production;
- reduction of wear or release of the equipment and the installations used for training and control of knowledge.

Unreliability of system should be considered exclusively as a consequence:

- errors entered into the system in development process which manifestation depends on input information and is found when processing of earlier not found sequences of data;
- mistakes entered in the course of broadcasting;
- receiving in case of data handling of the values exceeding admissible or an inappropriate look;
- inadmissible modifications of system in the course of its maintenance.

The quality of the e-learning system of safety of agro-industrial production including e-portfolios and models of competences of employees should be determined by the functional suitability, reliability, efficiency, applicability, maintenance, mobility. The functional suitability of the e-learning system it is necessary to connect to correctness, ability to interaction and security of system.

The correctness of the e-learning system should characterize ability of system to provide the correct or acceptable results and visual effects for users. The ability to interaction of the e-learning system should be defined by property of components of system to interact among themselves and an external environment due to unification of their interfaces. The security of the e-learning system should be described by labor input and duration of overcoming its subsystem of protection by potential hackers.

Reliability of the e-learning system of safety of agro-industrial production including e-portfolios and models of competences of employees should be evaluated by the time between failures, the average recovery time, the availability factor-the probability of finding the system in an efficient state. Reliability is determined by low probability of loss of operability of the e-learning system and fast restorability of working capacity after the happened failures or refusals.

Reliability of the e-learning system should be characterized:

- **Completeness:** Property not to get to conditions of refusals owing to mistakes and defects in programs and data;
- **Resistance to Defects and Mistakes:** Property to automatically maintain the set level of quality of functioning in the event of abnormal situations;
- **Restorability:** Property in case of refusal to renew the required level of quality of functioning;
- **Availability or Readiness:** Property to perform the required function at present of time under the set use conditions.

The lower bound of reliability should be determined by values of its attributes in case of which the functional suitability sharply decreases and use of the e-learning system becomes inconvenient, dangerous or unprofitable.

To qualitative characteristics of the e-learning system, e-portfolios and models of competences it is necessary to carry:

- applicability or practicality connected to labor intensity and duration which are necessary for a study and complete mastering of functions and technology of use of systems;
- clearness as a possibility of not labor-consuming assessment of applicability of systems for the purposes of users;
- ease of use providing to users convenience and comfort of operation and management of the system;
- a maintainability as the continuous user support by consultations, adaptations and adjustments of system, including changeover of versions;
- convertibility as simplicity of implementation of the specified changes and managing the configuration of the system;
- stability as minimizing influences of changes in system;
- testability as simplicity of quality check of changes of the modified components in system;
- mobility as readiness of the e-learning system to transfer from one hardware operating environment to another;
- adaptability as fitness of the e-learning system, databases and knowledge to modification when deployed in various hardware and operating environments;
- simplicity of installation as simplicity of placement of the e-learning system in new hardware and operational environment;
- suitability as the adaptability of the components of the e-learning system to simple enough use instead of replaceable ones.

Attributes of quality of the e-learning system of agro-industrial production have different measures and scales and in the majority are incomparable among themselves. For estimation of influence of the chosen attributes of quality on functional suitability of the e-learning system it is necessary to appropriate to each of them influence coefficient. Coefficients of influence should be formed isn't formalized – potential users and formalized – expert estimation and establishment of the fullest influence of each attribute of quality on functional suitability taking into account costs of realization of the corresponding attribute.

FUTURE RESEARCH DIRECTIONS

Future researches are planned to be conducted in the field of improvement of network intellectual e-learning system of safety of agro-industrial production, the personified components of the e-learning systems, including a portfolio and models and also in the field of national standardization of various aspects of network e-learning of safety of production. Perspective researches will be directed to improvement of structure and the user interface of systems of electronic portfolios and models of competences of employees, to ensuring completeness of their saturation with information, important for the safety of agro-industrial production, and depth of their immersion in procedures of e-learning.

CONCLUSION

The e-learning systems on safety of agro-industrial production should be carried out as multifunctional intellectual professional systems that allow those responsible for preparation and attestation of employees in concrete production structures to form educational programs that are adapted to real production, ensuring the acquisition of sustainable knowledge, skills and competencies in the safety of agro-industrial production.

The created system of national standards, including the requirements for the organization of network e-training of personnel, involving the use of interconnected web-applications on corporate portals, is considered as an effective tool for influencing the creation of safe agro-industrial production.

The use of e-portfolio systems and models of competences of employees in e-learning systems for safety of agro-industrial production makes it possible to achieve the necessary quality of training, the objectivity of certification of employees, to reduce production injuries and production the caused incidence for the reasons of untrained personnel, to improve work with personnel and to increase efficiency of personnel decisions, to provide direct participation in the formation of effective training programs of the trained employees, to take into account with fullness the necessary personal and professional characteristics both when training and at personnel appointments.

REFERENCES

GOST R 57099-2016. (2016). *Information and communication technologies in education. E-learning production safety. General provisions.* (in Russian)

GOST R 58099-2017. (2017). *Information and communication technologies in education. E-learning production safety. E-portfolio of employee production safety.* (in Russian)

Khalin, E.V. (2016). *Method and network intellectual system of safety production.* Patent Russian Federation N°2589302. Bulletin of inventions, N° 19. (in Russian)

Khalin, E.V., & Pozdneev, B. M. (2016). *E-portfolio of employee on safety of production and a method of his production.* Patent Russian Federation N°2603494. Bulletin of inventions, N° 33. (in Russian)

Khalin, E.V., Strebkov, D. S., & Mikhaylova, E. E. (2016). *A method and network system of safety of production using intellectual graphical descriptions of emergency situations.* Patent Russian Federation N° 2591008. Bulletin of inventions, N° 19. (in Russian)

Khalin, E.V. (2016a). *Method and system of e-learning production safety.* Patent Russian Federation N° 2591687. Bulletin of inventions, N° 20. (in Russian)

Khalin, E.V. (2016b). Systems of e-learning production safety of agricultural production. *Competitiveness in the Global World: Economy, Science, Technologies*, *5*, 206-209. (in Russian)

Khalin, E.V. (2017). *Electrical safety* (2nd ed.). NIIPF TECHINTELL. (in Russian)

Khalin, E.V., Pozdneev, B. M., & Mikhaylova, E. E. (2017). About the new national standard on electronic training of safety of production. [in Russian]. *Vestnik Svyazi*, *3*, 11–16.

Khalin, E.V., & Pozdneev, B. M. (2018). *Model of competences for safety of production and a way of formation of model.* Patent Russian Federation N°2641249. Bulletin of inventions, N° 2. (in Russian)

Khalin, E.V. (2018). Innovative methods and technologies for ensuring production safety. *Safety and labor protection on railway transport*, *1*, 22-29. (in Russian)

Khalin, E.V., Pozdneev, B. M., & Mikhaylova, E. E. (2018). National standard of the system e-portfolio of employees for production safety. [in Russian]. *Vestnik Svyazi*, *5*, 45–50.

Khalin, E.V., & Mikhaylova, E. E. (2018). About standardization of requirements to model of competences of *employee* on safety of production. [in Russian]. *Elektrotsekh*, *4*, 10–19.

Khalin, E.V. (2018a). The interface of the competence model for production safety. *Labor protection and safety at the industrial enterprises*, *3*, 25-33. (in Russian)

Khalin, E.V. (2018b). Requirements for designing the interface of the competence model for production safety. *Modern methods of technical diagnostics and non-destructive testing of details and assemblies*, *2*, 52-60. (in Russian)

Kharchenko, V., & Vasant, P. (2018). *Handbook of Research on Renewable Energy and Electric Resources for Sustainable Rural Development.* Academic Press. doi:10.4018/978-1-5225-3867-7

Vasant, P., Litvinchev, I., & Marmolejo, J. A. (Eds.). (2018). *Modeling, simulation, and optimization.* Academic Press. doi:10.1007/978-3-319-70542-2

Zelinka, I., Vasant, P., Duy, V. H., & Dao, T. T. (Eds.). (2018) Innovative Computing, Optimization and Its Applications: Modelling and Simulations. Academic Press. doi:10.1007/978-3-319-66984-7

Chapter 14
Research of Green Vegetable Cultivation Technology Under Photoculture Conditions in Irradiation Chamber

Igor Yudaev
Don State Agrarian University, Russia

Diana Charova
Volgograd State Agrarian University, Russia

Andrey Feklistov
Volgograd State Agrarian University, Russia

Sergey Mashkov
Samara State Agricultural Academy, Russia

Pavel Kryuchin
Samara State Agricultural Academy, Russia

Sergey Vasilyev
Samara State Agricultural Academy, Russia

Denis Morgunov
"ENERGOSTANDART" Group of Companies "Energospetsstroy", Russia

Yuliia Daus
https://orcid.org/0000-0001-9120-7637
Don State Agrarian University, Russia

Nikola Armenyanov
Ruse University, Bulgaria

ABSTRACT

The object of the research in the chapter is green vegetable crops grown in a manmade environment. The purpose of the presented research is to study the process of growing plants in a man-made environment, the main influencing factor of which is the man-made sources (LEDs) emission, as well as to design and construct the installation (irradiation chamber) for implementation of the studied technology. The chapter presents the results of experiments conducted by a team of researchers, which allowed to determine the technological process indicators of growing green vegetable crops in a man-made environment, and also to develop designs for small-sized irradiation chambers with automatic control, both for laboratory research and for small business and private use. Studies were conducted in different regions of the country and combined the efforts of several scientific groups of researchers. The chapter presents the results of researching these areas conducted by the authors, which are completely original.

DOI: 10.4018/978-1-5225-7573-3.ch014

INTRODUCTION

Nutritious and healthy diet can not be imagined today without a year-round presence of green vegetable crops, which are valuable sources of vitamins, pectins, dietary fiber and other bioactive substances, including antioxidants. Fresh green vegetable crops intake is one of the most effective ways of correcting and preventing damage to biological structures caused by oxidative stress (Zenkov, Lankin & Menshchikova, 2001). Studies show that intake of only 100 grams of fresh greens provides a significant part of the daily human need for vitamin C, carotenoids, phenolic substances, potassium, calcium, manganese and some other elements (Golovko et al., 2010).

To provide useful green products year round, regardless of climatic zones of residence, to as many people as possible, it is necessary to grow them not only using industrial technologies, but also in small-scale production, as well as in private households and shared apartments.

Ubiquitous, year round green vegetable crops growing technology can be implemented with small-sized modular-type cultivating facilities - irradiation chambers, small phytotrons and vertical greenhouses, compact greenhouses, grow boxes and similar structures (Berkovich et al., 2005; Zabel et al., 2014). In this case, the main technological parameter of the man-made plant growing environment is the "correct" irradiation of cultivated plants with the spectrum, radiation intensity and mode that is necessary for their development and growth (Mashkov, 1966; Leman, 1976; Protasova, 1987; Protasova et al., 1990; Tikhomirov, Sharupich & Lisovsky, 2000).

Plant cultivation in completely isolated rooms or in structures placed therein using intense electric lighting (photoculture), allows eliminating the external influence as much as possible, cultivating plants year-round in comfortable conditions for them, and thus getting the maximum return with high quality of production. Photoculture becomes relevant today due to the climate change, the presence of geographically remote regions with unfavorable and extreme climatic conditions, and rapidly growing logistics costs.

BACKGROUND

Light or natural optical radiation is the most important abiotic factor for plant life and its role is primarily determined by a special plants' position in the biosphere as autotrophs, forming organic matter from simple inorganic compounds, while using the solar radiation energy for synthesis (Tooming, 1977; McCree, 1981; Raven, Evert & Eichhorn, 2005).

Light is necessary for the normal development of plants for three reasons. First, it is needed for photosynthesis process, when CO_2 and water combine in the presence of light to form carbohydrates - food for plants. Second, light regulates the external appearance of plants, their constitution. Third, the relative duration of light and darkness periods controls a number of natural physiological processes in plant development and, in particular, blooming of a large number of them (Mashkov, 1966; McCree, 1981; Raven, Evert & Eichhorn, 2005).

Poor lightning can affect one, specific, or all of the plant's cellular reactions. Therefore, when man-made lighting is used to change the growing conditions, the nature of its effects on the reactions that arise should be taken into account. Life processes in a plant organism are closely related both to the intensity of light and the irradiation mode, and to the spectral composition (Protasova, 1987; Protasova et al., 1990; Tikhomirov, Lisovsky & Sidko, 1991; Kong et al., 2015; Ouzounis, Rosenqvist & Ottosen, 2015).

Light intensity affects the photosynthesis rate. The average daily natural light intensity is 100 W/m^2.

At low light intensity, plant respiration processes predominate, and when the light intensity increases, the role of photosynthesis increases linearly. With a further increase in intensity, photosynthesis rate increases more slowly, gradually shifting to the "saturation phase". Therefore, if we continue to increase the intensity of optical radiation, the photosynthesis process becomes less intense - its rate decreases.

At low light intensity, plants are elongated, they form flowered stems, the fruits are poorly formed and small with poor taste.

Intensive light allows to increase yield, obtaining high-quality commercial products, substantially reducing the vegetation timing. By varying light intensity (with optical radiation), one can coordinate the photosynthesis process, and, thus, plant growth and development. At the same time, intense light is harmful for green vegetable crops cultivation, as it leads to the leaf surface growth slowing down, decrease in the leaf quality, they turn yellow and become stiff (Canals et al., 2008).

According to agricultural technology standards, the optimum irradiation rate in an industrial greenhouse is 40 W/m^2 PAR with a photoperiod of 14 hours for growing seedlings, and 100 W/m^2 with a photoperiod of 16 hours for growing commercial products.

Green plants need dormancy and periods of darkness just as much as as light. Without six hours of darkness, they will start to shatter flowers, lag behind in development. The temperature difference between dormancy and waking should not exceed 8 °C.

Therefore, it can be said that the lightning intensity, daylight hours, and the spectral composition of the radiation flux incident on plants are the limiting parameters that strongly influence the growth rate and development. Light pattern is the fundamental contributing factor in the creation of man-made environmental conditions (microclimate) in under cover facilities, since it can not be replaced with any other agricultural technology method (Leman, 1976; Golovnev, Terent'ev, & Fedonkin, 1980; Protasova, 1987; Tikhomirov, Sharupich & Lisovsky, 2000).

Small cultivation facility structures with opaque covers are used not only where there is a serious lack of solar radiation, for example, in the northern territories, but also in small spaces that are not equipped and do not meet the traditional for greenhouse business agricultural technology plant cultivation requirements, and which are in supply in store rooms, cellars, garages, pantries and similar premises. To produce green products in the circumstances, it is necessary to create man-made conditions for plants duplicating the conditions of the natural vegetation environment.

As have already been mentioned above, creation of man-made light irradiation and illumination necessary for plant development allowing for photosynthesis therein, is the main contributing factor (Voskresenskaya, 1965; Protasova & Kefeli, 1982; Raven, Evert & Eichhorn, 2005). In the natural vegetation environment, plants get all the necessary substances for their growth and development from the environment: carbon dioxide - from the air, water and nutrients - from the soil, and they are powered by energy from the sun rays. The energy also triggers chemical reactions, during which carbon dioxide and water turn into glucose (food) and oxygen.

Plants absorb solar energy due to a special leaf structure, which includes green cells - chloroplasts, that have a special substance - chlorophyll. It is the plant pigment that makes the leaves green, and is responsible for absorbing the sunlight energy. Visible spectrum section (380-700 nm), absorbed by chloroplasts pigments was called photosynthetically active radiation (PAR).

Each section of the visible light spectrum plays a certain role in the vital activity of plants (Voskresenskaya, 1965; Voskresenskaya & Nechaeva, 1967; Mortensen & Stromme, 1987; Protasova et al., 1990; Tikhomirov, Sharupich & Lisovsky, 2000; Raven, Evert & Eichhorn, 2005):

- medium-wave ultraviolet rays (280-315 nm) are necessary for metabolic processes and plant growth - they slow down the stem stretching, increase the content of vitamins, facilitate the hardening process and increase plant cold resistance; practically do not affect the synthesis of chlorophyll;
- long-wave ultraviolet rays (320-400 nm) have a regulatory role; only a few percent needed for normal plant development;
- purple and blue rays (400-500 nm) are almost completely absorbed by chlorophyll, with maximum photosynthesis effectiveness. Influencing the plant development, this part of the light wave spectrum inhibits the growth of stems, laminas and chards, thus creating compact plants with thicker leaves that are adapted for better absorption and general use of light. Intensive photosynthesis process with protein synthesis occurs in the leaves. Due to cryptochrome pigment, blue light plays a significant role in plant life. Other special pigments, carotenoids, which make the leaves bright yellow and orange in fall, also take part in the absorption of blue-violet light. The essential role of blue light in the cultivation of green products and seedlings is that it limits the stretching of young plants stems, since the plant cells division is facilitated, but their stretching is inhibited, thereby promoting the development of strong plants and nursery plants. At the same time, blue light suppresses seed germination and causes the development of phototropism phenomenon (leaves and flowers turning towards a source of light), since cells growth slows down near the source of blue light and the stem bends towards the source of life;
- green and yellow rays (500-600 nm) are basically not absorbed by green-colored leaves, and therefore are completely reflected. Due to high penetrating power, they are useful for photosynthesis of optically-dense leaves, lower cover leaves, solid planting. With man-made monochrome irradiation, the leaves get thinner, the stem stretches and photosynthetic rate is low;
- red and orange rays (625-700 nm) are the suppliers of the main share of energy for photosynthesis and contribute to intensive growth of leaves and stems of plants. Chlorophyll almost fully absorbs red and orange color, thus facilitating intensive photosynthesis process with carbohydrate synthesis. This spectrum part of the light incident on the plant has a huge influence on all the physiological processes occurring therein. Low intensity red light (600-700 nm) actively influences the physiological processes in photoperiodic plants sensitive to interchange of light and darkness. So, for example, irradiation of tomatoes and cucumbers in the evening with such light from special lamps results in accelerated development and growth and a significant increase in yield.

Apart from chlorophyll, plants have another photosensitive pigment - phytochrome, which can have different properties under the exposure to near red light (660 nm) and far red light (730 nm). And the pigment "remembers" what light it was irradiated with the last time. Thanks to the phytochrome, the plant "knows" the time of day. Phytochrome ensures that the plant properly reacts to the time of day change - morning, evening, night or day, resulting in a corresponding change in the vital activity of the plant. The blooming period is also regulated by phytochrome.

Near red light stimulates the activity of vital processes and determines the beginning of the light day for the plant, while far red color alternatively signals the evening fall.

Summing up: red color activates germination and growth; when exposed to far red, hypocotyl growth rate decreases, the germ hook starts straightening, chlorophyll synthesis starts, the cotyledons begin to turn green; in the vegetation stage, blue spectral region light gains importance for the plant; in the blooming and fructification stages, red spectrum dominates again.

In addition to chlorophyll, other pigments that are part of the photosynthetic unit absorb the light energy; however, they do not significantly contribute to energy conversion for photosynthesis, since chlorophyll *a* (640 and 660 nm) and chlorophyll *b* (430 and 450 nm) have the highest degree of absorption (Protasova & Kefeli, 1982; Raven, Evert, & Eichhorn, 2005).

Light that is not absorbed to a large extent by chlorophyll *b*, for example, with a wavelength of 660 nm, is captured by chlorophyll *a*, which actively absorbs it at this wavelength specifically, and thus chlorophyll *a* and chlorophyll *b* complement each other in sunlight transformation. At the same time, there is a fairly wide spectral range, from 500 to 600 nm, where light absorption is relatively weak and occurs only due to photoactive pigments other than chlorophylls. Most plants, however, do not need to capture light precisely in this spectral region, since they have enough light absorbed in the blue and red regions (Protasova & Kefeli, 1982; Raven, Evert & Eichhorn, 2005).

Experiments have shown that light quanta from the PAR range are not equally effective for initiating photosynthesis in chloroplasts. In terms of the effect on the photosynthesis process, blue light quantum is equivalent to red light quantum with a lower energy. Therefore, from the photosynthesis process energy point of view, it is more economically viable to irradiate plants with red light than with blue light. But at the same time it is worth noting that all light is more or less important for plant lives. Blue light is important for plants, as it has a photomorphological and regulatory effect, but effective photosynthesis needs much less blue light than red light (Voskresenskaya & Nechaeva, 1967; McCree, 1972; Goins et al., 1997); Banerjee & Batschauer, 2005; Son & Oh, 2013; Lee, Lim, & Kim, 2014; Kopsell, Sams & Morrow, 2015).

Each particular main PAR area (blue, green and red) is marginally suitable for plant cultivation under the conditions of man-made radiation and premises closed off from the environment, and only the radiation generated in a particular ratio of energy across the spectrum can ensure proper plants cultivation (Protasova, 1987). It has also been experimentally proved that illumination sufficient for normal plant growth and development is a reliable obstacle to the accumulation of excessive nitrates in plants.

Thus, the spectral composition of light, as well as its intensity and the change of the light and dark phase, is a strong morphogenetic factor regulating both growth and photosynthetic responses in the life support system of the whole plant. In this case, to grow the majority of cultivated plants, it is recommended that man-made radiation sources have the following PAR spectrum energy ratio: 25-30% in the blue area (380-490 nm), 20% in green (490-590 nm) and 50% in the red area (600-700 nm) (Protasova, 1987). That is, optimization of the spectral composition of the optical radiation source is crucial for the predicted plant growth and normal development, as well as for obtainment of the necessary amount of plant products with the desired biochemical and flavor characteristics (Avercheva, 2014; Konovalova et al., 2016).

When plants are grown in light-tight facilities, the technological operation of plant irradiation is carried out with specialized light sources, which differ from the usual ones not only in the design, but also in the spectrum of radiation necessary for effective photosynthesis process. Mainly traditional (incandescent lamps, fluorescent lamps, etc.) or specialized light sources (phyto lamps with radiation spectrum that is the most acceptable for plants) are used for supplementary lighting and irradiation (Protasova et al., 1990; Prikupets & Tikhomirov, 1994).

But at the same time, many lamps - man-made radiation sources used in greenhouses and greenhouse complexes, are not always suitable for use in small-scale cultivating structures due to the size of the latter, the temperature regimes arising during their operation, electricity consumed, and no possibility to create the radiation intensity and spectral composition modes necessary for normal plant development.

At present, light sources based on light-emitting diodes (LEDs) are intensively used for the purposes. LED wavelength in the red spectrum region is 620-780 nm, in orange - 600-620 nm, in yellow - 585-595 nm, in green - 500-570 nm, in light blue - 465-490 nm and in blue - 430-465 nm. Therefore, by making combinations of different color groups LEDs, it is possible to obtain a light source with nearly any spectral composition in the visible range.

Low power consumption should be noted as another LED advantage. In this case, LED emission is direct, and this allows for a more efficient use of lamps on their bases for supplementary lightning and irradiation of plants. It should also be taken into account that LED lifetime is several-fold larger than fluorescent lamp lifetime. LED fixtures, unlike fixtures with other lamps, are not fragile; therefore, they can be resistant to mechanical and atmospheric impacts, and the possibility of low-voltage power supply makes their operation safe (Mitchell, 2012; Gary, 2015).

LED intensity depends on the current flowing through the chip. This allows for relatively easy control of radiation intensity of the fixture by changing the amperage. Using LEDs with different radiation wavelengths and varying the current for different LEDs, it is possible to obtain the radiation spectra of different composition and intensity, thus selecting the necessary spectrum and radiation intensity, depending on the specific period of plant development.

Small-scale cultivation facilities, despite their size, are quite sophisticated technical complexes that are easily automated and can be quite smoothly added with the lighting system control (both by intensity and spectral composition), organizing such control operations by programs, taking into account the plant development phase.

The main advantage of LED lamps used for plant lighting in cultivation facilities is the possibility of selecting the radiation spectrum that is near-perfect for plant growth: red light is needed for root system growth, fruit ripening and blooming, while blue is needed for leaves development and plant growth. In comparison with the radiation sources used in greenhouse complexes, the main part of a sodium lamp spectrum, for example, lies in the orange-red region and the blue light is clearly insufficient; this results in plants stretching up, becoming more fragile and badly transported (Protasova et al., 1990; Dzakovich, Gómez & Mitchell, 2015).

Another LED advantage over the lamps currently being used is the fact that almost all electrical energy supplied to the device can be converted into optical radiation energy that will be completely absorbed by the plants and will not be reflected in the surrounding space.

MAIN FOCUS OF THE CHAPTER

The object of the research are vegetable green plants illustrated by lettuce grown in an irradiation chamber with man-made environment with regulated irradiation modes created.

The purpose of the research presented is to study the plant growing process in a small-scale irradiation chamber with man-made environment and man-made sources (LEDs) impact on their growth and development, as well as to design and create a working installation (irradiation chamber) model for implementation of the technology offered.

Rationale for Supplementary Lightning and Irradiation Modes Necessary for Plant Growth and Development

Based on the analytical review carried out to determine the influence of the spectral characteristics of optical radiation sources on the plant physiological development, the following light pattern algorithm was offered and implemented for the chosen crop, leaf lettuce.

Given the lack of daylight and small size of the growing facilities, the light ratio algorithm will depend on the ontogeny stage.

After the seeds are sown in the ground and before the sprouts, irradiation is carried out directly with red light only. This will contribute to the intensive and active growth of the aboveground part, and establishment of the main root.

When the white sprout reaches the ground surface, a light-growth and photomorphogenetic response occurs: hypocotyl or mesocotyl growth is strongly inhibited, and the epicotyl (the first true internode) and leaves growth is intensified. The ethylene content in the hook area decreases and the hook straightens. The plant turns green and goes to the phototrophic mode of nutrition. When the first green leaf appears, an important stage in plant development comes up - the growth of the second and third leaves. All crops are the most sensitive to solar radiation in the first growing season (I and III organogenesis stages, 1-3 leaf stages). Violation of irradiation modes during these periods inhibits growth processes and disrupts consistency of the physiological functions that determine the morphogenesis processes. Leading shoot death is observed in all cases where irradiation is insufficient at this stage. Therefore, supplementary lightening in the following days is carried out with mixed red (600-760 nm) and blue (430-480 nm) light at their ratio of R/B = 3/1 (p.u.), since further intensive growth continues during the period and the main task of the period is to proportionally increase the height of the plant and the number of leaves thereon. It is necessary to eliminate plant elongation and stunting, weak bushiness and the presence of small area leaves. Therefore, the light pattern is mixed - red (600-760 nm) and blue (430-480 nm) light at a ratio of R/B = 3/1 (p.u.), which allows intensifying the herbage growth.

During the full maturity start period, yellow light (565-590 nm) is added to the indicated ranges, since at this plant growth stage the plant is blooming and the seed material is formed.

At the last stage (full maturity), the plants are predominantly illuminated by blue-green-red spectrum at a ratio of 1/0.5/0.5 (p.u.), ensuring a complete genetic material set up from the plant to the seed, if the plants are grown to get seed grains. If plants are grown to get herbage, this step is skipped (Sarychev, 2001).

The specified order (as an algorithm for including the corresponding optical radiation sources) is included in the program processed by automatic control controlling the supplementary lighting system. The operating time of optical radiation sources is 14-16 hours with an irradiation level of at least 17 mol/m^2/day. The need for the device to maintain the temperature of the irradiation chamber medium within 20-25 °C and timely arrange irrigation and full-scale fertilizing is an additional condition for ensuring growth and good harvesting.

Rationale for and Design Development of a Small Irradiation Chamber for Growing Plants Under Intense Photoculture

The following are the main requirements for small-scale structures for implementation of intense photoculture: these facilities should occupy as little space as possible and be located in the premises in the areas not used by the person or visited for a minimum amount of time; they must be technologically efficient with agricultural technology processes automated to the maximum extent; installations should consume minimum energy, which should be used with maximum efficiency in technological operations - organization of supplementary lighting, irrigation, maintaining temperature and humidity, etc.

Therefore, the use of closed-type irradiation facilities with an opaque cover is often grounded and advised, for example - irradiation chambers, boxes, rack systems with chambers, irradiation cabinets where the entire radiation flux generated by man-made light sources is directed at the plants, and internal irradiators surfaces are made of reflective materials.

The most perfect figure is a ball, since it has a center and a set of points creating the surface of the ball itself is equally spaced therefrom. If we consider this form as a structural embodiment of a small-scale cultivation facility, an optical radiation source may be located in the center, and the plants, located at an equal distance from the center, will be on the ball surface. If this structure is technically implemented, it will be the most energy-efficient design of a small greenhouse, since all photons from the light source will reach the cultivated plants.

Therefore, in designing a small cultivation facility (irradiation chamber), it was decided to make its arch shape as close to a ball as possible. Therefore, it was decided to produce the arch of the irradiation chamber (where LEDs are placed) in a half-cylinder shape, which would allow to evenly distribute LEDs arranged in the form of luminous matrix lines across its surface. It is also well-known from the building structures theory that light is distributed more evenly within such space, which maximally meets the environmental safety and energy efficiency requirements.

The structure of an irradiation facility (chamber) for growing green vegetable products with no daylight used is a design solution option of a small-scale cultivation facility with full, intense photoculture (Figure 1).

The developed facility for growing plants under the conditions of a completely man-made environment provides for the creation and possibility of control of the variable irradiation mode with specified LED spectral brightness ratios with simultaneous volumetric and uniform surface distribution of optical radiation in the interior volume of the irradiation chamber (Charova, Yudaev & Feklistov, 2016).

A structurally small facility for year-round growing of green vegetables consists of a microprocessor control unit and an irradiation chamber, which can optionally be divided into two compartments (boxes). Control circuit is equipped with a power supply unit with output voltage U = 12 V and power P = 250 W, assembled based on Microchip PIC16F1937 microcontroller. 32 COB (Chip-on-Board) RGB LEDs of 3 W power are selected as optical radiation sources, which are collected in luminous matrix lines and evenly located in each of the two boxes along the entire inner surface of the arch of the irradiation chamber. LEDs are connected in series through the resistors in the light lines that make up the electrical circuit. To provide more efficient LED cooling and to maintain the necessary microclimate parameters, depending on crop agricultural technology, a fan with a controlled air flow via control commands from the microprocessor unit is mounted in the end walls of the irradiation chamber.

During the seedlings growing, the air temperature in the irradiation chamber was maintained at +21 ... +22°C and humidity - 45 ... 55%, the soil temperature was kept within +22 ... +24 °C at the humidity

Figure 1. External appearance of a produced small-scale facility for year-round green vegetables cultivation (irradiation chamber)

a) b)

of 50 ... 60%. Dosed automatic watering of plants with warm water was organized, and the air mobility was carried out by forced ventilation system. DS1820 electronic sensor was used (Figure 2,a) to measure soil temperature, SHT10 (Figure 2,б) to measure the soil humidity and temperate, and DHT11 (Figure 2,в) to measure the air humidity and temperate All sensors worked in coordination with the control unit, allowing not only to measure, but to control the results on a digital indicator board (Figure 2,г). Testo 540 instrument (illuminometer) (Figure 2,д) was used to measure the light level.

The Study of Optical Radiation Sources (LEDs) Characteristics

32 CHANZON COB (Chip-on-Board) RGB LEDs with a capacity of 32 W were accepted to installation as the radiation sources, which were collected in LED matrix bands and evenly placed in each of the two units along the entire inner surface of the irradiator arch. LEDs in the matrix bands are connected in series via the resistors into light lines, representing an electrical circuit through which the current flows and which are located along the chamber arch perimeter.

The main electro-optical properties of LEDs selected for the research are shown in Table 1.

To study the photoculture process of Typhoon leaf lettuce, the manufactured irradiation chamber design was divided into two zones - two boxes (Figure 3).

LEDs with an adjustable spectrum (red, blue, green) are placed in the first box, which was installed and was not changed during the entire period of growth and development of the lettuce plants. The distribution of red, blue and green chip powers (R-B-G) corresponded to the following values: 0.76, 1.12, 1.12 W, with a total box capacity by the above colors - 12.16; 17.92; 17.92 W. The parameters were maintained throughout the growing season. A mode adjustable depending on the development period of the lettuce plants was installed in the second working box, and the power ratio of LEDs used was varied (Table 2). To control the brightness when mixing the required colors in the radiation spectrum, the irradiator has a control device, which operating principle is based on the pulse-width modulation (PWM).

Figure 2. Electronic sensors: a) DS1820 temperature sensor; b) SHT10 humidity and temperature sensor; c) DHT11 humidity and temperature sensor, as well as the scoreboard with digital display (d) and the external appearance of Testo Illuminometer 540 (e)

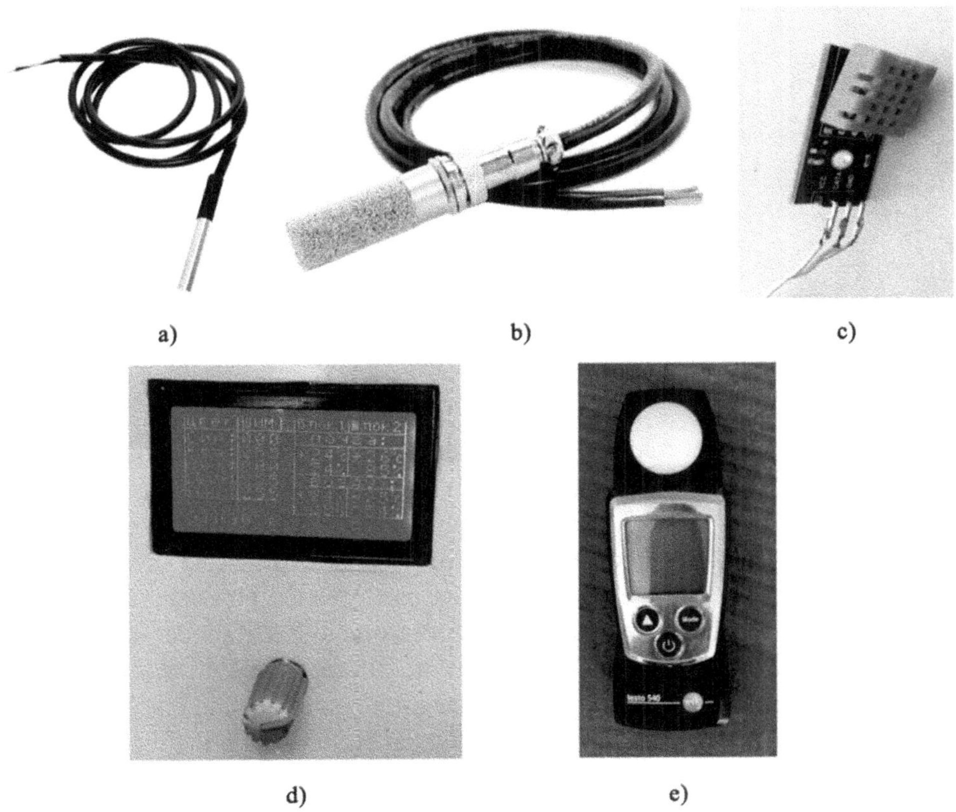

Table 1. LED electro-optical properties

Parameters	LEDs (RGB)		
	Red (R)	**Blue (G)**	**Green (B)**
Peak wavelength λ_{peak} (nm), at I = 350 Ma	632	520	465
Operating Temperature	-40°C/+100°C		
Luminous flux, Lm	40-50	30-40	70-80

The advantages of such systems are determined by the possibility of increasing interference immunity, reliability, accuracy of regulation, as well as simplicity in implementing the circuit solution, etc.

The diagram of the control block for individual processes and the irradiation chamber in general is shown in Figure 4. The diagram is implemented based on the PIC16F1937 (DD1) microcontroller, the following main automation elements are connected to: for collecting information about the environmental parameters - SHT10 (DD2) soil humidity and temperature sensors and DHT11 (DD3) air humidity and temperature sensors; for information output - WG128x64 (DD4) graphical LCD display with LED backlighting with a contrast range adjustable with the R2 divider; for block set-up - an incremental encoder

Figure 3. LEDs connection in irradiation chamber boxes diagram

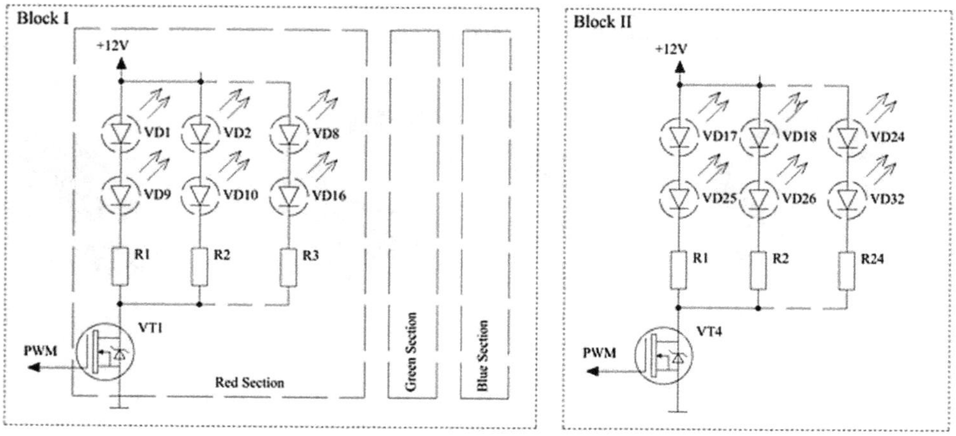

(BR1) with a (SB1) EC11 tact switch; for supplementary lightning and ventilation process control by pulse-width modulation (PWM) - IRL3705 (VT1-VT5) insulated-gate field-effect transistors. The whole diagram is powered by a 220 V static power supply unit with IC7805 (DA1) integral voltage regulator. DD1 clock source is built-in, with a frequency of 32 MHz.

One of the criteria for choosing the microcontroller used was the presence of 5 CCP(CAPTURE/COMPARE/PWM) modules, which are necessary for operation in PWM mode. At this microcontroller clock frequency, the PWM can operate at frequencies of 1.95 to 333.3 kHz, with a bit depth of 10 to 6.6 bits. Based on the sufficiency and convenience, the 8-bit resolution and the frequency of 125 kHz were chosen. The on-off time of each channel is set with an encoder, with the set-up value represented on the display.

The work program is written in low-level assembler programming language and has more than 3,500 commands because of the use of a graphic display.

Since the device operates under an 8-bit processor control, it was decided to numerically adjust the color relation in the range from 0 to 255 units, where 255 corresponds to 100% of the LED (chip) power, resulting in 16.5 million color combinations.

The daily operating time of the irradiation chamber in supplementary lightning mode was 16 hours.

Prior to experiments with the lettuce plants cultivation, LED emission sources operating modes were studied and characteristics for each of them were recorded - chromaticity coordinates and correlated color temperature; spectral characteristics; illumination; pulsation, etc. To study the characteristics of LEDs and LED matrix bands collected on their basis and used in the irradiation chamber, the latter were studied with the following instruments: TKA-VD spectral colorimeter, TKA-PKM illuminometer and Argus-07 illuminometer and flicker meter (Figure 5). Research was carried out for the irradiation chamber LEDs placed in its both boxes.

The results of the research on the characteristics of radiation sources in the irradiation chamber - LED matrix bands of the irradiation chamber are presented in Tables 3 and 4, and Figure 5.

Figure 4. Irradiation chamber process control block diagram

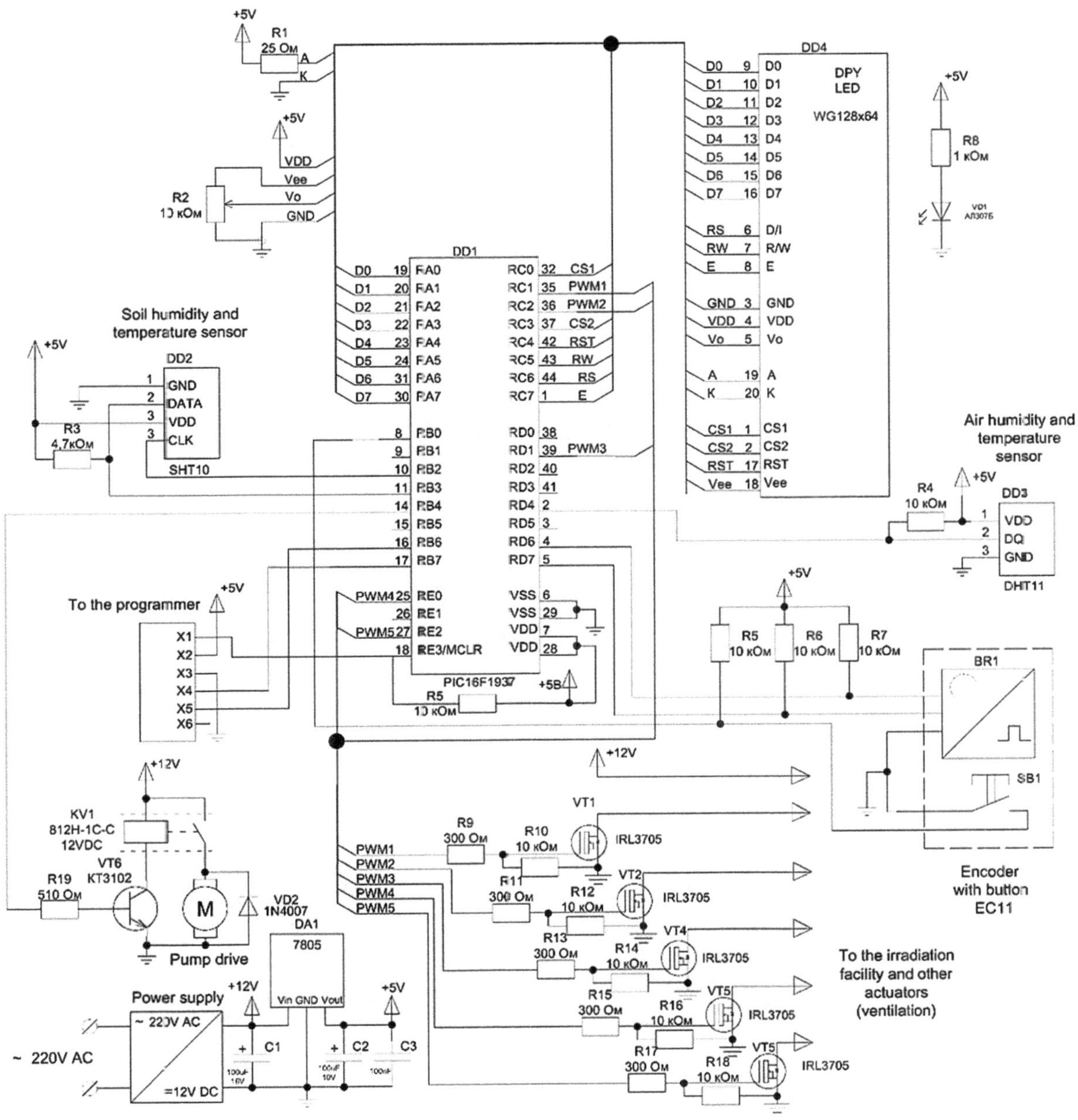

Research and Study of Lettuce Plants Growth and Development under the Conditions of a Small Irradiation Chamber

During the experimental research, the earlier defined theoretical assumptions about the influence of the light patterns chosen for lettuce cultivation were checked, and the operating parameters of the lightning system with the optimized spectral composition of the optical radiation sources and their influence on

Table 2. Color characteristics and relations in chips during experimental research

Date	Working Box	Color Relations R:B:G	Power Distribution in a Single Chip R:B:G (W)	Total Block Power R:B:G (W)	Illumination, lux
August 9-16, 2016	1	255:255:255	0.76:1.12:1.12	12.16:17.92:17.92	6,865
	2	255:0:0	0.76:0:0	12.16:0:0	2,260
August 17-24, 2016	1	255:255:255	0.76:1.12:1.12	12.16:17.92:17.92	6,865
	2	170:85:0	0.51:0.37:0	8.16:5.92:0	1,450
August 25, 2016	1	255:255:255	0.76:1.12:1.12	12.16:17.92:17.92	6,865
	2	170:85:85	0.51:0.37:0.37	8.16:5.92:5.92	2,220

Figure 5. Pictures of experiments on the characteristics of LED-based optical radiation sources with: a-c) TKA-VD Spectral Colorimeter; d) TKA-PKM Illuminometer, Argus-07 Illuminometer and Flicker Meter

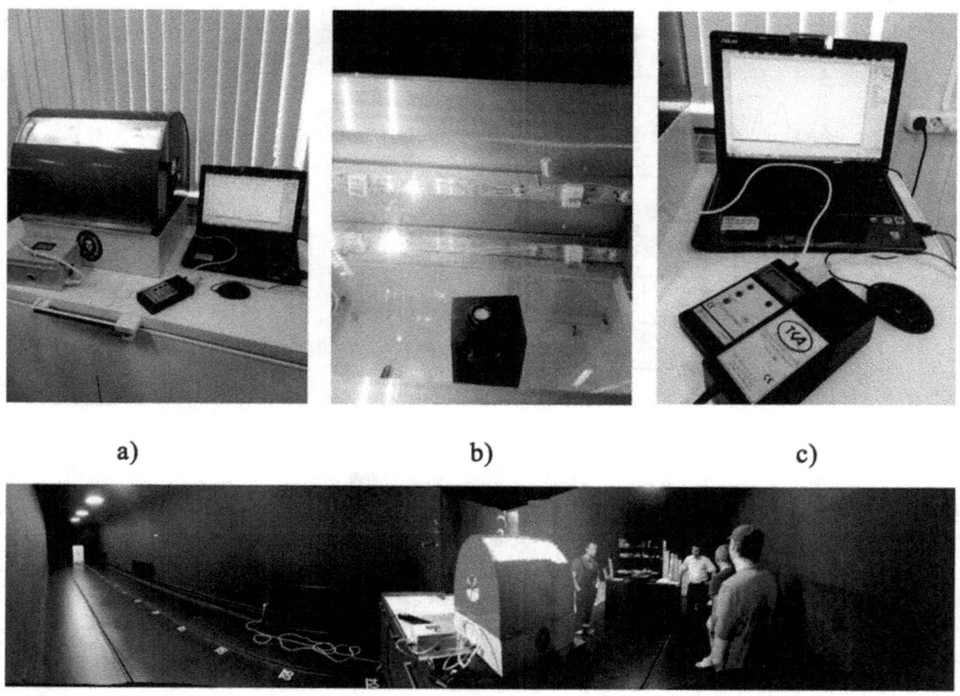

d)

plants grown in the absence of daylight in a man-made environment were estimated (Charova & Yudaev, 2014; Charova et al., 2017).

The choice of the object of the research was driven by leaf lettuce importance as a nutrient dense plant, and therefore it is among the top ten useful vegetables. Leaf lettuce is the absolute leader among the vegetables in vitamin K content and on top of that it contains organic acids, a large amount of vitamin A, C, as well as E, PP, iron, phosphorus, iodine, copper, sodium, potassium, zinc. Leaf lettuce is a source

Table 3. Results of the research and study of LED optical radiation source characteristics

Mode	Measuring Instruments					
	ARGUS-07 Illuminometer and Flicker Meter		TKA-VD Spectral Colorimeter			
	Illumination, lux	Ripple	Color Coordinates (X; Y; Z)	Chromaticity Coordinates (x; y; u'; v')	Illumination, lux	Correlated Color Temperature, K
Blue (255)	171.7	0.6	–	–	–	–
Red (255)	830	2.8	224; 100; 1 224; 100; 1 224; 100; 1	0.688; 0.307; 0.518; 0.521 0.688; 0.307; 0.519; 0.520 0.688; 0.307; 0.519; 0.521	841.3 832.3 820.0	unchanged
Green (255)	1,564	1.6	24; 100; 20 24; 100; 20 24; 100; 20	0.167; 0.690; 0.061; 0.567 0.167; 0.692; 0.061; 0.568 0.167; 0.690; 0.061; 0.567	1,540.0 1,536.9 1,539.1	unchanged
White (255)	2,122	1.2	134; 100; 277 132; 100; 279 131; 100; 281	0.262; 0.196; 0.218; 0.365 0.259; 0.195; 0.214; 0.364 0.257; 0.195; 0.213; 0.363	2,954.8 2,901.2 2,871.7	20,000 20,000 20,000
White (max)	–	–	100; 100; 100 100; 100; 100 99; 100; 100	0.334; 0.332; 0.211; 0.473 0.333; 0.333; 0.211; 0.473 0.332; 0.332; 0.210; 0.473	2,194.4 2,165.0 2,196.6	5,400 5,500 5,500

Table 4. Results of the research and study of the characteristics of the irradiation chamber LED matrices radiation source with TKA-VD Spectral Colorimeter

Mode (Red-Blue-Green)	Irradiation										
	Maximum at the Wavelength, kW/m²	Wave Band PAR 400-700 nm		Spectral/PPFD							
				Blue 400-500 nm		Green 500-600 nm		Red 600-700 nm		Far Red 700-790 nm	
		kW/m²	µmol/m²/s	kW/m²	µmol/m²/s	kW/m²	µmol/m²/s	kW/m²	µmol/m²/s	kW/m²	µmol/m²/s
R-B-G (255-255-255) Figure 3a	605	39.9	174.4	19.2	73.260	8.82	38.850	11.8	62.318	99.6	0.583
R-B-G (255-0-0) Figure 3б	582	15.3	79.4	494	1.874	745	3.475	14,1	74.025	254	1.541
R-B-G (170-85-0) Figure 3в	414	17.6	82.2	6.92	26.267	731	3.355	9.99	52.582	152	0.914
R-B-G (85-170-0) Figure 3d	2.9	1.6	79.5	485.0	50.712	570.2	3.131	655.4	25.652	884.8	0.254

Figure 6. Measurement of the spectral intensity of the irradiation chamber LED matrix bands taken with TKA-VD/04 Spectral Colorimeter, for the following operation modes (blue-red-green): a) 255-255-255; b) - 255-0-0; c) - 170-85-0; d) - 85-170-0

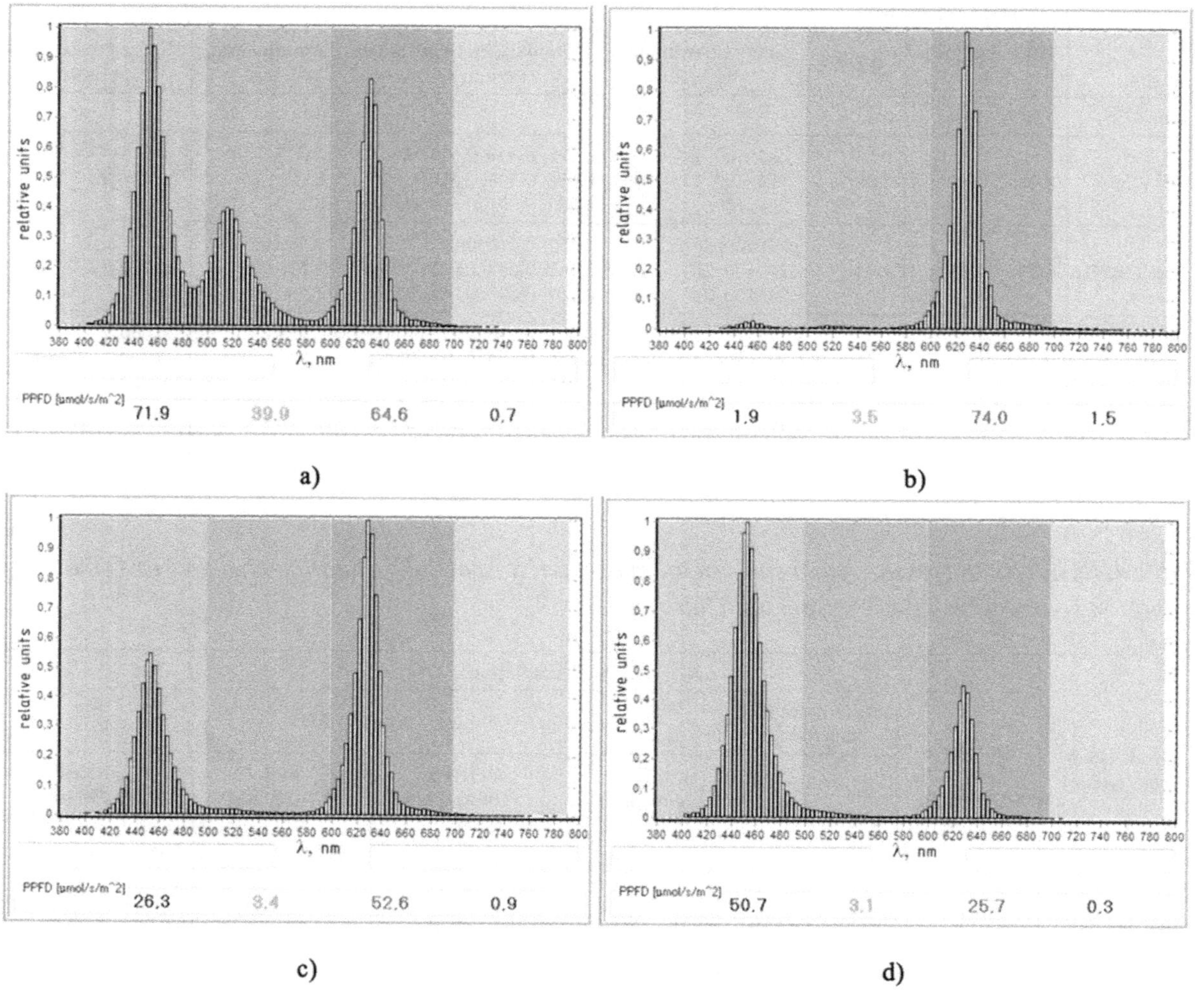

of folic acid, which is absolutely necessary for development and qualitative growth of such systems as the immune and circulatory, so important for the body.

Plant growth and development was observed and the observation results were recorded throughout the entire cultivation period. Let's review the research results and perform the analysis.

August 9, 2016 Typhoon leaf lettuce seeds sown. 4 seeds were manually sown in 180 ml pricking-up cups and placed in the irradiation chamber. The research was carried out with a leaf salad, which growing season, as declared by the manufacturer, was 30-40 days. Commercial certified peat mix with water suspension acidity of 5.5-7.0 and a mass fraction of moisture not exceeding 65% was used as the ground.

Figure 7. Pictures of lettuce plants of the first (a) and second (b) boxes (August 15, 2018), and the measured spectral intensity of the irradiation chamber LED matrix bands of the first (c) and second (d) boxes

a)

b)

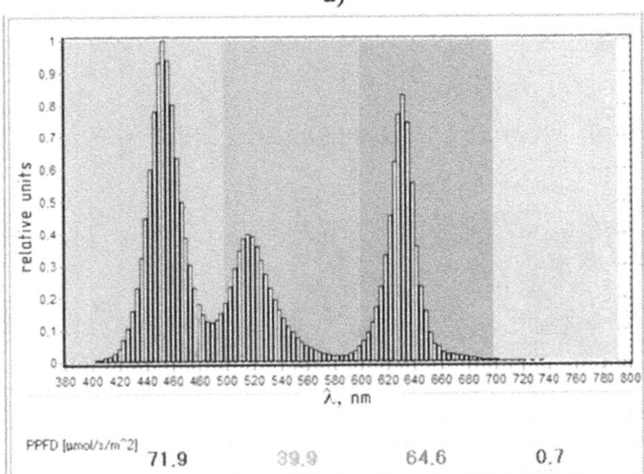

c)

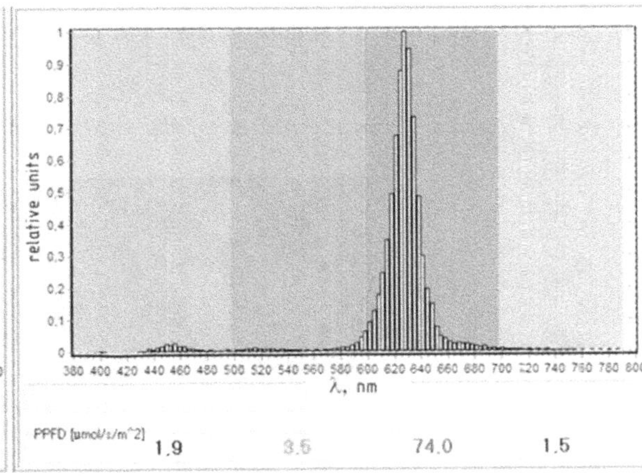

d)

Sprouts

In he first box, good sprouts were observed on August 11, while in the second box they were recorded on August 10. For comparison, seeds were sown in the conditions of the natural vegetation environment - first sprouts were seen on August 11.

August 12, 2016 Plant observations showed: in the first box, the plants had 2 leaves, while the plant height was 0.3 cm; in the second box - 2 leaves and 0.4 cm respectively.

August 15, 2016 Plant observations showed: in the first box, the plants had 3 leaves, while the plant height was 0.4 cm; in the second box - 2 leaves and 2.0 cm respectively, control figures were equal - 2 leaves and 1.5-2.0 cm.

August 22, 2016 Plant observations showed: in the first box, the plants had 4-5 leaves, while the plant height was 4.0-5.0 cm; in the second box - 2-3 leaves and 5.0-6.0 cm respectively, control figures were equal - 2 leaves and 1.5-2.0 cm.

September 1, 2016 Plant observations showed: in the first box, the plants had 5 leaves, while the plant height was 7.0 cm; in the second box - 3-4 leaves and 8.0 cm respectively.

September 12, 2016 Plant observations showed: In the first box, the plants had 5-6 leaves, while the plant height was 10.0 cm.The leaf was large, slightly alveolar, strongly repand, the tissue texture was crispy, juicy with tender taste.

In the second box - 4-5 leaves and 11.0 cm respectively. The leaf elongated, narrow, slightly repand, tissue texture was soft, fragile, not particularly juicy.

The obtained results of the conducted experiments allow to state the following:

1. there is a definite dependence of Typhoon lettuce salad development and harvested crop quality on the irradiation regimes;
2. today LEDs are the most promising optical radiation sources for organizing the photoculture of greenhouse plants, especially green vegetable crops.

Figure 8. Pictures of lettuce plants of the first (a) and the second (b) boxes (August 22, 2018)

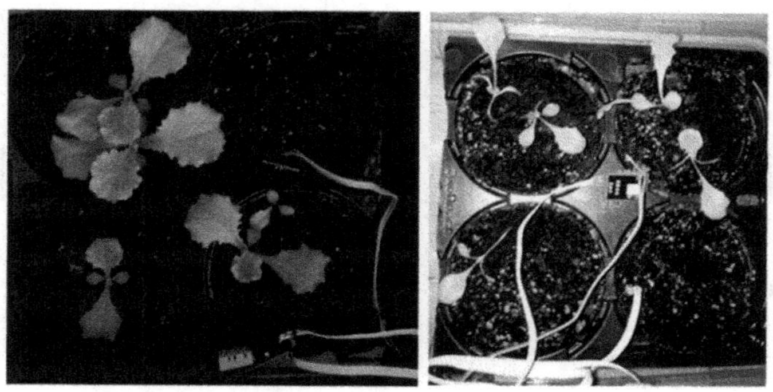

a) b)

Figure 9. Pictures of lettuce plants of the first (left) and the second (right) boxes (September 1, 2018)

Figure 10. Pictures of lettuce plants of the first (a) and the second (b) boxes (September 12, 2018)

a) b)

Figure 11. Changes in height of lettuce plants grown in the first and second boxes of the irradiation chamber

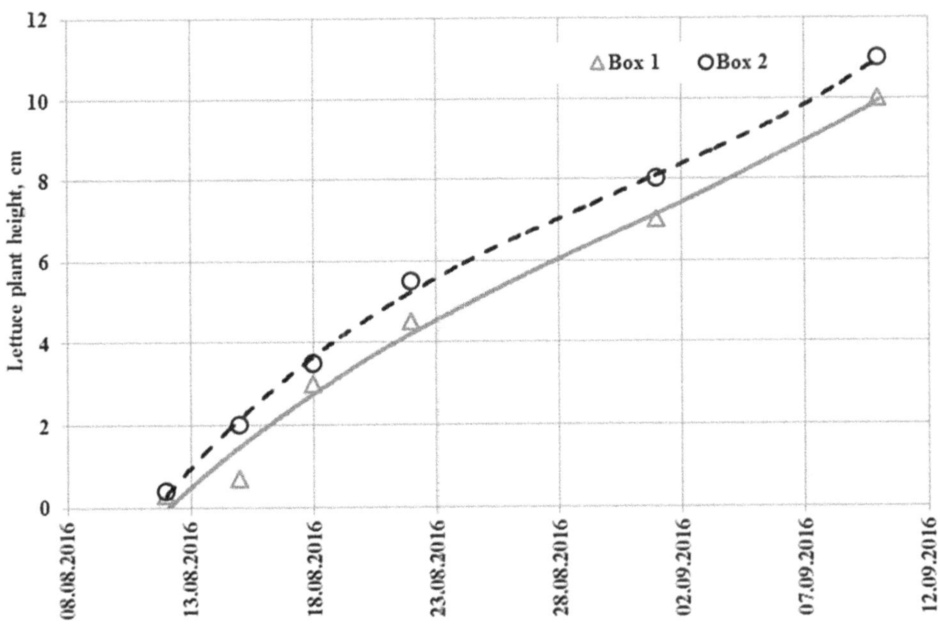

At the end of lettuce cultivation experiments, the organoleptic indicators of the lettuce were analyzed. Organoleptic indicators of two encoded samples of a lettuce plant were examined according to a 10-point rating scale developed by expert tasters. Average examination results are presented in Table 5.

Table 5 data suggests that the second sample is ahead in taste, but scored less in the remaining indicators. To keep the assessment impartial, a complex quality index was calculated, which amounted to 7.57 for the first sample, and 7.67 for the second sample. Therefore, it was concluded that the second sample would be preferable for the consumer, since its complex indicator was higher. Despite the low color, crisp and external appearance scores, the second sample had better taste qualities, as this indicator was a priority for vegetable products consumers.

Table 5. Average sample points

Sample	Indicator			
	Taste	Color	Crisp	External Appearance
1 - salad grown in box 1	6.67	8.56	8.00	8.44
2 - salad grown in box 2	8.89	6.67	5.89	7.44

Figure 12. Analysis of lettuce organoleptic properties

Based on experimental research results, the following parameters were identified to be crucial for yield assessment: the height of lettuce plants and the number of leaves harvested therefrom.

Based on the theory grounded by leading scientists in optical electrotechnology area and the results of triple-repeated experiments in 2016-2017, on the basis of organoleptic analysis results, the best-selling option of lettuce products grown was identified, the supplementary lightning mode of which are shown in Table 6.

The following operating factors were chosen for mathematical optimization: irradiation mode (the spectral source composition) with impact exposure. 3W RGB LEDs with a power ratio of R:B:G=255:255:255=0.77:1.12:1.1 (where 255 corresponds to the maximum chip power; R, B, G - red, blue and green LED color). In all the experiments, the exposure (irradiation time) was 16 hours.

Having processed the experimental data presented on the first graph, a linear relationship was obtained as a result of the regression-correlation analysis:

$$Y = 3.927 - 0.018 \cdot \mathrm{K} + 0 \cdot C + 0.0018 \cdot 3 + 0.285 \cdot T\,, \qquad (1)$$

where Y is the plant height; K is the red chip power; C is the blue chip power; 3 is the green chip power; T is the number of days from the experience start.

We determine the significance of the results obtained with the Fisher criterion. Let the significance level $\alpha = 0.05$, and the number of degrees of freedom ν_1 and ν_2, let ν_1 = m - 1 = 3 and ν_2 = n - m = 35 - 4 = 31 (where m is the number of variables, and n is the number of data points). $F_{крит}$ (0.05; 3; 31) = 0.116. Since we obtained the F-experimental equal to $F_{набл}$ = 672.46, which is much greater than F-critical, then with a probability of 95% we can state that the correlation found between the variables is significant.

We determine the significance of the regression equation coefficients with the Student's test. The critical value $t_{крит}$ (0.05; n - m) = 2.039. Experimental values: $t_1=m_1/se_1$=16.75; $t_2=m_2/se_2$=0.494; $t_3=0$; $t_4=m_4/se_4$=-5.726; it can be concluded that the first and fourth values are greater than the critical value in absolute value, which means that the plant height is more influenced by the red color power and the

Table 6. Results of experimental research with typhoon lettuce

Plant Height, cm	Number of Leaves, pcs.	Chip Power (Maximum Chip Power Is 255)			Number of Days from the Experiment Start
		Red (R)	Blue (B)	Green (G)	
0	0	255	0	0	1
0	0	255	0	0	2
0.2	2	255	0	0	3
0.4	2	255	0	0	4
0.8	2	255	0	0	5
0.8	2	255	0	0	6
1.5	2	170	85	0	7
2.4	2	170	85	0	8
3.5	2	170	85	0	9
3.5	2	170	85	0	10
4.2	2	170	85	0	11
4.8	2	170	85	0	12
5.0	2	170	85	0	13
5.6	3	170	85	0	14
5.6	3	170	85	0	15
5.6	3	170	85	0	16
6.2	3	170	85	85	17
6.3	4	170	85	85	18
6.8	4	170	85	85	19
6.9	4	170	85	85	20
7.4	5	170	85	85	21
7.6	5	170	85	85	22
7.9	6	170	85	85	23
8.0	6	170	85	85	24
8.0	7	170	85	85	25
8.0	7	170	85	85	26
8.4	8	170	85	85	27
8.4	8	170	85	85	28
9.0	8	170	85	85	29
9.4	9	170	85	85	30
9.4	9	170	85	85	31
10.0	9	170	85	85	32
11.0	9	170	85	85	33
11.0	9	170	85	85	34
11.0	9	170	85	85	35

Figure 13. Changes in height and leaf mass accumulation of lettuce plants grown in the second box of the irradiation chamber

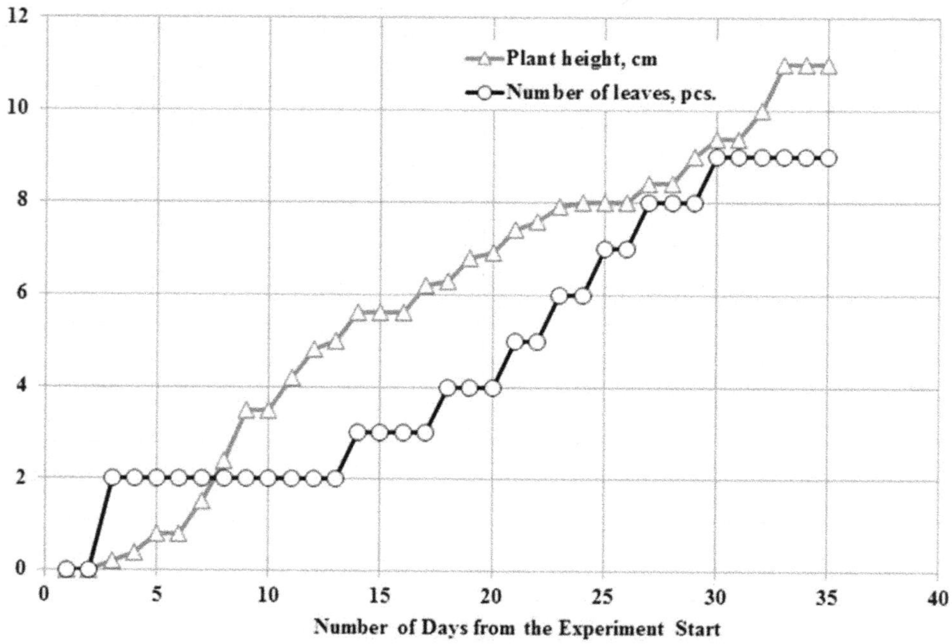

number of days from the experiment start, and we can state this with 95% confidence. Blue and green chip powers influence the plant height to a lesser extent.

Having similarly processed the experimental data from the second graph, the following linear relationship was obtained:

$$X = -4.981 + 0.02 \cdot K + 0 \cdot C - 0.004 \cdot 3 + 0.335 \cdot T\,, \qquad (2)$$

where *X* is the number of leaves.

We determine the significance of the results obtained with the Fisher criterion. Let the significance level $\alpha = 0.05$, and the number of degrees of freedom ν_1 and ν_2, let $\nu_1 = m - 1 =$ and $\nu_2 = n - m = 35 - 4 = 31$. $F_{крит}$ (0.05; 3; 31) = 0.116. Since we obtained the F-experimental equal to $F_{набл} = 257.45$, which is much greater than F-critical, then with a probability of 95% we can state that the correlation found between the variables is significant.

We determine the significance of the regression equation coefficients with the Student's test. The critical value $t_{крит}$ (0.05; $n - m$) = 2.039. Experimental values: $t_1 = m_1/se_1 = 14.32$; $t_2 = m_2/se_2 = 0.89$; $t_3 = 0$; $t_4 = m_4/se_4 = -4.703$; it can be concluded that the first and fourth values are greater than the critical value in absolute value, which means that the number of leaves of the plant is more influenced by the red color power and the number of days from the experiment start, and we can state this with 95% confidence. Blue and green chip powers influence the number of leaves of lettuce to a lesser extent.

Therefore, we get a functional dependence reflecting the relationship between the change in plant height and the number of leaves on one plant on LED spectral characteristics change over time.

The maximum accumulation of economic biomass does not necessarily result from the maximum photosynthesis of the green leaf; therefore, there is a need to determine the actual increase in the new plant biomass as a controlled parameter, depending on the spectral composition and intensity of irradiation in the plant development process.

To assess the economic efficiency of the projected installation implementation, there was performed assessment of the payback period and the efficiency of the installation's application within five years of its production.

In the complete list of equipment for the manufacture of individual components and the assembly of the entire installation there were included such items as: desk, shelving, soldering iron, soldering iron fan, screwdriver, digital oscilloscope, multimeter, numerical controlled milling machine, luxmeter with the total cost of 301,000 rubles.

To assemble an installation, one set of components is worth 12,880 rubles. The starting price of one finished unit at the beginning of 2018 is 50,000 rubles. Wages per employee per month are 7,500 rubles, which is indexed by years, taking into account the inflation rate and a multiple of the number of equipment produced per year. The rental of the premises for the production is 6,700 rubles. The cost of room renting was determined taking into account inflation, from 2021 the premise is planned to rent a larger one.

The project provides for the circulation of loan funds from the bank (Sberbank of Russia). The cost of loan for equipment is 301,000 rubles; the cost of loan for 20 sets of component parts is 257,600 rubles, and the total costs for loan is 558,600 rubles. The monthly payment is 19,612 rubles, the loan rate is 20%, the maturity is 36 months.

Taking into account these criteria, there were calculated the production costs, which are presented in Table 7.

There were defined the economic indicators of the project, that are presented in Table 8.

Table 7. Production costs

Indexes	Production Costs, rub.				
	2018	2019	2020	2021	2022
Wages	90000	144180	191160	119812.5	115020
Accessories	128800	206337.6	273571.2	342930	411516
Rent of premises	80400	85867.2	91190.97	97118.38	103431.07
Amortization of equipment	7525	7525	7525	7525	7525
Loan payment	235344	235344	235344	0	0
Other costs	9201.75	13317.29	16903.41	17021.58	19124.76
Total	553288.8	694590.09	817714.6	586428.5	658638.84

Table 8. Economic indicators of the project

Indexes	Economic Indicators of the Project				
	2018	2019	2020	2021	2022
Volume of products, pcs.	10	15	20	25	30
Revenues, ths rub	500	801	1134.22	1509.93	1929.68
Production costs, ths rub	553.29	694.59	817.72	586.43	658.64
Profit, ths rub	-53.29	106.41	316.51	923.49	1271.05
Level of total profitability,%	-9.63	15.32	38.71	157.48	192.98
Inflation, %	–	6.8	6.2	6.5	6.5
Price of installation, rub.	50,000	53,400	56,710.8	60,397	64,322.81

SOLUTIONS AND RECOMMENDATIONS

Based on the conducted research, the following recommendations can be drawn:

1. both intense photoculture modes of lettuce plants during cultivation, as with the invariable spectral intensity of the irradiation chamber LED matrix band throughout the entire cultivation period, and varied, depending on the plant development period, show basically the same quality marketable plant collection yield;
2. the mode with adjusting the spectral emission depending on the lettuce plant growth phase allows to obtain higher consumer quality products;
3. the developed design of a small irradiation chamber and intense photoculture modes can be recommended for everyday household use for year-round green vegetables cultivation, and also for use as modular elements for green vegetables production by small producers, for example, catering, cafes, restaurants and shops of green products.

FUTURE RESEARCH DIRECTIONS

As a future research perspective, it is necessary to clearly define the intense green vegetables photoculture technology, which can be done by performing a number of experimental studies, having studied not only consumer qualities and biochemical indicators of grown products, but spectral characteristics and intensity of LEDs in a confined small irradiation facilities space.

The presented technology for growing green vegetable plants is based not only on experiment results, but also on the known plant development data, their physiology and biochemistry, as well as the theory of rhythmicity of all biological processes occurring in plants.

When using a lighting installation (irradiation chamber) for green mass growing, standard green lettuce can be obtained twice a month with a marketability of about 88%, which fully covers the expenses for nearly several harvests.

The fact that products produced under man-made light are green is an essential factor that determines the possibility of technology implementation. The very fact of complete harmony of light and other conditions of growing plants makes it impossible to accumulate free and excessive nitrites and nitrates therein.

The neutrality and purity of water and mineral salts solutions used for fertilizing plants preclude the excessive accumulation of heavy metals or radionuclides; the absence of pests and plant diseases eliminates the need for pesticides and results in no traces thereof in the products. The optimal combination of all plant growth and development factors makes products dietary. Numerous biochemical analyzes carried out on different plants and in various organizations, explicitly confirm the dietary value and ecological purity of the products grown, their value for human life. Products have significantly increased content of sugars, vitamins, glycosides, fats, dry matter and other components that are exceptionally valuable for humans. (Konovalova et al., 2016)

The possibility of a well-founded, targeted influence on the biochemical composition of crop products produced in small cultivation structures allows to develop a concept for production of useful and therapeutic products with an increased content of organically bound trace nutrients, the lack of which in human food determines human diseases – microelementoses, i.e. the diseases of the endocrine system and prostate gland, diabetes, blood anemia, stroke, cancers and others. Growing plant food in a man-made environment allows to create a completely new approach to health improvement methods without drug prevention.

CONCLUSION

In conclusion it can be stated that:

1. small irradiation facilities for year-round green crops growing are necessary not only to receive products useful for human life all year long, but also to grow them under climatic restrictions when products can not be cultivated in the natural vegetation environment, as well as in the Arctic, Antarctic and long-term transport expeditions;
2. by varying the LED spectral emission and intensity in the irradiation chamber, depending on the lettuce plants growth phase, it is possible to obtain higher consumer and commercial quality products, with the necessary biochemical composition and characteristics.

REFERENCES

Avercheva, O., Berkovich, Y. A., Smolyanina, S., Bassarskaya, E., Pogosyan, S., Ptushenko, V., ... Zhigalova, T. (2014). Biochemical, photosynthetic and productive parameters of Chinese cabbage grown under blue-red LED assembly designed for space agriculture. *Advances in Space Research*, *53*(11), 1574–1581. doi:10.1016/j.asr.2014.03.003

Banerjee, R., & Batschauer, A. (2005). Plant blue-light receptors. *Planta*, *20*(3), 498–502. doi:10.1007C0425-004-1418-z PMID:15714356

Berkovich, Yu. A., Krivobok, N. M., Smolyanina, S. O., & Erokhin, A. N. (2005). Kosmicheskie oranzherei: nastoyashchee i budushchee [Space greenhouses: present and future]. Moscow: Slovo [Word].

Canals, L. M., Munoz, I., Hospido, A., Plassmann, K., & McLaren, S. (2008). *Life Cycle Assessment (LCA) of Domestic vs. Imported Vegetables. Case studies on broccoli, salad crops and green beans.* Centre for Environmental Strategy.

Charova, D.I., & Yudaev, I.V. (2014). Povyshenie urozhajnosti ovoshchnyh kul'tur za schet ispol'zovaniya SID dlya ehlektrodosvechivaniya rastenij v sooruzheniyah zashchishchennogo grunta [Increasing the yield of vegetable crops through the use of LEDs for electric lighting of plants in protected ground facilities]. In *Nauchnye osnovy strategii razvitiya APK i sel'skih territorij v usloviyah VTO. Materialy mezhdunarodnoj nauchno-prakticheskoj konferencii, posvyashchennoj 70-letiyu obrazovaniya VolGAU* [Scientific basis of the strategy of development of agriculture and rural areas in the WTO. Materials of the international scientific and practical conference dedicated to the 70th anniversary of the VolSAU]. Volgograd: VolGAU.

Charova, D.I., Yudaev, I.V., & Feklistov, A.S. (2016). Issledovaniya po vyrashchivaniyu listovogo salata v svetodiodnoj obluchatel'noj kamere [Studies on the cultivation of lettuce in the led irradiation chamber]. In *Materialy mezhdunarodnoj nauchno-prakticheskoj konferencii, posvyashchennoj pamyati vedushchego ehlektrotekhnologa Rossii akademika Ivana Fyodorovicha Borodina* [Materials of international scientific-practical conference dedicated to the memory of leading Electrotechnology Russia academician Ivan Borodin]. Volgograd: VolGAU.

Charova, D.I., Yudaev, I.V., Feklistov, A.S., Vorotnikov, I. N., & Gabrielyan, S. Z. (2017). Vyrashchivanie listovogo salata v svetodiodnoj obluchatel'noj kamere [Growing lettuce in the led irradiation chamber]. *Sel'skij mekhanizator* [Rural mechanic], *1*, 20-21.

Dzakovich, M. P., Gómez, C., & Mitchell, C. A. (2015). Tomatoes Grown with Light-emitting Diodes or High-pressure Sodium Supplemental Lights have Similar Fruit-quality Attributes. *HortScience*, *50*(10), 1498–1502.

Gary, W. S. (2015). Commercial Transition to LEDs: A Pathway to High-value Products. *HortScience*, *50*(9), 1297–1300.

Goins, G. D., Yorio, N. C., Sanwo, M. M., & Brown, C. S. (1997). Photomorphogenesis, photosynthesis, and seed yield of wheat plants grown under red light-emitting diodes (LEDs) with and without supplemental blue lighting. *Journal of Experimental Botany*, *48*(7), 1407–1413. doi:10.1093/jxb/48.7.1407 PMID:11541074

Golovko, T. K., Tabalenkova, G. N., Dalke, I. V., Zahozhi, I. G., Gregori, E. E., & Botkin, A. V. (2010). Produkcionnyj process i pishchevaya cennost' zelennyh kul'tur zashchishchennogo grunta na severe [Production and nutritional value of leaf vegetables of the protected ground in the North]. *Gavrish*, *5*, 32–35.

Golovnev, N. B., Terent'ev, V. M., & Fedonkin, D. V. (1980). Ob ocenke ehffektivnosti izlucheniya dlya vyrashchivaniya rastenij [On the evaluation of radiation efficiency for growing plants]. *Svetotekhnika*, *5*, 16–17.

Kong, Sh.-W., Chung, H.-Y., Chang, M.-Y., & Fang, W. (2015). The contribution of different spectral sections to increase fresh weight of boston lettuce. *HortScience*, *50*(7), 1006–1010.

Konovalova, I. O., Berkovich, Yu. A., Yerokhin, A. N., Smolyanina, S. O., Yakovleva, O. S., Znamensky, A. I., ... Tsvirko, V. I. (2016). Optimization of the LED lighting system for a vitamin spacegreenhouse. *Aerospace and Environmental Medicine*, *50*(3), 17–22. PMID:27344853

Kopsell, D. A., Sams, C. E., & Morrow, R. C. (2015). Blue Wavelengths from LED Lighting Increase Nutritionally Important Metabolites in Specialty Crops. *HortScience*, *50*(9), 1285–1288.

Lee, J. S., Lim, T. G., & Kim, Y. H. (2014). Growth and phytochemicals in lettuce as affected by different ratios of blue to red radiation. *Acta Horticulturae*, (1037): 843–848.

Leman, V.M. (1976). *Kurs svetokul'tury rastenij* [Course of plant light culture]. Moscow: Vysshaya shkola [Higher school].

Mashkov, B. S. (1966). *Vyrashchivanie rastenij pri iskusstvennom osveshchenii* [Growing plants under artificial light]. Leningrad: Kolos.

McCree, K. J. (1972). The action spectrum, absorptance and quantum yield of photosynthesis in crop plants. *Agricultural Meteorology*, *9*, 191–216. doi:10.1016/0002-1571(71)90022-7

McCree, K. J. (1981). Photosynthetically active radiation. In *Encyclopedia of Plant Physiology, 12A* (pp. 41–55). Berlin: Springer-Verlag.

Mitchell, C. A. (2012). Plant lighting in controlled environments for space and earth applications. *VII Intl. Symp. Light Hort. Systems*. 10.17660/ActaHortic.2012.956.1

Mortensen, L. M., & Stromme, E. (1987). Effects of light quality on some greenhouse crops. *Scientia Horticulturae*, *33*(1-2), 27–36. doi:10.1016/0304-4238(87)90029-X

Ouzounis, Th., Rosenqvist, E., & Ottosen, C.-O. (2015). Spectral effects of artificial light on plant physiology and secondary metabolism: A review. *HortScience*, *50*(8), 1128–1135.

Prikupets, L. B., & Tikhomirov, A. A. (1994). Optimization of lamp spectrum for vegetable growth. In T.W. Tibbitts (Ed.), *International Lighting in Controlled Environments Workshop*. Academic Press.

Protasova, N. N. (1987). Svetokul'tura kak sposob vyyavleniya potencial'noj produktivnosti rastenij [Light culture as a method of revealing potential productivity of plants]. *Fiziologiya rastenij. Plant Physiology*, *34*(4), 812–822.

Protasova, N. N., & Kefeli, V. I. (1982). Fotosintez i rost vysshih rastenij, ih vzaimosvyaz' i korrelyacii [Photosynthesis and growth of higher plants, their interconnection and correlation]. In Fiziologiya fotosinteza [Physiology of photosynthesis]. Moscow: Nauka [Science].

Protasova, N. N., Wells, J. M., Dobrovolsky, M. V., & Tsoglin, L. N. (1990). Spektral'nye harakteristiki istochnikov sveta i osobennosti rosta rastenij v usloviyah iskusstvennogo osveshcheniya [Spectral characteristics of light sources and features of plant growth in artificial lighting]. *Fiziologiya rastenij. Plant Physiology*, *37*(2), 386–396. PMID:16667287

Raven, P. H., Evert, R. F., & Eichhorn, S. E. (2005). Photosynthesis, Light, and Life. *Biologia Plantarum, 7*, 119–127.

Sarychev, G. S. (2001). Produktivnost' cenozov ogurcov i tomatov v funkcii spektral'nyh harakteristik OU [Productivity of cenosis of cucumbers and tomatoes in a function of the spectral characteristics of Oh]. *Svetotekhnika, 2*, 27–29.

Son, K.-H., & Oh, M.-M. (2013). Leaf shape, growth, and antioxidant phenolic compounds of two lettuce cultivars grown under various combinations of blue and red light-emitting diodes. *HortScience, 48*, 988–995.

Tikhomirov, A. A., Lisovsky, G. M., & Sidko, F. Y. (1991). *Spektral'nyj sostav sveta i produktivnost' rastenij* [Spectral composition of light and plant productivity]. Novosibirsk: Nauka.

Tikhomirov, A. A., Sharupich, V. P., & Lisovsky, G. M. (2000). *Svetokul'tura rastenij: biofizicheskie i biotekhnologicheskie osnovy* [Light culture of plants: biophysical and biotechnological bases]. Novosibirsk: Izdatel'stvo RAN.

Tooming, H. T. (1977). *Solnechnaya radiaciya i formirovanie urozhaya* [Solar radiation and yield formation]. Leningrad: Gidrometeoizdat.

Voskresenskaya, N. P. (1965). *Fotosintez i spektral'nyj sostav sveta* [Photosynthesis and spectral composition of light]. Moscow: Nauka.

Voskresenskaya, N. P., & Nechaeva, V. P. (1967). Dejstvie sinego, krasnogo i zelenogo sveta na soderzhanie belka, nukleinovyh kislot i hlorofilla v molodyh rasteniyah yachmenya [Effect of blue, red and green light on protein, nucleic acids and chlorophyll in young plants of barley]. *Fiziologiya rastenij. Plant Physiology, 32*(14), 299–307.

Zabel, P., Bamsey, M., Schubert, D., & Tajmar, M. (2014). Review and analysis of plant growth chambers and greenhouse modules for space. *44th Intl. Conf. Environ. Systems, 44*, 1-17.

Zenkov, N. K., Lankin, V. Z., & Menshchikova, E. B. (2001). *Okislitel'nyj stress: biohimicheskie i patofiziologicheskie aspekty* [Oxidative stress: biochemical and pathophysiological aspects]. Moscow: Nauka.

ADDITIONAL READING

Brazaitytė, A., Ulinskaitė, R., Duchovskis, P., Samuolienė, G., Šikšnianienė, J. B., Šabajevienė, G., ... Žukauskas, A. (2006). Optimization of lighting spectrum for photosynthetic system and productivity of lettuce by using light-emitting diodes. *Acta Horticulturae*, (711): 183–188. doi:10.17660/ActaHortic.2006.711.22

Bugbee, B. (2016). Towards an optimal spectral quality for plant growth and development: The importance of radiation capture. *Acta Horticulturae*, (1134): 1–12. doi:10.17660/ActaHortic.2016.1134.1

Gates, D. M., Keegan, H. J., Schleter, J. C., & Weidner, V. R. (1965). Spectral Properties of Plants. *Applied Optics, 4*(1), 11–20. doi:10.1364/AO.4.000011

Johkan, M., Shoji, K., Goto, F., Hahida, S., & Yoshihara, T. (2012). Effect of green light wavelength and intensity on photomorphogenesis and photosynthesis in *Lactuca sativa*. *Environmental and Experimental Botany*, *75*, 128–133. doi:10.1016/j.envexpbot.2011.08.010

Lin, C. (2000). Plant blue-light receptors. *Trends in Plant Science*, *5*(8), 337–342. doi:10.1016/S1360-1385(00)01687-3 PMID:10908878

Normy fiziologicheskih potrebnostej v ehnergii i pishchevyh veshchestvah dlya razlichnyh grupp naseleniya Rossijskoj Federacii. Metodicheskie rekomendacii (2009) [Norms of physiological needs in energy and nutrients for different groups of the population of the Russian Federation. Methodical recommendation]. MP 2.3.1.2432-08 от 18.12.2008. Federal'nyj centr gigieny i ehpidemiologii Rospotrebnadzora [Federal center of hygiene and epidemiology of Rospotrebnadzor].

Owen, W. G., & Lopez, R. G. (2015). End-of-production Supplemental Lighting with Red and Blue Light-emitting Diodes (LEDs) Influences Red Pigmentation of Four Lettuce Varieties. *HortScience*, *50*(5), 676–684.

Ritchie, R. J. (2010). Modelling photosynthetic photon flux density and maximum potential gross photosynthesis. *Photosynthetica*, *48*(4), 596–609. doi:10.100711099-010-0077-5

Terashima, I., Fujita, T., Inoue, T., Chow, W. S., & Oguchi, R. (2009). Green light drives leaf photosynthesis more efficiently than red light in strong white light: Revisiting the enigmatic question of why leaves are green. *Plant & Cell Physiology*, *50*(4), 684–697. doi:10.1093/pcp/pcp034 PMID:19246458

Yakovlev, A. N., Turanov, S. B., Kozyreva, I. N., & Starodubtseva, D. V. (2014). Sources with Different Spectra Radiation Influence on Plants Growth and Development. *Advanced Materials Research*, *1040*, 830–834. doi:10.4028/www.scientific.net/AMR.1040.830

KEY TERMS AND DEFINITIONS

Green Crops: Mean vegetable plants grown to produce fresh greens used for food fresh.

Light-Emitting Diode: Means a semiconductor device with a single p-n-junction, where the direct conversion of electrical energy into thermal energy occurs.

Optical Spectrum: Means the electromagnetic spectrum or the distribution of electromagnetic radiation intensity by frequencies or wavelengths.

Photoculture: Means the process of effective plant cultivation using man-made radiation sources, the energy of which is necessary for the photosynthesis process.

Photosynthetically Active Radiation (PAR): Means a part of the solar radiation reaching the biocenosis (plants) in the wavelength range from 400 to 700 nm, used by plants for photosynthesis.

Pulse-Width Modulation (PWM): Means the process of managing power supplied to a load, by changing the pulse ratio at a constant frequency.

Chapter 15
Electro-Impulse Irreversible Plant Tissue Damage as Highly Efficient Agricultural Technology

Igor Yudaev
Don State Agrarian University, Russia

Yuliia Daus
Don State Agrarian University, Russia

Ruslan Kokurin
Volgograd State Agrarian University, Russia

Petr Prokofyev
Volgograd State Agrarian University, Russia

Varvara Gamaga
Moscow State Pedagogical University, Russia

Nikola Armenyanov
Ruse University, Bugaria

ABSTRACT

The object of research in the chapter is the plant tissue of weeds, fruits, vegetables, melons, etc. The purpose of the presented research is to study the process of electro-impulsive irreversible damage of plant tissue as the basis for energy saving, efficient electro-technologies with minimal negative effects of chemical components on the environment and food products. The chapter presents the results of experiments carried out by the team of researchers that allowed it to determine the technological indices of the irreversible electro-impulse damage of plant tissue (weeds, fruits, vegetables, melons, etc.), which is the basis for developing the technology of electro-impulse destruction of weeds, electro-impulse plasmolysis of fruits and melons cultures. This research was conducted in the southern region of the country and combined the efforts of several scientific groups of researchers.

DOI: 10.4018/978-1-5225-7573-3.ch015

INTRODUCTION

The use of electroimpulse high-voltage actuations or impulsed electric fields with high intensity (high-intensity pulsed electric fields, HIPEF) is the basis of innovative technologies of plant raw materials in processing and food industry, and also cultivated crops and weed plants processing, hay grasses, etc. in the crop farming sector of agriculture. In most cases, the purpose of such technological operations is the achievement of irreversible damage to the internal (cellular) structures of processed objects plant tissues due to their nonthermal destruction, which is characterized by minimum energy costs and processing time, high processing efficiency, environmental and food safety.

BACKGROUND

In the technological processes of crop production the electroimpulse high-voltage actuations are applied in the weeds and uncontrolled vegetation destruction (Diprose & Benson, 1984; Vincent, Panneton, & Fleurat-Lessard, 2001; Yudaev & Brenina, 2008; Stankovic, Cvijanovic & Dukic, 2016; Toporkov & Korolev, 2017; Baev et al., 2018), thinning of agricultural crops (Vincent, Panneton & Fleurat-Lessard, 2001), sunflower and tobacco pre-harvest processing (Baev, Savchuk & Armianov, 1985; Antonov, Armianov & Dragiev, 1984; Baev & Borodin, 2002; Baev et al., 2018) hay grass pre-processing before drying, etc. (Chervyakov, 1979).

In the processing of agricultural raw materials and food production the electrical actuation is used to harmful microorganisms, bacteria and enzymes inactivation, (Toepfl, Heinz & Knorr, 2005). Electroimpulse high-voltage actuation with a field density more than 10 kV/cm is used in the fruit and vegetable juices production process, effectively providing to inactivate bacteria with a negligible flavor deterioration of the obtained products, and also their flavour, color and nutritional value deterioration (Aguiló-Aguayo et al., 2009; Yeom, Chism & Zhang, 2002). This electrotechnology application also allows to solve the technological tasks dealing with reversible and irreversible cells permeabilization of plant tissues (Knorr et al., 1994; Angersbach, Heinz & Knorr, 2000). Pulsed electric fields of high intensity with a field density less than 10 kV/cm allow us to obtain high-value metabolites in the food industry (Eshtiaghi & Knorr, 2002; Fincan, DeVito & Dejmek, 2004), increase the yield of fruit and vegetable juices (Bouzrara & Vorobiev, 2000; Ade-Omowaye et al., 2001; Popova, Papchenko & Bologa, 2014), advance the mass transfer in the food drying process, etc. (Rastogi, Eshtiaghi & Knorr, 1999; Ade-Omowaye et al., 2001).

MAIN FOCUS OF THE CHAPTER

All these mentioned applications of high-voltage electro-impulse actuations are based on the functional failure effect of plant tissues cell structures membrane shells, which is achieved due to their irreversible damage.

Therefore, the purpose of the research is the investigation of the actuation mechanism of electrical high voltage pulses on whole plants tissue, raw materials or products and also engineering assessment examination of caring process issues and explaining their subject matters.

PLANT CELL MEMBRANES FUNCTIONAL FAILURE AS THE BASIS OF THEIR IRREVERSIBLE DAMAGE BY HIGH VOLTAGE ELECTRO-IMPULSES

According to the theory of electrical conductivity plants and their tissues, in our opinion, are the second class conductors, since their intracellular and intercellular solutions contain many complex organic compounds and different substances that have the ability to conduct electrical current under the action of external applied voltage. This conclusion is verified, first of all, that environment inside the cells and between them is filled with a liquid fraction, which in its structure is associated with the electrolyte with the main charge carriers – ions. It is worth mentioning that biological environment differ from electrolytes only by the fact that their homogeneity is damaged by the existence of organelles variety, intracellular and intercellular environments containing dissolved organic matter and being distinguished by cell membranes. Under the influence of applied external electric field the ions, being oriented in it, move and thus cause the ion currents appearance. It is also known that biological tissues, including plant, are the ordered structures organized from a large number of elementary components – cells, which are separated from other cells and intercellular spaces by semi-permeable membranes. To maintain the viability and normal functioning of such elements certain amount of energy is required, its replenishment is carried out by a complex biochemical and bioenergetical exchange processes. The membranes permeability of undamaged cells is carried out by necessary for homeostasis ions moving from the extracellular environment to the cells inside and back under the influence of the membrane potential difference. Such interactions are determined by the ion- and mass-transfer processes phenomena, which are characterized by the presence in such structures ion flow or currents. Thus, it is possible to access with high confidence that the undamaged plant tissue, according to the electrical conductivity theory, is the conductor of the second kind.

To explain the damage mechanism one should take as a basis researches dealing with electroporation of cell membranes. The electroporation process is based on the fact that the high voltage pulses increase the permeability of biological membranes without any considerable output of thermal energy. By the effecting of pulses with a duration from tens of microseconds to tens of milliseconds on cells dispersion, creating in the environment electric field, which is characterized by the intensity from several hundred to several thousand volts per centimeter, it causes a sharp conductivity increase of cell membranes. It is due to the fact that the electrical potential difference between intracellular fluid and intercellular environment causes structural disruption of the cell membrane continuity or damage, that result to pores forming. (Zimmermann, 1986; Ho & Mittal, 1996). Depending on the electric field intensity value the membranes normal functioning and further permeabilization can be both reversible and irreversible (Benz & Zimmermann, 1980).

After moderate elektroprocessing, which does not damage the membrane shell, the increased cells conductivity is reduced to normal values over time from several seconds to several minutes, however, the increase in the amplitude of effecting impulses or processing duration leads to irreversible disruption of membranes cell normal functioning. According to modern opinions, the plant cells membranes are constructed mainly of proteins (50-70% of dry weight), lipids (38-17%) and carbohydrates (15-10%) and represent a complex organized system whose main barrier functions are organized by phospholipid bilayer permeated by proteins, performing the role of selective channels or active pumps for ions and metabolites, and which can be activated to ensure cell metabolism. The hypothetic cause of cell membranes conductivity increase may be both in lipid bilayer and proteins. The study of the processes occurring during electroporation indicates that the electric breakdown of particular composition bilayer lipid membrane can be reversible if the voltage on the membrane does not exceed values of the order of 0.2-0.5 V. Irreversible destruction, characterized by the disruption in the membrane its semi-permeable properties, occurs when the voltage value on it more than 1 V, which corresponds to electric field intensity $2 \cdot 10^6$ V/cm. Researches of biophysicists indicate that breakdown of the lipid component is responsible for the cells permeability increase. Experiments with bilayer lipid membrane showed that electrical breakdown occurs stochasticily, and the average lifetime of the membrane depends on voltage nonlinearly. These observations allow us to justify the theory of the pores formation and development in fluid lipid bilayers under the influence of an electric field, particularly and pulse electric fields of high intensity.

If the voltage drop on the membrane exceeds a critical value which equals 0.5 V and above, holes-pores begin to grow in the membrane body created by thermal fluctuations. The electroporation theory of bilayer lipid membranes suggests that the high voltage electric pulse action causes a local structure reorganization resulting to perforated water channels, increasing the whole cell electric permeability. In other words, we can say that under the electric field influence with a critical voltage value on the membrane there occurs irreversible electroporation with the cell membrane destruction, its permeability defect and loss of cell viability. All this discussed facts are defined by the appearance in the membrane body pores with a radius greater than the normal permissible value, which determines the increase of membranes permeability and, consequently, their irreversible damage (Weaver & Chizmadzhev, 1996; Neu & Krassowska, 1999; Kotnik & Miklavcic, 2006; Toepfl, 2006).

THE MECHANISM OF ELECTRO-IMPULSE DAMAGE TO PLANT TISSUE ACCORDING TO THE THEORY OF ELECTRICAL CONDUCTIVITY

To explain the subject matter of the processes occurring in the plant tissue under the high voltage electric pulses influence, it is necessary to know the construction of equivalent electric scheme of the plant tissue displacement, that makes this explanation more informative and accurately. As an electric model of plant origin objects tissue, it is common practice to use a circuit drawing structure of the biological (plant) cell displacement (Fig. 1), which consists of parallel connection of active resistance intercellular space (R_2) and the branch that contains the series connection of protoplasm active resistance (R_1) with complex membrane resistance, representing parallel switched capacitance ($C_м$) and active resistance (R_3) (Klimov, Savchuk & Baev, 1970; Zhang, Stout & Willison, 1990; Inaba et al., 1995; Bauchot, Harker & Arnold, 2000; Baev & Borodin, 2002; Yudaev et al., 2004; Beraa et al., 2016; Baev et al., 2018).

Figure 1. The fundamental equivalent electric scheme of a plant cell displacement

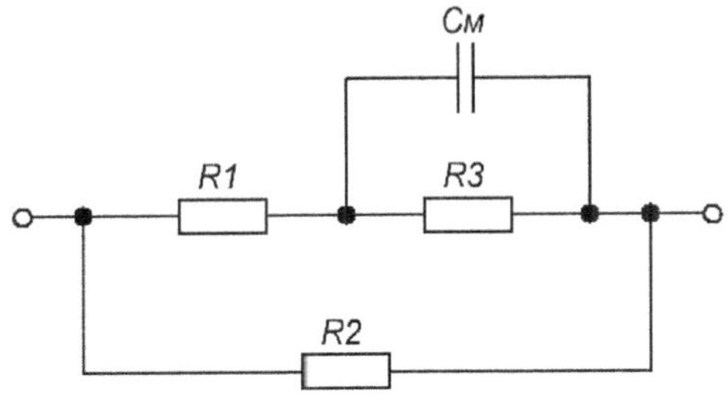

The study of anatomical and morphological structure of plant cell and simultaneous analysis, its already found electro-conductive properties, allows to determine the elements parameters values of equivalent electric scheme of displacement quantitatively. As already mentioned, the intracellular and intercellular environment is filled with a good conductive electric current "solution". Scientists cytologists established that in higher plants the protoplasm or intracellular content structure contains large amounts of organic compounds water inclusions, for example, vacuolar sap. The intercellular environment is filled with a large number of macromolecular units, salts and their solutions. According to the theory of electrical conductivity applied to the described cell structure, it comes out that the intercellular space resistance (R_2) is greater than the resistance of the intracellular "solution" (R_1). The structure of biological membranes and their organization are characterized by the presence of components with significant dielectric properties, which do not allow to conduct ions electricity freely except through intended for it special pores. Experimentally proved, that the basic structural elements of membranes, lipids and proteins, have a very low conductivity of the ionic current, that is, act as a high-quality dielectric. This seems to suggest, that membranes` electrical resistance (R_3) is much greater than the resistance of intercellular space (R_2) and protoplasm (R_1), and the conducted researches lead to the conclusion that the active components of the cell displacement differ from each other quantitatively, that is $R_3 >> R_2 >> R_1$ (Baev, & Borodin, 2002; Yudaev et al., 2004; Baev et al., 2018).

In plants and plant raw materials processing, there is not one single undamaged cell included in an electric circuit, but some of their number, meeting on the current flow way. Having considered the morphological structure of the plant tissue, you may notice that the cells in the current on its flow way creates complex, mixed electrical series-parallel connection including both little surface cells, and more massive, internal. It turns out that the equivalent circuit of tissue displacement, consisting of passive electrical elements is quite difficult one and has a bulky appearance, which makes difficult possibility to estimate electro conductive properties quantitatively and to determine the response (sensitivity) of tissue on the electro-impulse influence, and therefore, to perform theoretical calculations and analysis one need a more compact electric model.

Each individual cell is described by the same electrical components, characterized by different quantitative indicators, as it is well known that cells have different geometrical dimensions and therefore different electro conductivity, but the active-capacitive nature of the models does not change. Plant tissue contains a large number of parallel and series-connected basic circuits of cells displacements, and therefore, in our view, the final version of the displacement circuit of the researching tissue is a compound of the equivalent, the total active resistances and cells capacitance values that organize the construction of the treated area or volume of plant tissue. As evidence, the following logical reasoning are: if simplify or convolute electrical components connection (in series, parallel, series-parallel or parallel-serial), each of which is a simple elementary compound consisting of the same electrical components, the resulting equivalent circuit will represent the same connection as the original components, but different from elementary at some coefficient, let us call it the coefficient of the convolution. In parallel or parallel-serial connection this coefficient is less than the one element, and in series or series-parallel – it is more, but in any case, the resulting simplification will appeared as basic initial elementary connection from the same initial electric components, differing only quantitatively in the value of the convolution coefficient.

Therefore we can conclude that the electrical displacement circuit of the plant tissue is identical to fundamental equivalent electric circuit of the plant cell displacement, but at the same time its elements have the following parameters: R_1 is the total resistance of the protoplasm of those cells which organize the researched plant tissue along the processing flow; R_2 is the total active intercellular space resistance for the considered part of the plant tissue; and R_3 and $C_м$ are the equivalent active resistance and capacitance of cell membranes of the processed plant area or plant material volume.

It is impossible to conduct analysis of the processes occurring in plant tissues during electric pulse effects and leading to irreversible damage of the internal structure, having information only about the construction of equivalent electric circuit of the plant tissue displacement, it is also necessary to understand the quantitative components of this scheme and their changes in the process of electrical damage.

Since the total resistance of the plant tissue is active-capacitive, then the measurement current signal form will reflect the process of charge-discharge capacity of the cell membrane (Fig. 2) by submitting the unipolar voltage pulses of squared shape on a plant sample. Considering that during the existence of the squared voltage pulse all the elements of the displacement scheme are constant, and the front length of the voltage pulse is close to zero, it is easy to establish an analytic expression for the current scheme in the transient regime:

$$i(t) = i_{fr}(t) + i_c(t) = \frac{U \cdot (R_1 + R_2) \cdot e^{pt}}{R_n \cdot (R_1 + R_2) + R_1 \cdot R_2} + \frac{U \cdot (R_1 + R_2 + R_3)}{R_n \cdot (R_1 + R_2 + R_3) + R_2 \cdot (R_1 + R_3)} \times \left[\frac{C_{<} \cdot R_3 \cdot (R_n \cdot (R_1 + R_2) + R_1 \cdot R_2) - e^{pt}}{C_{<} \cdot R_3 \cdot (R_n \cdot (R_1 + R_2) + R_1 \cdot R_2)} \right], \quad (1)$$

where i_{fr}, i_n are the free and forced (involuntary) current components of the transient process, A; R_n is the resistance of shunt in Ohms; p is the radical of the characteristic equation:

$$p = -\frac{1}{\tau} = -\frac{R_n \cdot (R_1 + R_2 + R_3) + R_2 \cdot (R_1 + R_2)}{C_{<} \cdot R_3 \cdot (R_n \cdot (R_1 + R_2) + R_1 \cdot R_2)}. \quad (2)$$

To clarify the mechanism of the occurring phenomena and the grounding of the electro-impulse actuation process on plants, objects and materials of plant origin, let us analyze how are these quantitative indicators of the circuit elements of the plant tissue displacement changed. As an example, let us consider the numerical values change of the electrical equivalent circuit components, to determine them let us use the analytical equation (1-2) and the experimental researches results of electro conductive properties of the plant stem area *Amaranthus retroflexus L.*

To determine the parameters values of equivalent circuit elements we will use equation (1) and solutions of equation (2) for time points t = 0 и t =∞. (Klimov, Savchuk & Baev, 1970; Baev & Borodin, 2002; Yudaev et al., 2004) It is known that for living tissue $R_3 >> R_2$; $R_2 >> R_1$ and $C_м \approx 10^{-9}$-10^{-12} F and by adopting the assumptions that $(R_1+R_2+R_3) \approx R_3$; $(R_1+R_3) \approx R_3$; $(R_1+R_2) \approx R_2$ etc., Baev, V.I., et al., 2018), one can significantly simplify the equations (1-2) and obtain formulas for the determination of equivalent scheme parameters:

$$\left.\begin{aligned} &R_1 \approx \frac{U}{i_{fr}} - R_n; R_2 \approx \frac{U}{i_c} - R_n; \\ &R_3 \approx \frac{U \cdot (R_1 + R_2) - i_c \cdot (R_n \cdot (R_1 + R_2) + R_1 \cdot R_2)}{U - i_c \cdot R_2}; \\ &C_< \approx \frac{\tau \cdot (R_n + R_2)}{R_n \cdot (R_1 + R_2) + R_1 \cdot R_2} \end{aligned}\right\} \tag{3}$$

Assessing the results of the electrical model elements values calculations of plant tissue according to the formulas (3) and obtained data by computer simulation method, it can be noted that the calculating error does not exceed 10-20%.

Figure 2 shows oscillograms of the measuring voltage pulses (Fig. 2,a) and current for the undamaged stem plant tissue *Amaranthus retroflexus L.* in the phase of fruiting (Fig. 2, b) and the same signal during the electric pulse damage to the same tissue (Fig. 2, *c-f*).

The electrical voltage pulses of amplitude U = 3 kV effected on the stem plant tissue fragment with a length of 20.4 mm and equivalent diameter of 1.95 mm, storage capacity of the discharge circuit was C = 100 pF, and in one pulse the specific value energy $W_{1s} = 7.32 \cdot 10^{-3}$ J/cm^3 was applied.

While applying of the rectangular pulse voltage to the undamaged plant tissue (Fig. 2, a) oscillogram of measuring flowing through it current is a steeply falling exponential curve, due to the charging of the membrane capacitance $C_м$. The maximum value of the measuring current is almost in 4 times more than for the end of the transition process (Fig. 2,b), it is characterized by the completion of charging capacity $C_м$.

For the undamaged stem plant tissue *Amaranthus retroflexus L.* the values of the equivalent circuit parameters, calculated by the formulae (3) are equal to: R= 15.59 kOhm; R_2= 92.38 kOhm; R_3 = 853.42 kOhm; $C_м$ = 165.91 pF and the oscillogramm of the measuring current flowing through this area is shown in Figure 2 b.

After partial damage by electrical high voltage pulses, primarily, the resistance of cell membranes R_3, is decreased in the plant tissue, due to the loss of its semipermeable properties. This fact leads to the membrane capacitance shunt $C_м$ by the declined resistance R_3, which results to the loss of membrane capacitive properties. Escaping through the opened pores in the cell wall membranes the protoplasm sap increases the conductivity of the intercellular space. Now the oscillogram of the measuring current through the plant tissue from applied measurement of a rectangular voltage pulse is a weakly falling exponential curve. The highest value of the measuring current due to the active resistance reduction of intercellular space R_2 and of the membrane R_3 increases only in 1.14 times according to the original (before damage) value and its established value for the same time in 3.29 times (Fig. 2, c). The ratio of maximum value to the final value will decrease and will be 1.39. The following effect on plant sample by high voltage pulses caused the quantitative indicators change of the current oscillogram (Fig. 2, d-e), it showed a decrease in the total active resistance of the equivalent circuit to the protoplasm resistance value R_1 and the final loss of the semi-permeable properties of cell membranes. At a total damage of the plant tissue, the active resistance value of the membrane R_3 and protoplasm R_1 is almost equal, the capacitive properties of the membranes almost completely disappear, and the oscillogram of measuring current through the tissue will be close to the oscillogram type of measuring rectangular voltage pulse

Figure 2. Oscillograms of voltage pulses (a) and the current flowing through undamaged stem plant tissue Amaranthus retroflexus L (b), and under the action of electric pulses (c, d, e, f)

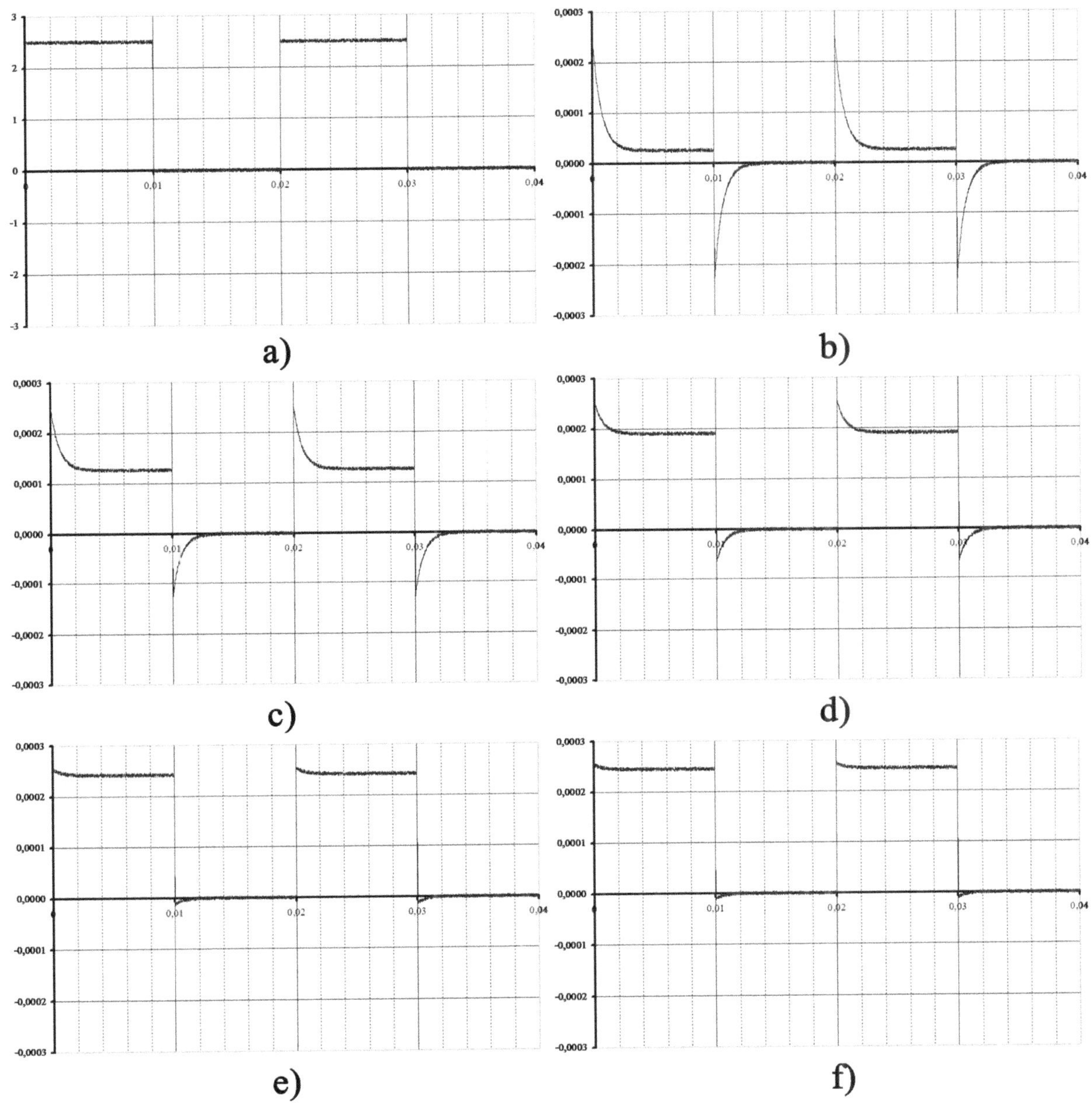

with weakly expressed by the exponential dependence on its roof (Fig. 2, f). The ratio of the maximum current value to the final current will be 1.05, which can be explained through the significantly decreased value or lack of the equivalent capacitive element in the circuit $C_м$.

It is possible to assess the parameters change of equivalent circuit elements, considering the plant tissue damage, by calculating their values in equation (3) and the calculate data should be taken from the oscillograms (Fig. 2,b, d, e, f). The results of the calculations are presented in table 1.

Table 1. The values of the stalk Amaranthus retroflexus L. plant tissue replacement parameters at actuating by high voltage electrical impulses

The Number of Actuating Impulses	R_1, Ohm	R_2, Ohm	R_3, Ohm	$C_м$, pF
m= 0	15595.24	92380.95	853423.81	165.93
m= 1	12395.83	21159.42	44772.11	380.48
m= 2	9907.41	13626.37	18567.79	523.13
m= 3	9545.45	12164.95	14798.59	598.65
m= 4	9026.55	11078.43	12273.16	686.16
m= 5	8695.65	9907.41	9299.06	746.03
m= 6	7916.67	8859.65	7849.29	840.14

The nature of the change in the values of the elements R_1, R_2, R_3 parameters of the plant tissue replacement electrical circuit corresponds to these components behavior shown above when it is electrically damaged. The change in membrane capacitance $C_м$ does not fully fit into the previously considered substantiation. Let us try to clarify how the damage to plant tissue goes on at the electrical actuation of high voltage on it, using a replacement circuit, current oscillograms while damage, as well as the nature of the changes in the replacement electrical circuit calculated parameters.

When high voltage electric impulses actuate on the plant tissue, the membrane voltage difference increases, which leads to the increased sizes membrane pores spontaneous formation - greater than the permissible critical radius. This circumstance is connected with the fact that the formed pores cannot immediately close ("heal") leaving the membrane intact, i.e. unhurt (Weaver & Chizmadzhev, 1996; Joshi & Schoenbach, 2000; Melikov et al., 2001). The intracellular solution is mixed with the contents of the intercellular substance through the opened pores with a large diameter, gradually leveling the resistance of the intercellular fluid R_2 and protoplasm R_1. At the same time, due to the solutions contained in the cell and the intercellular space mixing the resistance R_2 decreases to the value of R_1. A slight decrease in the resistance of R_1 can be explained by the fact that the cellular juice has a more acidic reaction than the intercellular solution, therefore, getting into the intercellular duct it causes proteins denaturation and enzymes inactivation, which leads to water release and increase in the received mixture electrical conductivity. The membrane R_3 resistance reducing occurs due to the fact that all the open pores in the membrane shell are filled with the got solution with a sufficiently high electrical conductivity.

A slight increase in the membrane capacity $C_м$ is determined by the process time constant increase, which is clearly seen in the Figures 2, b-f, but simultaneously with this the same capacity decrease was noted above. Analyzing the tissue replacement scheme from the point of view of theoretical electrical engineering, you can see that you can exclude the capacitance $C_м$ not only by removing it from the circuit, but also by shunting it with a small in the value active resistance R_3. This hypothesis is confirmed by the fact that in the case of electrical damage, the membrane resistance is significantly reduced by increasing the number of pores in the plant tissue cells membranes walls. A slight increase in the capacitance of the membrane $C_м$ does not contradict the biophysicists' researches. Chizmadzhev Yu.A. and the staff (Weaver & Chizmadzhev, 1996) give the formula for calculating the change in the membrane electrical capacitance when a structural defect (pore) appears in it, and shows that it depends on the capacity of a membrane area unit multiplied by the coefficient $\left(\frac{\varepsilon_s}{\varepsilon_m} - 1\right)$, whose value is greater than the one element:

$$\Delta ! = \pi \cdot r^2 \cdot C_0 \cdot \left[\frac{\varepsilon_s}{\varepsilon_m} - 1\right], \tag{4}$$

where ε_s is the water dielectric permeability, $\varepsilon_s = 80$; ε_m is the membrane lipids dielectric permittivity, $\varepsilon_m = 2\text{-}3$; C_0 is the capacity of a membrane area unit; r is the pore radius.

The processes occurring in plant tissue at its electrical damage and the evaluation of the change in its replacement scheme elements parameters analysis variant should be considered as a working hypothesis that allows describing the nature of changes in tissues at its electroimpulse irreversible damage with the high degree of reliability.

GENERAL TERMS EXPLAINING THE ESSENCE OF VEGETABLE TISSUE DAMAGE BY HIGH-VOLTAGE ELECTROIMPULSE ACTUATION

To evaluate and analyze the processes in plant tissues when processing them by electrical high-voltage impulses, it is necessary to study the questions connected with the plant-derived objects tissues sensitivity to the electrical actuation main damaging factor, the intensity of the electric field in the tissue or, equivalently, the conductivity current density and of the amount, absorbed by this tissue energy.

The plant tissue sensitivity to electrical actuation is suggested to be estimated with the help of such a quantitative index as the degree of its damage (S_n), which means the quantity, its numerical value is determined by the ratio of the tissue (or its active component) total resistance before the electrical treatment to tissue resistance during the processing at a measuring current fixed frequency, in our case at a frequency $f = 10$ kHz (Klimov, Savchuk & Baev, 1970; Baev & Borodin, 2002; Baev et al., 2018).

Applying the terms of the cellular damage to the plant origin objects theory modern interpretation of, we can say that under the influence of any damaging factor on plant tissue, there are always two generalized regularities (Baev & Borodin, 2002):

1. the interaction of the damaging factor and the object is of an energetic nature;
2. the resulting damage effect has a direct relationship with the amount of energy absorbed by the object.

Based upon the above terms, we can conclude the following: between the amount of damage, named, as stated above, the degree of damage, and the dose of the damaging factor, a well-known relationship that "the damaging agent dose increase by the certain number of times also increases the damaging effect size by the certain number of times" must be fulfilled (Baev & Borodin, 2002). This relationship can be expressed by the following formula:

$$S_? = k \cdot D^n , \tag{5}$$

where k is a constant that determines the curve position in the coordinate system; D is the dose of the damaging agent or intensity and time of the acting factor; n is a constant that determines the relationship between the agent dose change and the damage effect.

The dose or amount of the damaging agent is determined by the multiplying the intensity F by the time t of the damaging factor actuation, so the relationship (5) in a more general form can be represented as:

$$S_? = k_s \cdot F^p \cdot t^q , \tag{6}$$

where p and q are exponential indices, positive variables that depend on F and t; k_s is the proportionality coefficient.

In principle, the general equation (6) can have several particular cases. So, the same damage degree can be achieved by damaging factor intensity and duration various combinations, which is known from photochemical reactions under the name of the Bunsen-Roscoe law or the law of interchangeability.

With a constant actuating time to plant tissue, the damage will depend only on the intensity of the damaging factor.

Intermittent actuation can also have two variants:

1. the acting factor intensity and duration are unchanged, and the total processing time is made up of the individual actions duration;
2. the treatment is carried out by the same number of unit actuations m with damaging factor varying intensity.

To establish the analytical relationship between the damage degree to the processed raw materials or material plant tissue and the electric field intensity of the *(E)*, as well as the energy of the actuating impulses (*W*), we use the equation (6) connecting the depth (degree) of damage with the actuation intensity and the damaging factor actuation time (Baev & Borodin, 2002; Yudaev et al., 2004; Baev & Yudaev, 2004):

$$S_? = k_s \cdot E^n \cdot t^q , \tag{7}$$

where $E=U_0/l_{pt}$ is the average electric field strength in the treated plant tissue, *kV/m;* t is the actuation time or the impulse duration, s; n is the constant that determines the relationship between the change in the intensity of the field in plant tissue and the degree of its damage, it can take integral and fractional values; q is the constant, which determines the relationship between the actuation time to plant tissue and the degree of its damage, it can take integral and fractional values; U_0 is the initial voltage of the discharge circuit applied to the sample of plant tissue of the length l_{pt} and the cross-sectional area F_{pt},, kV; k_s is the proportionality coefficient, (cm/kV) $^{n} \cdot c^{-1}$.

ELECTRIC FIELD TENSION IN PLANT TISSUE INFLUENCE ON ITS DAMAGE DEGREE

Detailed researches and the study of the electroimpulse actuations influence were carried out with weeds, sunflower and tobacco plants, and therefore in the proposed analysis the bulk of the material will relate to irreversible damage to the plant tissues of the listed plants, although in experimental researches, vegetable raw materials, fruits, melons and gourds were also studied.

The investigated objects are plant origin parts and pieces samples of finite length l_{pt} and cross-sectional area F_{pt}. In analytical equation (7), for the convenience of the analysis carrying out, the actuation time t is taken in the first degree, which corresponds to the main linear portion of the dependence $S_n=f(E)$. This assumption is possible, since the investigated plant tissues were affected by single impulses that had the same duration in each series of experiments:

$$S_n = k_s \cdot E^n \cdot t = k_s \cdot \left(\frac{R_{pt} \cdot i}{l_{pt}}\right)^n \cdot t = k_s \cdot \left(\frac{R_{pt}^{n-1} \cdot i^{n-2}}{l_{pt}^n}\right) \cdot W = k_s \cdot \left(\frac{R_{pt}^{n-1} \cdot i^{n-2}}{l_{pt}^n}\right) \cdot m \cdot W_1 , \quad (8)$$
$$= k_s \cdot \left(\frac{R_{\blacklozenge t}}{l_{\blacklozenge t}^2}\right) \cdot m \cdot W_1 \cdot E^{n-2} = k_s \cdot \left(\frac{\rho_{\blacklozenge t}}{l_{\blacklozenge t} \cdot F_{\blacklozenge t}}\right) \cdot m \cdot W_1 \cdot \blacklozenge^{n-2}$$

or, if we take into account the parameters of the discharge circuit of the experimental installations impulsed voltage generators:

$$S_n = k_s \cdot \left(\frac{R_{pt}^{n-1} \cdot i^{n-2}}{l_{pt}^{n}}\right) \cdot m \cdot W_1 = k_s \cdot \left(\frac{R_{pt} \cdot E^{n-2}}{l_{pt}^2}\right) \cdot m \cdot \frac{C \cdot U_0^2}{2} = k_s \cdot \left(\frac{R_{pt} \cdot C}{2}\right) \cdot m \cdot E^n , \quad (9)$$

where i is the average value of the current flowing through the plant tissue, A; U_0 is the initial voltage of the discharge circuit applied to a sample of plant tissue of length l_{pt}, kV; k_s is the coefficient of proportionality, $(cm/kV)^n \cdot c^{-1}$; $R_{pt}=((\rho_{pt} \cdot l_{pt})/F_{pt})$ is the resistance of the investigated sample of plant tissue with specific electric resistance ρ_{pt}, the length l_{pt} and cross-section area F_{pt}, Ohm; $W=W_1 \cdot m$ is the total energy, J; $W_1=i^2 \cdot R_{pt} \cdot t=0,5 \cdot C \cdot U_0^2$ is the energy of a single impulse, J; C is the storage capacity of the discharge circuit, F; m is the number of actuating impulses.

The preliminary analysis of equations (7-9) shows that the degree of damage to plant tissues S_n depends: 1) on the strength of the electric field E in them, and the character of the dependence is determined by the value of some exponential index n; 2) on the total amount of the damaging energy W, which is determined by the multiplication the number of actuating impulses m by the energy of the unit impulse W_1.

Weeds, sunflower and tobacco plant tissue, objects of plant origin have the same structure as the object of electrical impact, and therefore the nature of the dependencies will be the same. As a confirmation, let us consider the behavior of the indicated dependences for sunflower (Figure 3), pumpkin (Figure 4) and some species of weeds (Fig. 5).

Quantitatively, the degree of damage, as well as the nature of the dependence (Figures 3-5) are determined by the value of the index n, which was found for the following plant tissues: for tobacco, $n = 1.26$; for sunflower $n = 1.5$; for the aloe plant $n = 0.75$; for apple pulp $n = 2.45$ (Baev & Borodin, 2002;

Figure 3. The sunflower plants (phase of development - "waxy ripeness") degree of damage to the stalk plant tissue (S_n) dependence on the intensity of the electric field (E) in the tissue with a different number of actuating impulses m=var and constant capacitances of the discharge circuit C = 47 pF, and also inductance L = 5 μH

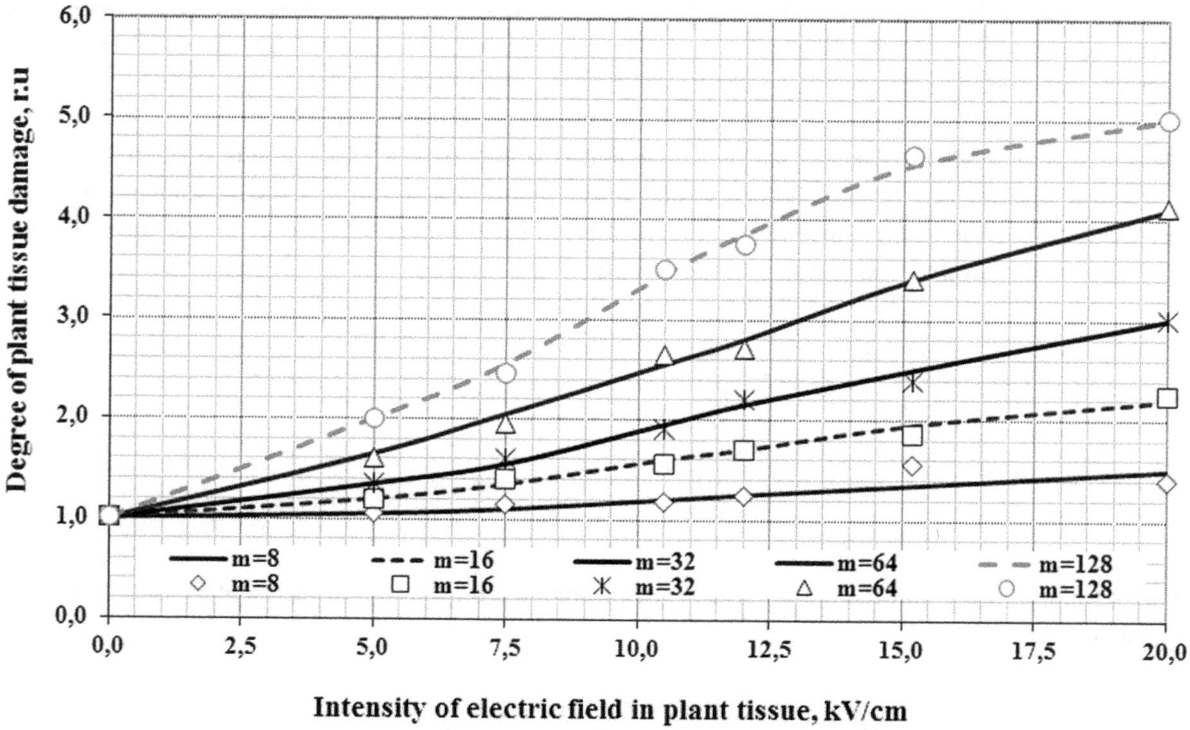

Yudaev et al., 2004). Here it should be noted the obvious fact - considering the electrical processing of various plant objects, you can see that the index n depends on the number of impulses m or the energy W applied to the processed object. Therefore, the values of index n given above should be characterized as averaged for the process of electroimpulse treatment.

Quantitatively, the index n was on the main linearly increasing portion of the dependence, which was constructed on a logarithmic scale, and the value of n was determined as the value of the tangent of the slope of the obtained lines to the abscissa axis. Each dependence is constructed for a specific number of actuating impulses with a constant value of the storage capacitance of the capacitor bank of the discharge circuit. The average values of the index n for some weeds are given in the Table 2.

Differences in the numerical values of the index n, most likely, are explained by the weeds plant tissue different sensitivity to the electrical impulses of high voltage influence on it.

It should be noted that there is a direct dependence between the value of the index n and the number of impulses m, and, consequently, on the energy absorbed by the plant tissue. The more actuating impulses are brought to the plant sample, the higher the value of the index n, for example, for the tissues of the plant *Euphórbia virgáta L.* at $m = 10$, the exponential index is $n = 0.63$, and for $m = 1000$ the index increases to $n = 1.796$.

Figure 4. The "Gribovskaya zimnyaya" pumpkin variety degree of damage (S_n) of the plant tissue dependence on the intensity of the electric field (E) in the tissue with a different number of actuating impulses m = var and constant capacitances of the discharge circuit C = 365 pF, and inductance L = 25 mcH

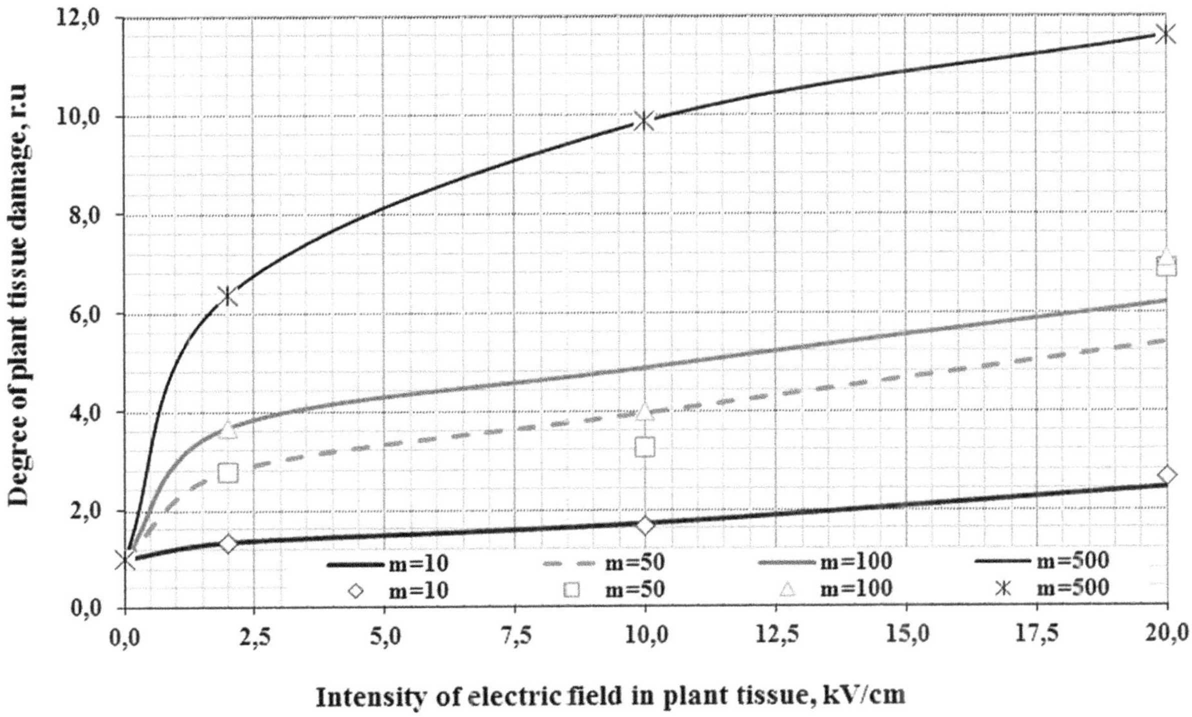

Table 2. The average values of the index n, which determines the relationship between the change in the intensity in plant tissue and the degree of damage for weeds various sections

Weed	The Average Values of the Index *n* for Different Weeds Species of and Plant Tissues Sectors			
	Stalk	Sector With Root Neck	Root	Average
Cirsium arvense L.	1.15	1.24	1.22	1.20
Sonchus arvensis L.	1.19	1.23	1.25	1.22
Euphórbia virgáta L.	1.06	1.21	1.14	1.14
Lactúca tatárica L.	1.17	1.25	1.26	1.23
Amaranthus retroflexus L.	1.22	1.31	1.21	1.25
Average value	1.16	1.25	1.22	1.21

To confirm the above assumption about the tissue different sensitivity, let us consider the dependences $S_n=f(E)$ constructed for weeds of different biological groups (Figure 5), which have a strongly clear S-shaped character, which emphasizes the presence of an initial section with an insignificant increase in the degree of damage with an increase in the tissue tension value and a section with a very high rate of change in the degree of damage, as well as the section where, as the tension increases, a significant

Figure 5. The weeds root system plant tissue damage degree (Sn): (a) - Euphórbia virgáta L.; b) Cirsium arvense L.; c) - Amaranthus retroflexus L. dependence on the intensity of the electric field (E) in plant tissue with a different number of impulses m = var and constant values of the capacity of the discharge circuit C = 1000 pF, and inductance L = 110 μH

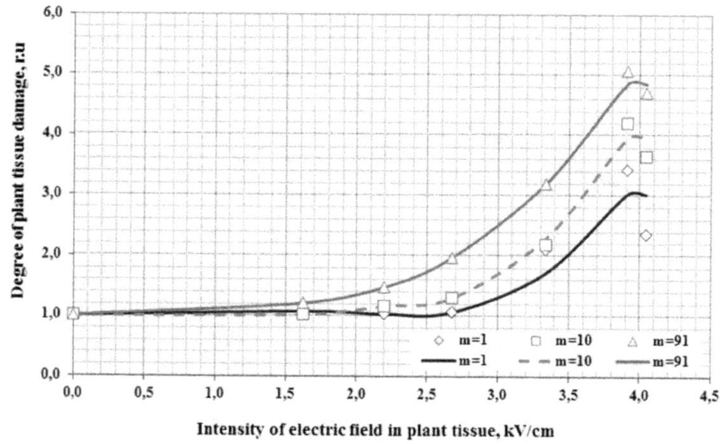

a)

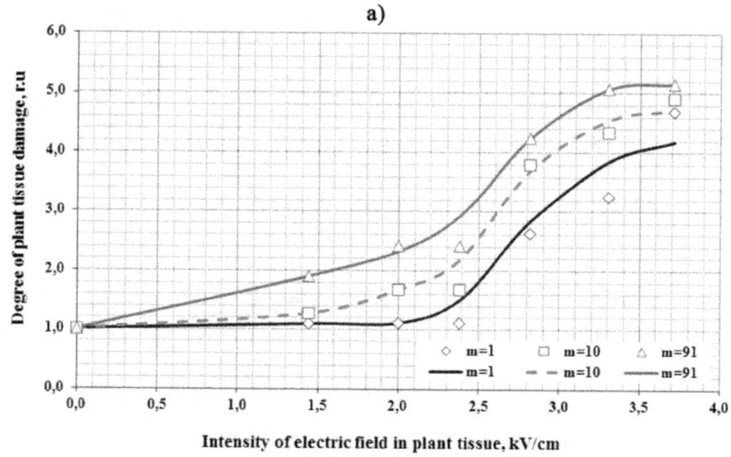

b)

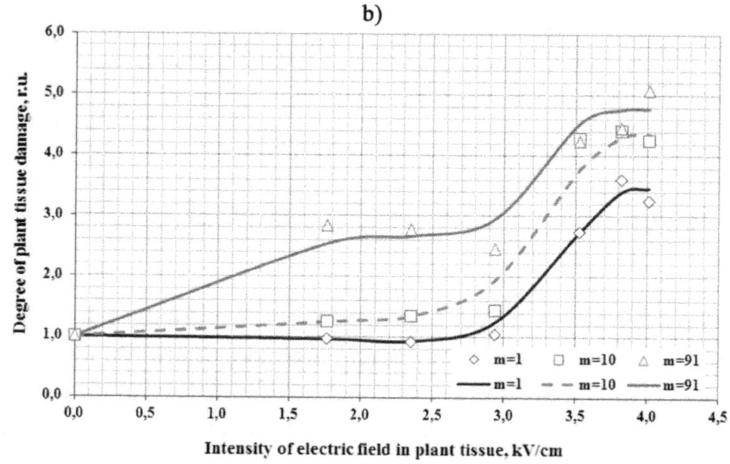

c)

change in the degree of damage is no longer observed, that is, the section where the tension reached its limiting value.

The nature of the curves presented in the Figure 5 makes it possible to draw the following conclusion: considering the dependence $S_n=f(E)$ at the number of actuating impulses equal to one element, it can be noted that the plant tissue of the root system of the *Amaranthus retroflexus L.* plant reaches a deeper damage than the *Euphórbia virgáta L.* and *Cirsium arvense L.* But at the same time, at $m = 91$, the value of the degree of damage to plant tissue practically becomes the same, and the behavior of the dependencies is very similar to each other.

In electrical processing, the applied initial voltage of the discharge circuit U_0 determines the value of the electric field E in the plant tissue. Of interest is its value, in which the degree of damage to the plant tissue reaches the limiting (maximum) value of $S_n=S_{nmax}$. It is the creation of such tension in plant tissue, or the current density flowing through it, leading to irreversible damage to the tissue, which was actuated upon by high-voltage impulses.

The analysis of the presented research results allows us to draw the following conclusions:

1. the degree of damage to plant tissue of plant origin various objects increases with increasing intensity of the electric field in the tissue, reaching its limiting value;
2. the exponential index n characterizes the sensitivity of the plant tissue to damage and, by its value, it can be concluded: what tissues and which plant objects are more sensitive to electroimpulse damage, and which are less sensitive;
3. with the increase in the energy supplied to the plant, a single impulse or a series of impulses, the degree of damage to weeds plant tissue increases;
4. for the reliable, irreversible damage to plant tissues of plant-derived objects, it is necessary to create sufficient electric field tension for the required technological purposes.

DAMAGE TO WEEDS PLANT TISSUE DEGREE DEPENDENCE ON ACTUATING ENERGY

Let us reveal the general conformities that determine the plant tissue sensitivity to the amount of damaging energy supplied when electric impulses of high voltage are applied to them.

The main element of the experimental installation is the discharge circuit of the impulse voltage generator with a storage capacitance C, where a certain specific amount of electrical energy is stored, and then directly during the plant sample processing is brought to it and absorbed by its tissue. Such an energy reserve determines, first of all, the energy of one impulse, which is supplied by electrical actuation to a plant or plant material. For irreversible damage to plant tissue, it is necessary to actuate on it with such an amount of energy that at the cellular level leads to irreversible damage to the vital functions of the tissues cell membranes.

The main dependence that was built based on the results of the experiments was the damage to plants plant tissue or plant material degree dependence on the number of impulses m: $S_n=f(m)$ or, which is the same, on the actuating energy W: $S_n=f(W)$ at $U_0= var$ or $C= var$. All the experimental dependences $S_n=f(m)$ или $S_n=f(W)$ have clearly S-shaped character, which fully confirms their theoretical substantiation, which is represented in the papers (Baev & Borodin, 2002; Yudaev et al., 2004). The got graphical dependencies are characterized by two main sections: on which a sharp increase in the degree of damage occurs (S_n) with the increase of the actuating energy applied to the plant *(W, m)*; and on which the degree of damage, having reached its limiting value $S_n=S_{nmax}$, remains constant despite the continuing increase in the total actuating energy. The first section steepness index or (what is the same) the rate of increase in the damage degree characterizes the plant tissue being treated sensitivity to the damaging effect of electrical energy. The more the rate of increase in the damage degree or the steeper the shape of the curve initial part $S_n=f(W)$, the more plant tissue, in our opinion, is susceptible to the electrical influence. According to the processes taking place in the plant tissue, the limiting (maximum) value of S_{nmax} is the quantitative index determining the maximum possible damage to the treated tissue when it is actuated by electrical impulses with the same dose of energy. From the energetical and technological points of view, when the degree of damage reaches the limit value, the electrical influence on the plant tissue must end, since its further processing does not increase the damage, but only leads to an increase in energy costs. The amount of energy at which the degree of damage reaches its limiting value can be called a lethal or damaging energy dose for a particular plant or plant material that is treated by electric impulses.

The equations (8-9) determine the proportional increase in the value of S_n with increasing damage energy. The amount of this energy supplied to the plant or material by a single actuating impulse depends on the value of the circuit C storage capacitance and the square of the initial voltage U_0 applied to the object being processed:

$$W_1 = 0,5 \cdot k_u \cdot C \cdot U_0^2, \tag{10}$$

Table 3. The range of values of total specific actuating energy on various weed sections plant tissue, corresponding to their damage limiting degrees

Weed	Section	Range of Damage Limiting Degrees S_{nmax}, r.u.		Range of Specific Cost of Electric Energy W_s, J/cm³	
		On Sections	Average	On Sections	Average
Cirsium arvense L.	Stalk	5.6-7.4	5.4-7.0	2.4-8.1	2.7-10.9
	Root neck	5.2-6.3		2.8-10.3	
	Root	5.3-7.3		2.9-14.3	
Sonchus arvensis L.	Stalk	4.8-7.7	4.7-7.5	3.3-15.2	3.7-15.8
	Root neck	4.6-7.3		4.1-16.4	
Euphórbia virgáta L.	Stalk	4.8-4.7	4.5-4.3	4.9-24.7	5.2-17.5
	Root neck	4.5-3.9		5.6-12.7	
	Root	4.3-4.3		5.1-15.2	
Lactúca tatárica L.	Stalk	5.6-7.4	5.5-7.3	3.8-6.8	3.3-8.1
	Root neck	5.8-6.9		3.3-8.4	
	Root	5.0-7.7		2.6-8.7	
Amaranthus retroflexus L.	Stalk	4.1-7.1	4.0-6.8	3.4-8.2	3.5-7.7
	Root neck	4.4-6.7		3.1-7.2	
	Root	3.6-6.6		3.9-7.8	

where k_u is the degree of the capacity discharge (by voltage), usually taken as 0,9-0,95.

With each actuating impulse in the plant tissue irreversible impairment of its normal functioning occurs, as a result the tissue gradually accumulates damage, reaching a critical value of the damage degree $S_n=S_{nmax}$. The behavior of the curves $S_n=f(W)$ or $S_n=f(m)$ has a strongly marked S-shaped character, which is manifested by the presence of a monotonically increasing section on these dependencies and the section at which the damage degree has a limiting value that practically does not change with an increase in the total actuating energy. It also should be noted that the speed of reach limit values of the damage degree S_{nmax} depends on the energy of one actuating impulse W_1: the higher it is, the higher the speed and the faster the damage to plant tissue will reach its maximum value.

Of practical interest are also data on the doses of energy at which the degree of damage to plant tissue reaches its limiting (maximum) value. The knowledge of these actuating energy doses values will make it possible to estimate the total energy costs for the proposed electrotechnological operation and develop technological requirements for the process of electrical impulse damage to plant tissue. For convenience of the generalizing analysis carrying out, we shall use not the values of the total actuating energy, but its specific index, which we define by the formula:

$$W_{1s} = \frac{W_1}{V_@}, \tag{11}$$

where V_{pt} is the volume of the plant sample processed by electrical impulses, cm^3.

When calculating the total energy impact, it is necessary to take into account the number of actuating impulses m with the specific energy of one impulse W_{1s}. The results of the carried out experimental researches with weeds are given in the Table 3.

The data presented in the Table 3 are the results of researches carried out during five years at two experimental installations. The spread of the results is determined by the fact that the installations had different parameters of the discharge circuits - in one case, the processing was carried out at voltages from 3 to 7.5 kV, and in the other one, from 5 to 12.5 kV, in addition, capacitors also had a difference in capacitance values of almost two orders of magnitude.

According to the electrical impulse damage of plant tissue technological process optimization, the effect S_n of the initial voltage of the discharge circuit U_0 on the degree of damage is of interest. The voltage applied to samples of plant tissue or plants determines the strength of the electric field in their tissues, and ultimately the degree of damage (8-9).

To elucidate the nature of this effect, several series of experimental researches were carried out, in which the capacitance of circuit C remained constant, and the processing voltage U_0 changed. As the result of these researches, a family of S-shaped dependency curves $S_n=f(m)$ was obtained.

In the case when weed plants with different initial resistances were subjected to electrical treatment, the limiting degree of damage increased with increasing of initial voltage of the discharge circuit U_0. In this case, the following dependence was fairly clearly in evidence: with increasing energy of the actuating impulses, the speed of reach limit values of the damage degree also increased (Fig. 6). In the example under consideration, tissue samples of the *Cirsium arvense L.* plant root system were subjected to electrical treatment, which were affected by electric impulses with specific energy W_{1s} = 2.0; 3.5; 6.8; 7.1 J/cm^3, and the discharge circuit had the following parameters: $C = 8.615 \cdot 10^{-9}$ F; $L = 110 \cdot 10^{-6}$ H; $U_0 = 6 \cdot 10^3$-$12 \cdot 10^3$ V.

Figure 6. The plant tissue of the Cirsium arvense L. root degree of damage (Sn) dependence on the number of actuating impulses (m) at different values of the initial voltage Uo = var the discharge circuit C = 8.615 nF storage capacitance constant value, as well as the inductance L = 110 μH (provided that the initial resistance values of the treated sectors are different)

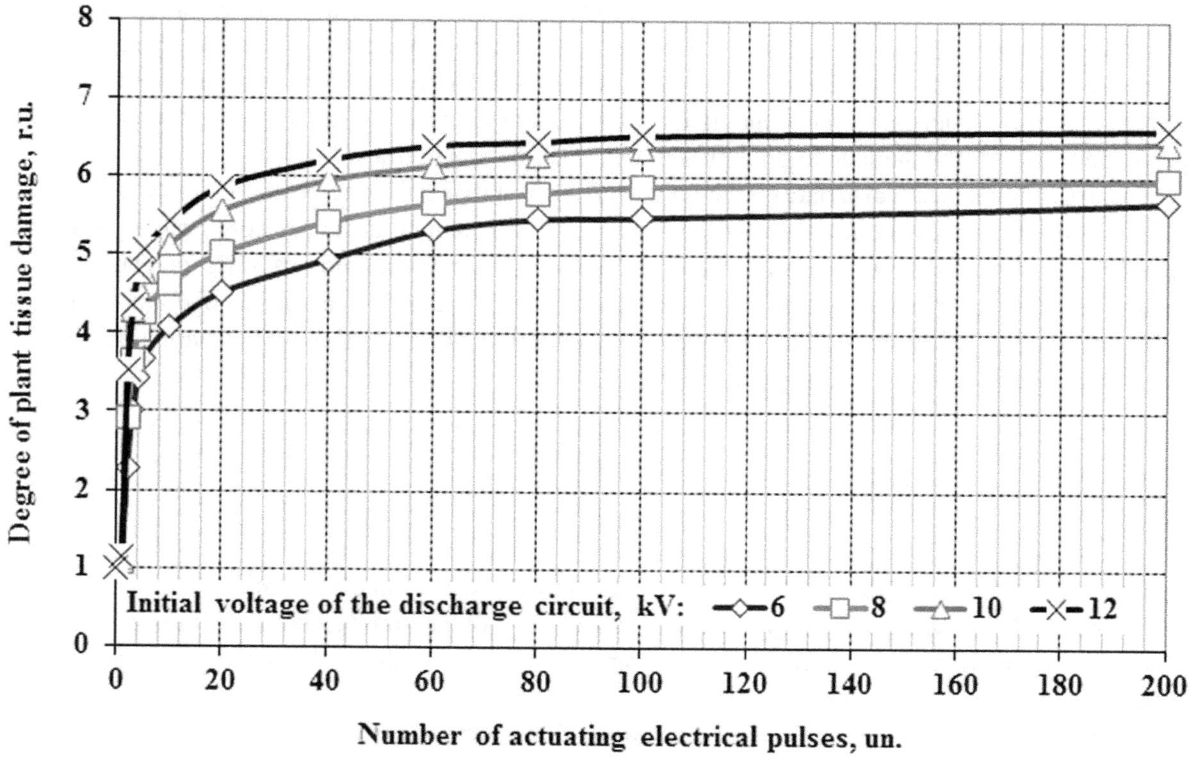

As one of the hypotheses of electroimpulse irreversible damage to the plant tissue, one can propose a version that if the initial values of the plant tissue investigated samples resistances are the same, the values of the limiting degrees of damage should also be equal in value. To confirm this thesis, appropriate experiments were carried out. Electrical actuation was on the tissue fragments of the *Cirsium arvense L.* plant root system, which were selected with the same initial resistance. Plant samples were processed by electric impulses of different energies so that the specific energies were equal to W_{1s} = 0.879; 1.13; 2.09; 2.45 J/cm^3. The parameters of the discharge circuit had the following values: $C = 8.615 \cdot 10^{-9}$ F; $L = 110 \cdot 10^{-6}$ H; $U_0 = 6 \cdot 10^3 - 12 \cdot 10^3$ V. The analysis of the dependencies behavior (Fig. 6) confirms the hypothesis.

In the case of electrical treatment along the plant stalks or in the longitudinal direction of the plant material fibers, the value of the initial voltage of the discharge circuit U_0 determines the value of the electric field E in the plant tissue. According to the technological requirements for the process of electroimpulse irreversible damage to plant tissues, the most interesting is the value of the intensity at which the degree of damage to the plant tissue reaches its limiting (maximum) value or, equivalently, the damaging value of the current density flowing through the treated plant. It is the creation of such tension in the plant tissue, or the current density in it, that leads the weeds to destruction, and the plant material to destruction.

Figure 7. The plant tissue of the Cirsium arvense L. root degree of damage (Sn) dependence on the number of actuating impulses (m) at different values of the initial voltage Uo = var the discharge circuit C = 8.615 nF storage capacitance constant value, as well as the inductance L = 110 μH (provided that the initial resistance values of the treated sectors are the same)

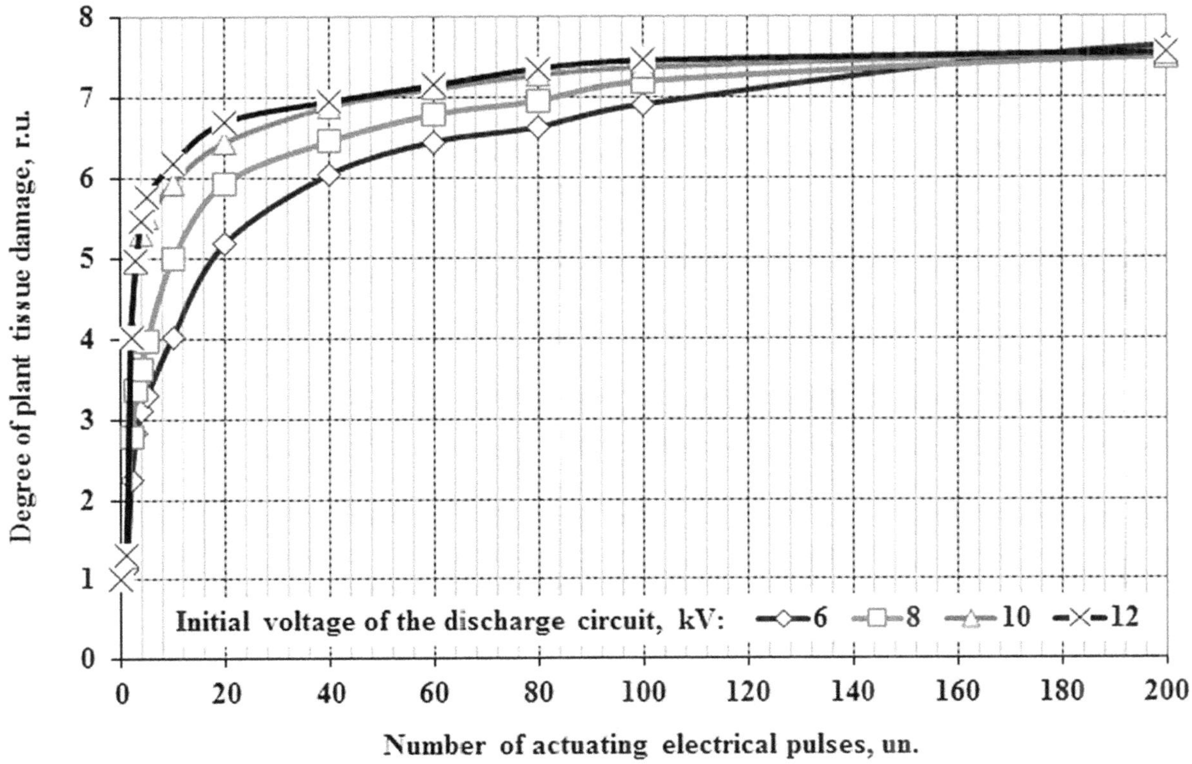

The analysis of the experiments results, their processing and theoretical representations allows us to state the following.

The various parts vegetative tissue of weeds such as *Cirsium arvense L.* and *Sonchus arvensis L.*, *Lactúca tatárica L.* is most sensitive to electrical actuations (this has already been emphasized since the value of the index *n* is the largest of the examined ones). This is characterized by the fact that the degree of tissue damage increases more intensively with each impulse and reaches the limiting value S_n earlier than in plant tissues of *Euphórbia virgáta L.* with equal energy actuations.

For a reliable (irreversible) damage to weeds tissues, the specific energy consumption is necessarily not less than 17.5 J/cm^3, with a maximum degree of damage not less than 7.5.

In addition to studying the actuation of high voltage electrical impulses on weeds plant tissue, the researches were carried out with samples of pumpkin and apples.

As an example, let us consider the results of experimental researches on the plant samples obtained from apples of the “Gala” variety, cut into samples of equal diameter, height and mass. The processing of apples was carried out at different values of the processing stress and the number of impulses. The weight of all the processed samples was in the range of 7.03-7.10 g. After the processing, a fixed pressure of 218.62 kPa was applied to each sample and the juice yield was estimated. The mass of the juice obtained by the pressing from a sample without electroimpulse treatment was 0.22 g. The results of the experiments are given in the form of the dependencies in Fig. 8.

Pumpkin samples of equal diameter, height and weight were also processed at different parameters of the electroimpulse actuation (voltage, number of actuating impulses) to obtain the pumpkin plant tissue damage degree dependence on the absorbed by tissue electroimpulse actuation energy (Fig. 9).

The the speed of reach limit values of the damage degree depends on the initial voltage U_0 of the discharge circuit: the greater the value of the voltage, the faster the plant tissue will reach the ultimate damage degree S_{nmax}.

Thus, by the type of the graph of the initial change in the damage degree to plant tissue after the electroimpulse treatment, one can judge the depth of tissue damage, the state of the plant or plant material, and subsequent changes therein. This technique can significantly shorten the time in determining and finding the regimes of irreversible damage to plant tissues with a minimum value of voltage and the expenditure of actuating energy.

ELECTROIMPULSE IRREVERSIBLE DAMAGE TO PLANT TISSUE PROCESS OPTIMAL PARAMETERS

Having carried out experimental researches on the high-voltage electrical impulses influence on plant tissues, where they are irreversibly damaged, and summarizing the obtained results it is possible to identify the optimal parameters of the proposed electrotechnological operations - weeds electroimpulse

Figure 8. The juice yield, g, obtained by the method of pressing from the apples of the "Gala" variety dependence on the number of actuating impulses (m), pieces, at different values of the electric field strength (E) in the plant tissue of the samples, and the constant capacitance values of the discharge circuit C = 365 pF and inductance L = 25 μH

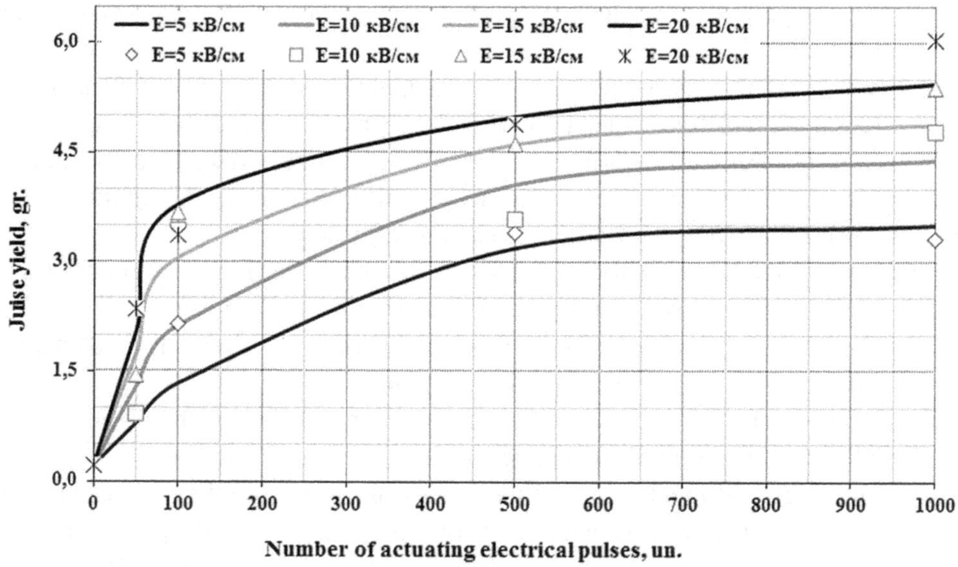

Figure 9. The plant tissue of the pumpkin "Gribovskaya zimnyaya" degree of damage (Sn) dependence on the number of actuating impulses (m) at different values of the electric field strength (E) in the plant tissue of the samples, and the constant capacitance values of the discharge circuit Ck = 365 pF and inductance L = 25 μHY

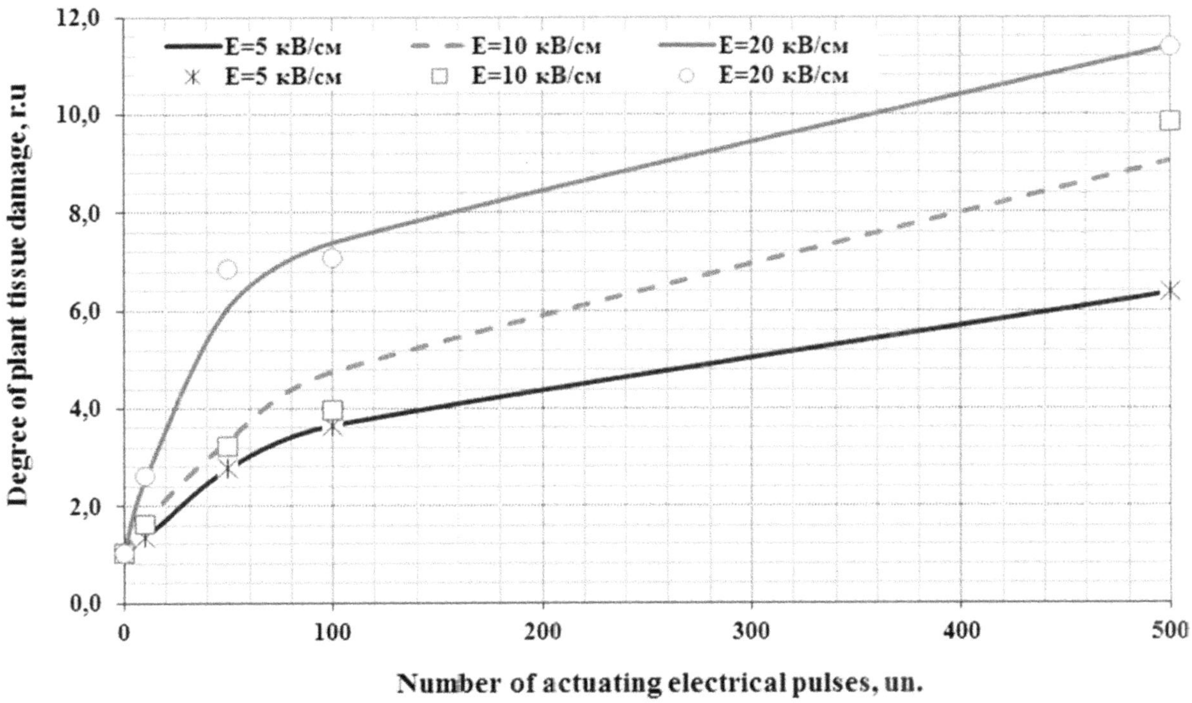

destruction, pre-harvesting, fruits and vegetables electroimpulse plasmolysis, melons and gourds, fruit and hay grasses processing before drying, etc.

The optimization under consideration is based on the following assumptions:

- minimal costs of electric energy were reached during the processing, which led to the maximum (limiting) value of the damage degree to plant tissues;
- optimal combination of the parameters of the impulse voltage discharge circuit (C, U_0) is realized at which the greatest actuating technological effect is achieved.

The analytical equation was obtained in the papers (Baev & Borodin, 2002; Yudaev & Baev, 2006) for determining the minimum energy expenditure W_{pt} at tobacco and sunflower plant tissues electroimpulse treatment, as well as weeds in dependence on the initial voltage of the discharge circuit U_0, the resistance of the plant tissue R_{pt}, its damage degree S_n, the index n, and also on the geometric parameters that determine the path of the processing current flow - the length of the current path through the plant l_{pt}. This dependence is developed from the formulas (8-9):

$$W_{@} = \frac{S_?}{k_s \cdot \frac{R_@^{n-1} \cdot i^{n-2}}{l_@^n}} = \frac{S_?}{k_s \cdot \frac{R_@}{l_@^n} \cdot U_0^{n-2}}. \qquad (12)$$

Analyzing the equation (12), it is possible to justify the theoretical conditions for the optimal operation of the discharge circuit of the impulse voltage discharge circuit loaded on a plant object (sunflower, tobacco, weeds, plant material). Using the computations of earlier researches (Baev & Borodin, 2002; Yudaev et al., 2004), we obtain the dependence of the minimum energy expenditure for electroimpulse damage to plant tissue from corresponding to this applied voltage values processing mode.

The analysis is based on the assumptions and speculations that the treated plant tissues reach the same value of the damage degree S_n, and the samples have equal lengths of treated sections l_{pt}, over which the current flows, and practically the same initial resistance R_{pt}. Since the average index of degree *n* for various plant materials lies in the range 0 <n <2, we suppose while the analysis that the range of the considered value *n* variation can take values from zero and up to 3-4. The minimum value of electric energy W_{pt}, spent for irreversible damage to weed plant tissues, varies depending on the value of the applied voltage U_0 at the following values of *n*:

- at n ≥ 3 the value $U_0^{n-2} \geq U_0$ and the dependence $W_{pt}=f(U_0)$ can be approximately described by an analytical equation $W_@ = a_1 \cdot U_0^{-k}$, where *k* is any integer or fractional number; a_1 is a constant coefficient. In the case under consideration, an increase in the processing voltage U_0 leads to a decrease in W_{pt};
- at 2 <n <3 the value $U_0^{n-2} \prec U_0$, and the dependence $W_{pt}=f(U_0)$ can approximately be represented by an analytical equation of the form $W_@ = a_2 \cdot U_0^2 - b_2 \cdot U_0 + c_2$, where a_2, b_2, c_2 are constant coefficients. In the case under consideration, an increase in the applied voltage U_0 results in a less dramatic decrease in W_{pt}, than in the previous paragraph;
- at n = 2, the value $U_0^{n-2} = 1$, and the dependence $W_{pt}=f(U_0)$ can approximately be described by an analytical equation of the form $W_@ = a_3$, where a_3 is a constant coefficient. In this case, the energy W_{pt} does not depend on the change in the processing voltage;
- at 1 <n <2 the value $U_0^{n-2} \succ U_0$, and the dependence $W_{pt}=f(U_0)$ approximately corresponds to an analytical equation of the form $W_@ = -a_4 \cdot U_0^2 + b_4 \cdot U_0 + c_4$, where a_4, b_4, c_4 are constant coefficients. In the case under consideration, an increase in the processing voltage U_0 leads to an increase in the value of W_{pt};
- at n = 1 the quantity $U_0^{n-2} = \frac{1}{U_0} \prec U_0$, and the dependence $W_{pt}=f(U_0)$ can approximately be written down as an analytical equation of the form $W_@ = a_5 \cdot U_0 - c_5$, where a_5, c_5 are constant coefficients. In this case, an increase in the voltage U_0 applied to the plant leads to a linear increase in the value of W_{pt};

- at 0 <n <1 the quantity $U_0^{n-2} \prec U_0$, and the dependence $W_{pt}=f(U_0)$ can approximately be described by an analytical equation of the form $W_{@} = a_6 \cdot U_0^2 + b_5 \cdot U_0 - c_5$, where a_6, b_6, c_6 are constant coefficients. In this case, an increase in the processing voltage U_0 leads to a more rapid increase in the value of W_{pt}, than in the fourth case.

In order to look over the change nature in the processing energy W_{pt} as a function of the initial voltage on the discharge circuit U_0, the corresponding experimental researches were carried out, which results are shown in the form of graphical dependences in Fig. 10, using the example of the stalk sample of the plant *Euphórbia virgáta L.* In the analysis, it was assumed that the change in the treatment energy W_{pt} can be identified with the duration of the electrical actuation on plant tissue, since the actuation was carried out by the impulses identical in amplitude and duration, and the parameters of the discharge circuit did not change during the experiments. The analysis of the data obtained by the experiment made it possible to ascertain that depending on the index *n* value in order to obtain the electroimpulse actuation on plant tissues planned technological effect, their treatment must be carried out either by impulses with a large voltage amplitude but with a short actuation time, or with a small amplitude of the actuating impulses, but with a sufficiently long processing time.

Having carried out a joint analysis of the obtained graphical dependences $W_{pt}=f(n)$ at $U_0=const$ and theoretical conclusions about the influence of the initial voltage U_0 on the processing energy W_{pt} with the corresponding changes in the index *n*, it can be stated that for the plant tissue electroimpulse ir-

Figure 10. The relative dose of the actuating energy (Wpt) on the Euphórbia virgata L. stalks plant tissue dependence on the index (n) at different initial voltage of the discharge circuit (U_0=var)

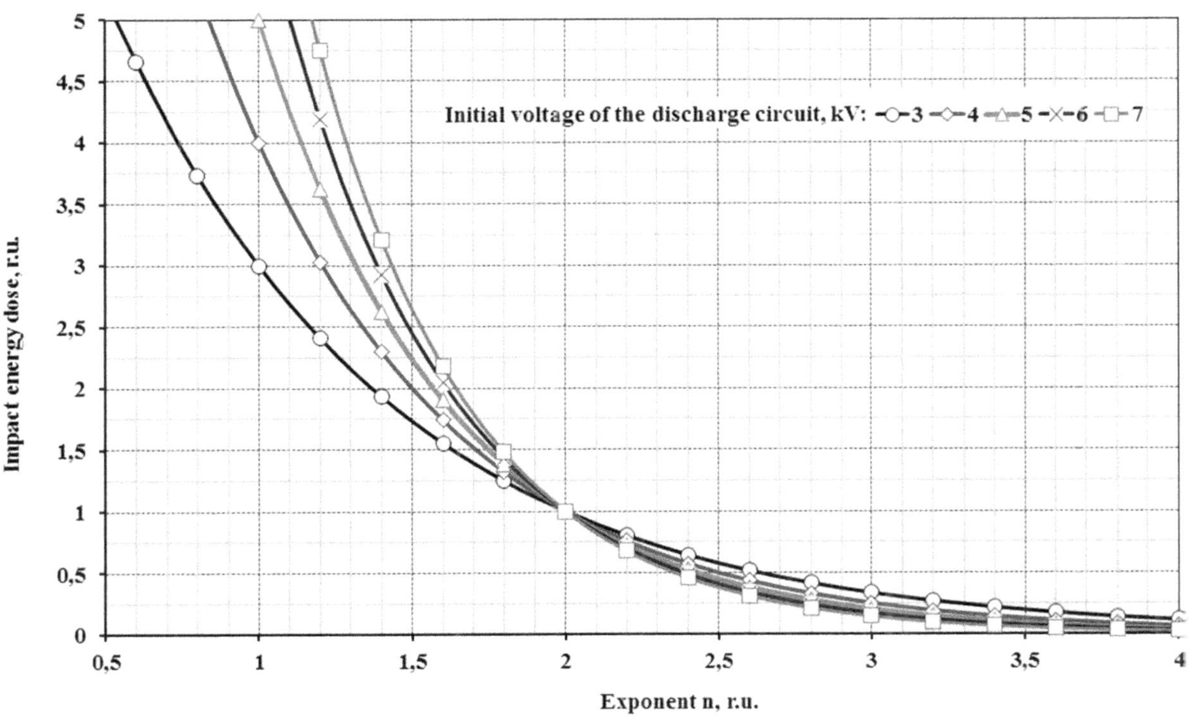

reversible damage at the lowest energy consumption, the operating range should be in the area of high tension for tissues with the index *n> 2* and in the area of lower tension values for tissues - *n <2*. In the cases of weeds electroimpulse damage *(n = 1,14-1,29)*, the discharge circuit of the impulse voltages source should be set to work with a relatively low processing

In the plants and plant material electroimpulse treatment, the time characteristics of the actuation - the duration of the voltage impulse and the time of the impulse front accumulation - are also influential factors. The carried out experimental researches allow to estimate these parameters by processing of oscillograms of voltage and current flowing through plant tissues influencing impulses.

The impulse duration was experimentally determined for the treatment of weeds certain species (pigweed, city goosefoot, canada thistle) with using different storage capacitances of circuit *C*. As the result, the range of initial actuating impulses $t_{pul} \approx$ 16-65 μs, the processing with their help leads to irreversible damage to plant tissues. In this case, single impulses in plant tissue caused damage, the degree of which was equal to S_n = 1.15-1.31, and with prolonged influence the degree of damage reached its maximum value after which the tissue was damaged irreversibly and the plant died.

The impulse duration can be estimated by calculation using the technique known from high voltage technique (Kuffel, Zaengl & Kuffel, 2000; Mytnikov, 2012), which can be valid for our case provided that $R_{pt} >> R_k$ and $C > C_м$. In this case, the impulse lifetime can be estimated as:

$$t_{pul} \approx 0.7 \cdot C \cdot R_{@} . \tag{13}$$

Comparing the results of the experimental researches based on the analysis of oscillograms and measurement data by instruments, with the calculation performed on the basis of equation (13), it can be stated that the experiment and calculation mismatching error is 17 ... 20%. In our opinion, the calculation formula can be used for a general estimation of the impulse time.

The analysis of the presented formula (13) shows that the impulse duration at $R_{pt} >> R_k$ and $C > C_м$ can be regulated by changing the capacitance value of the discharge circuit *C*. At the same time, by decreasing the duration of t_{pul} by decreasing the value of the capacitance *C*, it is necessary to carry out the treatment, as already mentioned, by voltages of a large amplitude, and with an increase in t_{pul} and an increase in the capacitance *C* - at a lower value of the voltage U_0.

The impulse front accumulation time, which also determines the treatment technological efficiency, can be estimated from the equation of high voltage technology ((Kuffel, Zaengl & Kuffel, 2000; Mytnikov, 2012) and is valid for case $R_{pt} >> R_k$ and $C > C_м$:

$$t_{ed} \approx 2.3 \cdot \mathrm{C}_{<} \cdot R_k , \tag{14}$$

where R_k is the total resistance of the discharge circuit and the connecting wires, Ohm.

The impulse front accumulation time is determined by the electroimpulse treatment of certain weed plants species (pigweed, city goosefoot, canada thistle). As the result of the performed calculations, the range in which the rise time of the front of the initial actuating impulses lies $t_{ed} = 0.1 \cdot 10^{-6}$-$0.2 \cdot 10^{-6}$ s was set.

The analysis of equation (14) allows us to say that to adjust the rise time of the pulse front, it is necessary to change the discharge circuit R_k resistance, which is quite feasible under real production conditions.

In this case, the resistance values can reduce the accumulation time, and, consequently, increase the rate of change in the processing voltage along the front.

It should be noted that the obtained values of t_{pul} and t_{ed} do not contradict the results of calculations of the transient processes in the discharge circuit, as well as the data of the current and voltage impulses oscillogram analysis in experimental researches.

To finish the discussion on the electroimpulse damage to plant tissues process parameters optimization, one cannot but analyze the dependence of such an important indicator as the degree of tissue damage depending on the initial tension treatment U_0 and storage capacity of the discharge circuit C. Since the main purpose of the treatment is irreversible tissue damage and its subsequent death, as the main indicator for the analysis should be taken the limiting (maximum) degree of damage to weeds plant tissue S_{nmax}.

Having processed the data of the carried out researches the dependences were constructed: $S_{nmax}=f(U_0)$ at $C = const$; $S_{nmax}=f(C)$ at $U_0 = const$; $U_0=f(C)$ at $S_n = S_{nmax}$. As the example, let us consider the nature of the change in the limiting degree of damage to the *Amaranthus retroflexus L.* stalk plant tissue, depending on the initial processing tension (Figure 11) and the capacitance of the discharge circuit (Figure 12).

The analysis of the dependencies $S_{nmax}=f(U_0)$ at $C = const$ (Figure 11) shows that the maximum degree of damage to plant tissue is achieved by using the storage capacity of the contour C = 470; 1100; 2200; 4400 pF and the voltage range U_0 = 8.5-12.0 kV. Here it should be noted that the rate of reaching the limiting value of the tissue damage degree is different - the greater the discharge circuit capacitance value, and hence the energy of the actuating impulse, so that the plant tissues are more quickly irreversibly damaged and the extent of their damage becomes the limiting (maximum).

Figure 11. The limiting degrees of damage (Snmax) of the Amaranthus retroflexus L. stalks plant tissues dependence on the initial tension of the contour (U_0) at different storage capacitances of the circuit (C)

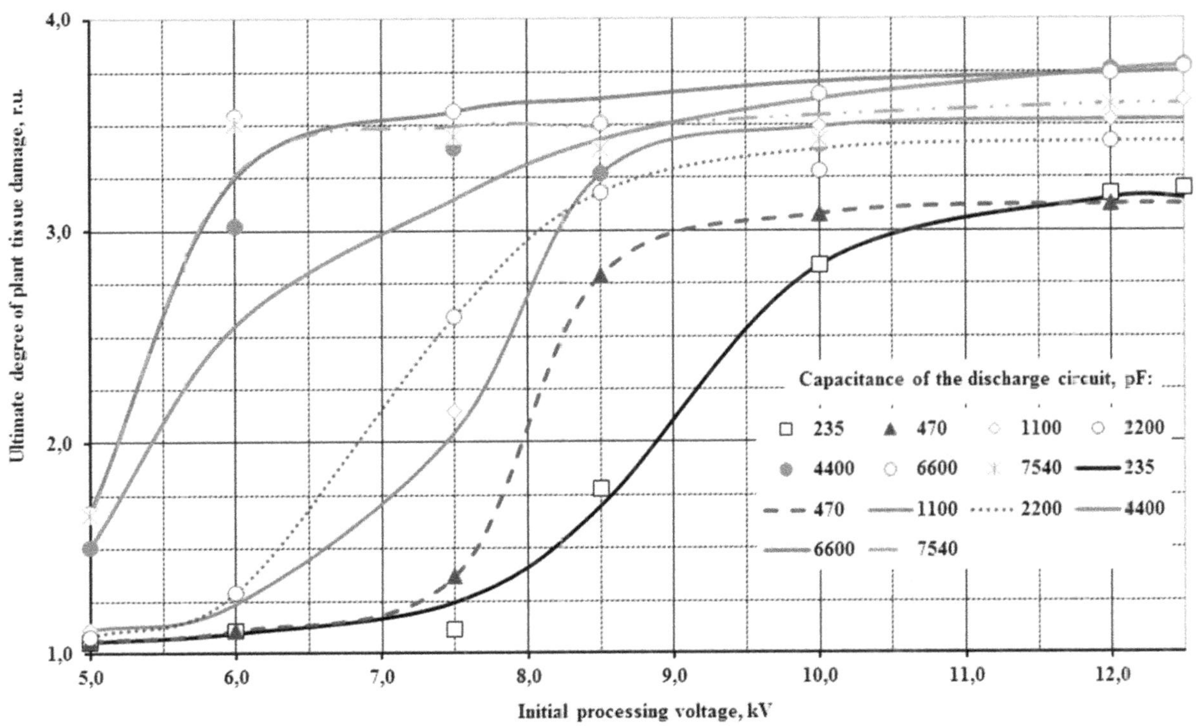

The analysis of the dependencies $S_{nmax}=f(C)$ at $U_0 = const$ (Figure 12) shows that the obtained curves gradually reach the maximum value of the limiting degree of damage at C = 1100-4400 pF and more, at all values of the initial circuit voltage. This behavior of these dependencies suggests that in order to achieve the maximum values of the limiting damage degrees to plant tissues, the storage capacity of the discharge circuit C must correspond to a value of at least 4400 pF.

In order to determine the parameters of the discharge circuit at which the desired technological effect is achieved, that is, irreversible damage is created in plant weed tissues with a limiting value of S_{nmax} at the actuating energy W_{pt} minimum cost, it is necessary to carry out a joint analysis of the contour U_0 and the storage capacity C initial voltage dependence on the limiting degree of damage tissues of various weeds S_{nmax}. As the result, a family of curves for different types of weeds was obtained (Figure 13).

Analysis of the presented dependences (Figure 13) allows us to state that in order to achieve the ultimate degree of damage to plant tissues, the treatment must be performed by adjusting the discharge circuit for a larger value of the voltage U_0 with a small capacitance C, which, as mentioned earlier, determines the duration of the actuating impulse. In the case of setting the circuit to a lower value of the processing voltage U_0, the processing time will increase, characterizing by the increase in the value of the storage capacitance C.

Figure 12. The limiting degrees of damage (Snmax) of the Amaranthus retroflexus L. stalks plant tissues dependence on the storage capacity of the contour (C) for different values of the initial circuit voltage (U_0)

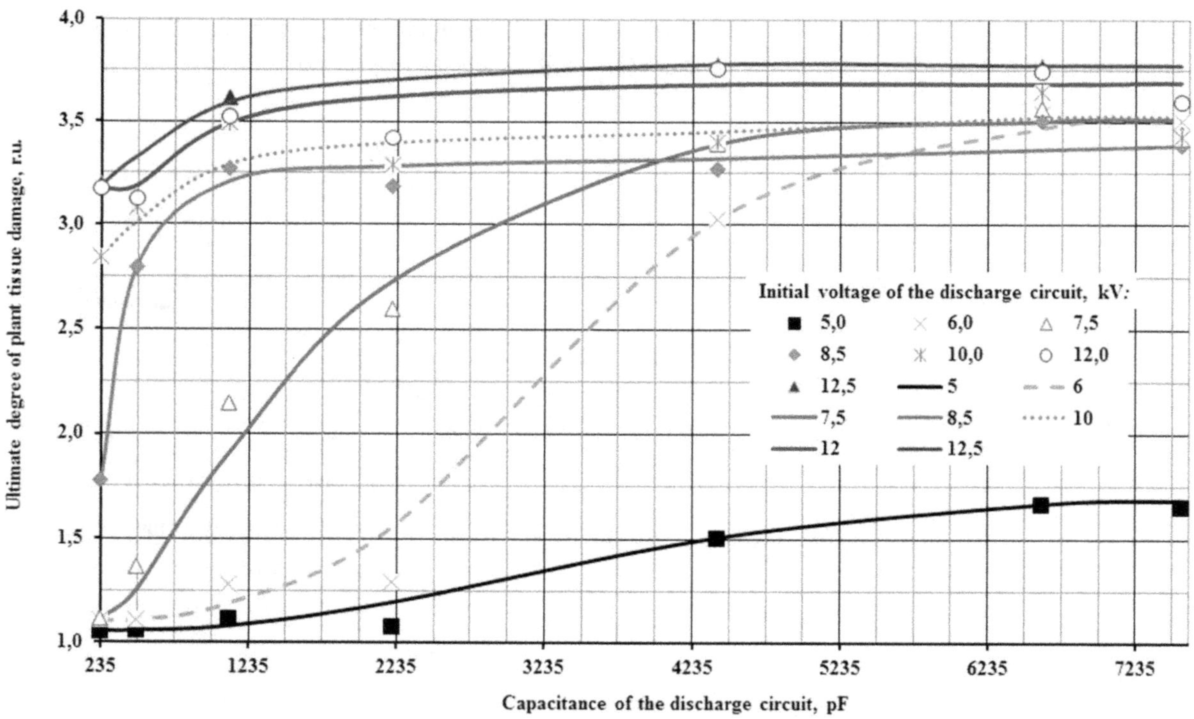

Figure 13. The initial processing voltage (U_0) dependence on the capacitance of the discharge circuit (C) with the minimum of the actuating energy, capable irreversibly damage the various weeds plant tissue

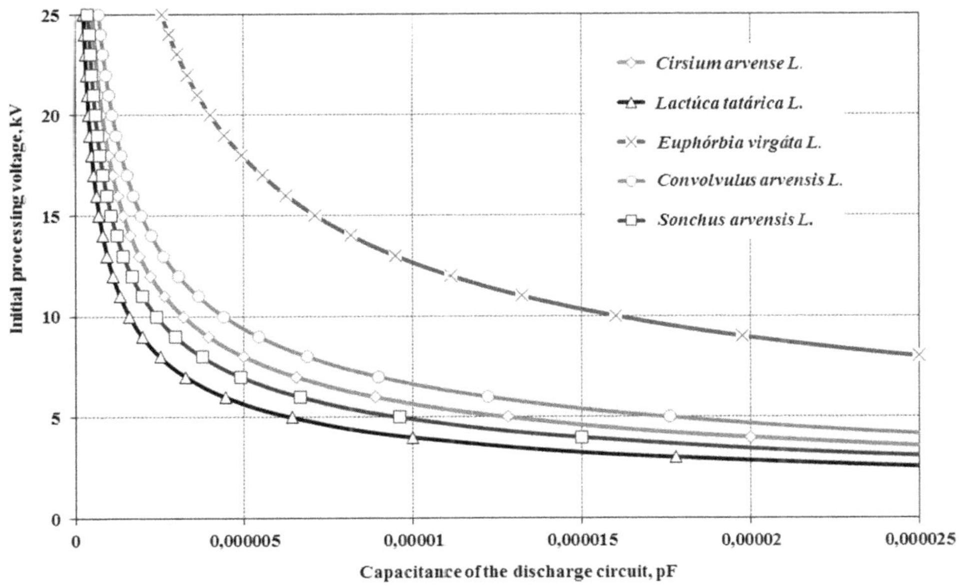

SOLUTIONS AND RECOMMENDATIONS

It is necessary to achieve the damage to plant tissues degree limiting value at the lowest possible consumption of electrical energy. Such a result can be obtained by affecting on the plant or plant material being treated, depending on its sensitivity, either by voltage impulses with a large amplitude and short duration or by impulses with a longer duration but by a small amplitude.

FUTURE RESEARCH DIRECTIONS

As a perspective for the scientific research and practical application continuation of the terms presented in the chapter on various plant-based objects irreversible tissue damage, we see the solution of the following tasks:

1. search, substantiation and identification of plants, fruit samples, vegetables, melons and gourds, etc. electroimpulse treatment optimal regimes, which unambiguously lead to irreversible damage to plant tissue;
2. substantiation and design of a reliable, highly effective circuit solution of the electrical part for electrotechnological installations that implement the principle of irreversible damage to processed objects plant tissue;
3. substantiation of the proposed electrotechnological operation application into technological processes existing in the processing industry and agriculture.

CONCLUSION

As the result of the carried out theoretical analysis and experimental researches results analysis it is possible to formulate requirements to the process of plant tissues irreversible electroimpulse damage that can form the basis for the requirements for electrotechnological operations in agriculture and processing plant raw materials industry, namely, the weeds electroimpulse destruction, sunflower and tobacco plants pre-harvesting treatment, thinning of agricultural crops thick plantings, fruits, vegetables, melons and gourds electroplasmolysis before extraction or drying, etc.

REFERENCES

Ade-Omowaye, B. I. O., Angersbach, A., Taiwo, K. A., & Knorr, D. (2001). Use of pulsed electric field treatment to improve dehydration characteristics of plant based foods. *Trends in Food Science & Technology*, *12*(8), 285–295. doi:10.1016/S0924-2244(01)00095-4

Ade-Omowaye, B. I. O., Angersbach, A., Taiwo, K. A., & Knorr, D. (2001). The use of pulsed electric field in producing juice from paprika (*Capsicum Annuum L.*). *Journal of Food Processing and Preservation*, *25*(5), 353–365. doi:10.1111/j.1745-4549.2001.tb00466.x

Aguiló-Aguayo, I., Oms-oliu, G., Soliva-fortuny, R., & Martin-Belloso, O. (2009). Flavour retention and related enzyme activities during storage of strawberry juices processed by high-intensity pulsed electric fields or heat. *Food Chemistry*, *116*(1), 59–65. doi:10.1016/j.foodchem.2009.02.007

Angersbach, A., Heinz, V., & Knorr, D. (2000). Effects of pulsed electric fields on cell membranes in real food systems. *Innovative Food Science & Emerging Technologies*, *1*(2), 135–149. doi:10.1016/S1466-8564(00)00010-2

Antonov, G.M., Armianov, N.K., & Dragiev, D.I. (1984). Elektroiskrovite razrjadi tjutjuno proshhvodstvo [Electric spark discharge in tobacco industry]. *Mekhanizaciya na sel'skogo stopanstvo* [Mechanization in agriculture], *12*, 48-51.

Baev, V. I., & Borodin, I. F. (2002). E`lektroimpul`snaya preduborochnaya obrabotka podsolnechnika i tabaka: monografiya [The elektroimpulse preharvest processing of sunflower and tobacco plants: the monography]. Volgograd: Page-2.

Baev, V. I., Savchuk, V. N., & Armianov, N. K. (1985). Eksperimental'nye issledovaniya parametrov elektroimpul'snoy obrabotki tabaka [Experimental research of parameters of tobacco electric spark treatment]. Sbornik trudov VSKHI [Collected Works of VSHI], 92, 44-45.

Baev, V.I., & Yudaev, I.V. (2004). Chuvstvitel'nost' tkani sornykh rasteniy k elektroimpul'snomu vozdeystviyu [Sensitivity of weed plants tissue to electrical impulse action]. *Mehanizacija i elektrifikacija sel'skogo hozjajstva* [Mechanization and electrification of agriculture], *3*, 5-8.

Baev, V. I., Yudaev, I. V., Petrukhin, V. A., Baev, I. V., Prokofyev, P. V., & Armianov, N. K. (2018). Electrotechnology as One of the Most Advanced Branches in the Agricultural Production Development. In Handbook of Research on Renewable Energy and Electric Resources for Sustainable Rural Development. Hershey, PA: IGI Global. doi:10.4018/978-1-5225-3867-7.ch007

Bauchot, A. D., Harker, F. R., & Arnold, W. M. (2000). The use of electrical impedance spectroscopy to assess the physiological condition of kiwifruit. *Postharvest Biology and Technology*, *18*(1), 9–18. doi:10.1016/S0925-5214(99)00056-3

Benz, R., & Zimmermann, U. (1980). Relaxation studies on cell membranes and lipid bilayers in the high electric field range. *Bioelectrochemistry and Bioenergetics*, *7*(4), 723–739. doi:10.1016/0302-4598(80)80037-7

Beraa, T. K., Berab, S., Karc, K., & Mondalc, S. (2016). Studying the Variations of Complex Electrical Bio-Impedance of Plant Tissues During Boiling. *Procedia Technology*, *23*, 248–255. doi:10.1016/j.protcy.2016.03.024

Bouzrara, H., & Vorobiev, E. (2000). Beet juice extraction by pressing and pulsed electric fields. *International Sugar Journal*, *102*(1216), 194–200.

Chervyakov, D. M. (1979). Mexanizm vozdejstviya e'lektricheskix razryadov na rasteniya [Mechanism of influence of electric discharges on plants]. *Elektronnaya Obrabotka Materialov*, *5*, 70–71.

Diprose, M. F., & Benson, F. A. (1984). Electrical methods of killing plants. *Journal of Agricultural Engineering Research*, *30*, 197–209. doi:10.1016/S0021-8634(84)80021-9

Eshtiaghi, M. N., & Knorr, D. (2002). High electric field pulse treatment: Potential for sugar beet processing. *Journal of Food Engineering*, *53*(2), 265–272. doi:10.1016/S0260-8774(01)00114-5

Fincan, M., DeVito, F., & Dejmek, P. (2004). Pulsed electric field treatment for solid-liquid extraction of red beetroot pigment. *Journal of Food Engineering*, *64*(3), 381–388. doi:10.1016/j.jfoodeng.2003.11.006

Ho, S. Y., & Mittal, G. S. (1996). Electroporation of cell membranes: A review. *Critical Reviews in Biotechnology*, *16*(4), 349–362. doi:10.3109/07388559609147426 PMID:8989868

Inaba, A., Manabe, T., Tsuji, H., & Iwamoto, T. (1995). Electrical Impedance Analysis of Tissue Properties Associated with Ethylene Induction by Electric Currents in Cucumber (*Cucumis sativus L.*). *Fruit. Plant Physiology*, *107*(1), 199–205. doi:10.1104/pp.107.1.199 PMID:12228354

Joshi, R., & Schoenbach, K. (2000). Electroporation dynamics in biological cells subjected to ultrafast electrical pulses: A numerical simulation study. *Physical Review E: Statistical Physics, Plasmas, Fluids, and Related Interdisciplinary Topics*, *62*(1), 1025–1033. doi:10.1103/PhysRevE.62.1025 PMID:11088559

Klimov, A. A., Savchuk, V. N., & Baev, V. I. (1970). O nekotorykh elektrofizicheskikh parametrakh i svoistvakh rastitel'noi tkani kak ob"ekta elektroiskrovogo vozdeistviya [About some electrophysical parameters and properties of plant tissue as an object of electric spark action]. *Elektronnaya Obrabotka Materialov*, *1*, 66–71.

Knorr, D., Geulen, M., Grahl, T., & Sitzmann, W. (1994). Food application of high electric field pulses. *Trends in Food Science & Technology*, *5*(3), 71–75. doi:10.1016/0924-2244(94)90240-2

Kotnik, T., & Miklavcic, D. (2006). Theoretical evaluation of voltage inducement on internal membranes of biological cells exposed to electric fields. *Biophysical Journal*, *90*(2), 480–491. doi:10.1529/biophysj.105.070771 PMID:16239325

Kuffel, E., Zaengl, W. S., & Kuffel, J. (2000). *High Voltage Engineering. Fundamentals*. Butterworth – Heinemann.

Melikov, K. C., Frolov, V. A., Shcherbakov, A., Samsonov, A. V., Chizmadzhev, Y. A., & Chernomordik, L. V. (2001). Voltage-Induced Nonconductive Pre-Pores and Metastable Single Pores in Unmodified Planar Lipid Bilayer. *Biophysical Journal*, *80*(4), 1829–1836. doi:10.1016/S0006-3495(01)76153-X PMID:11259296

Mytnikov, A. V. (2012). *High Voltage Engineering*. Tomsk: TPU. Neu J., & Krassowska W. Asymptotic model of lectroporation. *Physical Review. E*, *59*(3), 3471–3482.

Popova, N. A., Papchenko, A. Ya., & Bologa, M. K. (2014). Ehlektroplazmoliz vinograda s primeneniem bipolyarnyh impul'sov [Electroplasmolysis of grapes using bipolar pulses]. *Elektronnaya Obrabotka Materialov*, *50*(6), 83–91.

Rastogi, N. K., Eshtiaghi, M. N., & Knorr, D. (1999). Accelerated mass transfer during osmotic dehydration of high intensity electrical field pulse pretreated carrots. *Journal of Food Science*, *64*(6), 1020–1023. doi:10.1111/j.1365-2621.1999.tb12272.x

Stankovic, M., Cvijanovic, M., & Dukic, V. (2016). Ecological importance of electrical devices innovative in the process of anti *Ambrosia artemisiifolia L. Economics of Agriculture*, *3*, 861–870.

Toepfl, S. (2006). *Pulsed Electric Fields (PEF) for Permeabilization of Cell Membranes in Food-and Bioprocessing: Applications, Process and Equipment Design and Cost Analysis*. Ph.D. Thesis.

Toepfl, S., Heinz, V., & Knorr, D. (2005). Overview of pulsed electric field processing for food. In D. W. Sun (Ed.), *Emerging Technologies for Food Processing* (pp. 67–97). Oxford, UK: Elsevier. doi:10.1016/B978-012676757-5/50006-2

Toporkov, V.N., & Korolev, V.A. (2017). *Ehlektroimpul'snaya ustanovka dlya bor'by s sornyakami: monografiya* [Electropulse installation for weed control: Monograph]. Moskva: Federal'nyj nauchnyj agroinzhenernyj centr VIM [Federal agricultural research centre VIM].

Vincent, C., Panneton, B., & Fleurat-Lessard, F. (2001). Electrical weed control: Theory and applications. In *Physical control methods in plant protection*. Springer-Verlag Berlin Heidelberg. doi:10.1007/978-3-662-04584-8

Weaver, J., & Chizmadzhev, Y. (1996). Theory of electroporation: A review. *Bioelectrochemistry and Bioenergetics*, *41*(2), 135–160. doi:10.1016/S0302-4598(96)05062-3

Yeom, H. W., Chism, G. W., & Zhang, Q. H. (2002). Inactivation of pectin methyl esterase in orange juice by pulsed electric fields. *Food Engineering and Physical Properties*, *67*(6), 2154–2159.

Yudaev, I.V., & Baev, V.I. (2006). Optimal'nye parametry elektroimpul'snogo unichtozheniya sornykh rasteniy [Optimal parameters for electrical impulse destroying of weeds]. *Vestnik Volgogradskoy sel'skokhozyaystvennoy akademii. Nauka i vysshee professional'noe obrazovanie* [Herald of Volgograd State Agricultural Academy. Science and higher professional education], *2*(2), 71-78.

Yudaev, I. V., Baev, V. I., Eliseev, D. S., & Brenina, T. P. (2004). Sornye rasteniya kak ob'ekt elektricheskoy propolki: biologicheskie osobennosti i elektrofizicheskie svoystva: monografiya [Weed plants as an object of electrical weeding: Biological features and electrophysical properties: Monograph]. Volgograd: Stanitsa-2.

Yudaev, I. V., & Brenina, T. P. (2008). The definition of electro impulses used in weed control. *Journal of agricultural sciens*, *53*(1), 37–44.

Zhang, M. I. N., Stout, D. G., & Willison, J. H. M. (1990). Electrical Impedance Analysis in Plant Tissue: Symplastic Resistance and Membrane Capacitance. In the Hayden Model. *Journal of Experimental Botany*, *41*, 371–380. doi:10.1093/jxb/41.3.371

Zimmermann, U. (1986). Electric breakdown, electropermeabilization and electrofusion. *Physiological Biochemical Pharmacology*, *105*, 175–256. doi:10.1007/BFb0034499

ADDITIONAL READING

Antonov, G. M., Armianov, N. K., Dragiev, D. I., Baev, V. I., & Savchuk, V. N. (1986). Samokhoden agregat za elektroiskrova obrabotka na tyutyun [Mobile facility for electric spark treatment of tobacco]. *Sel'skostonanskaya tekhnika* [Agricultural machinery]. *22*, *2*, 57–61.

Baev, V. I. (1972). *Issledovanie vliyanie formy iskrovogo impul'sa na povrezhdenie rastitel'noy tkani* [Research of spark impulse form influence to plant tissue damage]. (pp. 45–50). Volgograd, Russia: Trudy Volgogradskogo SHI. [Works of Volgograd agricultural institute]

Baev, V. I., & Armianov, N. K. (1996). Analiz elektricheskogo soprotivleniya kornevoy sistemy sornykh rasteniy [Analysis of electrical resistance of weed`s root system]. Sbornik: Uchenye Volgogradskogo sel'khozinstituta - nauke i praktike. Materialy nauchnoy konferencii, posvyashchennoy 50-letiyu vuza [Digest: Scientists of Volgograd Agricultural Institute - to science and practice. Materials of science conference devoted to 50th anniversary of institute]. Volgograd, Russia, 189-190.

Baev, V. I., & Savchuk, V. N. (1974). Deistvuyushchie faktory elektroiskrovogo razryada pri obrabotke rastenii [Actual factors of electric spark discharge during the plants processing] [Electronic materials processing]. *Elektronnaya Obrabotka Materialov*, *1*, 70–72.

Baev, V. I., Savchuk, V. N., Antonov, G. M., Armianov, N. K., & Dragiev, D. I. (1984). Otnosno izbora na napryazhenie pri naprechnata obrabotka na tyutyunevy rasteniya s elektricheski razryadi [About voltage selection for transversal processing of tobacco plants with electrical discharges.]. *Nauchni trudove na VTU "Angel Kynchev". [Scientific works of Highest Technical School "Angel Kanchev"].* Ruse, Bulgaria. *XXVI, 7,* 33–38.

Baev, V.I., & Yudaev, I.V. (2001). Tekhnologicheskaya effektivnost' elektroimpul'snoi obrabotki sornyakov [Technological efficiency of weeds electro-impulse processing]. *Mekhanizatsiya i elektrifikatsiya sel'skogo khozyaistva [Mechanization and electrification of agriculture], 10,* 17-19.

Baev, V. I., & Yudaev, I. V. (2004). Udel'nye pokazateli processa elektricheskogo povrezhdeniya sornoy rastitel'nosti [Specific indexes of the process of electrical weed damaging]. *1-ya Rossiyskaya nauchno-prakticheskaya konferenciya Fiziko-tekhnicheskie problemy sozdaniya novykh tekhnologiy v APK [First Russian science-practice conference Physical and technical problems of new technologies creation in agribusiness industry].* – Stavropol, Russia, *2,* 265-268.

Baev, V.I., Yudaev, I.V., Baev, I.V., & Prokofiev, P.V. (2006). Tekhnologicheskoe obespechenie elektroimpul'snoy kul'tvacii [Technological support of electrical impulse cultivation]. *Vestnik Moskovskogo gosudarstvennogo oblastnogo universiteta. Seriya "Estestvennye nauki" [Moscow state region university herald. "Natural sciences" series], 3,* 19-23.

Diprose, M. F., Benson, F. A., & Turner, N. W. (1980). The use of high voltage electricity for weed beet control. *Proceedings British Crop Protection Conference, 2,* 545-548.

Diprose, M. F., Hacham, R., & Benson, F. A. (1978). Weed control by high voltage electric shocks. *Proceedings British Crop Protection Conference Weeds.* Brighton, England, *2,* 443-450.

Dykes, W. Q. (1980). Principles and practices of electrical weed control. American Society of Agricultural Engineers, 80-1007, 9.

Evreinov, M. G. (1954). *Elektrifikacija sel'skogo hozjajstva* [Electrification of agriculture]. Moscow: Goskultprosvetizdat.

Kaufman, K. R., & Schaffner, L. M. (1979). *Lasco electric discharge system weeder. Excerpts from Lasco EDS Lightning Weeder.* Fargo: North Dakota State University.

Klimov, A. A., & Baev, V. I. (1970). Analiticheskoe issledovanie parametrov vysokovol'tnogo razryadnogo kontura [Analytic research of high voltage discharge circuit parameters] [Works of Volgograd agricultural institute]. *Trudy Volgogradskogo SHI, 34,* 181–190.

Nedjalkov, N. (1996). Optimizarane na konstruktivnite paramrtri na elementite na navesnija elektrod za elektroiskrovata obrabotka na pleveli [Optimization of design parameters and elements of hinged electrodes for electric spark treatment of weeds] *Sel'skostonanskaya tekhnika [Agricultural machinery], 33, 2,* 16-18.

Slesarev, V.N., Gubanova, N.Ju., & Nechaev, B.V. (1970). Unichtozhenie sornjakov jelektricheskim tokom [Destruction of weeds by electric current]. *Mehanizacija i elektrifikacija sel'skogo hozjajstva [Mechanization and electrification of socialist agriculture], 12,* 45-46.

Spirin, A. A. (1953). *Jelektricheskie impul'sy vysokogo naprjazhenija v probleme bor'by s sornjakami: avtoref. dis. ... dokt. tehn. nauk* [Electrical high voltage impulses for the problem of controlling weeds: author's abstract of the thesis of the Doctor of Technical Sciences]. Baku.

Vigoureux, A. (1982). Mécanisation de la destruction des montées à graines en culture betteravière. *Publ. Trimest. Inst. R. Belge L'Amélior. Betterave, 50,* 3–36.

Yudaev, I. V. (2004). Izmenenie parametrov ekvivalentnoy skhemy zameshcheniya rastitel'noy tkani pri elektricheskom povrezhdenii [Changing of plant tissue equivalent circuit parameters after electrical damaging] *Power supply and energy saving in agriculture: Works of 4th International science-practice conference All-Russian Research Institute of Electrification in Agriculture*. Moscow, Russia, *2,* 142-149.

Yudaev, I. V., Baev, I. V., Brenina, T. P., Krivoshchapov, V. A., & Prokofiev, P. V. (2010). Elektroimpul'snyy propol'shchik [Electrical impulse weeder]. Dostizheniya nauki v Volgogradskoy oblasti 2004-2009 [Science achievements at Volgograd area in 2004-2009]. Volgograd, Russia, 317-320.

Yudaev, I.V., Baev, V.I., Baev, I.V., Prokofiev, P.V., & Sokolov A.V. (2002). Metodika ocenki sostoyaniya sornoy rastitel'nosti pri elektricheskoy oborabotke [Method of the condition assessing of weeds during electrical weeding]. *Elektricheskie apparaty i elektricheskie tekhnologii sel'skogo khozyaystva: sbornik nauchnykh trudov MGAU [Electrical apparatus and electrical technologies in agriculture: science works of MSAU bulletin], 10,* 71-73.

Yudaev, I.V., Baev, V.I., Brenina, T.P., & Eliseev, D.S (2002). Sornye rasteniya kak ob"ekt elektricheskogo vozdeystviya [Weed plants as an objects of electrical action]. *Elektricheskie apparaty i elektricheskie tekhnologii sel'skogo khozyaystva: sbornik nauchnykh trudov MGAU. [Electrical apparatus and electrical technologies in agriculture: science works of MSAU bulletin], 10,* 68-71.

Yudaev, I. V., Baev, V. I., Brenina, T. P., & Eliseev, D. S. (2003). Soprotivlenie uchastkov cepi protekaniya toka pri elektricheskom unichtozhenii sornyakov [The resistance of current flow circuit portion during electrical weed destroying]. *Fiziko-tekhnicheskie problemy sozdaniya novykh tekhnologiy v APK: II-ya Rossiyskaya nauchno-prakticheskaya konferenciya* [Physical and technical problems of new technologies creation in agribusiness industry. Second Russian science-practice conference: scientific works digest]. Stavropol, Russia, *1,* 11-14.

KEY TERMS AND DEFINITIONS

Damaging Degree: Determined as plant tissue resistance ratio before and after the electric processing, calculated at frequency 10 kHz.

Electric Field Strength: Is a quantitative expression of the intensity of an electric field at a particular location. A field strength of 1 V/m represents a potential difference of one volt between points separated by one meter.

Electrotechnology in Agriculture: Complex of methods and techniques of electrophysical influence on technological processes in order to obtain agricultural products of specified properties with a minimum of labor and energy costs, as well as to create comfortable conditions and stimulate the development of living organisms for increasing quantity and quality of agricultural products without expanding the acreage, livestock and poultry, as well as increasing the shelf life of finished products and protecting it from pests and diseases.

High-Voltage Electroimpulse Action: Effect of the electrophysical nature on the processed material, which is carried out by means of successively generated high-tension voltage pulses (at given frequency), for example, in the discharge circuit of the impulse voltage generator.

Irreversible Damage to Plant Tissue: An irreversible damage to plant tissue cells, leading to a large degree and persistent disturbances of intracellular homeostasis, which cannot be eliminated even by the maximum activation of protective-compensatory-adaptive mechanisms in still viable intracellular and extracellular structures, and eventually leading to cells and the plant tissue itself death.

Reversible Damage to Plant Tissue: A short-term disruption of the intracellular homeostasis of the cells that make up plant tissue, which is eliminated by activating intra- and extracellular protective-compensatory-adaptive mechanisms, thereby facilitating the tissue cells vital activity restoration.

Chapter 16
Control of Advanced Fodder Disinfection in Terms of Economic Criteria

Alexandr Dubrovin
Federal Scientific Agroengineering Center VIM, Russia

ABSTRACT

Upon long-term storage, germinated grain with 50-60% humidity begins to mold and rot. Therefore, it should be fed off to livestock as soon as possible after its germination. To extend storage life of germinated grain, it is necessary either to dry it to 14% relative humidity, or to carry out its disinfection. Grain forms the basis of combined fodder in agriculture. Some technological solutions are available to provide fodder disinfection optimized in technological or economic terms, as well as rational energy use, including disinfection by high-speed electron bunches, ultra-high frequency electric waves, combined IR and conductive (contact) electric heating, and toxic ozone gas. There also exists a similar solution for fodder premises disinfection through controlling excessive pressure of their internal air. In control by technological (production) criterion, the mode of minimal integrated cost of calculated loss of quality (expenditures) of products being disinfected (fodder) is determined, so the highest livestock productivity in the current context is obtained.

INTRODUCTION

Upon long-term storage (over 4…5 hours) germinated grain with 50…60% humidity begins to mold and rot. Therefore, it should be fed off to livestock as soon as possible after its germination. To extend storage life of germinated grain it is necessary either to dry it to 14% relative humidity (Vendin S.V. et al., 2015), or to carry out its disinfection. The Grain is the basis of combined fodder in agriculture.

Some technological solutions are available to provide fodder disinfection optimized, for example (Vasant, P., & Voropai, N., 2016; Vasant P. et al., 2018), in technological or economic terms, as well as rational energy use (Kharchenko, V. & Vasant P., 2018; Vasilyev A. N., 2013; Vasiliev A.N. et al., 2016), including disinfection by high-speed electron bunches (Dubrovin A.V., 20.11.2014), ultra-high frequency electric

DOI: 10.4018/978-1-5225-7573-3.ch016

waves (Mohapatra, G. et al., 2014; Budnikov D.A. et al., 2018; Dubrovin A.V.,03.07.2017; Syrovatka V.I., 2014), combined infrared and conductive (contact) electric heating (Dubrovin A.V., 31.05.2017; Shuvalov A.M. et al., 2014), and toxic ozone gas (Granella, S. et al., 2018; Abhijit, S, Sambit D. & Pooja S., 2017; Mason, L.J. et al., 2014; Dubrovin A.V., 19.01.2017; Pershin A.F. et al., 2014). There also exists a similar solution for fodder premises disinfection through controlling excessive pressure of their internal air (Dubrovin A.V., 16.03.2017; Zhiltsov V.I., 1979).

The minimal sum of the cost of calculated loss of fodder being disinfected, and of operational energy costs for radiation processing and transportation of fodder and other agricultural products, is automatically determined. In terms of radiation doze value, functional dependence of expenditures Z_{prod1} on losses of livestock and poultry productivity because of fodder contamination with microorganisms and fungi in the absence of radiation or at its low level, are artificially developed (Dubrovin A.V., 20.11.2014). Dependences of economic expenditures Z_{ost1} on losses of fodder and other products because of their excessive radiation with high-speed electron bunches interacting with the cellular structure of fodder and other products biomass, are also determined.

The first of these dependences Z_{prod1} is nonlinearly decreasing with irradiance radiation P_{obl} growth, starting from a specific level of biomaterial contamination with microorganisms and fungi, known in advance on the results of measuring hygienic quality of materials to be sterilized by nanosecond electric impulses.

The second dependence Z_{oost1} is nonlinearly increasing starting from a minimal value of radiation threshold sufficient for the manifestation of the first irreversible changes in biological products of vegetable and animal origin.

Acceptable level of expenditures for losses of animal and poultry products because of changed fodder characteristics is also determined in the process of specific experimental work.

They also model similar cost dependences of economic expenditures for electric energy for fodder transportation Z_{tran1}, as well as for their radiation $Z_{obluc1,}$ on the radiation dose value.

The third dependence Z_{tran1} is a constant value at permanent velocity of movement of the conveyor operating element and at constant mass flow of products, changing in value proportionally to the velocity of movement of the conveyor operating element and the product mass.

The fourth dependence Z_{obluc1} is linearly increasing with the radiation dose P_{obl} growth.

The obtained four expenditure functions are summed up in the range of variation of artificially formed signal of a radiation dose to determine the minimum value of this target function (the criterion of optimization by the minimal sum of the specified expenditures) $Z_{\Sigma 1} = Z_{prod1} + Z_{ost1} + Z_{obluc1} + Z_{tran1}$. Thus, precise, economically optimal and energy-saving disinfection of each product (its whole mass) is carried out.

In control by technological (production) criterion, the mode of minimal integrated cost of calculated loss of quality (expenditures) of products being disinfected (fodder) is determined, so the highest livestock productivity in the current context is obtained. In control by economic criterion, they determine the minimal sum of the cost of calculated loss of quality of fodder being disinfected, and of operational energy costs for radiation processing, heating, ozonation and fodder transportation in the disinfection area.

The paper objective is to substantiate the process of automated search and achievement of the two corresponding modes of fodder disinfection by nanosecond electric impulses, optimal in technological and economical terms and providing rational energy use, that differ in efficiency (performance) characteristics.

The first objective is achieved through determination of the first energy mode of fodder disinfection corresponding to technological minimum of the first integrated cost of animal and poultry products losses while feeding livestock with disinfected fodder, because of it bacteriological and microbial contamination and expenditures caused by its quality loss due to overexposure to nanosecond electric impulses. The highest animal and poultry productivity is provided.

The second objective is achieved through determination of the second energy mode of fodder disinfection. This mode corresponds to the economic minimum of the second sum of the cost of livestock and poultry losses and of considered operational energy costs for fodder treatment with nanosecond superhigh frequency impulses, as well as for fodder transportation in disinfection area. Livestock is fed with fodder disinfected in an economically optimal mode, whose quality is partly lowered in comparison with the technologically optimal disinfection mode, because of fodder bacteriological and microbiological contamination or overexposure to nanosecond electric impulses in the process of control of economically optimal disinfection mode. The highest economic result in monetary terms (maximum profit or its increment, minimal production cost or its chosen component as an adopted control characteristic) is provided.

RESEARCH METHOD

Radiation evoked by nanosecond electric impulses is absorbed by biological materials causing *impact* action of electromagnetic field on intercellular biological ties, mostly of animal origin, thus destroying microbes, bacteria, moldand fungi residing on these vegetable materials, for example, on fodder.

The range of radiation dose values required for a specific type of fodder being disinfected, is established. The values of radiation dose are to be determined in advance during test disinfecting of preproduction fodder lots with nanosecond electric impulses. The amount of harmful biological components is proportional to the volume, that is, the mass of material being disinfected. Radiation dose (1) is radiation energy absorbed by a unit of fodder mass, providing its disinfection depending on biological characteristics of these harmful components, the disinfecting chamber design and energy radiators in the form of nanosecond electric impulses.

$$D_{spec} = \frac{E_{spec}}{M_{prod}} \tag{1}$$

where D_{spec} – is the specified radiation dose for the product, J/kg; E_{spec} – is the radiation energy absorbed by the product, J; M_{prod} – is the product mass, kg. For the sake of simplicity, weight factors in the mathematical model of the type (1) and others are not shown.

Specified energy of radiation absorbed E^{absor}_{spec} 2) is proportional to the product mass:

$$E^{absor}_{spec} = D_{spec} \times M_{prod} \tag{2}$$

Energy of absorbed radiation (3) is proportional to the product mass end to the radiation dose:

$$E^{absor} = D \times M_{prod} \tag{3}$$

EXPERIMENTAL PROCEDURE

At first, cost dependence of expenditures on losses of animal and poultry productivity because of their contamination with microorganisms and fungi in the absence of radiation at its low level, must be artificially modeled in terms of radiation doze value. It is also necessary to determine cost dependence on losses of animal and poultry productivity because of excessive radiation interacting with the cellular structure of harmful biomass in livestock and poultry fodder.

The first of these dependences is nonlinearly decreasing with radiation growth, starting from a specific level of biomaterial contamination with microorganisms, fungi, known in advance on the results of measuring hygienic quality of materials to be processed by nanosecond electric impulses.

The second dependence is nonlinearly increasing starting from a minimal value of radiation threshold sufficient for the manifestation of the first irreversible changes in biological products of vegetable origin, that is, in fodder. Acceptable level of expenditures for losses of animal and poultry products because of changed fodder characteristics is also determined in the process of experimental work.

It is also advisable to model similar cost dependences of energy consumption for fodder radiation on the radiation dose value, as well as on energy consumption for fodder transportation.

The third dependence of electric energy consumption for radiation is linearly increasing with the radiation dose growth.

The fourth dependence of energy consumption for fodder transportation is a constant value at permanent velocity of the conveyor operating element and at constant mass flow of fodder over time. In control in economic terms and at rational energy consumption for fodder disinfection, it is required to sum up the obtained four expenditure functions in the range of variation of artificially formed signal of a radiation dose and to determine the minimum of this second sum (figure 1).

RESULTS AND THEIR DISCUSSION

The unit (figure2) operates in the following way. Fodder to be disinfected is loaded onto the measuring conveyor wherefrom it is fed to the main conveyor and thereon to the disinfection chamber. It is advisable to use a metal screw conveyor, as helical surface of a metal screw conveyor to a great extent prevent losses of electromagnetic radiation field from the disinfection chamber.

The unit 9 calculates two target functions of total expenditures depending on radiation dose. The first is the function of total losses of cost of livestock and poultry productivity because of microbes and bacteria presence in fodder at low radiation doses and of overheating of fodder at high radiation doses. The second function reflects total losses of cost of livestock and poultry productivity and operating energy cost for fodder disinfection and transportation in the form of their total sum. The unit 10 determines the minimal value of a target function of total expenditures, selected with the use of the element for selecting a criterion of optimization of disinfection mode 13, that is, yields the optimal value of the disinfection mode in terms of radiation dose, corresponding to this minimum. In calculation, temperature of fodder supplied for disinfection and relative humidity of fodder supplied for disinfection are considered with the use of the sensor 11 and the sensor 12 correspondingly.

Therefore, with changing of the above fodder input characteristics, dependencies of costs of productivity loss and operating expenditures for disinfection, as well as settings of parameters of technologically and economically optimal disinfection modes in terms of radiation dose are also changed in the form

Figure 1. Representation of technologically or economically optimal process of disinfection of livestock and poultry fodder with nanosecond electric impulses

D – radiation dose for material being disinfected, J/kg; ($D_{technol}^{max}$ – $D_{technol}^{min}$) – specified technological process range of radiation doses for specific material types, J/kg; ΔP – cost of loss of livestock and poultry productivity caused by high humidity B_f (microbes, bacteria, mold and fungi, etc.) and by over radiation (loss of vitamins, destruction of protein molecules, etc.), ruble/time unit; Z – economic expenditures, ruble/time unit; $Z_{transport}$ – total costs of fodder transport, ruble/time unit; $\Delta P_{humidity}$ – value of livestock productivity losses caused only by high humidity, ruble/time unit; $\Delta P_{overexposure}$ – value of livestock and poultry productivity loss caused only by over radiation (excessive radiation), ruble/time unit; $\Delta P_{technol}$ – cost of total loss of livestock and poultry productivity because of unconditional (non-disinfected or ill-disinfected) fodder, ruble/time unit; $Z_{energy1}$ – radiation energy costs at high temperature T_f and low humidity B_f of fodder to be disinfected, ruble/time unit; $Z_{energy2}$ – radiation energy costs at low temperature T_f and high humidity B_f of fodder to be disinfected, ruble/time unit; Z_{econ1} – total costs of disinfection and of productivity loss at high temperature T_f and low humidity B_f of fodder to be disinfected, ruble/time unit; Z_{econ2} – total costs of disinfection and livestock productivity loss at low temperature T_f and high humidity B_f of fodder to be disinfected, ruble/time unit; $D_{technol}^{wholesale}$ – technologically optimal radiation dose at which the cost of loss of livestock and poultry productivity $\Delta P_{technol}^{min}$ is the lowest in the result of feeding disinfected fodder, J/kg; $D_{econ1}^{wholesale}$ – economically optimal radiation dose Z_{econ1}^{min} at which the sum of the cost of loss of livestock and poultry productivity due to feeding disinfected fodder, at high temperature T_f and low humidity B_f of fodder to be disinfected, and of the cost of energy consumption for radiation is the lowest, J/kg; $D_{econ2}^{wholesale}$ – economically optimal radiation dose Z_{econ2}^{min} at which the sum of the cost of loss of livestock and poultry productivity due to feeding disinfected fodder at low temperature T_f and high humidity B_f of fodder to be disinfected, and of the cost of energy consumption for radiation is the lowest, J/kg; ΔP_1 – the cost of additional productivity loss because of the radiation mode $D_{ekon1}^{wholesale}$ deviation from the optimal $D_{technol}^{wholesale}$ in pursuit of saving energy spent on drying $\Delta Z_{energy1}$, ruble/time unit; ΔP_2 – the cost of additional productivity loss because of the radiation mode $D_{ekon2}^{wholesale}$ deviation from the optimal $D_{technol}^{wholesale}$ in pursuit of saving energy spent on disinfection $\Delta Z_{energy2}$, ruble/time unit; $\Delta P_{technol}(D_{technol}^{wholesale}) + Z_{energy1}(D_{technol}^{wholesale})$ – the cost of total value of productivity losses and of radiation energy for disinfection in the technologically best disinfection mode $D_{technol}^{wholesale}$ at high temperature T_f and low humidity B_f of fodder to be disinfected, ruble/time unit; $\Delta P_{technol}(D_{technol}^{wholesale}) + Z_{energy2}(D_{technol}^{wholesale})$ – the cost of total value of productivity losses and of radiation energy for disinfection in the technologically best disinfection mode $D_{technol}^{wholesale}$ at low temperature T_f and high humidity B_f of fodder to be disinfected, ruble/time unit; $\Delta F_{radiation}$ – the highest technological advantage in monetary terms when passing from standard control of disinfection process in the standard process range of radiation doses ($D_{technol}^{max}$ – $D_{technol}^{min}$) to advanced fine control of the disinfection mode $D_{technol}^{wholesale}$, ruble/time unit; ΔZ_{econ1} – the cost of reduction of total productivity loss and of the costs of energy for disinfection in economically optimal control mode at high temperature T_f and low humidity B_f of fodder to be disinfected, in comparison with the technologically best mode, ruble/time unit. The cost of energy saving $\Delta Z_{energy1}$ exceeds the cost of additional productivity losses ΔP_1 by the value of reduction of total productivity loss and of costs of energy spent on disinfection ΔZ_{econ1}; ΔZ_{econ2} – the cost of reduction of total productivity loss and of the cost of energy spent on disinfection in the economically optimal control mode at low temperature T_f and high humidity B_f of fodder to be disinfected in comparison with the technologically best mode, ruble/time unit. The cost of energy saving $\Delta Z_{energy1}$ exceeds the cost of additional productivity losses ΔP_2 by the value of reduction of total productivity loss and of costs of energy spent on disinfection ΔZ_{econ2}

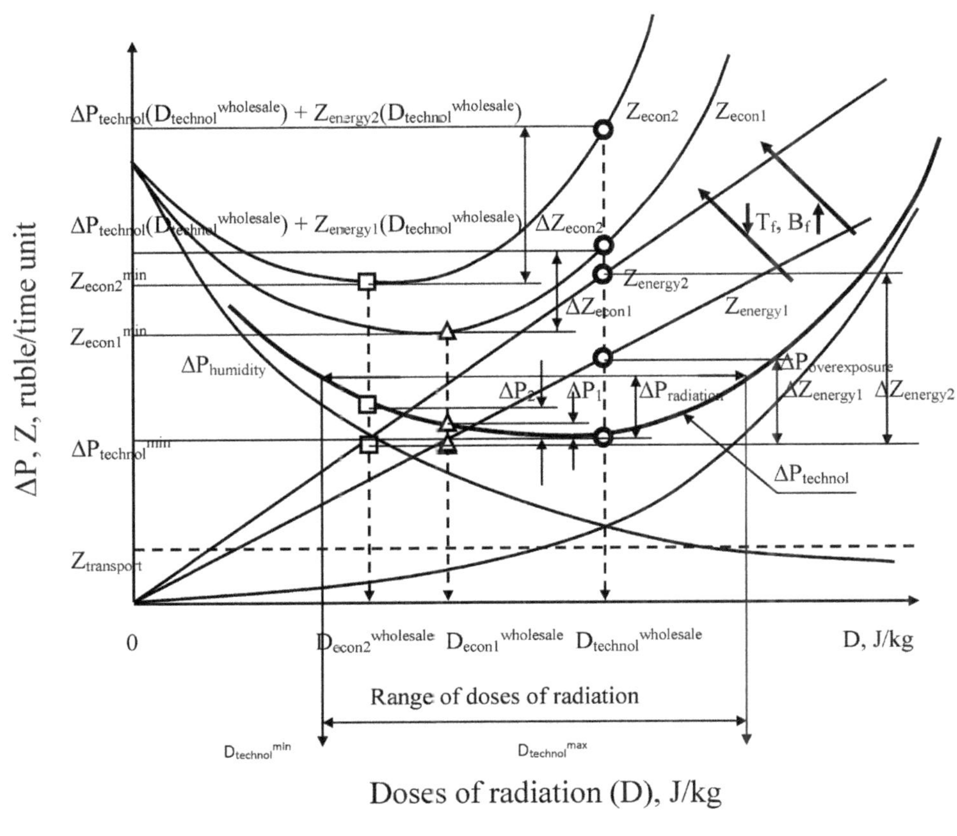

Figure 2. Shows a general diagram of the automated device for control of fodder disinfection with nanosecond electric impulses, functional diagram of the device for technologically and economically optimal disinfection of fodder for livestock and poultry with nanosecond electric impulses

1 – sensor of motion speed of fodder supplied for disinfection in the conveyor operating elements; 2 – meter of flow of fodder supplied for disinfection; 3 – unit of controllers for setting values of artificial signals of irradiance in the range of changing of radiation dose from zero to its limit value, minimal and maximal technological radiation dose, time signal, distance from the output cross section of the meter of flow of fodder supplied for disinfection to the reference axis of the radiator horn in the disinfection chamber, minimal and maximal radiation dose, sweep waveforms for radiation dose in time, regional unit prices of livestock and poultry products, of fodder and electricity, constants and coefficients of mathematical models of control of the mode of irradiation rate of bulk fodder; 4 – irradiation power meter; 5 – radiation power controller; 6 – fodder irradiator; 7 – controller on the basis of a magnetron or a klystron, nanosecond impulses generator; 8 – radiator horn in the disinfection chamber; 9 – valuator of the two target functions of total expenditures or a valuator of the first function of total loss of livestock and poultry productivity losses and of the second function of total productivity losses and expenditures for fodder disinfection; 10 – unit for determining minimal value of the selected target function of total expenditures (or a unit for optimization of the disinfection mode considering radiation dose); 11 – sensor of temperature of fodder supplied for disinfection; 12 – sensor of relative humidity of fodder supplied for disinfection; 13 – element for selecting a criterion of disinfection mode optimization; 14 – multiplicative identity; 15 – unit for controlled time delay; 16 – generator of nanosecond electric impulses

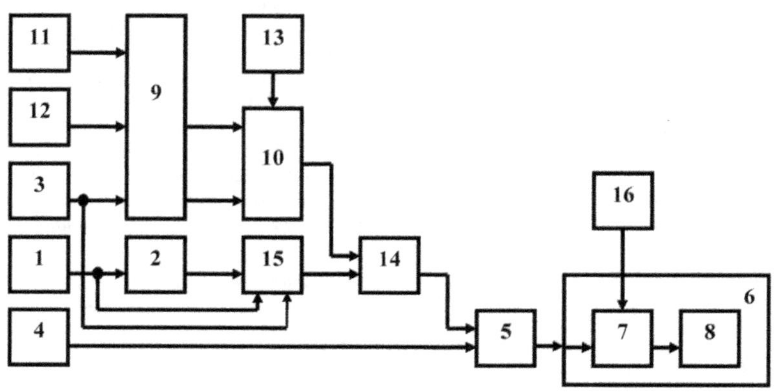

of a signal of required radiation dose at the unit 10 output. To consider relative position of the output cross section of the flow meter 2 and the reference axis of the radiator horn in the disinfection chamber, time delay is implemented in the unit of controlled time delay 15. A signal of prescribed distance from the output cross section of meter of flow of fodder supplied for disinfection to the reference axis of the radiator horn in the disinfection chamber from the corresponding output of the unit of controllers 3 is divided in the unit 5 by a signal of motion speed of the operating element of fodder supplied for disinfection in the conveyor operating elements of the flow meter 2 from the output of the speed sensor 1.

Consequently, a signal from the flow meter 2 is delayed by time T, sec (4):

$$T = \frac{S}{V} \tag{4}$$

where S distance along movement of bulk fodder from the outputcross section of flow meter 2 to the direction of axis of the radiator horn in the disinfection chamber, prescribed in the unit of controllers 3, m; V measured motion speed of an operating element of the flow meter 2 or speed of fodder movement, m/s.

Figure 3. Shows the general diagram of the arrangement of technical means for technically and economically optimal fodder disinfection with nanosecond electric impulses, or general arrangement of technical means for technologically and economically optimal disinfection of livestock and poultry fodder
1 – sensor of motion speed of fodder supplied for disinfection in the conveyor operating elements; 2 – meter of flow of fodder supplied for disinfection; 4 – irradiation power meter; 5 – irradiation power controller; 6 – fodder irradiator; 7 – controller on the basis of a magnetron or a klystron, nanosecond impulses generator; 8 – radiator horn; 16 – generator of nanosecond electric impulses; 17 – disinfection chamber with walls metalized on the inside, with a conveyor (a version of metal screw conveyor); 18 – input bin for fodder supplied for disinfection; 19 – output bin for disinfected fodder

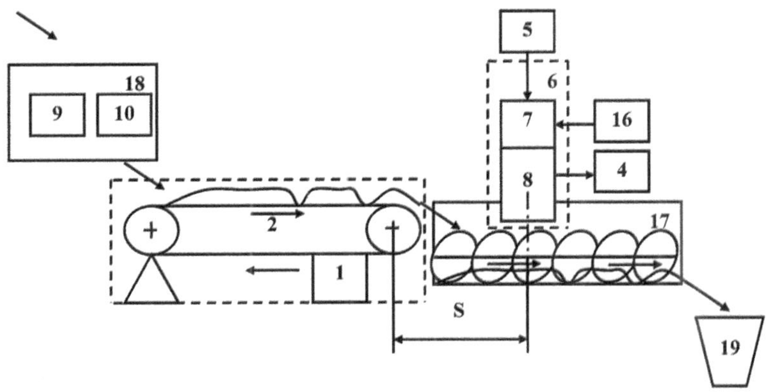

Distance, specified in the unit of controllers 3 is divided by the speed of fodder movement measured by the speed sensor 3. In the result, delay of time of signal of fodder floe is obtained, equal to the time of fodder movement from the flow meter 2 to the radiator horn 8. Multiplication of a signal of required technologically or economically optimal radiation dose by a signal of fodder flow from the output of the fodder flow meter 2 into the disinfection zone, delayed with respect to time of delay of entry of the cross section of the "fodder tube" into radiation zone, that is, multiplication of required optimal value of energy of radiation of fodder mass unit by fodder mass per time unit in the multiplication element 14, at its output gives a signal of required disinfection mode, with the use of – element a criterion of optimization of a disinfection mode 13, corresponding to an optimal value of radiation power at the instant when fodder flow measured in the meter 2, reaches the place where the radiator horn 8 is located in the disinfection chamber.

CONCLUSION AND FINDINGS

A specified disinfection capacity calculated according to the method illustrated at figure1, is compared with a measured value in the form of output signal of a radiation power meter. Radiation modes are adjusted with the use of a super-high frequency radiation power controller and a radiation device made on the basis of a magnetron or a klystron modulating oscillations with the use of a nanosecond impulse generator.

The radiating horn of the electronic device is incorporated into the fodder disinfection chamber to which bulk fodder to be disinfected is supplied by a metallic screw conveyor. In the inlet bin for fodder to be disinfected its physical characteristics are measured with the use of temperature and relative humidity meters. To the outlet bin bulk fodder is fed which is disinfected considering a specified production efficiency criterion, either technological (for example, for fodder to be immediately fed tolivestock), or economical (for fodder to be stored).

The process of the best possible fodder disinfection for production or economic purposes is carried out in automatic mode.

REFERENCES

Abhijit, S., Sambit, D., & Pooja, S. (2017). Ozone Pollution, Agriculture, and Food Security. In Environmental Issues Surrounding Human Overpopulation. IGI Global.

Budnikov, D. A., & Vasiliev, A. N. (2018). The Use of Microwave Energy at Thermal Treatment of Grain Crops. In Handbook of Research on Renewable Energy and Electric Resources for Sustainable Rural Development. Hershey, PA: IGI Global.

Dubrovin, A. V. (2014). Osnovy avtomatizirovannogo upravleniya tekhnologicheskimi protsessami v ptitsevodstve po ekonomicheskomu kriteriyu [Basic principles of automated control of technological processes in poultry breeding by economic criterion]. Moscow: FGBNU VIESH. (in Russian)

Granella, S., Christ, D., Werncke, I., Bechlin, T. R., & Machado Coelho, S. R. (2018). Effect of drying and ozonation process on naturally contaminated wheat seeds. *Journal of Cereal Science*, *80*, 205–211. doi:10.1016/j.jcs.2018.03.003

Kharchenko, V., & Vasant, P. (2018). *Handbook of Research on Renewable Energy and Electric Resources for Sustainable Rural Development*. IGI Global. doi:10.4018/978-1-5225-3867-7

Mason, L. J., Woloshuk, C. P., Mendoza, F., Maier, D. E., & Kells, S. A. (2014). Ozone: A new control strategy for stored grain. *9th International Working Conference on Stored Product Protection*, 904-907.

Mohapatra & Kar. (2014). Effect of Microwave Aided Disinfestation of Callosobruchus Maculatus on Green Gram Quality. *International Journal of Agriculture and Food Science Technology*, *5*(2), 55–62.

Pershin, A.F., & Smirnov A.A. (2014). Opredeleniye optimalnoi dozy ozona pri obrabotke zernovykh materialov [Determination of optimal ozone dose in processing grain materials]. *Tekhnika v sel'skom khozyaistve* [Equipment in agriculture], *6,* 9-11. (in Russian)

RF Patent of invention N°2533585. (2014). *Device for economic and energy-saving disinfection of fodder and animal and poultry products* / Dubrovin A.V. // Issued 20.11.2014, Bulletin N° 32. (in Russian)

RF Patent of invention N°2608532. (2017). *Method and device for specified technologically and economically optimal ozonation of moving bulk fodder for livestock and poultry breeding* / Dubrovin A.V. // Issued 19.01.2017, Bulletin N°2. (in Russian)

RF Patent of invention N°2613453. (2017). *Method and device of determination of economically optimal excessive air pressure for control of harmful microorganisms in ambient air of livestock and poultry houses* / Dubrovin A.V. // Issued 16.03.2017, Bulletin N°8. (in Russian)

RF Patent of invention N°2621140. (2017). *Method and device for specified technologically and economically optimal combined infrared and conductive drying of moving bulk fodder for livestock and poultry breeding*/ Dubrovin A.V. // Issued 31.05.2017. Bulletin N°16. (in Russian)

RF Patent of invention N°2624199. (2017). *Method and device of technologically and economically optimal super-high frequency drying of bulk fodder for livestock and poultry breeding* / Dubrovin A.V. // Issued 03.07.2017, Bulletin N° 19. (in Russian)

Shuvalov A.M., Mashkov A.N., Chernov D.S., Kalinin V.F., & Nabatov K.A. (2014). Raschet pokazateley protsessa nagreva zerna soi [Calculation of process performance of soybeans heating]. *Tekhnika v sel'skom khozyaistve* [Equipment in agriculture], *6,* 7-9. (in Russian)

Syrovatka, V.I. (2014). Sovershenstvovaniyetekhnologicheskikhprotsessovproizvodstvakombikormov v khozyaistvakh [Improvement of processes of combined fodder (mixed feed) production at farms]. *Vestnik VNIIMZH, 1*(13), 4-11. (in Russian)

Vasant, P., Alparslan-Gok, S. Z., & Weber, G.-W. (2018). *Handbook of Research on Emergent Applications of Optimization Algorithms*. IGI Global. doi:10.4018/978-1-5225-2990-3

Vasant, P., & Voropai, N. (2016). *Sustaining Power Resources through Energy Optimization and Engineering*. IGI Global. doi:10.4018/978-1-4666-9755-3

Vasiliev, A.N., Ospanov, A.B., Budnikov, D.K., Karmanov, D.K, Salginbayev, D.B., & Vasilyev, A.A. (2016). Controlling reactions of biological objects of agricultural production with the use of electrotechnology. *International Journal of Pharmacy & Technology, 8*(4), 26855-26869.

Vasilyev, A. N. (2013). Elektrotechnologies as controlling influence on biological objects in agricultural production. *Research in Agricultural Electric Engineering*, *3*(3), 96–102.

Vendin, S.V., Bulavin, S.A., & Saenko, U.V. (2015). Opredeleniye parametrov sushilki proroschennogo zerna [Determination of parameters of the germinated grain dryer]. *Mekhanizatsiya i elektrifikatsiya selskogo khozyaistva* [Mechanization and electrification of agriculture], *1,* 8-10. (in Russian)

Zhiltsov, V. I. (1979). *Issldovaniye rezhimov raboty elektrifitsirovannykh ustanovok dlya obespecheniya staticheskogo davleniya vozdukha v ptichnike* [Reseach of operation modes of electrified units providing static air pressure in poultry houses] (Unpublished PhD dissertation). Moscow: VIESH. (in Russian)

Compilation of References

"OZUF" na ob'yektakh veterinarnogo nadzora. (n.d.). *Trudy mezhdunarodnoy nauchno-tekhnicheskoy konferentsii Energoobespecheniye i energosberezheniye v sel'skom khozyaystve, 3*, 247-249.

Abhijit, S., Sambit, D., & Pooja, S. (2017). Ozone Pollution, Agriculture, and Food Security. In Environmental Issues Surrounding Human Overpopulation. IGI Global.

Ade-Omowaye, B. I. O., Angersbach, A., Taiwo, K. A., & Knorr, D. (2001). The use of pulsed electric field in producing juice from paprika (*Capsicum Annuum L.*). *Journal of Food Processing and Preservation, 25*(5), 353–365. doi:10.1111/j.1745-4549.2001.tb00466.x

Ade-Omowaye, B. I. O., Angersbach, A., Taiwo, K. A., & Knorr, D. (2001). Use of pulsed electric field treatment to improve dehydration characteristics of plant based foods. *Trends in Food Science & Technology, 12*(8), 285–295. doi:10.1016/S0924-2244(01)00095-4

Adler, Y. P., Markova, E. V., & Granovskiy, Y. V. (1976). Planirovaniye eksperimenta pri poiske optimal'nykh usloviy [Planning an experiment when searching for optimal conditions]. Moskow: Nauka.

Aguiló-Aguayo, I., Oms-oliu, G., Soliva-fortuny, R., & Martin-Belloso, O. (2009). Flavour retention and related enzyme activities during storage of strawberry juices processed by high-intensity pulsed electric fields or heat. *Food Chemistry, 116*(1), 59–65. doi:10.1016/j.foodchem.2009.02.007

Aldibekov, I.T. (2010). *Vybor i primenenie ehnergosberegayushchih sistem teploobespecheniya molochnyh ferm i fermerskih hozyajstv na baze ehlektrodnyh teplovyh ustanovok (rekomendacii)* [Selection and application of energy-saving heat supply systems for dairy farms and farms based on electrode heat installations (recommendations)]. Almaty, Kazakhstan: Inzhu-Marzhan.

Alferova, L.K., & Yuferev, L.Y. (2002). Bakteritsidnaya ustanovka povyshennoy effektivnosti. *Tekhnika v sel'skom khozyaystve, 2*, 29-31.

Alferova, L.K.,, Kozlov, A.I., & Yuferev, L.Y. (2000). Ul'trafioletovyy obluchatel'-ozonator. *Sel'skiy mekhanizator, 6*, 35-36.

Algirdas, R. (2006). Application of ozone for reduction of mycological infection in wheat grain. *Annals of Agricultural and Environmental Medicine, 13*(2), 287–294. PMID:17196003

Amaya, J. (1996). Effects of stationary magnetic fields on germination and growth of seeds. *Hortic. Abst, 68*, 1363.

Angersbach, A., Heinz, V., & Knorr, D. (2000). Effects of pulsed electric fields on cell membranes in real food systems. *Innovative Food Science & Emerging Technologies, 1*(2), 135–149. doi:10.1016/S1466-8564(00)00010-2

Anisimova, L. V. (1998). *The Russian patent N° 2121782*. Moscow, Russia: Federal Institute of Industrial Property.

Aniskin, V. I., & Chizhikov, A.G. (2002). Tekhnologicheskie priemy i tekhnicheskoe obespechenie sushki semyan i zerna v Rossii [Technological receptions and technical support of drying of seeds and grain in Russia]. *1st International Scientific and Practical Conference "Modern Energy-Saving Thermal Technologies (Drying and Thermal Handling of Materials)" (SETT-2002): Proceedings of the Conference, 4*, 7-12.

Aniskin, V. I., Okun', G. S., & Vercman, I. I. (1988). Modelirovanie processa nizkotemperaturnoj sushki zerna v sloe s pomoshch'yu EHVM [Modeling the process of low-temperature drying of grain in a layer with the help of a computer]. Collection of scientific papers VIM, 115, 25-29.

Aniskin, V. I. (1972). *Teoriya i tekhnologiya sushki i vremennoj konservacii zerna aktivnym ventilirovaniem* [Theory and technology of drying and temporary preservation of grain by active ventilation]. Moscow, Russia: Kolos.

Antonov, G.M., Armianov, N.K., & Dragiev, D.I. (1984). Elektroiskrovite razrjadi tjutjuno proshhvodstvo [Electric spark discharge in tobacco industry]. *Mekhanizaciya na sel'skogo stopanstvo* [Mechanization in agriculture], *12*, 48-51.

Antonov, N. M. (1978). *Issledovanie nagruzhennosti transmissii traktora Kirovets na pahote* [Investigation of the loading of the transmission of the tractor "Kirovets" on the plowing] (PhD dissertation). Saint Petersburg, Russia: SPBGAU

Antropov, L. I. (1984). Teoreticheskaya elektrokhimiya [Theoretical Electrochemistry]. Moskow: Vysshaya shkola.

Avdeev, A. V., Nachinov, D. S., & Nachinova, A. A. (2009). Modelirovanie processa i rezhimov konvektivnoj sushki zerna [Modeling of the process and regimes of convective grain drying]. Vestnik Orlovskogo gosudarstvennogo agrarnogo universiteta, 16(1), 39-43.

Avercheva, O., Berkovich, Y. A., Smolyanina, S., Bassarskaya, E., Pogosyan, S., Ptushenko, V., ... Zhigalova, T. (2014). Biochemical, photosynthetic and productive parameters of Chinese cabbage grown under blue-red LED assembly designed for space agriculture. *Advances in Space Research*, *53*(11), 1574–1581. doi:10.1016/j.asr.2014.03.003

Babanin, N. V. (2015). Uluchshenie plavnosti hoda mashinno-traktoronogo agregata na baze traktora klassa 1.4 s uprugodempfiruyushchim privodom na vedushchih kolesah [Improvement of the smoothness of the machine and tractor unit on the basis of a class 1.4 tractor with an elastic damping mechanism on driving wheels]. *Current Problems of Science and Education*, *22*, 46.

Babanin, N. V., & Polivaev, O. I. (2015). Eksperimentalnie issledovaniya na plavnost hoda, proizvoditelnost i toplivnuyu ekonomichnost mashinno-traktornogo agregata na baze traktora klassa 1.4, oborudovannogo gazogidravlicheskim uprugodempfiruyushchim privodom [Experimental researches on smoothness of a course, productivity and fuel economy of the machine-tractor unit on the basis of a class 1.4 tractor, equipped with a gas-hydraulic elastic-damping drive]. *Bulletin of the Voronezh State Agrarian University*, *3*(46), 112–118.

Baev, V. I., & Borodin, I. F. (2002). E`lektroimpul`snaya preduborochnaya obrabotka podsolnechnika i tabaka: monografiya [The elektroimpulse preharvest processing of sunflower and tobacco plants: the monography]. Volgograd: Page-2.

Baev, V. I., Savchuk, V. N., & Armianov, N. K. (1985). Eksperimental'nye issledovaniya parametrov elektroimpul'snoy obrabotki tabaka [Experimental research of parameters of tobacco electric spark treatment]. Sbornik trudov VSKHI [Collected Works of VSHI], 92, 44-45.

Baev, V. I., Yudaev, I. V., Petrukhin, V. A., Baev, I. V., Prokofyev, P. V., & Armianov, N. K. (2018). Electrotechnology as One of the Most Advanced Branches in the Agricultural Production Development. In Handbook of Research on Renewable Energy and Electric Resources for Sustainable Rural Development. Hershey, PA: IGI Global. doi:10.4018/978-1-5225-3867-7.ch007

Baev, V.I., & Yudaev, I.V. (2004). Chuvstvitel'nost' tkani sornykh rasteniy k elektroimpul'snomu vozdeystviyu [Sensitivity of weed plants tissue to electrical impulse action]. *Mehanizacija i elektrifikacija sel'skogo hozjajstva* [Mechanization and electrification of agriculture], *3,* 5-8.

Bakhmutov, V.A. (1980). Razmeshcheniye semyan po ploshchadi pri ryadovykh posevakh [seeds placement on the area at ordinary crops]. *Mekhanizatsiya i elektrifikatsiya sel khozyaistva, 5,* 24-27.

Bandi, B. (1988) *Metody optimizacii. Vvodnij kurs* [Methods of optimization. Introductory course]. Radio i svyaz'. (in Russian)

Band, S. (1901). *Die Eirkungen des Eichtes Sikreorganiscen.* Milt, Finsen.

Banerjee, R., & Batschauer, A. (2005). Plant blue-light receptors. *Planta, 20*(3), 498–502. doi:10.100700425-004-1418-z PMID:15714356

Baranov, L.A. & Zakharov, V.A. (2006) *Svetotechnika i electrotechniologia* [Lighting engineering and electrotechnology]. Kolos. (in Russian)

Barkov, V. I. (2006). *Vybor i postroenie sistem ehlektroteploobespecheniya v zhivotnovodstve* [Selection and construction of electric heating systems in animal husbandry] (Doctoral dissertation). Almaty, Kazakhstan.

Basiuk, O.I. (2006). Pro efektyvnist zastosuvannia mahnitnoi obrobky vody [On the effectiveness of the application of magnetic water treatment]. *Nova tema, 2,* 36.

Batalin, Y. V., & Svidzinsky, S. A. (2001). Perspektivy osvoeniya kalijnyh mestorozhdenij Volgogradskoj oblasti i osnovnye problemy razrabotki brom-magnievogo syr'ya – bishofita [Prospects for the development of potash deposits in the Volgograd region and the main problems of developing bromine-magnesium raw materials – bischofite]. *Osvoenie i ispol'zovanie prirodnyh resursov Volgogradskoj oblasti – put' ustojchivogo razvitiya regiona: materialy nauchno-prakticheskoj konferencii 15-16 dekabrya 2000* [Development and use of natural resources of the Volgograd region - the way of sustainable development of the region: materials of the scientific and practical conference. December 15-16, 2000], 63-64.

Bauchot, A. D., Harker, F. R., & Arnold, W. M. (2000). The use of electrical impedance spectroscopy to assess the physiological condition of kiwifruit. *Postharvest Biology and Technology, 18*(1), 9–18. doi:10.1016/S0925-5214(99)00056-3

Belitskaya, M. N., & Gribust, I. R. (2009). Stabilizaciya i produktivnost' zernovyh agrocenozov pri ispol'zovanii nanostrukturirovannogo rastvora bishofita [Stabilization and Productivity of Grain Agrocenoses with the Use of a Nanostructured Bischofite Solution]. *Materialy 2-oj Vserossijskoj nauchno-tekhnicheskoj konferencii «Nanotekhnologii i nanomaterialy: sovremennoe sostoyanie i perspektivy razvitiya v usloviyah Volgogradskoj oblasti»* [Proceedings of the 2nd All-Russian Scientific and Technical Conference «Nanotechnologies and Nanomaterials: Current State and Prospects of Development in Volgograd Region»], 53-58.

Belitskaya, M.N., & Gribust, I.R. (2012). Razrabotka tekhnologii primeneniya ehlektroliticheski nanostrukturirovannogo rastvora na osnove bishofita [Development of technology for the application of electrolytically nanostructured solution based on bischofite]. *Ehkologiya Rossii: na puti k innovaciyam* [Ecology of Russia: on the way to innovation], *6,* 192-194.

Belitskaya, M. N., & Kryukova, E. A. (2010). Bishofit pomogaet zashchitit' posevy [Bishofit helps protect crops]. *Zemledelie [Agriculture], 5,* 41–43.

Beno, Iu. I., Dyka, M. V., Skvarko, K. O. (2013). Doslidzhennia vplyvu postiinoho mahnitnoho polia na prorostannia nasinnia Physostegia virginiana Alba (L.) benth [The influence of constant magnetic field on seed germination of Physostegia virginiana Alba (L.) benth]. *Biolohichni Studii / Studia Biologica, 7*(3), 107-114.

Benz, R., & Zimmermann, U. (1980). Relaxation studies on cell membranes and lipid bilayers in the high electric field range. *Bioelectrochemistry and Bioenergetics*, *7*(4), 723–739. doi:10.1016/0302-4598(80)80037-7

Beraa, T. K., Berab, S., Karc, K., & Mondalc, S. (2016). Studying the Variations of Complex Electrical Bio-Impedance of Plant Tissues During Boiling. *Procedia Technology*, *23*, 248–255. doi:10.1016/j.protcy.2016.03.024

Berkovich, Yu. A., Krivobok, N. M., Smolyanina, S. O., & Erokhin, A. N. (2005). Kosmicheskie oranzherei: nastoyashchee i budushchee [Space greenhouses: present and future]. Moscow: Slovo [Word].

Bespamyatnova, N. M. (2002). Nauchno-metodicheskiye osnovy adaptatsii pochvoobrabatyvayushchikh i posevnykh mashin [Scientific and methodical bases of adaptation of tillage and sowing machines]. Rostov-on-Don, Russia; NPK.

Bespamyatnova, N. M. (2010). Vibratsii v tekhnologicheskikh protsessakh [Vibrations in technological processes]. Zernograd: Russia, VNIPTIMESKh.

Bespamyatnova, N.M. (2002). *Nauchno-metodicheskie osnovi adaptatsii pochvoobrabativayushchih mashin* [Scientific and methodological fundamentals of adaptation of soil cultivating and sowing machines]. Rostov-on-Don, Russia: LLC "Terra".

Boltinskiy, V. N. (1949). *Rabota dvigatelya tractora pod nagruzkoy* [The work of the tractor engine under unsteady load] Moscow, Russia: OGIZ-SELHOZGIZ.

Boltinsky, V. N. (1959). Moshcnost traktornogo dvigatelya pri rabote s neustanovivsheysya nagruzkoy i eye opredelenie [Tractor engine power when working with unsteady load and its determination]. *Mechanization of Socialist Agriculture*, *4*, 13–16.

Bondarenko, Ya. N. (2001). Material'no-syr'evaya baza Volgogradskoj oblasti kak osnova ustojchivogo razvitiya regiona [Material and raw materials base of the Volgograd region as a basis for sustainable development of the region]. Osvoenie i ispol'zovanie prirodnyh resursov Volgogradskoj oblasti – put' ustojchivogo razvitiya regiona [Development and use of natural resources of the Volgograd region - the way of sustainable development of the region], 11-21.

Borodin, I. F. (2005). Nanoehlektrotekhnologii v sel'skom hozyajstve [Nanoelectrotechnologies in agriculture]. Mekhanizaciya i ehlektrifikaciya sel'skogo hozyajstva, 10, 2–5. (in Russian)

Bourgin, M., Violleau, F., Debrauwer, L., & Albet, J. (2011). Ozonation of imidacloprid in aqueous solutions: Reaction monitoring and identification of degradation products. *Journal of Hazardous Materials*, *190*(1-3), 60–68. doi:10.1016/j.jhazmat.2011.02.065 PMID:21524846

Bouzrara, H., & Vorobiev, E. (2000). Beet juice extraction by pressing and pulsed electric fields. *International Sugar Journal*, *102*(1216), 194–200.

Brandt, A. B., & Taheeva, S. V. (1967). *Optical parameters of plant organisms*. Moscow: Science.

Budnikov, D. A. (2015). Issledovanie raspredelenija naprjazhennosti SVCh polja v zernovom sloe [Investigation of the distribution of the microwave field intensity in the grain layer]. *Inzhenernyj vestnik Dona*, 3. Retrieved from http://www.ivdon.ru/ru/magazine/archive/n3y2015/3234

Budnikov, D. A., & Vasiliev, A. N. (2018). The Use of Microwave Energy at Thermal Treatment of Grain Crops. In Handbook of Research on Renewable Energy and Electric Resources for Sustainable Rural Development. Hershey, PA: IGI Global.

Budnikov, D. A., & Vasiliev, D. A. (2018). The Use of Microwave Energy at Thermal Treatment of Grain Crops. In Handbook of Research on Renewable Energy and Electric Resources for Sustainable Rural Development. Hershey, PA: IGI Global. Doi:10.4018/978-1-5225-3867-7.ch020

Bychkov, N. I., & Yakulin, V. G. (2003). Vibor tractor dlya vspashki [Choice of a tractor for plowing]. *Mechanization and Electrification of Agriculture, 3*, 11–12.

Bystrickij, D.N., Kozhevnikova, N.F., Lyamtsov, A. K. & Murugov, V. P. (1981). *EHlektricheskie ustanovki infrakrasnoro izlucheniya v zhivotnovodstve* [Electrical installations of infrared radiation in animal husbandry]. Moscow, Russia: EHnergoizdat.

Canals, L. M., Munoz, I., Hospido, A., Plassmann, K., & McLaren, S. (2008). *Life Cycle Assessment (LCA) of Domestic vs. Imported Vegetables. Case studies on broccoli, salad crops and green beans*. Centre for Environmental Strategy.

Chang, C. S., Converse, H. H., & Steele, J. L. (1993). Modeling of temperature of grain during storage with aeration Trans. ASAE. -St.Joseph(Mich.), 36(2), 509-519.

Chaplin, M. F. (2007). The memory of water: An overview. *Homeopathy, 96*(3), 143–150. doi:10.1016/j.homp.2007.05.006 PMID:17678809

Charova, D.I., & Yudaev, I.V. (2014). Povyshenie urozhajnosti ovoshchnyh kul'tur za schet ispol'zovaniya SID dlya ehlektrodosvechivaniya rastenij v sooruzheniyah zashchishchennogo grunta [Increasing the yield of vegetable crops through the use of LEDs for electric lighting of plants in protected ground facilities]. In *Nauchnye osnovy strategii razvitiya APK i sel'skih territorij v usloviyah VTO. Materialy mezhdunarodnoj nauchno-prakticheskoj konferencii, posvyashchennoj 70-letiyu obrazovaniya VolGAU* [Scientific basis of the strategy of development of agriculture and rural areas in the WTO. Materials of the international scientific and practical conference dedicated to the 70th anniversary of the VolSAU]. Volgograd: VolGAU.

Charova, D.I., Yudaev, I.V., & Feklistov, A.S. (2016). Issledovaniya po vyrashchivaniyu listovogo salata v svetodiodnoj obluchatel'noj kamere [Studies on the cultivation of lettuce in the led irradiation chamber]. In *Materialy mezhdunarodnoj nauchno-prakticheskoj konferencii, posvyashchennoj pamyati vedushchego ehlektrotekhnologa Rossii akademika Ivana Fyodorovicha Borodina* [Materials of international scientific-practical conference dedicated to the memory of leading Electrotechnology Russia academician Ivan Borodin]. Volgograd: VolGAU.

Charova, D.I., Yudaev, I.V., Feklistov, A.S., Vorotnikov, I. N., & Gabrielyan, S. Z. (2017). Vyrashchivanie listovogo salata v svetodiodnoj obluchatel'noj kamere [Growing lettuce in the led irradiation chamber]. *Sel'skij mekhanizator* [Rural mechanic], *1*, 20-21.

Chernovolov, V. A. (2007). Proyektirovaniye rabochego protsessa i modernizatsiya sborochnoy edinitsy mashiny dlya vneseniya mineralnykh udobreniy [Design of the working process and modernization of the Assembly unit of the machine for the application of mineral fertilizers]. In Uchebnoye posobiye po diplomnomu i kursovomu proyektirovaniyu. Zernograd, Russia: Academic Press.

Chervyakov, D. M. (1979). Mexanizm vozdejstviya e'lektricheskix razryadov na rasteniya [Mechanism of influence of electric discharges on plants]. *Elektronnaya Obrabotka Materialov, 5*, 70–71.

Chizhikov, A. G., & Golubkovich, A. V. (2005). *Tekhnologicheskie i tekhnicheskie priemy ehnergosberezheniya pri sushke zerna v sel'skom hozyajstve* [Technological and technical methods of energy saving during grain drying in agriculture]. The second international scientific and practical conference "Modern energy-saving thermal technologies (drying and thermal processes) SETT-2005", Moscow, Russia.

Cho, Y. S., Song, K. B., & Yamda, K. (2010). Effect of Ultraviolet Irradiation on Molecular Properties and Immunoglobulin Production-regulating Activity of β-Lactoglobulin. *Food Science and Biotechnology, 19*(3), 595–602. doi:10.100710068-010-0084-0

Coruzzi, G., & Bush, D. (2001). Nitrogen and Carbon Nutrient and Metabolite Signaling in Plants. *Plant Physiology, 125*(1), 61–64. doi:10.1104/pp.125.1.61 PMID:11154297

D'yakonov, V. P. (2004). *Matlab 6.0/6.1/6.5/6.5 + SP1 + Simulink 4/5. Obrabotka signalov i izobrazhenij*. Moscow: Solon-Press.

De Freitas. (2014). Development of a solid-liquid extraction method with low-temperature partitioning for the determination of insecticides in ozonized maize grain. *Quimica Nova, 37*(2), 238–243.

De Freitas. (2017). Degradation kinetics of pirimiphos-methyl residues in maize grains exposed to ozone gas. Journal of Stored Products Research, 1-5.

De Vita, P., Colecchia, S. A., Pecorella, I., & Saia, S. (2017). Reduced inter-row distance improves yield and competition against weeds in a semi-dwarf durum wheat variety. *European Journal of Agronomy, 85*, 69–77. doi:10.1016/j eja.2017.02.003

Deboli, R., Calvo, A., & Preti, C. (2017). Whole-body vibration: Measurement of horizontal and vertical transmissibility of an agricultural tractor seat. *International Journal of Industrial Ergonomics, 58*, 69–78. doi:10.1016/j.ergon.2017.02.002

Delyagin, V.N. (2004). Obosnovanie racional'nyh temperaturno-vlazhnostnyh rezhimov zhivotnovodcheskih pomeshchenij [Substantiation for rational temperature and humidity conditions of livestock buildings]. *EHnergoobespechenie i ehnergosberezhenie v sel'skom hozyajstve, 3*, 250-255.

Denisov, V. P. (2005). *Optimizatsiya rabochego protsessa zemleroyno-transportnyh mashin s uchetom sluchaynogo haraktera nagruzok* [Optimization of the working process of earth-moving machines taking into account the random nature of the loads] (PhD dissertation). SibADI, Omsk, Russia.

Derevyagin, B.C., Sedlechkiy, V.N., & Ermakov, V.A. (1989). *Bishofity Nizhnego Povolzh'ya* [Bishofits of the Lower Volga Region]. Rostov: Izd-vo Rostovskogo universiteta [Publishing house of the Rostov University].

Derevyagin, V. S., Makarov, A. S., & Kovalsky, F. I. (1982). Geologiya i kalienosnost' nizhnepermskoj galogennoj formacii Severnogo Prikaspiya [Geology and Kalianosity of the Lower Permian Halogen Formation of the Northern Caspian Region]. Stroenie i uslovie obrazovaniya solenosny`x formacij [The structure and conditions of formation of saline formations], 98-109.

Dimitrov, N. (2014). *Effect of the storage duration and moisture on the temperature pattern of respiration of soft wheat grain*. University of Food Technology.

Diprose, M. F., & Benson, F. A. (1984). Electrical methods of killing plants. *Journal of Agricultural Engineering Research, 30*, 197–209. doi:10.1016/S0021-8634(84)80021-9

Dospekhov, B. A. (1985). Metodika polevogo opyta [Technique of field experience]. Moskow: Agropromizdat.

Duan, T., Chapman, S. C., Guo, Y., & Zheng, B. (2017). Dynamic monitoring of NDVI in wheat agronomy and breeding trials using an unmanned aerial vehicle. *Field Crops Research, 210*, 71–80. doi:10.1016/j.fcr.2017.05.025

Dubrovin, A. V. (2014). Osnovy avtomatizirovannogo upravleniya tekhnologicheskimi protsessami v ptitsevodstve po ekonomicheskomu kriteriyu [Basic principles of automated control of technological processes in poultry breeding by economic criterion]. Moscow: FGBNU VIESH. (in Russian)

Duke, M., & Goss, G. (2007). Investigation of tractor driver seat performance with non-linear stiffness and on-off damper. *Biosystems Engineering, 94*(4), 477–486. doi:10.1016/j.biosystemseng.2007.01.005

Dyakonov, V. P. (2008). *SIMULINK 5/6/7. Samouchitel* [SIMULINK 5/6/7. Tutorial]. Moscow, Russia: DMK-Press.

Dzakovich, M. P., Gómez, C., & Mitchell, C. A. (2015). Tomatoes Grown with Light-emitting Diodes or High-pressure Sodium Supplemental Lights have Similar Fruit-quality Attributes. *HortScience*, *50*(10), 1498–1502.

Egorov, G. A. (1973). *Vliyanie tepla i vlagi na processy hraneniya i pererabotki zerna* [Influence of heat and moisture on the processes of storage and processing of grain]. Moscow, Russia: Kolos.

Eikevik, T. M. (2012). Microwave-Assisted Atmospheric Freeze Drying of Green Peas: A Case Study, Drying Technology. *International Journal (Toronto, Ont.)*, *30*(14), 1592–1599.

Electric, R. C. T. (2016). Utility Patent No. 2576209 registered in the State Register of Inventions of the Russian Federation 02.02.2016.

Ermakov, V. A., & Grebennikov, N. P. (1985). Zakonomernosti bishofitovyh zalezhej solenosnoj tol [Regularities of bischofite deposits of saline tol]. In Problemy solenakopleniya [Problems of salt accumulation]. Moscow: Nauka [Science].

Ermokhyn, Yu. Y., & Krasnytskyi, V. M. (2009). *Modern approaches to the diagnosis of nitrogen nutrition and the use of nitrogen fertilizers: methodical recommendations*. Omsk: Variant-Omsk.

Eshtiaghi, M. N., & Knorr, D. (2002). High electric field pulse treatment: Potential for sugar beet processing. *Journal of Food Engineering*, *53*(2), 265–272. doi:10.1016/S0260-8774(01)00114-5

Eviev, V. N., Jabborov, N. I., & Ochirov, N. G. (2014). Povishenie effektivnosti gusenichnyh traktorov klassa 4, osnashchennyh dvigatelyami postoyannoy moshchnosty [Increase of efficiency of caterpillar tractors of class 4, equipped with constant power engines]. *Technology of Wheeled and Caterpillar Vehicles: Research Institute of Automotive Design*, *6*, 51–56.

Ewbank, N., & Buckby, L. (2002). Lynx engineering enters seed drill market with the seedline system. *Landwards, 82*(1), 153-155.

Fernandes, F. A. N., Rodrigues, S., García-Pérez, J. V., & Cárcel, J. A. (2016). Effects of UltrasoundAssisted Air-Drying on Vitamins and Carotenoids of Cherry Tomatoes. *Drying Technology*, *34*(8), 986–996. doi:10.1080/07373937.2015.1090445

Feuerhuber, K., Lindert, S. O., & Schlacher, K. (2013). *Vibration attenuation by semi-active dampers.* Paper presented at 14th International Conference on Computer Aided Systems Theory (EUROCAST), Las Palmas, Spain.

Feynman, R., Leighton, R., & Sands, M. (1977) Feinmanovskie lekcii po fizike v 8 t [Lectures on physics by R. Faiyman: in 8 vol.], Vol. 5. Elektrichestvo i magnetism [Electricity and magnetism]. Mir. (in Russian)

Filatov, A. S., & Paulys, V. Y. U. (2017) Ustrojstva aktivnogo ventilirovaniya zernovyh mass pri hranenii [Devices of active ventilation of cereal masses during storage]. In The *collection of topical issues of science and economy: new challenges and solutions Collection of materials of the LI International Student Scientific and Practical Conference* (pp. 204-206). Academic Press.

Filin, V. A., Pronin, V. Y., Kryuchkov, S. V., & Filin, A. V. (2016). Obosnovanie stepeni zagruzki traktornogo dvigatelya pri vypolnenii selskokhozyastvennykh rabot [Substantiation of the degree of loading of the tractor engine in the performance of agricultural works]. In Proceedings of Energy-Efficient and Resource-Saving Technologies and Systems. Saransk, Russia: FSBEU of Higher Professional Education of the Moscow State University named after N.P. Ogaryova.

Fincan, M., DeVito, F., & Dejmek, P. (2004). Pulsed electric field treatment for solid-liquid extraction of red beetroot pigment. *Journal of Food Engineering*, *64*(3), 381–388. doi:10.1016/j.jfoodeng.2003.11.006

Flóreza, M., Carbonella, V., & Martínez, E. (2007). *Exposure of maize seeds to stationary magnetic fields: Effects on germination and early growth.* Academic Press.

Fomichev, V.T., & Filimonova, N.A. (2010). Poluchenie fungicidov na osnove minerala bishofita [Preparation of fungicides based on bischofite mineral]. *Internet-Vestnik VolgGASU* [Internet-Bulletin of VolgGASA], *1*, 4-8.

Fomichev, V.T., Drevin, V.E., Filimonova, N.A., Kulikova, I.A., Kamkova, S.V., & Yudaev I.V. (2016). *Perspektivy osvoeniya i tekhnologiya fiziko-himicheskoj aktivacii mineral'nogo syr'ya Volgogradskih mestorozhdenij* [Perspectives of development and technology of physical and chemical activation of mineral raw materials of Volgograd deposits]. Volgograd: VolGAU.

Fomichev, V.T., Naumova, G.A., & Akhremenko, O.Y. (2013). Innovacionnaya tekhnologiya povysheniya biologicheskoj aktivnosti rastvora minerala bishofita [Innovative technology of increasing the biological activity of a solution of bischofite mineral]. *Vestnik VolGU. Seriya 10. Innovacionnaya deyatel'nost'* [Bulletin of the VolSU. Series 10. Innovative activity], *1*(8), 131-138.

Fomichev, V.T., Yankovskaya, M.B., & Abyzov, V.V. (2014). Izuchenie rezul'tativnosti dejstviya kompleksnogo preparata «Bilator-6» na porazhaemost' gribnymi zabolevaniyami rastenij zemlyaniki sadovoj i barbarisa obyknovennogo [The study of the effectiveness of the complex preparation «Bilator-6» on the damage caused by fungal diseases of strawberry and barberry plants of the common]. *Vestnik VolGU. Seriya 11. Estestvennye nauki* [Bulletin of the VolSU. Series 11. Natural Sciences], *2*(8), 36-42.

Fritz, N., & Elsner, W. (2002). *Non linear numerical torsional vibration analysis of the JD 6920 tractor drive train.* Paper presented at Conference on Agricultural Engineering, Halle, Germany.

Gabdrafikov, F. Z., & Safina, R. R. (2017). Povyshenie tehniko-ekonomicheskih pokazateley traktornyh dizeley na chastichnyh i neustancvivshihsya rezhimah raboty [Increase of technical and economic indicators of tractor diesel engines on partial and unsteady operating modes]. *Russian Electronic Scientific Journal*, *2*(24), 95–109.

Gallyamova, T. R., & Shirobokova, T. A. (2016). *Svetodiodny svetil'nik dlya ravnomernogo osveschenia rabochei poverhnosti ptichnika* [A led lamp for the uniform illumination of the working plane in the poultry house]. Pat. 159705 RF, MPK F21S 13/00 Publ. 20.02.2016. Bulletin 5. (in Russian)

Gallyamova, T. R., Novosyolov, I. M., & Shirobokova, T. A. (2012) Opredelenie optimal'nogo svetoraspredelenia dlya obespechenia ravnomernosti osveschenia pticevodcheskih pomeschenij [Determination of optimal light spreading for providing uniformity of illumination of the poultry facilities]. *FGBOU VPO Izhevskaya gosudarstvennaya sel'skohozaistvennaya academia, 3,* 271-274. (in Russian)

Gallyamova, T. R., Shirobokova, T. A., & Kochetkov, N. P. (2015). Experimental'nye issledovania osveschyonnosty, sozdavaemoi opytnym obrazcom svetodiodnogo svetil'nika s lineikamy kruglosimmetrichnyh svetodiodov [Experimental studies of illumination created by an experimental sample of a led lamp with the lines of round symmetric leds]. In *Mezhdunarodnaya prakticheskaya konferencia "Teoria i praktika - ustoichivomu razvitiu agropromyshlennogo kompleksa"* [International practical conference "Sustainable development of agroindustrial complex: theory and practice]. FGBOU VPO Izhevskaya GSHA. (in Russian)

Gallyamova, T.R. & Shirobokova, T.A. (2015). Svidetel'stvo o gosudarstvennoi registratsii programmy dlya EVM 2015617923 [Certificate of registration 2015617923 for the computer programme]. In *Yevribrid svetodiodov v sisteme osveschenia ptichnika* [Led eurobreed in the system of the poultry house illumination]. Zayavitel' i patentoobladatel' FGBOU VPO Izhevskaya GSHA, 2013116823/07. Publ. 24.07.2015. (in Russian)

Gary, W. S. (2015). Commercial Transition to LEDs: A Pathway to High-value Products. *HortScience*, *50*(9), 1297–1300.

Gates, F. L. (1928). On nuclear derivatives and the lethal action of ultraviolet light. *Science*, (68): 259–275. PMID:17758687

Ginzburg, A.C. (1973). *Osnovy teorii i tekhniki sushki pishchevyh produktov* [Fundamentals of the theory and technology of food drying]. Moscow, Russia: Pishchevaya promyshlennost.

Gobelev, S.N. & Lyamcov A.K. (2006). Rezul'taty ispytanij infrakrasnyh lamp na srok sluzhby [The results of the tests of infrared lamp service life]. *Tekhnika v sel'skom hozyajstve, 6*, 39–41.

Goins, G. D., Yorio, N. C., Sanwo, M. M., & Brown, C. S. (1997). Photomorphogenesis, photosynthesis, and seed yield of wheat plants grown under red light-emitting diodes (LEDs) with and without supplemental blue lighting. *Journal of Experimental Botany*, *48*(7), 1407–1413. doi:10.1093/jxb/48.7.1407 PMID:11541074

Golovko, T. K., Tabalenkova, G. N., Dalke, I. V., Zahozhi, I. G., Gregori, E. E., & Botkin, A. V. (2010). Produkcionnyj process i pishchevaya cennost' zelennyh kul'tur zashchishchennogo grunta na severe [Production and nutritional value of leaf vegetables of the protected ground in the North]. *Gavrish*, *5*, 32–35.

Golovnev, N. B., Terent'ev, V. M., & Fedonkin, D. V. (1980). Ob ocenke ehffektivnosti izlucheniya dlya vyrashchivaniya rastenij [On the evaluation of radiation efficiency for growing plants]. *Svetotekhnika*, *5*, 16–17.

Goroshkov, Y. I., & Bondarev, N. A. (1981). Contact Network, Moscow. *Transport*, 397.

Gorskij, I. V. (2004). *Obrabotka semyan pshenicy ozonirovannym vozduhom [*Processing of wheat seeds with ozonized air]. Dissertation of the candidate of technical sciences, Russia, Moscow.

Goryachkin, V. P. (1965). *Zemledel'cheskaya mekhanika. Sobraniye sochineniy. Tom 2* [Agricultural mechanics. Collected Works: Volume 2]. Moscow, Russia: Kolos.

GOST 12.1.005-88. (1988). *Obshchie sanitarno-gigienicheskie trebovaniya k vozduhu rabochej zony* [General sanitary and hygienic requirements for the air of the working area]. Moscow: Izd-vo standartov.

GOST R 57099-2016. (2016). *Information and communication technologies in education. E-learning production safety. General provisions.* (in Russian)

GOST R 58099-2017. (2017). *Information and communication technologies in education. E-learning production safety. E-portfolio of employee production safety.* (in Russian)

Gracheva, N.N., Rudenko, N.B., Kononenko, A.F., & Litvinov, V.N. (2017). *Primenenie EHVM v agronomii. Laboratornyj praktikum* [The use of computers in agronomy. Laboratory Practice]. Zernograd, Russia: Donskoj gosudarstvennyj agrarnyj universitet.

Granella, S. J., Christ, D., Werncke, I., Bechlin, T. R., & Machado Coelho, S. R. (2018). Effect of drying and ozonation process on naturally contaminated wheat seeds. *Journal of Cereal Science*, *80*, 205–211. doi:10.1016/j.jcs.2018.03.003

Grebennikov, N. P., & Ermakov, V. A. (1982). Bishofit: problemy i perspektivy [Bishofit: problems and prospects] *Tekhnika i nauka. Engineering and Science*, *4*, 56–64.

Grebnev, V. P., Poliaev, O. I., & Vorokhobin, A. V. (2011). *Traktory i avtomobili: teoriya i ekspluatatsionniye svoystva: uchebnoye posobie dlya studentov vuzov, obuchaushchihsya po napravleniyu «Agroingeneriya»* [Tractors and automobiles: theory and operational properties: a teaching aid for university students studying in the direction of "Agroengineering"]. Voronezh, Russia: KnoRus.

Grigor'yeva, O. (2014). Sposoby podgotovki semyan k posevu [Methods for preparing of seeds for sowing]. *LesProm, 6*(104), 176-177.

Grishchenko, F. V. (1970). Teoreticheskoye i eksperimentalnoye obosnovaniye seyalok-kultivatorov [Theoretical and experimental rationale of grain drills-cultivators]. In Materialy Nauchno-tekhnicheskogo soveta. Vypusk 28. Sostoyaniye i perspektivy razvitiya seyalok-kultivatorov. Moscow, Russia: Academic Press.

Grishchenko, F. V. (1963). Osnovy teorii dvizheniya semyan i raspredeleniye ikh v pochve pri bezryadkovom poseve [Fundamentals of the theory of seeds movement and their distribution in the soil at sowing without rows]. *Ryazanskiy SKhI, 10*, 31–33.

Gritsenko, A. I., & Ovchinnikov, A. D. (1986). Nekotorye voprosy ustojchivosti bishofita v pristvol'noj zone skvazhiny [Some problems of bischofite stability in the near-well zone of a well]. Neftepromyslovaya geologiya i stroitel'stvo skvazhin v Prikaspijskom regione [Oilfield geology and well construction in the Caspian region], 75-84.

Gulyaev, G. A. (1990). *Avtomatizaciya processov posleuborochnoj obrabotki i hraneniya zerna* [Automation of post-harvest processing and storage of grain]. Moscow, Russia: Agropromizdat.

Gurhan, R., & Cay, C. I. (2008). Performances of tractor's driver seats with passive damping system under simulated test conditions. *Tarim Bilimleri Dergisi-Journal of Agricultural Sciences, 14*(4), 401–408.

Gurov, I. P. (2000). *Osnovy teorii informacii i peredachi signalov* [The fundamentals of information theory and signaling]. St. Petersburg, Russia: BHV-Peterburg.

Guttsait, E. M., Krasnopol'skij, A. E., & Milyutin, D. V. (2007). Paschyoty svetodiodnyh modulei dlya mestnogo osveschenia [Calculations of led modules for local illumination] [in Russian]. *Svetotehnika, 4*, 52–56.

Hammer, W.M. (n.d.). Diurnal rhythm and photoperiodism in testicular recrundes sense of the house finch. Poultry Science, 42(5), 1294–1295.

Hansson, P. A. (1994). *Gain scheduling based controller for active cab suspension on agricultural tractors.* Paper presented at 2nd IFAC/ISHS Workshop on Mathematical and Control Applications in Agriculture and Horticulture, Silsoe, England.

Hansson, P. A. (2002). Working space requirement for an agricultural tractor axle suspension. *Biosystems Engineering, 81*(1), 57–71. doi:10.1006/bioe.2001.0007

He, Z., Zhang, Y., Wang, Z., Zhao, Z., & Yi, S. (2016). Reducing Wood Drying time by Application of Ultrasound Pretreatment. *Drying Technology, 34*(10), 1141–1146. doi:10.1080/07373937.2015.1099107

Hilmi, D. (2010). *The Fundamental Properties of Information-Carrying Relations. In Thinking Machines and the Philosophy of Computer Science: Concepts and Principles.* Hershey, PA: IGI Global. doi:10.4018/978-1-61692-014-2.ch002

Hizhnyi Yu, A., Nedilko, S. G., Chornii, V. P., Slobodyanik, M. S., Zatovsky, I. V., & Terebilenko, K. V. (2014). Electronic structures and origin of intrinsic luminescence in Bi-containing oxide crystals BiPO4, K3Bi5(PO4)6, K2Bi(PO4)(MoO4), K2Bi(PO4)(WO4) and K5Bi(MoO4)4. *Journal of Alloys and Compounds, 614*, 420–435. doi:10.1016/j.jallcom.2014.06.111

Horodnii, M. M. (Ed.). (2006). Agrochemicals (4th ed.). View. LLC "Ariste".

Horodnii, M. M., Melnychuk, S. D., & Honchar, O. M. (2006). Applied biochemistry and quality management of crop production. Ariste.

Ho, S. Y., & Mittal, G. S. (1996). Electroporation of cell membranes: A review. *Critical Reviews in Biotechnology, 16*(4), 349–362. doi:10.3109/07388559609147426 PMID:8989868

Hospodarenko, H. M. (2016) Fertilizer of agricultural crops. Kiev: LLC "SIC GRUP UKRAINE".

Hyland, G. J. (2003). Bio-Electromagnetism. In Integrative Biophysics-Biophotonics. Kluwer Academic.

Ilyuhin V.V., Noskov B.E. & Babakin B.S. (1983). *Sposob sushki materialov* [Method of drying materials]. Author's certificate 1016643 USSR, MKI5 F 26 V 3/34.

Inaba, A., Manabe, T., Tsuji, H., & Iwamoto, T. (1995). Electrical Impedance Analysis of Tissue Properties Associated with Ethylene Induction by Electric Currents in Cucumber (*Cucumis sativus L.*). *Fruit. Plant Physiology*, *107*(1), 199–205. doi:10.1104/pp.107.1.199 PMID:12228354

Intellectual Robotized Agricultural System for Food Production. (2012). Utility Patent No. 2462853 registered in the State Register of Inventions of the Russian Federation 10.10.2012.

Iofinov, S. A. (1974). *Ekpluatatsiya mashinno-traktornyh agregatov* [Operation of machine-tractor units]. Moscow, Russia: Kolos.

Iofinov, S. A., & Mikuberg, B. K. (1971). Opredelenie ekspluatatsionnih parametrov i pokazateley raboty agregatov pri veroyatnostnom haraktere issleduemyh velichin [Determination of operational parameters and performance indicators of units for the probabilistic nature of the investigated quantities]. *Mechanization and Electrification of Social Agriculture*, *12*, 42–46.

Izenberg, Y. B. (2006). Spravochnaya kniga po svetotehnike [Reference book on lighting engineering] (3rd ed.). Znak. (in Russian)

Janulevičius, A., & Giedra, K. (2008). Analysis of the main dynamic parameters of a split power transmission. *Transport*, *23*(2), 112–118. doi:10.3846/1648-4142.2008.23.112-118

Jerzy, R. (2011). Changes in Microwave-treated Wheat Grain Properties. *Advances in Induction and Microwave Heating of Mineral and Organic Materials*.

Joshi, R., & Schoenbach, K. (2000). Electroporation dynamics in biological cells subjected to ultrafast electrical pulses: A numerical simulation study. *Physical Review E: Statistical Physics, Plasmas, Fluids, and Related Interdisciplinary Topics*, *62*(1), 1025–1033. doi:10.1103/PhysRevE.62.1025 PMID:11088559

Jyun-Ping, J., Jiann-Yeou, R., & Haala, N. (2018). Robust and adaptive band-to-band image transform of UAS miniature multi-lens multispectral camera. *ISPRS Journal of Photogrammetry and Remote Sensing*, *137*, 47–60. doi:10.1016/j.isprsjprs.2017.12.009

Kaasová, J., Hubáčková, B., Kadlec, P., Příhoda, J., & Bubník, Z. (2002). Chemical and biochemical changes during microwave treatment of wheat. *Czech Journal of Food Sciences*, *20*(2), 74–78. doi:10.17221/3513-CJFS

Kagan, N.B., Kaufman, V.G., Pron'ko, M.G., & YAnevskij G.D. (1980). *EHlektrotermicheskoe oborudovanie dlya sel'skohozyajstvennogo proizvodstva* [Electrothermal equipment for agricultural production]. Moscow, Russia: EHnergiya.

Kalmykov, A. V. (2014). *Snizheniye dinamicheskoy nagruzhennosty silovoy peredachi traktora za schet izmeneniya krutilnoy zhestkosti reaktivnogo zvena* [Reduction of the dynamic loading of the power transmission of the tractor due to the change in the torsional stiffness of the reactive link] (PhD dissertation). Volgograd State Agrarian University, Volgograd, Russia.

Kalyanina. (2017). We Just Have To Stop And Think: What Is Our National Interest. *Expert Magazine, 49*, 22-27.

Kanakin, N.S. & Kogan, Y. M. (1986). *Tekhniko-ehkonomicheskie voprosy ehlektrifikacii sel'skogo hozyajstva* [Technical and economic issues of agriculture]. Moscow, Russia: EHnergoatomizdat.

Karabanitsky, A.P., &Levshukova, O.A. (2014). *Teoreticheskoe obosnovanie parametrov energosberegaushchikh mashinno-traktornykh agregatov: uchebnoe posobie* [Theoretical justification of the parameters of energy-saving machine-tractor units: manual]. Krasnodar, Russia: KubSAU.

Kavtarashvili, A.S. & Ridzhal, P. (2001). Preryvistoye osveschenie kur vo vtorom cikle produktivnosti [Intermittent illumination of the chickens during the second cycle of their productivity] *Peredovoi nauch.-proizv. opyt v ptitsevodstve: Ekspress-inform. Vseros. N.-i. i tehnol. in-t ptitsevodstvc, 2,* 10-12. (in Russian)

Kavtarashvili, A. S., Novorotov, E. N., Gladin, D., & Kolokol'nikova, T. N. (2012). Svetodiodnoye osveschenie pri soderzhanii roditel'skogo stada [Led illumination during the parents flock keeping]. *Pticevodstvo, 5,* 15–17.

Kavtarashvili, A. Sh., Novorotov, E. N., & Kolokol'nikova, T. N. (2014). Racional'ny vozrast kur sovremennyh yaichnyh krossov pri nachale svetovoi stimulyacii [Rational age of chickens at the launch of illuminating stimulation]. *Pticevodstvo, 2,* 43–46.

Kazancev S.P. & Ignatkin I.Y. (2013). Rekuperator teploty dlya svinovodcheskogo kompleksa [The heat exchanger of the heat for a pig-breeding complex]. *Mekhanizaciya i ehlektrifikaciya sel'skogo hozyajstva, 4,* 17-18.

Keey, R. B. (2011). *Dryers*. Thermopedia. doi:10.1615/AtoZ.d.dryers

Khalin, E.V. (2016). *Method and network intellectual system of safety production.* Patent Russian Federation N°2589302 Bulletin of inventions, N° 19. (in Russian)

Khalin, E.V. (2016a). *Method and system of e-learning production safety.* Patent Russian Federation N° 2591687. Bulletin of inventions, N° 20. (in Russian)

Khalin, E.V. (2016b). Systems of e-learning production safety of agricultural production. *Competitiveness in the Global World: Economy, Science, Technologies, 5,* 206-209. (in Russian)

Khalin, E.V. (2017). *Electrical safety* (2nd ed.). NIIPF TECHINTELL. (in Russian)

Khalin, E.V. (2018). Innovative methods and technologies for ensuring production safety. *Safety and labor protection on railway transport, 1,* 22-29. (in Russian)

Khalin, E.V. (2018a). The interface of the competence model for production safety. *Labor protection and safety at the industrial enterprises, 3,* 25-33. (in Russian)

Khalin, E.V. (2018b). Requirements for designing the interface of the competence model for production safety. *Modern methods of technical diagnostics and non-destructive testing of details and assemblies, 2,* 52-60. (in Russian)

Khalin, E.V., & Pozdneev, B. M. (2016). *E-portfolio of employee on safety of production and a method of his production.* Patent Russian Federation N°2603494. Bulletin of inventions, N° 33. (in Russian)

Khalin, E.V., & Pozdneev, B. M. (2018). *Model of competences for safety of production and a way of formation of model.* Patent Russian Federation N°2641249. Bulletin of inventions, N° 2. (in Russian)

Khalin, E.V., Strebkov, D. S., & Mikhaylova, E. E. (2016). *A method and network system of safety of production using intellectual graphical descriptions of emergency situations.* Patent Russian Federation N° 2591008. Bulletin of inventions, N° 19. (in Russian)

Khalin, E.V., & Mikhaylova, E. E. (2018). About standardization of requirements to model of competences of employee on safety of production. [in Russian]. *Elektrotsekh, 4,* 10–19.

Khalin, E.V., Pozdneev, B. M., & Mikhaylova, E. E. (2017). About the new national standard on electronic training of safety of production. [in Russian]. *Vestnik Svyazi, 3*, 11–16.

Khalin, E.V., Pozdneev, B. M., & Mikhaylova, E. E. (2018). National standard of the system e-portfolio of employees for production safety. [in Russian]. *Vestnik Svyazi, 5*, 45–50.

Kharchenko, V., & Vasant, P. (2018). Handbook of Research on Renewable Energy and Electric Resources for Sustainable Rural Development. Academic Press. doi:10.4018/978-1-5225-3867-7

King, D. (1958). Get more eggs with new lighting plan. *Poultry Tribune, 64*, 26–28.

Kireev, M. V., Grigor'ev, S. M., & Koval'chuk, Y. U. K. (1981). *Posleuborochnaya obrabotka zerna v hozyajstvah* [After-harvesting of grain in farms]. Leningrad: Kolos.

Kirtbai, Yu. K. (1976). *Rezervy v ispolzovanii mashinno-traktornogo parka* [Reserves in the use of machine and tractor fleet]. Moscow, Russia: Kolos.

Kiva, O. V., & Khodurskyi, V. Y. (2010). Doslidzhennia ta rozrobka prystroiu dlia peredposivnoi obrobky nasinnia tsukrovoho buriaku [Research and development of the device for pre-sowing processing of seeds of sugar beet]. *Visnyk Poltavskoi derzhavnoi ahrarnoi akademii, 4*, 176 – 178.

Klassen, V. I. (1982). Omagnichivaniye vodnykh system [Magnetization of water systems]. Moskow: Khimiya.

Klimov, A. A., Savchuk, V. N., & Baev, V. I. (1970). O nekotorykh elektrofizicheskikh parametrakh i svoistvakh rastitel'noi tkani kak ob"ekta elektroiskrovogo vozdeistviya [About some electrophysical parameters and properties of plant tissue as an object of electric spark action]. *Elektronnaya Obrabotka Materialov, 1*, 66–71.

Knorr, D., Geulen, M., Grahl, T., & Sitzmann, W. (1994). Food application of high electric field pulses. *Trends in Food Science & Technology, 5*(3), 71–75. doi:10.1016/0924-2244(94)90240-2

Knorring, G.M., Fadin, I.M., & Sidorov, V.N. (1992). *Spravochnaya kniga dlya proyektirovschikov elektricheskogo osveschenia* [Reference book for the designers of electrical lighting]. Sankt Petersburg: Energoatomizdat. (in Russian)

Kolin, A. R., Sergeyev, V. V., & Gorbatsevich, N. A. (2008). *Vozdeystviye gradiyentnym magnitnym polem na posadochnyy material i vegetiruyushchiye kartofel'nyye rasteniya* [Effect of a gradient magnetic field on planting material and vegetative potato plants]. Russkiye vysokiye tekhnologii. Available at: http://skutis.ucoz.ru/publ/26-1-0-13

Kondratev, K. Ia., & Fedchenko, P. P. (1982). *Spectral reflectivity and vegetation recognition.* Gidrometeoizdat.

Kondratieva, N. P., Baranov, S. A., & Kondratiev, R. G. (2012). Energoresursosberegayuschaya systema osveschenia myasnogo krossa "Smena 7" [Energy resources conservation system of illumination of the meat cross "Smena 7"]. *Trudy mezhdunarodnoi nauchno-tehynicheskoi konferencii "Energoobespechenie i energosberezhenie v sel'skom hoziastve, 3,* 177-179. (in Russian)

Kong, Sh.-W., Chung, H.-Y., Chang, M.-Y., & Fang, W. (2015). The contribution of different spectral sections to increase fresh weight of boston lettuce. *HortScience, 50*(7), 1006–1010.

Konovalova, I. O., Berkovich, Yu. A., Yerokhin, A. N., Smolyanina, S. O., Yakovleva, O. S., Znamensky, A. I., ... Tsvirko, V. I. (2016). Optimization of the LED lighting system for a vitamin spacegreenhouse. *Aerospace and Environmental Medicine, 50*(3), 17–22. PMID:27344853

Kopsell, D. A., Sams, C. E., & Morrow, R. C. (2015). Blue Wavelengths from LED Lighting Increase Nutritionally Important Metabolites in Specialty Crops. *HortScience, 50*(9), 1285–1288.

Korobiichuk, I., Lysenko, V., Opryshko, O., Komarchyk, D., Pasichnyk, N., & Juś, A. (2018). Crop Monitoring for Nitrogen Nutrition Level by Digital Camera. *Automation 2018, AISC, 743*, 595–603. Retrieved from https://link.springer.com/chapter/10.1007/978-3-319-77179-3_56

Kotelnikov, A. V. (1986). Earth Currents in Electric Railway Transport. Moscow. *Transport*, 279.

Kotnik, T., & Miklavcic, D. (2006). Theoretical evaluation of voltage inducement on internal membranes of biological cells exposed to electric fields. *Biophysical Journal, 90*(2), 480–491. doi:10.1529/biophysj.105.070771 PMID:16239325

Kovalenko, O. Y. (2009). *Svetotehnicheskie ustanovki dlya povyshenia produktivnosti sel'skohozyaystvennyh zhivotnyh* [Light engineering installations for increasing the productivity of farm animals] (Unpublished doctoral dissertation). Saransk. (in Russian)

Kovrikov, I.T. (1976). Obosnovaniye nekotorykh parametrov raspredelitelya semyan i soshnika bezryadkovoy zernovoy seyalki [The rationale for some of the parameters of the distributor of seeds of scattering sowing coulters] *Traktory i selskokhozyaystvennyye mashiny, 4*, 26 - 28.

Kovrikov, I.T. (1983). Osnovnyye printsipy razrabotki raspredelitelnykh ustroystv podpochvenno-razbrosnykh soshnikov zernovykh seyalok [Basic principles of development of distribution devices of coulters of grain seeders subsurface-scattering]. *Traktory i selkhozmashiny, 5*, 13-14.

Kovrikov, I.T., & Skvortsov, G.V. (1977). Sovershenstvovaniye soshnikov dlya bezryadkovogo poseva po sternevym fonam [Improvement of coulters for non-row sowing on stubble backgrounds]. *Traktory i selskokhozyaystvennyye mashiny, 5*, 16 – 18.

Kovrikov, I.T., Nogtikov, A.A., & Sazonov, S.N. (1995). Povysheniye Effektivnosti ispolzovaniya kombinirovannykh pochvoobrabatyvayushche-posevnykh mashin [Increasing the efficiency of combined tillage and sowing machines using]. *Sovershenstvovaniye tekhnologiy i tekhn. sredstv dlya mekhanizatsii s.-kh. protsessov.-Vseros.N.-I. i proyekt.-tekhnol. In-t po ispolz. Tekhniki i nefteproduktov v sel. khoz.-ve, 82*, 80 – 81.

Kozinski, V.A. (1991). *Elektricheskoye osveschenie i obluchenie* [Electrical lighting and radiation]. Agropromizdat. (in Russian)

Kozyrskyi, V., Savchenko, V., & Sinyavsky, O. (2018). *Presowing Processing of Seeds in Magnetic Field. In Handbook of Research on Renewable Energy and Electric Resources for Sustainable Rural Development* (pp. 576–620). IGI Global. doi:10.4018/978-1-5225-3867-7.ch024

Krasnoshchekov, N. V. (2002). O proyektirovanii tekhnologiy proizvodstva selsko-khozyaystvennoy produktsii [About design of production of agricultural products technologies]. *Prioritety mekhanizatsii rasteniyevodstva i zhivotnovodstva, 138*, 11–20.

Krausp, V. R. (2010). Scientific Methods and Experience in Computerization of Innovative Project Management in Industrial Agriculture until 2020. Youth Scientific School. The Internet- and Nanotechnologies. Automated Enterprises.

Krausp, V. R. (2014). Industrialization and Electric Robotization of Industrial Agriculture for Organic Food Production in Full Cycle "Field-Store" under the slogan "Science is a Productive Force and Mother Earth." In *1ST All-Russian Forum for Russia's Technological Leadership. Book of Reports.* "Technological Doctrine 2014," MOO RUSTO.

Krausp, V. R. (1975). *Automation of Grain Postharvest Treatment.* Moscow: Machine Building.

Krausp, V. R. (1980). *Complex Automation in Livestock Industry.* Moscow: Machine Building.

Krausp, V. R. (2008). *Strategy for Automation and Informatization of the Agricultural Production Management*. Moscow: Theory. The Internet- and Nanotechnologies. Practice.

Kravchenko, V. A. (2010). *Povishenie dinamicheskih i ekspluatatsionnyh pokazateley selskohozyastvennyh mashinno-traktornyh agregatov* [Increase of dynamic and operational parameters of agricultural machine and tractor units] (PhD dissertation). FSEE HE ABSSAEA, Zernograd Russia.

Kravchenko, V. A., Senkevich, A. A., & Senkevich, S. E. (2003). *Povysheni eeffektivnosti funktsionirovaniya mashinno-traktornogo agregata na baze traktora klassa 1.4 putem vvedeniya v transmissiyu uprugo-dempfiruyushchego elementa* [Increase of efficiency of functioning of the machine-tractor unit on the basis of a tractor of a class 1.4 by introduction in transmission of an elastic-damping element]. Paper presented at Materials of the XLII Scientific and Technical Conference of the Chelyabinsk SAU, Chelyabinsk, Russia.

Kravchenko, V. A., Senkevich, S. E., Senkevich, A. A., Goncharov, D. A., & Duryagina, V. V. (2008). *Ustroystvo dlya snizheniya zhestkosti transmissii mashinno-traktornogo agregata [Device for reducing the stiffness of the transmission of the machine-tractor unit*]. RU Patent 2398147 F16H47/04.

Kravchenko, V. A., Senkevich, S. E., Senkevich, A. A., Yarovoy, V. G., & Tolstoukhov, Y. S. (2004). *Ustroystvo dlya snizheniya zhestkosti transmissii transportnogo sredstva* [Device to reduce the rigidity of the vehicle's transmission]. RU Patent 2222440, B60K 17/10

Kravchenko, V. A., Duryagina, V. V., & Gamolina, I. E. (2014). Matematicheskoe modelirovanie tyagovoy nagruzki MTA [Mathematical modeling of traction load MTU] [Kuban State University Scientific Journal]. *Polytematic Network Electronic Scientific Journal of the Kuban State University*, *100*, 459–472.

Kravchenko, V. A., Duryagina, V. V., & Gamolina, I. E. (2017). Analiz resultatov analiticheskih issledovaniy pahotnogo agregata na baze traktora klassa 1.4 s UDM v transmissii [Analysis of the results of analytical studies of arable unit on the basis of a class 1.4 tractor with EDM in transmission]. *Bulletin of the Don Agrarian Science*, *37*, 36–42.

Kravchenko, V. A., Oberemok, V. A., & Kravchenko, L. V. (2014). Povishenie effektivnosti mashinno-traktornyh agregatov na baze kolesnyh traktorov [Increase of efficiency of machine-tractor units based on wheeled tractors]. *Technology of Wheeled and Caterpillar Machines*, *6*, 45–50.

Kravchenko, V. A., Senkevich, S. E., & Senkevich, A. A. (2007). Issledovanie vliyaniya uprugodempfiruyushchego mehanizma na pokazateli raboti posevnogo mashinno-traktornogo agregata [Investigation of the influence of an elastic-damping mechanism on the performance of a seed machine-tractor unit]. *Mechanization and Electrification of Agriculture*, *11*, 21–23.

Krejmeris I.B., Tryukas K.C. & Tryukene V.A. (1991). Sposob sushki semyan zernovyh kul'tur [Method for drying seeds of cereals]. *Author's certificate N°1630661 SSSR, MKI A01 F25/08.*

Krozyer, Dzh. Hekhl, Khardy, & Kheinyher. (2014). Nutrition management of maize at high plant density in broad-leaved and narrow-row crops. *Plant Nutrition, 2*, 17-22.

Kuffel, E., Zaengl, W. S., & Kuffel, J. (2000). *High Voltage Engineering. Fundamentals*. Butterworth – Heinemann.

Kulikova, I. A. (2003). Poluchenie dezinfekanta na osnove prirodnogo syr'ya dlya processov obezzarazhivaniya prirodnyh vod [Preparation of disinfectant based on natural raw materials for disinfection of natural waters]. In *Ehkologiya. Ohrana sredy. Stroitel'stvo: doklady VIII Regional'noj konferencii molodyh issledovatelej Volgogradskoj oblasti* [Ecology. Protection of the environment. Construction: reports of the VIII Regional Conference of Young Researchers of the Volgograd Region]. Volgograd: VolgGASA.

Kulikova, I.A. (2004). Ehkologicheskaya bezopasnost' processa hlorirovaniya prirodnyh i stochnyh vod [Ecological Safety of the Process of Chlorination of Natural and Wastewater]. *Vestnik VolgGASU. Seriya Estestvennye nauki* [Vestnik VolgGASU. Series of Natural Sciences], *3*(10), 159-160.

Kuperman, F. M., & Andryenko, S. S. (1959). *Physiology of corn*. Moscow: Publishing House of the Moscow Unitary.

Kutateladze S.S. (1990). *Teploperedacha i gidrodinamicheskoe soprotivlenie*. [Heat transfer and hydrodynamic resistance]. Moscow, Russia: EHnergoatomizdat.

Kutateladze, S. S., & Borishanskij, V. M. (2012). Spravochnik po teploperedache [Handbook of heat transfer]. Moscow, Russia: Kniga po Trebovaniyu.

Kutkov, G. M. (1980). *Tyagovaya dinamika traktorov* [Traction dynamics of tractors]. Moscow, Russia: *Mechanical Engineering (New York, N.Y.)*.

Kutkov, G. M., Puchkov, V. S., & Kholin, A. I. (1975). Analiz istochnikov genaratsii kolebaniy nagruzki na dvigatel selskohozyastvennyh traktorov [Analysis of the sources of oscillation of the load on the engine of agricultural tractors]. *Tractors and Agricultural Machinery*, *6*, 9–10.

Kuzma-Kichta, Yu. A. (1994). *Metody intensifikacii teploobmena* [Methods of heat transfer intensification]. Moscow, Russia: MPEI.

Kuzmichyov, A. V. (2017). Opredelenie i ocenka teplovyh uslovij v zone obogreva infrakrasnymi obogrevatelyami molodnyaka zhivotnyh [Determination and evaluation of thermal conditions in the heating zone by infrared heaters of young animals]. *Vestnik VIEHSKH*, *4*(29), 26–32.

Kuzmichyov, A. V., Lyamcov, A. K., & Tikhomirov, D. A. (2015). Teplotekhnicheskie pokazateli IK obluchatelej dlya molodnyaka zhivotnyh [Thermal performance of IR irradiators for young animals]. *Svetotekhnika*, *3*, 57–58.

Kuzmichyov, A. V., Malyshev, V. V., & Tikhomirov, D. A. (2011). Efficiency of the combined pasteurization of milk using UV and IR irradiation. *Light & Engineering*, *19*(1), 74–78.

Kuznetsov, N. G. (2006). *Stabilizatsiya rezhimov raboty skorostnyh mashinno-traktornyh agregatov* [Stabilization of the operating modes for high-speed machine-tractor units] (PhD dissertation). Volgograd State Agricultural Academy, Volgograd, Russia.

Kuznetsov, N. G., Gapich, D. S., & Shiryaeva, E. V. (2013). Vliyanie neustanovivshegosya haraktera nagruzhenuya kolesnogo traktora krukovim usiliem na koeffitsient buksovaniya [Influence of the unsteady nature of loading of the wheeled tractor by hook force on the coefficient of slipping]. *Tractors and Agricultural Machinery*, *10*, 25–26.

Labetskas, G. S., & Slavinkas, S. S. (2002). Kak ekonomit pri ekspluatatsii traktora MTZ-82 [How to save while operating the tractor MTZ-82]. *Tractors and Agricultural Machinery*, *9*, 9–10.

Lachuga, Yu. F., & Ksendzov, V. A. (2000). *Teoreticheskaya mekhanika* [Theoretical mechanics]. Moscow, Russia: Kolos.

Lauva, A., Aboltins, A., Palabinskis, J., & Karpova-Sadigova, N. (2006). *Grain drying experiments with unheated air*. University of Agriculture.

Lavrikova, N. A., & Fomichev, V. T. (2009). Ehkologobiologicheskaya bezopasnost' stroitel'nyh materialov i izdelij [Ecological and Biological Safety of Building Materials and Products]. *Kachestvo vnutrennego vozduha i okruzhayushchej sredy: materialy VII Mezhdunar. nauch. konf* [Quality of Internal Air and the Environment: Materials VII Intern. sci. Conf.], 8-12.

Lavrukhin, P. V. (2000). *Sovershenstvovaniye protsessa podpochvennogo razbrosnogo raspredeleniya semyan* [Improvement of the process subsurface distribution of seeds] (PhD dissertation). Zernograd, Russia.

Lavrukhin, P. V. (2003). Issledovaniye sovremennykh sposobov i skhem razmeshcheniya semyan po ploshchadi polya [Study of modern methods and schemes of seed placement on the area of the field]. Otchet o NIR. / FGOU VPO ACh-GAA; NoGR01.20.0303698.

Lavrukhin, P. V., & Ivanov, P. A. (2009). *The coulter for scattering sowing*. RU patent No 89324 U1 A 01C 7/20. Moscow, Russia: Federal Institute of Industrial Property.

Lebedev, D.P., Rasstrigin, V.N., & Tikhomirov D.A. (2000). Teploobmen v ehlektricheskih parovodonagrevatelyah [Heat exchange in electric steam and water heaters]. *Promyshlennaya teplotekhnika, 4*(22), 34–39.

Lee, J. S., Lim, T. G., & Kim, Y. H. (2014). Growth and phytochemicals in lettuce as affected by different ratios of blue to red radiation. *Acta Horticulturae*, (1037): 843–848.

Leman, V.M. (1976). *Kurs svetokul'tury rastenij* [Course of plant light culture]. Moscow: Vysshaya shkola [Higher school].

Lisoval, A. P., Makarenko, V. M., & Kravchenko, S. M. (Eds.). (2002). Fertilizer application system. Vyshcha shkola.

Lobyshev, V. I. (2005). Water is a sensor to weak forces including electromagnetic fields of low intensity. *Electromagnetic Biology and Medicine*, *24*(3), 449–461. doi:10.1080/15368370500382248

Lohinova, I. V., & Smyk, S. I. (2012). Prognosis of fertilizer efficiency for corn on grain according to soil diagnostics. Scientific reports of NUBiP, 3(32).

Lurie, A. B. (1969). *Dinamika regulirovaniya navesnykh selskokhozyaystvennykh agregatov* [The dynamics of regulation of mounted agricultural units]. Leningrad, Russia: Mechanical Engineering.

Lurye, A. B. (1979). *Modelirovaniye selskokhozyaystvennykh agregatov i ikh sistem upravleniya* [Modeling of agricultural units and their control systems]. Leningrad, Russia: Kolos.

Lykov, A. V. (1968). *Teoriya sushki* [Theory of drying]. Moscow: Ehnergiya.

Lysakov, A. A., & Ivanov, R. V. (2014). Vliyaniye magnitnogo polya na sokhrannost' kartofelya [The influence of the magnetic field on the conservation of potatoes]. *Uspekhi sovremennogo estestvoznaniya, 8*, 103-106.

Lysenko, V., Opryshko, O., Komarchuk, D., Pasichnyk, N., Zaets, N., & Dudnyk, A. (2017). Usage of Flying Robots for Monitoring Nitrogen in Wheat Crops. *The 9th IEEE International Conference on Intelligent Data Acquisition and Advanced Computing Systems: Technology and Applications*, 30–34. 10.1109/IDAACS.2017.8095044

Lysov, A. M. (1966). Issledovanie vliyaniya tipa transmisii na proizvoditelnost selskohozyaystvennykh traktorov, provedennoe na analogovoy vychislitelnoy mashine [Investigation of the effect of transmission type on the performance of agricultural tractors, conducted on an analog computer]. In Proceedings of Hydrostatic and Hydromechanical Tractor Transmissions. Moscow, Russia: Works of SATI.

Lyubushko, N.I., & Zvolinskiy, V.N. (2001). Zernovyye seyalki na rubezhe XXI veka [Grain drills at the turn of the XXI century]. *Traktory i selskokhozyaystvennyye mashiny, 2*, 4 -7.

Machihina, L. I., Ushakov, T. I., Lvova, L. S., & Denisova, E. V. (2012). Nauchnyj analiz hraneniya zerna v metallicheskih silosah [Scientific analysis of grain storage in metal silos]. *Hleboprodukty, 9*, 54–59.

Machnev, A.B. (2012). Soshnik s napravitelem-raspredelitelem semyan dlya poseva zernovykh kultur [Coulter with guide-distributor of seeds for sowing of crops]. *Traktory i selkhozmashiny, 7*, 42-43.

Madden, C. (1987). *Electrical process to improve manufacturing efficiency* (Vol. 5). Technology Ireland.

Mahnitna obrobka vody pered posadkoiu. (2018). [Magnetic water treatment before planting]. Available at: http://budivnik.in.ua/magnitna-obrobka-nasinnya-pered-posadkoyu.html

Makushin, A.N. (2012). Izmenenie himicheskogo sostava zerna i tekhnologicheskih dostoinstv krupy pri hranenii zerna sortov prosa [Change in the chemical composition of grains and technological advantages of cereals during storage of grains of millet varieties]. *Izvestiya Samarskoj gosudarstvennoj sel'skohozyajstvennoj akademii, 4,* 82-86.

Malev, M.K. (1971). *Obosnovaniye parametrov rabochikh organov seyalok- kultivatorov dlya poseva na pochvakh* [Substantiation of parameters of working bodies of seeders-cultivators for sowing on soils exposed to wind erosion]. Kazan, Russia: Tr. Kaz. NII mekhanizatsii sel. khoz-va.

Malkin, Y.S., Furtat, I.Y., Zhuravska, N.Y., & Usachov, V.P. (2014). Perspektyvy stvorennia resursozberihaiuchykh tekhnolohii shliakhom mahnitnoi obrobky vody ta vodnykh rozchyniv [Prospects for resource-saving technologies through magnetic treatment of water and water solutions]. *Ventyliatsiia, osvitlennia ta vodopostachannia, 17,* 120 – 127.

Malkin, Y.S., Zhuravska, N.Y., & Kovalenko, N.O. (2015). Protses obrobky vody v mahnitnykh poliakh [The process of water treatment in magnetic fields]. *Ventyliatsiia, osvitlennia ta vodopostachannia, 18,* 70 – 74.

Marchuk, I., Makarenko, V., Rozstalnyi, V., & Savchuk, A. (2002). *Fertilizers and their use.* Kiev: Directory.

Marjoram, R., St-Clair, K., McMahon, B., Goelz, A., & Achen, A. (2008). *Integrated MR cabin suspension for reduction of human exposure to whole-body vibration in heavy-duty vehicles.* Bremen, Germany: Messebremen-hvg hanseatische veranstaltungs-gmbh.

Markelova, E. K., Petrova, L. N., Uhanova, V. Yu., Tikhomirov, A. V., Tikhomirov, D. A., & Pershin, A. F. (2013). Pat. RF MPK A01K1/03. Sposob i ustrojstvo ochistki vozdushnoj sredy zhivotnovodcheskih i pticevodcheskih pomeshchenij [Method and device of air cleaning of livestock and poultry premises]. Moscow, Russia: Academic Press.

Martinez, F. R., Pacheco, A. D., Aguilar, C. H., Pardo, G. P., & Ortiz, E. M. (2014). Effects of magnetic field irradiation on broccoli seed with accelerated aging. *Acta Agrophysica, 21*(1), 63–73.

Mashkov, B. S. (1966). *Vyrashchivanie rastenij pri iskusstvennom osveshchenii* [Growing plants under artificial light]. Leningrad: Kolos.

Mashukov, A. S., & Safonov, Y. A. (2017) Sushka zerna aktivnym ventilirovaniem s ispol'zovaniem reguliruemogo ehlektroprivoda ventilyatorov [Drying the grain with active ventilation using an adjustable electric fan drive]. In The collection of Achievements of Science - Agriculture Materials of the All-Russian Scientific and Practical Conference (correspondence) (pp. 317-322). Academic Press.

Mason, L. J., Woloshuk, C. P., Mendoza, F., Maier, D. E., & Kells, S. A. (2014). Ozone: A new control strategy for stored grain. *9th International Working Conference on Stored Product Protection*, 904-907.

McCree, K. J. (1972). The action spectrum, absorptance and quantum yield of photosynthesis in crop plants. *Agricultural Meteorology, 9*, 191–216. doi:10.1016/0002-1571(71)90022-7

McCree, K. J. (1981). Photosynthetically active radiation. In *Encyclopedia of Plant Physiology, 12A* (pp. 41–55). Berlin: Springer-Verlag.

Mehdi, K. (2017). *Encyclopedia of Information Science and Technology* (4th ed.). Hershey, PA: IGI Global. doi:10.4018/978-1-5225-2255-3

Melikov, K. C., Frolov, V. A., Shcherbakov, A., Samsonov, A. V., Chizmadzhev, Y. A., & Chernomordik, L. V. (2001). Voltage-Induced Nonconductive Pre-Pores and Metastable Single Pores in Unmodified Planar Lipid Bilayer. *Biophysical Journal*, *80*(4), 1829–1836. doi:10.1016/S0006-3495(01)76153-X PMID:11259296

Melynik, B. E. (1986). *Aktivnoe ventilirovanie zerna: Spravochnik* [Active grain ventilation: Handbook]. Moscow, Russia: Agropromizdat.

Men'shova, E. A., & Nizharadz, T. S. (2012). Vliyaniye predposevnoy obrabotki semyan yachmenya na ego ustoychivost' k boleznyam i urozhaynost' [Effect of presowing treatment of barley seeds on its resistance to disease and yield]. *Izvestiya Samarskogo nauchnogo tsentra Rossiyskoy akademii nauk, 14*(5), 241-244.

Mikhalyov, A. A. (2014). *Razrabotka sistemy osvescheniya pticevodcheskih pomeshchenij na osnove svetodiodnyh svetil'nikov i rezonansnogo istochnika pitania* [Development of a lighting system for poultry houses based on LED lamps and a resonant power source]. Dissertation Abstracts PhD in Engineering. (in Russian)

Mironyuk, S.S. (2015). The effect of antimicrobic treatment on the quality and storage duration of aubergine fruits. *Collection of scientific works of Vinnitsa National Agrarian University*, *1*, 111-116.

Mitchell, C. A. (2012). Plant lighting in controlled environments for space and earth applications. *VII Intl. Symp. Light Hort. Systems*. 10.17660/ActaHortic.2012.956.1

Mochunova, N. A. (2015). Vliyanie chastoti kolebaniy tyagovoy nagruzki na soprotivleniye perekativaniyu kolesnogo trakrora [Influence of the oscillation frequency of the traction load on the rolling resistance of a wheeled tractor]. *Nature Management*, *4*, 85–87.

Mohapatra, D., & Mishra, S. (2011). Current trends in Drying and Dehydration of Foods. In *Food Engineering*. NOVA Science Publishers Inc.

Mohapatra, Giri, & Kar. (2014). Effect of Microwave Aided Disinfestation of Callosobruchus Maculatus on Green Gram Quality. *International Journal of Agriculture and Food Science Technology*, *5*(2), 55–62.

Moldau, X. (1965). *Vegetative course of the brightness coefficients of leaves of some plants. Issues of the radiation regime of vegetation cover*. Tar-tu. Institute of Physics and Astronomy of the Academy of Sciences of the Estonian SSR.

Morozov, N. M. (2011). Organizacionno-ehkonomicheskie i tekhnologicheskie osnovy mekhanizacii i avtomatizacii zhivotnovodstva [Organizational, economic and technological basis of mechanization and automation of livestock]. Moscow, Russia: FGBNU «Rosinformagrotekh».

Mortensen, L. M., & Stromme, E. (1987). Effects of light quality on some greenhouse crops. *Scientia Horticulturae*, *33*(1-2), 27–36. doi:10.1016/0304-4238(87)90029-X

Mytnikov, A. V. (2012). *High Voltage Engineering*. Tomsk: TPU. Neu J., & Krassowska W. Asymptotic model of lectroporation. *Physical Review. E*, *59*(3), 3471–3482.

Mzhel'ski, B.I., & Mzhel'skaya, E.B. (2001). *Ocnovy teorii optimizatsii. Uchebnoye posobie* [Fundamentals of the theory of optimization]. MEI. (in Russian)

Nakasono, S., & Saiki, H. (2000). Effect of ELF magnetic fields on protein synthesis in Escherichia coli K12. *Radiation Research*, *154*, 208–216.

Nejadi, Mohammad, & Nikbakht. (2017). Numerical Simulation of Corn Drying in a Hybrid Fluidized BedâInfrared Dryer. *Journal of Food Process Engineering*.

Nekhoroshev, D. A. (2014). *Stabilizatsiya rezhimov MTA izpolzovaniem pnevmogidravlicheskoy mufty stsepteniya* [Stabilization of MTU modes using pneumatic hydraulic clutch] (PhD dissertation). Volgograd State Agrarian University, Volgograd, Russia.

Nekhoroshev, D. D. (2004). *Povishenie ekspluatatsionnykh pokazateley MTA putem ispolzovaniya traktorov s pnevmogidravlicheskoy muftoy* [Increase of operational indicators of MTU by using tractors with pneumatic hydraulic clutch] (PhD dissertation). Volgograd State Agrarian University, Volgograd, Russia.

Nekhoroshev, D. A., & Nekhoroshev, D. D. (2009). Planetarnaya uprugaya mufta v transmissii kolesnogo tractor klassa 1.4 [Planetary elastic coupling in the transmission of class 1.4 wheeled tractor]. *Achievements of Science and Technology of Agro-Industrial Complex*, *12*, 60–61.

Nekhoroshev, D. D., & Nekhoroshev, D. A. (2015). Energeticheskoe sredstvo v transmissii kolesnogo traktora [Power means in the transmission of the wheeled tractor]. *Unique Research of the XXI Century*, *12*(12), 40–41.

Nelson, S.O. (2004). RF and microwave permittivities of insects and some applications. *URSI EMTS*, 1224-1226.

Nelson, S. O. (2008). Dielectric properties of agricultural products and some applications. *Research in Agricultural Engineering*, *54*(2), 104–112. doi:10.17221/1000-RAE

Nelson, S. O. (2015). *Dielectric Properties of Agricultural Materials and Their Applications*. Academic Press.

Nogtikov, A.A., & Golivets, V.A. (1997). Obosnovaniye parametrov bokovykh po-verkhnostey rassekatelya [Justification of parameters of the side surfaces of the divider]. *Traktory i selskokhozyaystvennyye mashiny*, *11*, 31 - 32.

Nogtikov, A.A., & Sazonov, S. N. (1995). Obosnovaniye parametrov semyaprovoda dlya vnutripochvenno - razbrosnykh soshnikov [parameters Justification of the seed pipe for inside scattering soil sowing coulters]. *Tekhnika v selskom khozyaystve*, *3*, 29-31.

Nogtikov, A.A., &Sazonov, S.N. (1995). Povysheniye effektivnosti raboty ras-predeliteley vnutripochvenno-razbrosnykh soshnikov [The efficiency of the distributors of subsoil-broadcast opener]. *Inzh.-tekhn. obespecheniye APK. 4*, 13-14.

Nogtikov, A.A., Glotov, A.L., & Sazonov, D.S. (1996). Soshnik dlya vnutripochvennogo razbrosnogo poseva [The opener for subsurface broadcast seeding]. *Mekhanizatsiya i elektrifikatsiya sel. khoz.- va*, *2*, 29-30.

Novosyolov, I. M. (2011). *Obosnovanie effektivnosti tehnologicheskogo svetodiodnogo osveschenia promyshlennogo stada kur-nesushek krossa "Highsex – Brown"* [Substantiation of the effectiveness of an industrial flock of laying hens of the cross "Highsex – Brown"]. Dissertation Abstracts PhD in Engineering. Izhevsk. (in Russian)

O'Leary, G. J. (2018). Challenges and Responses to Ongoing and Projected Climate Change for Dryland Cereal Production Systems throughout the World. *Agronomy (Basel)*, *8*(4).

Okhotnikov, B.L. (2014). *Ekspluatatsiya dvigateley vnutrennego sgoraniya: uchebnoye posobie* [Operation of internal combustion engines: a training manual]. Ekaterinburg, Russia: Publishing house Ural.

Okonov, M. M., Khalgaeva, K. E., & Unkanzhinov, G. D. (2012). Urozhajnost' i kachestvo zerna ozimoj pshenicy pri ispol'zovanii Binorama, Biosila Bishofita na fone mineral'nyh udobrenij [Yield and quality of winter wheat grain using Binorama, Biosila Bishofit against mineral fertilizers]. *Plodorodie [Fertility]*, *1*(64), 15–16.

Onhonova, L. O., & Skryabina, I. Y. U. (2017). Ustrojstvo dlya aktivnogo ventilirovaniya i pnevmovygruzki sypuchego materiala [Device for active venting and pneumatic unloading of bulk material]. *Vestnik VSGUTU*, *3*(66), 50–64.

Ospanov, A. B., & Vasilev, A. N. (2016). Changing parameters of the microwave field in the grain layer. *J. Eng. Applied Sci.*, *11*(S1), 2915-2919.

Ospanov, A. B., & Vasilev, A. N. (2016). Controlling reactions of biological objects of agricultural production with the use of electrotechnology. *International Journal of Pharmacy and Technology, 8*(4), 26855-26869.

Ospanov, A.B., Vasilyev, A.N., Budnikov, D.A., Karmanov, D.K. & Vasilyev, A.A. (2017). *Improvement of grain drying and disinfection process in the microwave field. Monography.* Almaty, Kazakhstan: Nur-Print.

Ospanov, A. B., Vasilyev, A. N., Budnikov, D. A., Karmanov, D. K., & Vasilyev, A. A. (2017). *Improvement of grain drying and disinfection process in the microwave field. Monography.* Almaty: Nur-Print.

Ouzounis, Th., Rosenqvist, E., & Ottosen, C.-O. (2015). Spectral effects of artificial light on plant physiology and secondary metabolism: A review. *HortScience, 50*(8), 1128–1135.

Panova, T. V. (2011). *Razrabotka i obosnovanie parametrov pritochno-vytyazhnogo utilizatora tepla dlya uluchsheniya mikroklimaticheskih uslovij v zhivotnovodcheskih pomeshcheniyah* [Development and justification of parameters of the heat supply and exhaust heat exchanger to improve the microclimate conditions in livestock buildings] (Doctoral dissertation).

Parshin, I. V. (1978). *Issledovanie raboty traktornogo dvigatelya s garoturbinnym nadduvom v usloviyah selskokhozyastvennogo mashinno-traktornogo agregata* [Investigation of the operation of a tractor engine with gas turbine supercharging in the conditions of an agricultural machine-tractor unit] (PhD dissertation). FSEE HE ABSSAEA, Zernograd Russia.

Pershin, A.F., & Smirnov A.A. (2014). Opredeleniye optimalnoi dozy ozona pri obrabotke zernovykh materialov [Determination of optimal ozone dose in processing grain materials]. *Tekhnika v sel'skom khozyaistve* [Equipment in agriculture], *6,* 9-11. (in Russian)

Petraq, P., & François, P. (Eds.). (2015). *International Journal of Agricultural and Environmental Information Systems, 7*(2). doi:10.4018/IJAEIS

Petrik, G. G. (2011). O sistemnom podhode k modelirovaniyu ob"ektov i ih svojstv na molekulyarnom i termodinamicheskom urovnyah [On the system approach to the modeling of objects and their properties at the molecular and thermodynamic levels]. *Monitoring. Science and Technology, 2,* 86–101.

Philips, N.V. (1979). Gloeilampenfabrieken. *Germicidal lamps and applications.*

Pil'schikova, Y. A. (2014). *Povyshenie effektivnosti obluchatel'noy ustanovki dlya pticevodstva* [Increasing the efficiency of the irradiation plant for poultry farming]. Dissertation Abstracts PhD in Engineering. Saransk. (in Russian)

Pis'mennyj, E.N. (1999). Konstruktivnye metody povysheniya teploaehrodinamicheskoj ehffektivnosti trubchatyh poperechno-orebrennyh poverhnostej teploobmena [Design methods for increasing heat and aerodynamic efficiency of tubular transversely finned heat transfer surfaces]. *Promyshlennaya teplotekhnika, 2-3*(21).

Pobedin, A. V., Dolotov, A. A., & Shekhovtsov, V. V. (2016). *Decrease of the vibration load level on the tractor operator working place by means of using of vibrations dynamic dampers in the cabin suspension.* Paper presented at 2nd International Conference on Industrial Engineering (ICIE), Chelyabinsk, Russia. 10.1016/j.proeng.2016.07.136

Podleoeny, J., Pietruszewski, V., & Podleoena, A. (2004). Efficiency of the magnetic treatment of broad bean seeds cultivated under experimental plot conditions. *International Agrophysics, 18,* 65–71.

Polifke, G. (2016). *Hydrodamp - hydraulic torsional vibration damper for tractors.* Paper presented at Conference on Agricultural Engineering, Koln, Germany.

Polifke, G. (2017). *Hydrodamp - hydraulic torsional vibration damper for tractors and construction machines design for a race-sensitive drive train.* Paper presented at VDI Conference on Couplings and Clutch Systems in Drives 2017 / 2nd VDI Conference on Vibration Reduction in Mobile Systems, Ettlingen, Germany.

Polivaev, O. I., Zhidenko, K. S., Maksimov, I. I., & Levkin, I. G. (2017). *Uprugo-dempfiruyushchiy privod vedushchih koles traktora* [Elastic-damping drive of the tractor's driving wheels]. Paper presented at Modern Trends in the Development of Technology and Technical Means in Agriculture, Voronezh, Russia.

Polivaev, O.I., & Kutkov, A. Y. (2010). Effektivnost ispolzovaniya transportnykh sredstv s uprugodempfiruyushchim privodom vedushchikh koles pri tormozhenii [Efficiency of the use of transport units with an elastic-damping drive of driving wheels during braking]. *Machinery in Agriculture, 5*, 30-32.

Polivaev, O. I., Pilayev, S. N., & Kutkov, A. Y. (2011). Obosnovanie parametrov i effektivnosti ispolzovaniya uprugodempfiruyushchego privoda vedushchikh koles traktorno-transportnogo agregata v rezhime tormozheniya [Substantiation of parameters and efficiency of using the elastic-damping drive of the driving wheels of the tractor-transport unit in braking mode]. *Bulletin of the Voronezh State Agrarian University, 3*, 48–52.

Polivaev, O. I., & Vedrinsky, O. S. (2013). *Vliyanie uprugodempfiruyushchego privoda vedushchikh koles traktora na dinamicheskie nagruzki v transmissii pri razgone MTA* [Influence of the elastic-damping drive of the tractor's driving wheels on the dynamic loads in the transmission when accelerating the MTU]. Innovative Technologies and Technical Means for Agroindustrial Complex.

Polivaev, O. I., & Vedrinsky, O. S. (2014). Analiz vliyaniya uprugodempfiruyushchego privoda koles na dinamicheskie nagruzki v transmissii traktora pri razgone [Analysis of the effect of the elastic-damping drive of the wheels on the dynamic loads in the tractor's transmission during acceleration]. *The Bulletin of the Don Agrarian Science, 28*, 5–9.

Polivaev, O. I., Vedrinsky, O. S., & Derkanosova, N. M. (2016). Povishenie dolgovechnosti stsepleniya traktorov za schet uprugofriktsionnogo dempfera [Increasing the durability of tractor traction due to elastic-friction damper]. *Science and Education in Modern Conditions, 12*, 226–230.

Popova, N.V., Tokar, A.Y., & Mironyuk, S. S. (2015). Influence of treatment of antimicrobial agents with microbial damage of eggplant fruits. *Scientific works of the National University of Food Technologies, 3*, 236-242.

Popova, N. A., Papchenko, A. Ya., & Bologa, M. K. (2014). Ehlektroplazmoliz vinograda s primeneniem bipolyarnyh impul'sov [Electroplasmolysis of grapes using bipolar pulses]. *Elektronnaya Obrabotka Materialov, 50*(6), 83–91.

Prikupets, L. B., & Tikhomirov, A. A. (1994). Optimization of lamp spectrum for vegetable growth. In T.W. Tibbitts (Ed.), *International Lighting in Controlled Environments Workshop*. Academic Press.

Prokopenko, A.A., & Yuferev, L.Y. (2004). *Effektivnost' primeneniya obluchateley-retsirkulyatorov*. Academic Press.

Prokopenko, A. A. (1978). Influence of different doses of ultraviolet rays of unshielded bactericidal irradiators on the microclimate and productivity of the laying hens. *Proceedings of VNIIVSGE, 61*, 139–141.

Protasova, N. N., & Kefeli, V. I. (1982). Fotosintez i rost vysshih rastenij, ih vzaimosvyaz' i korrelyacii [Photosynthesis and growth of higher plants, their interconnection and correlation]. In Fiziologiya fotosinteza [Physiology of photosynthesis]. Moscow: Nauka [Science].

Protasova, N. N. (1987). Svetokul'tura kak sposob vyyavleniya potencial'noj produktivnosti rastenij [Light culture as a method of revealing potential productivity of plants]. *Fiziologiya rastenij. Plant Physiology, 34*(4), 812–822.

Protasova, N. N., Wells, J. M., Dobrovolsky, M. V., & Tsoglin, L. N. (1990). Spektral'nye harakteristiki istochnikov sveta i osobennosti rosta rastenij v usloviyah iskusstvennogo osveshcheniya [Spectral characteristics of light sources and features of plant growth in artificial lighting]. *Fiziologiya rastenij. Plant Physiology, 37*(2), 386–396. PMID:15667287

Prystash, I. V. (2003). Yield and quality of corn grain depending on fertilizer system on meadow-chernozem ground. *Collection of scientific works of the Institute of Agriculture of the Ukrainian Academy of Agrarian Sciences, 3*, 58–63.

Rachkulyk, V. Y., & Sytnykova, M. V. (1981). *Reflective properties and the state of vegetation cover*. Gidrometeoizdat.

Ralph, A. (1963). *Lighting programmes for replacement pullets. In Extension Poultry Specialist* (p. 95616). Davis, CA: Animal Science Department, University of California.

Ranjbaran, M., & Zare, D. (2013). Simulation of energetic- and exergetic performance of microwaveassisted fluidized bed drying of soybeans. *Energy*, *59*, 484–493. doi:10.1016/j.energy.2013.06.057

Rasstrigin, V. N. (1988). *Osnovy ehlektrifikacii teplovyh processov v sel'skohozyajstvennom proizvodstve* [Fundamentals of electrification of thermal processes in agricultural production]. Moscow, Russia: Agropromizdat.

Rastimeshin, S. A. (1990). *Tekhnicheskie sredstva lokal'nogo obogreva* [Technical means for local heating]. Moscow, Russia: Rosagropromizdat.

Rastimeshin, S. A. (1991). *Lokal'nyj obogrev molodnyaka zhivotnyh (teoriya i tekhnicheskie sredstva)* [Local heating of young animals (theory and technical means)]. Moscow, Russia: Agropromizdat.

Rastogi, N. K., Eshtiaghi, M. N., & Knorr, D. (1999). Accelerated mass transfer during osmotic dehydration of high intensity electrical field pulse pretreated carrots. *Journal of Food Science*, *64*(6), 1020–1023. doi:10.1111/j.1365-2621.1999.tb12272.x

Raven, P. H., Evert, R. F., & Eichhorn, S. E. (2005). Photosynthesis, Light, and Life. *Biologia Plantarum*, *7*, 119–127.

Reference Book for Hay Meadows and Pasture Fields. (1986). Moscow: Rosselkhozizdat.

RF Patent of invention N°2533585. (2014). *Device for economic and energy-saving disinfection of fodder and animal and poultry products* / Dubrovin A.V. // Issued 20.11.2014, Bulletin N° 32. (in Russian)

RF Patent of invention N°2608532. (2017). *Method and device for specified technologically and economically optimal ozonation of moving bulk fodder for livestock and poultry breeding* / Dubrovin A.V. // Issued 19.01.2017, Bulletin N°2. (in Russian)

RF Patent of invention N°2613453. (2017). *Method and device of determination of economically optimal excessive air pressure for control of harmful microorganisms in ambient air of livestock and poultry houses* / Dubrovin A.V. // Issued 16.03.2017, Bulletin N°8. (in Russian)

RF Patent of invention N°2621140. (2017). *Method and device for specified technologically and economically optimal combined infrared and conductive drying of moving bulk fodder for livestock and poultry breeding*/ Dubrovin A.V. // Issued 31.05.2017. Bulletin N°16. (in Russian)

RF Patent of invention N°2624199. (2017). *Method and device of technologically and economically optimal super-high frequency drying of bulk fodder for livestock and poultry breeding* / Dubrovin A.V. // Issued 03.07.2017, Bulletin N° 19. (in Russian)

Rozenberh, H. V. (1959). Absorption spectroscopy of dispersed substances. Uspekhy fyz. *Nauk.*, *69*(1), 57–104.

Rozhenau, U.M. (1966). *Teploobmen pri kipenii* [The boiling heat transfer]. Moscow, Russia: Sb. Sovremennye problemy teploobmena, EHnergiya.

Rvachëv, V. L. (1987). *Methods of optics of light scattering media in physics and biology*. Minsk: Belarusian Publishing House.

Saberioona, M., Amina, M., Anuarb, A., Gholizadehc, A., & Wayayokd, A. (2014). Assessment of rice leaf chlorophyll content using visible bands atdifferent growth stages at both the leaf and canopy scale. *International Journal of Applied Earth Observation and Geoinformation*, *32*, 35–45. doi:10.1016/j.jag.2014.03.018

Sabiev, U. K., & Shonov, A. E. (2017). Obzor i analiz bunkerov aktivnogo ventilirovaniya zerna [Review and analysis of active grain ventilation bins]. In *Collection of materials of the XXIII scientific and technical conference of students of the FGBOU VO OMSK GAU* (pp. 104-107). Academic Press.

Sakun, V. A. (1974). *Sushka i aktivnoe ventilirovanie zerna i zelenyh kormov* [Drying and active ventilation of grain and green forages]. Moscow: Kolos.

Sarkar. (2017). Ozone Pollution, Agriculture, and Food Security. In *Environmental Issues Surrounding Human Overpopulation.* IGI Global. doi:10.4018/978-1-5225-1683-5.ch01

Sarychev, G. S. (2001). Produktivnost' cenozov ogurcov i tomatov v funkcii spektral'nyh harakteristik OU [Productivity of cenosis of cucumbers and tomatoes in a function of the spectral characteristics of Oh]. *Svetotekhnika, 2*, 27–29.

Semenov, A. N. (1959). *Zernovyye seyalki* [Grain drills.]. Moscow, Russia: Mashgiz.

Severinov, O. V. (2016). *Sovershenstvovanie processa sushki zerna v bunkerah aktivnogo ventilirovaniya s ispol'zovaniem modelirovaniya* [Improvement of the process of grain drying in active ventilation bins using simulation] (Dissertation of the candidate of technical sciences). VIESH, Moscow, Russia.

Severinov, O.V., Rudenko, N.B., Budnikov, D.A. & Macynin, S.N. (2012). Bunker dlya hraneniya i ventilirovaniya zerna [Bunker for storing and venting grain]. *Patent for utility model RUS 113631 byul., 6.*

Shand, M. A. (2006). *The chemistry and technology of magnesia.* Wiley. doi:10.1002/0471980579

Shekhovtsov, K. V. (2014). *Snizhenie urovnya vibronagruzhennosti rabochego mesta operatora traktora za schet primeneniya dinamicheskikh gasiteley kolebaniy v sisteme podressorivaniya kabiny* [Reduction in the vibration level of the operator's workplace due to the use of dynamic vibration dampers in the cabin suspension system] (PhD dissertation). Volgograd State Technical University, Volgograd, Russia.

Shekhovtsov, V. V., Shevchuk, V. P., & Zelenko, S. V. (2002a). Rasprostranenie krutilnyh kolebaniy v valoprovode silovoy peredachi traktora VT-100 [Dissemination of torsional oscillations in the power transmission line of the power transmission of the tractor VT-100]. *Tractors and Agricultural Machinery, 8*, 10–11.

Shekhovtsov, V. V., Shevchuk, V. P., & Zelenko, S. V. (2002b). Issledovaniya rezonansnykh rezhimov silovoy peredachi traktora VT-100 [Investigation of the resonance modes of the power transmission of the tractor VT-100]. *Tractors and Agricultural Machinery, 7*, 5–6.

Shekhovtsov, V. V., Sokolov-Dobrev, N. S., & Potapov, P. V. (2016). Decreasing of the Dynamic Loading of Tractor Transmission by means of Change of the Reactive Element Torsional Stiffness. *Procedia Engineering, 150*, 1239–124[illegible]. doi:10.1016/j.proeng.2016.07.129

Shirobokova, T.A., Kochetkov, N.P. & Gallyamova, T.R. (2014). Modelirovanie svetodiodnogo svetil'nika dlya osveschenia ptichnikapri pri napol'nom soderzhanii roditel'skogo stada [Modeling of a led lamp for illumination of the poultry house with floor maintenance of the parent flock]. *Agrarny naychny zhurnal, 9,* 52-54. (in Russian)

Shpaar, D., Hinapp, K., Dreher, D., Zakharenko, A., & Kalenska, S. (2009). *Growing, harvesting, preserving and using.* Alfa-Steeve LTD.

Shuvalov A.M., Mashkov A.N., Chernov D.S., Kalinin V.F., & Nabatov K.A. (2014). Raschet pokazateley protsessa nagreva zerna soi [Calculation of process performance of soybeans heating]. *Tekhnika v sel'skom khozyaistve* [Equipment in agriculture], *6,* 7-9. (in Russian)

Sidortsov, I. G. (2007). Ustanovka dlya predposevnoy obrabotki semyan [Plant for presowing seed treatment]. *Tekhnika v sel'skom khozyaystve, 3,* 61-62.

Sinitsky, S. A. (2005*). Vliyanie nagruzki mashinno-traktornogo agregata na pokazateli dvigatelya v usloviyakh ekspluatatsii* [Influence of the load of the machine-tractor unit on the engine performance under operating conditions] (PhD dissertation). Kazan State Agricultural Academy, Kazan, Russia.

Siracusa, L., Gresta, F., & Sperlinga, E. (2017). Strip tillage and sowing: Is precision planting indispensable in silage maize. *Journal of Food Composition and Analysis, 62,* 1–7. doi:10.1016/j.jfca.2017.04.005

Skolzayev, V. A., & Chernovolov, V. A. (1969). Metodika tekhnologicheskogo rascheta tsentrobezhnykh diskovykh apparatov [Method of technological calculation of centrifugal disk devices]. Mekhanizatsiya i elektrifikatsiya selskokhozyaystvennogo proizvodstv, 12, 135-140.

Skolzayev, V.A., &Chernovolov, V.A. (1969). Elementy teorii raspredeleniya udobreniy diskovym tsentrobezhnym apparatom [Theory elements of the fertilizer distribution disc by centrifugal apparatus]. *Traktory i selkhozmashiny, 2,* 15-17.

Smith, W. L., Lagunas-Solar, M. C., & Cullor, J. S. (2002). Use of pulsed ultraviolet laser light for the cold pasteurization of bovine milk. U.S. *Journal of Food Protection, 65*(9), 1480–1482. doi:10.4315/0362-028X-65.9.1480 PMID:12233862

Solodova, E. A. (2002). *Vliyanie aehroionnoj obrabotki na mikroorganizmy, pishchevuyu i biologicheskuyu cennost' produkcii iz morskih gidrobiontov* [Influence of aeroion treatment on microorganisms, food and biological value of products from marine hydrobionts]. Dissertation of the candidate of technical sciences, Russia, Vladivostok.

Son, K.-H., & Oh, M.-M. (2013). Leaf shape, growth, and antioxidant phenolic compounds of two lettuce cultivars grown under various combinations of blue and red light-emitting diodes. *HortScience, 48,* 988–995.

Spalart, P. R., & Rumsey, C. L. (2007). Effective inflow conditions for turbulence models in aerodynamic calculations. *AIAA Journal, 45*(10), 2544–2553. doi:10.2514/1.29373

Spasov, A. A. (2000). *Magnij v medicinskoj praktike* [Magniy in medical practice]. Volgograd: Otrok.

Spiridonov, A. A. (1981). *Planirovanie eksperimenta pri issledovanii tehnologicheskikh protsessov* [Planning an experiment in the study of technological processes]. Moscow, Russia: Mechanical Engineering.

Stankovic, M., Cvijanovic, M., & Dukic, V. (2016). Ecological importance of electrical devices innovative in the process of anti *Ambrosia artemisiifolia L. Economics of Agriculture, 3,* 861–870.

Svidzinsky, S.A. & Moscow, G.A. (2008). *Metodika poiskovo-razvedochnyh rabot na mineral'nye soli v usloviyah solyanokupol'nyh struktur* [Methods of prospecting for mineral salts in salt-dome structures]. Saratov: Izdatel'stvo Saratovskogo universiteta [Saratov University Press].

Svistula, A. E., & Matievsky, G. D. (2011). Issledovanie optimizatsionnoy skorostnoy kharakteristiki dvigatelya postoyannoy moshchnosty [Investigation of the optimizing speed characteristic of a constant-power motor]. Internal Combustion Engines: National Technical University "Kharkov Polytechnic Institute, 2, 46-49.

Syrovatka, V.I. (2014). Sovershenstvovaniyetekhnologicheskikhprotsessovproizvodstvakombikormov v khozyaistvakh [Improvement of processes of combined fodder (mixed feed) production at farms]. *Vestnik VNIIMZH, 1*(13), 4-11. (in Russian)

System of Management of Robotized Mixer Feed Dispenser. (2016). Utility Patent No. 2585043 registered in the State Register of Inventions of the Russian Federation 28.04.

Tanimura Y., Hirotsuji J., Tanaka K. & Asakura T. (2003). Preserving fruits from putrefying by using negative air ions and ozone J.Japan. *Soc.Agr.Mach., 65*(1), 115-129.

Tarasenko, F.P. (1963). *Vvedenie v kurs teorii informacii* [Introduction to the course of information theory]. Tomsk, Russia: Izdatel'stvo Tomskogo universiteta.

Tarasik, V. P. (1984). *Proektirovanie kolesnykh tyagovo-transportnykh mashin* [Design of wheel traction-transport machines]. Minsk, Belarus: High. Sc.

Tarnizhevsky, M. V., & Tomlyanovich, D. K. (1986). Design of Electric Power Supply Devices for Trams and Trolleybuses. Moscow. *Transport*, 376.

Tihen'kij, V. I. (1993). *Metody i sredstva povysheniya ehffektivnosti processa sushki semyan pri ehlektrorazryadnom vozdejstvii na potok vozduha* [Methods and means of increasing the efficiency of the process of drying seeds in the discharge of electric discharge]. Dissertation of the candidate of technical sciences, Russia, St. Petersburg-Pushkin: NPO «Nechernozemagromash». (in Russian)

Tikhomirov D.A. & Vasiliev A.N. (2016). Programmnyj proekt rascheta teploehnergeticheskih parametrov ob"ektov zhivotnovodstva [Program project for calculation of heat and power parameters of livestock facilities]. *Mezhdunarodnyj nauchno-issledovatel'skij zhurnal, 8*(50), 95-98.

Tikhomirov D.A. (2013a). Teploehnergeticheskie parametry sistemy goryachego paro- i vodoobespecheniya ob"ektov zhivotnovodstva [Heat and power parameters of the system of hot steam and water supply of livestock]. *Mekhanizaciya i ehlektrifikaciya sel'skogo hozyajstva, 3*, 17–19.

Tikhomirov D.A. (2013b). Metodika teploehnergeticheskogo rascheta ehnergosberegayushchej ventilyacionno-otopitel'noj ustanovki dlya zhivotnovodcheskih ferm [Method of heat power calculation of energy-saving ventilation and heating plant for livestock farms]. *Al'ternativnaya ehnergetika i ehkologiya, 2*(1), 125–131.

Tikhomirov D.A. (2014). Fizicheskaya model' dinamiki teploobmena v ehlektricheskom paroperegrevatele [The physical model of the dynamics of heat transfer in electrical superheater]. *Tekhnika v sel'skom hozyajstve, 2*, 13–15.

Tikhomirov D.A. (2015). Opredelenie koehfficienta teplootdachi pri kipenii vody v ehlektricheskom parovodonagrevatele ehksperimental'nym putem [Determination of the heat transfer coefficient at boiling water in the electric steam heater by experiment]. *Innovacii v sel'skom hozyajstve, 3*(13). 85-89.

Tikhomirov, D. A. (2002). Povyshenie ehffektivnosti teploobmena v ehlektricheskih parovodonagrevatelyah [Improving the efficiency of heat transfer in electric steam heaters]. In Sovremennye ehnergosberegayushchie teplovye tekhnologii (sushka i termo-vlazhnostnaya obrabotka materialov): Trudy 1-j mezhdunarodnoj nauchno-prakticheskoj konferencii (pp. 213–216) Moscow, Russia: MGAU.

Tikhomirov, D.A. (2017). EHlektricheskie IK-obogrevateli: sostoyanie i perspektivy innovacionnogo razvitiya [Electric IR heaters: status and prospects of innovative development]. *Tekhnika i oborudovanie dlya sela, 1*, 2-7.

Tikhomirov, A. A., Lisovsky, G. M., & Sidko, F. Y. (1991). *Spektral'nyj sostav sveta i produktivnost' rastenij* [Spectral composition of light and plant productivity]. Novosibirsk: Nauka.

Tikhomirov, A. A., Sharupich, V. P., & Lisovsky, G. M. (2000). *Svetokul'tura rastenij: biofizicheskie i biotekhnologicheskie osnovy* [Light culture of plants: biophysical and biotechnological bases]. Novosibirsk: Izdatel'stvo RAN.

Tikhomirov, D. A. (2016). EHnergoehffektivnye ehlektricheskie sredstva i sistemy teploobespecheniya tekhno-logicheskih processov v zhivotnovodstve [Energy-efficient electric means and systems of heat supply of technological processes in animal husbandry]. *Vestnik VNIIMZH, 4*(24), 15–23.

Tireki, Sumnu, & Esin. (2006). Effect of Microwave, Infrared and Infrared-assisted Microwave Heating on the Drying Rate of Bread Dough. *American Journal of Food Technology*. Doi:10.3923/ajft.2006.82.93

Tkachenko, Y. (1998). *Method of crops growing*. Patent of Russian Federation A01G7/04, A01C1/00. N°2110910 declared. 25.10.96; published 20.05.98, N° 10.

Tkachev, R. V. (2000). *Ehlektroaktivirovanie processa sushki semyan* [Electroactivation of the process of drying the seeds]. Dissertation of the candidate of technical sciences, Russia, Moscow.

Toepfl, S. (2006). *Pulsed Electric Fields (PEF) for Permeabilization of Cell Membranes in Food-and Bioprocessing: Applications, Process and Equipment Design and Cost Analysis*. Ph.D. Thesis.

Toepfl, S., Heinz, V., & Knorr, D. (2005). Overview of pulsed electric field processing for food. In D. W. Sun (Ed.), *Emerging Technologies for Food Processing* (pp. 67–97). Oxford, UK: Elsevier. doi:10.1016/B978-012676757-5/50006-2

Tokar, A. Yu., & Mironyuk, S. S. (2015). Preservation of eggplant fruits provided processing of antimicrobial substances. *Progressive technique and technologies of food production of restaurant and trade*, *1*, 400–409.

Tolubinskij, V. I. (1989). Teploobmen pri kipenii [The boiling heat transfer]. Kiev, Ukraine: Naukova dumka.

Tooming, H. T. (1977). *Solnechnaya radiaciya i formirovanie urozhaya* [Solar radiation and yield formation]. Leningrad: Gidrometeoizdat.

Toporkov, V.N., & Korolev, V.A. (2017). *Ehlektroimpul'snaya ustanovka dlya bor'by s sornyakami: monografiya* [Electropulse installation for weed control: Monograph]. Moskva: Federal'nyj nauchnyj agroinzhenernyj centr VIM [Federal agricultural research centre VIM].

Tuvalev, Z. I. (1987). *Issledovaniye parametrov zernovykh sternevykh seyalok v usloviyakh Zauralia Bashkirii* [Study of parameters of grain stubble seeders in the conditions of the trans-urals Bashkiria] (PhD dissertation). Chelyabinsk, Russia.

Valyashko, M.G. (1976). *Brom v solyanyh otlozheniyah i rassolah kak geohimicheskij indikator ih genezisa, istorii i poiskovyj priznak* [Bromine in salt deposits and brines as a geochemical indicator of their genesis, history and search sign]. Academic Press.

Vanckul, A. S., Goryacheva, E. M., Garyaev, A. B., & Sorochinskij, V. F. (2017). Issledovanie vliyaniya razlichnyh faktorov na vremya ohlazhdeniya sloya zerna v metallicheskih silosah [Investigation of the influence of various factors on the cooling time of the grain layer in metal silos]. In The *collection Modern problems of engineering sciences, Collection of scientific works of the VI International Scientific and Technical Symposium, International Scientific and Technical Forum* (pp. 101-103). Academic Press.

Vasant, Zeynep, & Gerhard-Wilhelm. (2018). Handbook of Research on Emergent Applications of Optimization Algorithms. Academic Press. Doi:10.4018/978-1-5225-2990-3

Vasant, P., Alparslan-Gok, S. Z., & Weber, G.-W. (2017). *Handbook of Research on Emergent Applications of Optimization Algorithms*. IGI Global.

Vasant, P., Barsoum, N., & Webb, J. (Eds.). (2011). *Innovation in Power, Control, and Optimization: Emerging Energy Technologies*. IGI Global.

Vasant, P., Litvinchev, I., & Marmolejo, J. A. (Eds.). (2018). *Modeling, simulation, and optimization*. Academic Press. doi:10.1007/978-3-319-70542-2

Vasant, P., & Voropai, N. (2016). *Sustaining Power Resources through Energy Optimization and Engineering*. IGI Global. doi:10.4018/978-1-4666-9755-3

Vasant, P., Weber, G.-W., & Dieu, V. N. (2016). *Handbook of Research on Modern Optimization Algorithms and Applications in Engineering and Economics*. IGI Global. doi:10.4018/978-1-4666-9644-0

Vashkov, V. I. (1973). *Means and methods of sterilization, used in medicine*. Moscow: Medicine.

Vasiliev, A. N. (2018). Increasing Efficiency of Grain Drying With the Use of Electroactivated Air and Heater Control. In Handbook of Research on Renewable Energy and Electric Resources for Sustainable Rural Development. Hershey, PA: IGI Global. Doi:10.4018/978-1-5225-3867-7.ch011

Vasiliev, A.N., Ospanov, A.B., Budnikov, D.K., Karmanov, D.K, Salginbayev, D.B. & Vasilyev, A.A. (2016). Controlling reactions of biological objects of agricultural production with the use of electrotechnology. *International Journal of Pharmacy & Technology, 8*(4), 26855-26869.

Vasiliev, A.N., Ospanov, A.B., Budnikov, D.K., Karmanov, D.K, Salginbayev, D.B., & Vasilyev, A.A. (2016). Controlling reactions of biological objects of agricultural production with the use of electrotechnology. *International Journal of Pharmacy & Technology, 8*(4), 26855-26869.

Vasilyev, A. N. (1986). Statisticheskie harakteristiki parametrov atmosfernogo vozduha dlya aktivnogo ventilirovaniya semyan [Statistical characteristics of atmospheric air parameters for active seed ventilation]. Automation and computer facilities in agricultural production - Collection of scientific papers MIISP, 77-82.

Vasilyev, A. N. (2005). Kosvennyj kontrol' vlazhnosti i temperatury zerna pri sushke v plotnom sloe [Indirect control of moisture and temperature of grain during drying in a dense layer]. Mechanization and Electrification of Agriculture, 3, 5-7.

Vasilyev, A. N. (2006). Kriterij optimal'nosti i granichnye usloviya v modeli sushki plotnogo sloya [The optimality criterion and boundary conditions in the dense layer drying model]. Energy-saving technologies in the agroindustrial complex (All-Russian scientific-practical conference, collection of articles), 14-17.

Vasilyev, A. N. (2009). *EHnergosberegayushchie ehlektrotekhnologii sushki i predposevnoj obrabotki zerna aktivnym ventilirovaniem* [Energy-saving electrotechnology for drying and presowing grain processing by active ventilation] (Doctoral Dissertation). Moscow, Russia.

Vasilyev, A. N., & Rudenko, N. B. (2009). Model' teplovlagoobmena pri recirkulyacii zerna [Model of heat and moisture exchange during grain recycling]. Vestnik agrarnoj nauki Dona, 37-41.

Vasilyev, A. N., & Severinov, O. V. (2016). Ocenka ehffektivnosti usovershenstvovannogo algoritma upravleniya sushkoj zerna v bunkerah aktivnogo ventilirovaniya [Evaluation of the effectiveness of the improved algorithm for controlling grain drying in active storage bins]. *Theoretical and Scientific and Practical Journal Innovations in Agriculture, 3*(18), 165-171. Retrieved from http: //ej.viesh.ru/wp-content/uploads/2016/03/insel16.pdf

Vasilyev, A.N. & Severinov, O.V. (2014). Matematicheskaya model' teplo- i vlagoobmena v ehlementarnom sloe pri sushke zerna aktivnym ventilirovaniem [Mathematical model of heat and moisture exchange in the elementary layer during grain drying by active ventilation]. *AgroEHkoInfo, 2.*

Vasilyev, A.N., & Udincova, N.M. (2000). Analiz parametrov vozdushno-teplovoj obrabotki dlya uluchsheniya posevnyh kachestv semyan [Analysis of the parameters of air-heat treatment to improve the seed quality of seeds]. *Mekhanizaciya i ehlektrifikaciya sel'skogo hozyajstva, 1,* 9-10.

Vasilyev, A. N. (2008). *Ehlektrotekhnologiya i upravlenie pri intensifikacii sushki zerna aktivnym ventilirovaniem* [Electrotechnology and management in the intensification of grain drying by active ventilation]. Rostov-on-Don, Russia: Terra Print.

Vasilyev, A. N. (2008). Informacionnyj podhod k opisaniyu povedeniya zernovogo sloya pri sushke [Information approach to the description of grain layer behavior during drying]. *Vestnik FGOU VPO MGAU im. V.P. Goryachkina. Agroinzheneriya, 3*, 28–31.

Vasilyev, A. N. (2013). Elektrotechnologies as controlling influence on biological objects in agricultural production. *Research in Agricultural Electric Engineering, 3*(3), 96–102.

Vasilyev, A. N., Budnikov, D. A., Gracheva, N. N., & Severinov, O. V. (2016). *Sovershenstvovanie tehnologii sushki zerna v plotnom sloe s ispolzovaniem elektrotehnologiy, ASU i modelirovaniya protsessa* [Improvement of the technology of drying grain in a dense layer with the use of electrotechnologies, automated control systems and modeling of the process]. Moscow, Russia: FSBSI FSAC VIM.

Vasilyev, A. N., Severinov, O. V., & Makarova, Yu. M. (2016). Razrabotka komp'yuternoj modeli teplo- i vlagoobmena v plotnom sloe zerna [Development of a computer model of heat and moisture exchange in a dense layer of grain]. *Vestnik NGEHI, 4*(59), 63–71.

Vedenyapin, G. V. (1967). Obschaya metodika eksperimental'nogo issledovania i obrabotki opytnyh dannyh [General methodology of experimental studies and of processing the experimental data] (2nd ed.). Kolos. (in Russian)

Vedenyapin, G. V., Kirtbaya, Yu. K., & Sergeev, M. P. (1968). *Ekspluatatsiya mashinno-traktornogo parka* [Operation of the machine and tractor fleet]. Moscow, Russia: Kolos.

Vendin, S.V., Bulavin, S.A., & Saenko, U.V. (2015). Opredeleniye parametrov sushilki proroschennogo zerna [Determination of parameters of the germinated grain dryer]. *Mekhanizatsiya i elektrifikatsiya selskogo khozyaistva* [Mechanization and electrification of agriculture], *1*, 8-10. (in Russian)

Verbickij, V. V., Maratov, B. K., Timoshenko, A. M., & Novoselov, S. V. (1989) Instrukciya po aktivnomu ventilirovaniyu zerna i maslosemyan (tekhnika i tekhnologiya) [Instruction on active ventilation of grain and oilseeds (machinery and technology)]. Moscow, Russia.

Vershinin, V.S. (2011) Izmenenie harakteristik vlazhnosti vozduha [Change of characteristics of humidity of air]. *Santekhnika, otoplenie, kondicionirovanie, 3*(111), 84-86.

VIESH. (1997). *Metodicheskie rekomendacii po matematicheskomu modelirovaniyu processa sushki i ohlazhdeniya zerna v ustanovkah plotnogo sloya* [Methodical recommendations on mathematical modeling of the process of grain drying and cooling in installations of a dense layer]. Moscow, Russia: VIESH.

Vincent, C., Panneton, B., & Fleurat-Lessard, F. (2001). Electrical weed control: Theory and applications. In *Physical control methods in plant protection.* Springer-Verlag Berlin Heidelberg. doi:10.1007/978-3-662-04584-8

Vodyanikov, V.T. (2002). *Organizacionno- ekonomicheskie osnovy sel'skoi elektroenergetiki. Uchebnoye posobie dlya vuzov po agroinzhenernym special'nostyam* [Organizational and economic fundamentals of rural electroenergineering. A tutorial for colleges of agroengineering specialties]. Academic Press. (in Russian)

Voskresenskaya, N. P. (1965). *Fotosintez i spektral'nyj sostav sveta* [Photosynthesis and spectral composition of light]. Moscow: Nauka.

Voskresenskaya, N. P., & Nechaeva, V. P. (1967). Dejstvie sinego, krasnogo i zelenogo sveta na soderzhanie belka, nukleinovyh kislot i hlorofilla v molodyh rasteniyah yachmenya [Effect of blue, red and green light on protein, nucleic acids and chlorophyll in young plants of barley]. *Fiziologiya rastenij. Plant Physiology, 32*(14), 299–307.

Voyevodin, V. V., Zhumatij, S.A., Sobolev, S.I., Antonov, A.S., Bryzgalov, P.A., Nikitenko, D.A., & Stefanov, K.S. (2012). Practice of "Lomonosov" Supercomputer. *Open Systems J., 7,* 36-39. Retrieved from http://www.osp.ru/os/2012/07/13017641 (in Russian)

Wang, Y. (2011). Review of dielectric drying of foods and agricultural products. *International Journal of Agricultural and Biological Engineering, 4*(1). Retrieved from http://www.ijabe.org

Warchalewski, J. R., & Gralik, J. (2010). Influence of microwave heating on biological activities and electrophoretic pattern of albumin fraction of wheat grain. *Cereal Chemistry, 87*(1), 35–41. doi:10.1094/CCHEM-87-1-0035

Weaver, J., & Chizmadzhev, Y. (1996). Theory of electroporation: A review. *Bioelectrochemistry and Bioenergetics, 41*(2), 135–160. doi:10.1016/S0302-4598(96)05062-3

Wilson, W., Abplanalp, H., & Atrington, L. (1962). Sexual development of coturmix as affected by changes in photoperiods. *Poultry Science, 41*(1), 17–22. doi:10.3382/ps.0410017

Yagodin, B. A., Zhukov, Y. P., & Kobzarenko, V. I. (2002). Agrokhimiya [Agrochemistry]. Moskow: Kolos.

Yakovlev, A. A. (2005). O sistemnom podhode k formirovaniyu mnozhestv tekhnicheskih reshenij preobrazovatelej ehnergii [About the system approach to the formation of sets of technical solutions for energy converters]. *News of higher educational institutions. Mechanical Engineering (New York, N.Y.), 7,* 44–50.

Yara N-Tester™. (n.d.). Retrieved from http://www.yara.ua/crop-nutrition/tools-and-services/n-tester

Yeom, H. W., Chism, G. W., & Zhang, Q. H. (2002). Inactivation of pectin methyl esterase in orange juice by pulsed electric fields. *Food Engineering and Physical Properties, 67*(6), 2154–2159.

Yozef, M. (2011). OSRAM o buduschem biznesa v svetotehnicheskoi otrasli [OSRAM about the future of business in lighting industry]. *Svetotehnika, 1,* 59–61.

Yudaev, I. V., Baev, V. I., Eliseev, D. S., & Brenina, T. P. (2004). Sornye rasteniya kak ob"ekt elektricheskoy propolki: biologicheskie osobennosti i elektrofizicheskie svoystva: monografiya [Weed plants as an object of electrical weeding: Biological features and electrophysical properties: Monograph]. Volgograd: Stanitsa-2.

Yudaev, I.V., & Baev, V.I. (2006). Optimal'nye parametry elektroimpul'snogo unichtozheniya sornykh rasteniy [Optimal parameters for electrical impulse destroying of weeds]. *Vestnik Volgogradskoy sel'skokhozyaystvenncy akademii. Nauka i vysshee professional'noe obrazovanie* [Herald of Volgograd State Agricultural Academy. Science and higher professional education], *2*(2), 71-78.

Yudaev, I.V., & Kostychev, K.V. (2013). Obrabotka ehlektricheskim tokom zernofurazha, kak process uluchsheniya ego perevarimosti pri otkorme cyplyat-brojlerov [Electricity grain feed processing as a process of improving its digestibility in fattening chickens-broilers]. In *Integraciya nauki i proizvodstva – strategiya ustojchivogo razvitiya APK Rossii v VTO. Materialy mezhdunarodnoj nauchno-prakticheskoj konferencii* [The integration of science and production is the strategy of sustainable development of Russia's agro-industrial complex in the WTO. Materials of the international scientific-practical conference]. Volgograd: VolGAU.

Yudaev, I.V., Rodionov, S.N., Gamaga, V.V., Zotov, V.M., Kostychev, K.V., & Panteleev, A.K. (2016). Analiz harakteristik ehlektrolitov ispol'zuemyh pri ehlektricheskoj obrabotke rastitel'nyh kormov [Analysis of the characteristics of electrolytes used for electrical processing of plant feeds]. *Mekhanizaciya i ehlektrifikaciya sel'skogo hozyajstva* [Mechanization and electrification of agriculture], *11,* 28-30.

Yudaev, I. V., & Brenina, T. P. (2008). The definition of electro impulses used in weed control. *Journal of agricultural sciens, 53*(1), 37–44.

Yuferev, L. Y. (2016). *Energoresursosberegayuschie osvetitel'nye i obluchatel'nye systemy i ustanovki sel'skohozyastvennogo naznachenia na osnove rezonansnoi systemy elektropitania* [Energy resources saving illumination and radiation systems, and agricultural installations on the bases of a resonant power supply system] (Unpublished doctoral dissertation). Moskva. (in Russian)

Yuferev, L.Y., Alferova, L.K., & Baranov, D.A. (2008). Avtomatizatsiya obezzarazhivaniya ul'trafioletovym oblucheniyem v pomeshcheniyakh dlya vyrashchivaniya molodnyaka ptitsy. *Sbornik nauchnykh dokladov VIM*, *1*, 448-456.

Yuferev, L.Y., Alferova, L.K., & Prokopenko, A.A. (2007). Elektrofizicheskoye obezzarazhivaniye i dezodoratsiya vozdukha v zhivotnovodcheskikh pomeshcheniyakh. *Nauchnyye trudy GNU VNIIMZH Rossel'khozakademii*, *17*(4), 63-68.

Yuferev, L.Y., Alferova, L.K., Baranov, D.A., & Prokopenko, A.A. (2010). Obluchatel'nyy pribor na baze amal'gamnykh UF lamp dlya ob'yektov veterinarnogo nadzora. *Trudy mezhdunarodnoy nauchno-tekhnicheskoy konferentsii Energoobespecheniye i energosberezheniye v sel'skom khozyaystve*, *3*, 292-296.

Yuferev, L. Y. (2006). Increase the operational and energy characteristics of irradiators. *Proceedings of the International Scientific and Technical Conference Energy supply and energy saving in agriculture*, *3*, 270-275.

Yuferev, L. Y., & Alferova, L. K. (2016). *Svetotekhnika v sel'skom khozyaystve*. Moscow: FSAC VIM.

Zabel, P., Bamsey, M., Schubert, D., & Tajmar, M. (2014). Review and analysis of plant growth chambers and greenhouse modules for space. *44th Intl. Conf. Environ. Systems*, *44*, 1-17.

Zborovsky, A. B. (2001). Itogi i perspektivy ispol'zovaniya prirodnyh faktorov Volgogradskoj oblasti v medicine [Results and perspectives of the use of natural factors in the Volgograd region in medicine]. *Osvoenie i ispol'zovanie prirodnyh resursov Volgogradskoj oblasti – put' ustojchivogo razvitiya regiona: materialy nauchno-prakticheskoj konferencii 15-16 dekabrya 2000 [*Development and use of natural resources of the Volgograd region - the way of sustainable development of the region: materials of the scientific and practical conference December 15-16, 2000], 94-99.

Zelenko, V.N. (1988). *Konvektivnaya sushka sel'skohozyajstvennyh materialov v plotnom sloe. Osnovy teorii* [Convective drying of agricultural materials in a dense layer. Fundamentals of the theory]. Tver', Russia: Tverskoe oblastnoe knizhnoe izdatel'stvo.

Zelinka, I., Vasant, P., Duy, V. H., & Dao, T. T. (Eds.). (2018) Innovative Computing, Optimization and Its Applications: Modelling and Simulations. Academic Press. doi:10.1007/978-3-319-66984-7

Zenkov, N. K., Lankin, V. Z., & Menshchikova, E. B. (2001). *Okislitel'nyj stress: biohimicheskie i patofiziologicheskie aspekty* [Oxidative stress: biochemical and pathophysiological aspects]. Moscow: Nauka.

Zepeda-Bautista, R., Hernandez-Aguilar, C., Suazo-Lopez, F., Dominguez-Pacheco, A. F., Virgen-Vargas, J., Perez-Reyes, C., & Peon-Escalante, I. (2014). *Electromagnetic field in corn grain production and health*. Academic Press.

Zhang, M. I. N., Stout, D. G., & Willison, J. H. M. (1990). Electrical Impedance Analysis in Plant Tissue: Symplastic Resistance and Membrane Capacitance. In the Hayden Model. *Journal of Experimental Botany*, *41*, 371–380. doi:10.1093/jxb/41.3.371

Zhao, Y., Wang, W., Xie, J., Zheng, B., Miao, S., Lo, Y. M., ... Tian, Y. (2017). Microwave Vacuum Drying of Lotus Seeds: Effect of a Single-Stage Tempering Treatment on Drying Characteristics, Moisture Distribution and Product Quality. *Drying Technology*, *35*(13), 1561–1570. doi:10.1080/07373937.2016.1261890

Zheng, E. L., Fan, Y. D., Zhu, R., Zhu, Y., & Xian, J. Y. (2016). Prediction of the vibration characteristics for wheeled tractor with suspended driver seat including air spring and MR damper. *Journal of Mechanical Science and Technology*, *30*(9), 4143–4156. doi:10.100712206-016-0826-x

Zhidko, V. I., Rezchikov, V. A., & Ukolov, V. S. (1982). *Zernosushenie i zernosushilki* [Drying of grain and grain dryer]. Moscow, Russia: Kolos.

Zhiltsov, V. I. (1979). *Issldovaniye rezhimov raboty elektrifitsirovannykh ustanovok dlya obespecheniya staticheskogo davleniya vozdukha v ptichnike* [Reseach of operation modes of electrified units providing static air pressure in poultry houses] (Unpublished PhD dissertation). Moscow: VIESH. (in Russian)

Zielinska, M. (2012). A multi-stage combined heat pump and microwave vacuum drying of green peas. *Journal of Food Engineering*. doi:10.1016/j.jfoodeng.2012.10.047

Zielinska, M., & Markowski, M. (2017). The effect of microwavevacuum, ultrasonication and freezing on mass transfer kinetics and diffusivity during osmotic dehydration of cranberries. *Drying Technology*. doi:10.1080/07373937.2017.1390476

Zielinska, M., Sadowski, P., & Błaszczak, W. (2015). Thawing and Microwave-Assisted Drying of Blueberries (Vaccinium Corymbosum L.). *Lebensmittel-Wissenschaft + Technologie*, *62*(1–2), 555–563. doi:10.1016/j.lwt.2014.08.002

Zimagulov, A. Kh., & Nurulin, R. G. (2003). Trebovaniya k tyagovym MES dlya vypolneniya tehnologicheskih operatsiy [Requirements for traction MEM for the execution of technological operations]. *Tractors and Agricultural Machinery*, *4*, 8–9.

Zimmermann, U. (1986). Electric breakdown, electropermeabilization and electrofusion. *Physiological Biochemical Pharmacology*, *105*, 175–256. doi:10.1007/BFb0034499

Zoz, F., & Grisso, R. (2003). Traction and Tractor Performance. Paper presented at *Agricultural Equipment Technology Conference*, Louisville, KY.

Zuyev, F. G., Levachev, N. A., & Lotkov, N. A. (1987). Mekhanizatsiya pogruzochno-razgruzochnykh. transportnykh i skladskikh rabot [Mechanization of loading and unloading, transport and storage works]. Moscow. Russia: VO Agropromizdat.

Zyudkamp, H. & Bakalets, A. (n.d.). Viper-kompiuter dlya pticefabrik. [A Viper-computer for the poultry houses]. *Zhivotnovodstvo Rossii, 2,* 74-75. (in Russian)

About the Contributors

Valeriy Kharchenko's date of birth 18 December 1938. The researcher is a consultant in a wide range of energy, energy policy, RUE and RES. Doctor of Engineering, professor. He currently holds the position of Chief Scientific Officer of the Federal Scientific Agroengineering Center. He is the official reviewer of a number of international journals, including such as Solar Energy, Deputy Editor-in-Chief of "European Journal of Renewable Energy", member of the editorial boards of several international journals, honorary professor of the Institute of Hydropower and Renewable Energy of the National Research University" MEI. Under his edition a number of books was published by IGI Global (USA).

Pandian Vasant is a senior lecturer at Department of Fundamental and Applied Sciences, Faculty of Science and Information Technology, Universiti Teknologi PETRONAS in Malaysia. He holds PhD (UNEM, Costa Rica) in Computational Intelligence, MSc (UMS, Malaysia, Engineering Mathematics) and BSc (2nd Class Upper-Hons in Mathematics, UM, Malaysia) in Mathematics. He has co-authored research papers and articles in national journals, international journals, conference proceedings, conference paper presentation, and special issues lead guest editor, lead guest editor for book chapters, edited books, and keynote lecture. (218 publications indexed in SCOPUS). In the year 2009, Dr. Pandian Vasant was awarded top reviewer for the journal Applied Soft Computing (Elsevier), awarded outstanding reviewer in the year 2015 for ASOC (Elsevier) journal and Top reviewer for Sentinels of Science: Computer Science (Oct. 2015 - Sept. 2016). He has 26 years of working experience at the various universities from 1989-2017. Currently he is Editor-in-Chief of IJCO and IJEOE, Member of American Mathematical Society, MERLIN, and NAVY Research Groups.

* * *

Nikola Armenyanov's date of birth is 1947. In 1973 received the diploma of specialist in Electrical Engineering in Kyiv Polytechnic Institute. In the period 1991 – 2009 worked at the same university as an assistant of the Power supply of industrial enterprises department. In 1988 got degree of Doctor of Technical Sciences. Has got more than 200 scientific works.

Vladimir Baltachev was born in 1950, Izhevsk, Udmurt Republic, Russia. He graduated from the Foreign languages Department, Udmurt state University, 1977 as an interpreter and translator in the sphere of professional communication. In 1990, he defended his Candidate thesis on the history of professional English, Leningrad state university, Russia. Since 2010, V. Baltachev has been working for the Izhevsk state agricultural academy as an Associate Professor, Foreign languages Department. His primary scientific and professional interests and activities lie in the fields of contemporary animal husbandry issues.

Aleksandr Belov was born on June 14, 1985 in the village of. Tenevo Alikovsky district of the Chuvash Republic. From 1992 to 2002 studied at the school N°48 Cheboksary of the Chuvash Republic, graduated from school with a silver medal. From 2002 to 2007 He studied at the Chuvash State Agricultural Academy at the Faculty of Engineering, received a higher professional education in engineering "Mechanization of agricultural processing." products ". In 2017 he defended his doctoral dissertation. I am an employee of the FBNU FNAC VIM.

Dmitry Budnikov is the Head of the Laboratory of Electrophysical Impacts FSAC VIM. Research interests: "Intensification of technological processes in agriculture due to the electrical effects and improved control algorithms." The bulk of research is made for drying grain in the dense layer using microwave fields.

Diana Charova, in 2014, graduated from the Electric Power Department of the Volgograd State Agrarian University. Since September 2014 - assistant of the "Electrotechnology and electrical equipment in agricultural production" department of the Electrical engineering faculty of the Volgograd State Agrarian University. Main scientific fields: electrotechnology and lighting engineering. Has 15 publications.

Yuliia Daus's date of birth is 1986. In 2008 received the diploma of specialist in Energy Management in Zaporizhzhya National Technical University. In the period 2008 – 2014 worked at the same university as an assistant of the Power supply of industrial enterprises department. From 2012 to 2015 studied at post graduate school on Electrotechnical complexes and systems specialty. Now work as an editor of the scientific journal " Don Agricultural Science Bulletin ". In 2017 got master's degree in Electrical Engineering and power.

Valerij Drevin graduated from the Natural-geographical faculty at Volgograd State Pedagogical University named after A. S. Serafimovich. 2010-2012 –head of the Chemistry faculty of Volgograd State Agrarian University. Since January 2013 – Dean of the "Processing technologies and commodity science" faculty of Volgograd State Agrarian University. In 1999 defended thesis on "Synthesis and properties of peroxides based on poly-and perfluorinated carbonyl compounds". Main scientific field: chemistry of organometallic compounds; chemistry and study of properties of natural minerals; chemistry and study of properties of food products. He has 113 publications, including 1 scientific work in foreign countries.

Stepan Dudin in 2011 graduated from Izhevsk state Agricultural Academy, faculty of "Electrification and automation of agriculture", qualification / specialty electrical engineer. Since 2012 - graduate student of Federal State Budgetary Scientific Institution "Federal Scientific Center Agroengineering VIM". Research interests: electrification of thermal processes in agriculture. Author of 15 scientific articles and 3 patents.

Andrey Feklistov, in 2013, graduated from the faculty of electrification and automation of agriculture at the Volgograd state agrarian University. 2013-2018-assistant Professor at the faculty of electric power engineering of the Volgograd state agrarian University. Main scientific fields: microprocessor technology, Biophysics. He has 11 publications, including 1 scientific work in foreign countries.

Valerij Fomichev, in 1965 graduated from the Physics and mathematics faculty at Volgograd State Pedagogical Institute. 1987-2016 – head of the "General and applied chemistry" department of Volgograd State University of Architecture and Construction. Since January 2017 – Professor of the "General and inorganic chemistry" department of Volgograd State Technical University. In 1996 defended doctoral thesis on " Electrodeposition of chromium from a standard electrolyte with organic additives." Main scientific fields: chemistry of organoelement compounds; chemistry and research of natural minerals properties; electrochemistry. He has 362 publications, including 3 scientific works in foreign countries.

Tatyana Gallyamova was born in 1971, Izhevsk, Udmurt Republic, Russia. She graduated from the Physics Department of the Udmurt state University in 1994 with a degree in Physics. In 2012, she completed the correspondence master's degree in "Agroengineering". Presently, as a teacher of fundamental mathematics at the State agricultural academy in Izhevsk, she is interested in the problems of updating a contemporary system of higher education. In 2016, she defended her Candidate thesis on improving the efficiency and uniformity of led lighting of agricultural premises, by the example of a poultry house. Thus, her scientific concerns are primarily related to the problems of energy conservation at the objects of the APK (agro-industrial complex). Mathematical modeling, physical modeling, optimization of energy consumption of led lighting also lie in the sphere of her professional interest.

Varvara Gamaga is a Candidate of Biological Sciences, associate professor of the Ecology and Nature Management Department, deputy dean of the Geography Faculty on scientific work. Professional interests are problems and ensuring environmental safety of urban areas.

Gulnara Gizzatova, in 2001, graduated from the Natural Geography faculty of the Volgograd State Pedagogical University. Since September 2015 – associate Professor of the Chemistry department. In 2014 defended thesis in technical sciences at the Volgograd State University of Architecture and Civil Engineering on «Development of the method of water purification in swimming pools from surface active substances». The main scientific fields are: chemical, ecological, pedagogical and food. She has got 30 publications, including one scientific work in foreign countries.

Zakhid Godzhaev is deputy director in FSAC VIM,Moscow, Russian Federation. In 1979 he received PhD (Candidate of Engineering Sciences) in Moscow. After that, in 1995 he became a doctor of Engeniiring Sciences. Finally, in 2007 he reached a status of "Full Professor". Scientific interests: Creation of mobile agricultural machines and tractors. Also, professor at Bauman MOSCOW STATE TECHNICAL UNIVERSITY and The Educational Scientific-and-technical Center "MAMI". Have more than 150 publications, scientific works and state awards.

Pavel Ivanov 1998-2003 student of Azov-Black Sea state Agroengineering Academies. 2003-2006 post-graduate student of the Department "tractors and cars". 2006-2010 high school teacher at the Department "theoretical mechanics". 2010-2012high school teacher at the Department "quality management in of Agroengi-neering". Since 2012 teacher of higher school at the Department "technical ser-vice in the agro-industrial complex". In 2012 defended thesis on competition for the degree of candidate of technical Sciences. Dissertation topic -Substantiation of parameters of the working body of the sub- soil sowing of cereal grains in the conditions of pneumatic transportation of seeds in to coulters.

Svitlana Kalenska is a Prof., Doctor, Corresponding Member of National Academy of Agriculture Scienses Head of Plant Science Department of National University of Life and Enviromental Sciences of Ukraine Education: 1982 - graduated from the Ukrainian agricultural academy, specialty "Agricultural chemistry and soil sciences"; 1991 – degree of the Candidate of Agricultural sciences; 2001 – degree of the Doctor of Agricultural sciences Work position:1991-2001 - Institute of Agriculture of NAASU, Scientific research in the field of plant production of cereal crops; 2002 – until present National University of Life and Environmental Sciences of Ukraine: Professor of the Plant Production Department; Head of the Plant Production Department Scientific and educational activities: Fields of scientific research are biology of plant and crops production, seeds science, bioenergetika.

Evgeny Khalin is doctor of technical sciences, the full member of Academies of electrotechnical sciences of the Russian Federation, conducts more than forty five years researches in the field of electrical safety and safety of production. Head of the laboratory of the Federal State Budget Scientific Institution "Federal Scientific Agro-Engineering Center VIM" (FGBNU FNAC VIM). Specialist in the field of intellectual information and communication technologies in safety of production, the e-learning systems, engineering means of ensuring of electrical safety. The author more than 340 scientific publications, including 5 monographs, 48 patents of the Russian Federation, the head of development of 15 existing national standards.

Nicolai Khokhriakov obtained a degree in physics at Udmurt State University, USSR, in 1992, and PhD in the field of theoretical physics at the same University in 1996. Since 2000, he is an associate professor at Izhevsk State Agricultural Academy. He has more than 70 scientific publications. His research interests include mathematical modelling, information technologies, nanotechnologies.

Dmytro Komarchuk is an associate professor of the Department of Automation and Robotic Systems of NULES of Ukraine, PhD. He was born on 07.10.1987 in Mala Vilshanka, Bila Tserkva district, Kyiv region. He graduated in 2009 from National University of Bioresources and Natural Resources of Ukraine, majoring in electrification and automation of agriculture, engineer-researcher. In 2014 he defended his thesis in specialty "Electrical Engineering Systems and Systems". He published 34 scientific papers, including 18 articles, 6 declarative patents for invention, 10 abstracts at scientific conferences.

Valeriya Komarova is an associate professor of the Chemistry department, Volgograd State Agrarian University. She has more than 40 scientific publications.

Volodymyr Kozyrskyi is doctor of Engineering, Professor, Honored Scientist of Ukraine. He was graduated department of the electrification of Ukrainian Agricultural Academy (now The National University of Life and Environmental Sciences of Ukraine) in 1978. He was appointed director of the research institute of engineering and technology in September 2005. He was appointed director of the Education and Research Institute of Energetics, Automation and Energy Efficiency in 2009.

Valentin Krausp is a Graduate from Azovo-Chernomorskiy Institution of Agricultural Mechanization in 1954. Received the degree of doctor of philosophy in 1960 in Moscow. Received the degree of doctor of doctor of engineering sciences in 1997 in Moscow. Received the title of Professor in 1998 in Moscow.

Pavel Kryuchin, in 2009 graduated from the Engineering faculty at the Samara State Agricultural Academy. In 2013 defended the master's thesis on the subject "Improvement of Quality of Seeding of Hardly Loose Cultures Development and Use of the Universal Disk and Brush Sowing Device of the Pneumatic Selection Seeder". 2015 – 2016 – the associate professor of the "Electrification and automation of agrarian and industrial complex" department of the Engineering faculty of the Samara State Agricultural Academy. 2016 – 2017 – the deputy dean for study at the Engineering faculty of Samara State Agricultural Academy. Since September, 2017 – the deputy dean for scientific work at the Engineering faculty of Samara State Agricultural Academy. Main scientific areas: "Electrotechnologies and electric equipment in agriculture". Has 37 scientific publications.

Inna Kucherova, in 2011 graduated from Volgograd State Agricultural Academy. In 2014 defended thesis on "The use of camelina meal as a filler pre-mixes in the feeding of calves." Since 2015 associate professor of Chemistry, food and sanitary micro-biology department. Main scientific fields are: animal husbandry, animal feeding, chemistry. She has 22 publications, scientific works in foreign countries are absent.

Tatyana Kudinova is an associate professor of the Foreign Languages Department at the Orel State Agrarian University named after N.V. Parakhin and specializes in teaching English to bachelors studying in the specialty "Power Supply", "Electrical Equipment". Tatyana has been studied at the Orel State University from 1994 to 1999 and received the qualification "Teacher of English and German". In 2006, Tatyana defended her thesis on specialty 10.02.19 "Language Theory". She has great professional experience in teaching English and working with scientific terminology in various fields of science and technology. Tatyana has been working at the Orel State Agrarian University since 1999.

Pavel Lavrukhin is a 1976 -1981 student of Azov-Black sea Institute of agricultural mechanization. 1981 -1985 engineer of the North-Caucasus machinery testing station. 1985 – 1994 research associate, Department of mechanization field crops 1994 - 2001 senior researcher of the scientific-production company "Agrotehnic» in 2000 defended thesis for the degree of candidate of technical Sciences. 2001 - 2005 associate Professor of the Department "Theoretical mechanics» 2005-2018 associate Professor of " Tractors and cars»

Vitalii Lysenko is a Head of the Department of Automation and Robotic Systems of NULES of Ukraine. PhD., Professor. He was born on February 11, 1950 in the village of. Mlyev, Gorodischensky district of Cherkasy region. He is an electrical engineer, graduated from the faculty of electrification Ukrainian Labor Red Banner of the Agricultural Academy in 1972. In 1979 he defended his Ph.D. thesis in the specialty 05.13.07 – Automation of control processes on the theme "Development and research of circuits and technical means for electric charging and automatic aerosol spraying of medicinal preparations of poultry complexes". He has 263 publications, including 12 textbooks and manuals, 6 monographs, 15 copyright certificates and patents, over 100 scientific articles in professional journals.

Sergey Mashkov, in 2005 graduated from at the Engineering faculty of Samara State Agricultural Academy. In 2009 defended the master's thesis on a subject "Economic assessment of agricultural machinery in the production technology of crop production (on materials of the Samara region)". 2009-

2016 – the deputy dean for study at the Engineering faculty of Samara State Agricultural Academy. Since July, 2016 – the Dean of the Engineering faculty of Samara State Agricultural Academy. Main scientific areas: "Electrotechnologies and electric equipment in agriculture", Agroinzheneriya. Has 90 publications, including: 2 publications in foreign countries.

Elena Mikhaylova was born in 1989, the research associate of the Federal State Budget Scientific Institution "Federal Scientific Agro-Engineering Center VIM" (FGBNU FNAC VIM). The coauthor 55 scientific publications, including 4 patents of the Russian Federation, responsible executor in the development of 7 national standards.

Lyubov Minchenko is an associate professor of the Chemistry department, Volgograd State Agrarian University Birth - 1978, Ternovo, Volgograd region, Russian Federation 2012-2013 – Deputy dean for educational work of the Zooveterinary faculty of the Volgograd State Agrarian University In 2011, defended thesis on " Technology and mode of irrigation of onions with a sprinkler machine "Bauer" on light chestnut soils of the Lower Volga region ". Main scientific fields: agriculture, processing of agricultural products. He has more than 40 scientific publications.

Ihab Mohamed received his BSc from the Institute of Aviation Engineering and Technology (IAET) in 2009, Egypt. He also received his MSc in Electrical Engineering from Cairo University, Egypt, in 2014. From 2009 to 2015, he was a Teaching Assistant at Electronics and Communication Department, IAET. During 2015 to 2017, he was an MSc student of "European Master in Advanced Robotics (EMARO+)". He is currently pursuing a PhD degree in Robotics Engineering at INRIA Sophia Antipolis, France, under the supervision of Prof. Philippe Martinet, Guillaume Allibert, and Paolo Salaris. Ihab's research interests include predictive control, power electronics, robotics, computer vision, and machine learning.

Konstantin Morenko is a senior researcher of the Technological Systems of Renewable Energy Sources Department. The dissertation on usage of a special generator in small wind power plant for cand. of tech. sciences scientific degree was defended in 2014.

Denis Morgunov, in 2003 graduated from Samara State University. In 2008 - 2011 became Foreman of the Energospetsstroy company. Since 2011 is the director of LLC PTP Energostandart. Since 2015 is the graduate student of the Power Supply of Railway Transport department of SAMGUPS. Main scientific areas: "Electrotechnologies and electric equipment in agriculture", Agroinzheneriya. Has 5 publications.

Oleksii Opryshko is an associate professor of the Department of Automation and Robotic Systems of NULES of Ukraine. He was born on April 27, 1974 in Kyiv. He graduated from the State Academy of Light Industry of Ukraine in 1996. In 2003 he defended his Ph.D. thesis on the theme "Increase of operational reliability of the system of protective destruction of electrical installations of livestock farms". Since 2001 he has been working at the Center for Computer Technologies Training and Information Center of the University, since 2003 – at the Department of Automation. He is an author of 5 patents, 25 articles, 5 methodical developments, a mentor of the academic group. Scientific interests – information and computer-integrated technologies, ecological systems.

Nataliia Pasichnyk is an associate professor of the chair of Agricultural Chemistry and Quality of Plant Products of NULES of Ukraine, PhD. She graduated from National Agrarian University, specialty – agroecology, qualification – scientist agronomist, 1998. The theme of the candidate's thesis: "Efficiency of using coal-fox ("Tecos-3") in growing carrot dining on dark gray podzolized soils of the Northern Forest-steppe of Ukraine", specialty 06.01.04 – Agrochemistry, 2002. Sphere of scientific interests: remote monitoring of the crops condition and mineral nutrition using UAV, optimization of nutrition and fertilization of corn, application of water-soluble fertilizers.

Svetlana Ponomaryova graduated from the physics-and-mathematics faculty, Udmurt state university, Izhevsk, Russia, 1978. After three years of teaching mathematics at the physics-and-mathematics lyceum has proceeded with her carrier at the Izhevsk state agricultural academy. Defended a Candidate thesis on indicative targeted planning of grain production in a region, 2003. Her research interests lie in the field of mathematical module application in agricultural production. Along with that, problems of optimization of mathematics teaching in agrarian higher education is also of her great interest. S. Ponomaryova authored and co-authored more than 30 scientific publications.

Vitaliy Savchenko graduated with honors in 2008 Department of Energy and Automation, National University of Life and Environmental Sciences of Ukraine in specialty "Automated process control." In 2012 he defended his thesis for the degree of candidate of technical sciences, specialty 05.09.03 - Electrotechnical complexes and systems on "electro-technological complex for potato pre-treatment in a magnetic field." Has over 130 scientific and educational works.

Sergey Senkevich is a 1997-2002 student of the Agroengineering University. 2002-2005 postgraduate-course student of the Department of Tractors and Cars. 2006 defended thesis for the degree of candidate of technical sciences. 2009 awarded the title of associate professor of the Department of Tractors and Cars. 2005-2013 high school teacher at the Department of Tractors and Cars. 2013-now senior researcher, Laboratory of Mobile Energy Facilities.

Nikolay Sergeev graduated from the Azov-Black Sea Institute of Agricultural Mechanization in 1987, specializing in mechanization of agriculture. In 2009 I defended a thesis on the topic "Reduction of energy costs in the operation of a tilled aggregate due to changes in the structure of tires of a class 1.4 tractor". In 2012 I was awarded with the academic rank of associate professor at the Tractors and automobiles department. Currently I work as an assistant professor of the Tractors and automobiles department in the Don State Agrarian University.

Tatyana Shipaeva, in 1994 graduated from the Chemistry and technology faculty of the Volgograd State Technical University. In 1998 defended thesis on "Synthesis and study of the properties of multifunctional additives based on chlorinated paraffins, salts of organic acids and Group II metals for polyvinyl chloride". Since 2011, the senior lecturer of the Chemistry department. Main scientific fields: chemical, food and pedagogical. She has 59 publications.

Oleksandr Sinyavsky graduated in 1985 Faculty of Agricultural automation of Ukrainian Agricultural Academy. He worked at the department of electricdrive and electrotechnologies from 1992. In 1992 he defended his thesis on "Automated electrical preparation of nutrient solution in hydroponic greenhouses". Has more than 210 scientific and educational works.

Dmitry Tikhomirov is doctor of technical sciences, professor. He is a graduate of the Faculty of Electrification and Automation for Industry and transport of the Moscow Power Engineering Institute (MPEI). Science 1991 he is working in the All-Russian Institute of Agriculture Electrification, which entered the Federal Scientific Agroengineering Centre VIM in 2016. Head of the Department of Energy supply of agriculture, chief researcher. Direction of scientific work: "electrical and heat supply of technological processes in agriculture." Scientific research is aimed at reducing the energy intensity of agricultural production, energy saving and improving the reliability of energy supply of the agro-industrial complex. He is the author of more than 150 scientific works.

Aleksey Vasiliev is a doctor of technical sciences, professor, head of the scientific direction for energy supply of the agro-industrial complex of the Federal State Budget Scientific Institution "Federal Scientific Agro-Engineering Center VIM" (FSAC VIM).Prior to that, he worked as a pro-rector of the Azov-Black Sea State Agroengineering Academy, deputy director for scientific work at the FGBNU VIESH.

Evgeniy Vasilev graduated from Bauman Moscow State Technical University with bachelor's and master's degree in Engineering. 2017 - now PhD student at All-Russian National Research Institute for Farm Mechanization (VIM). The topic of my thesis is Theoretical and experimental researches of Elastic Damping Mechanism in the transmission of low propulsion tractor. 2015 - now working as design engineer at All-Russian National Institute of Radio Engineering.

Sergey Vasilyev, in 2003 graduated from the Engineering faculty at the Samara State Agricultural Academy. In 2007 defended the master's thesis on the subject "Improvement of Method and Technical Means for Horizontal Measurement of Hardness of the Soil at Introduction of Technology of Coordinate Agriculture". Since 2009 –associate professor of "Electrification and Automation of Agrarian and Industrial Complex" department of the Engineering faculty of the Samara State Agricultural Academy. Main scientific areas: "Electrotechnologies and electric equipment in agriculture". Has 65 publications.

Aleksey Vozmischev was born September 15, 1972 in Glazov, Udmurt Republic, Russia. Received a diploma of higher education in information technologies in power engineering. A. Vozmischev was awarded the title of Laureate of the state prize of the Udmurt Republic for technical re-equipment of production at the OOO "Udmurt poultry farm", 2011. Since 2013, he has worked as a chief power engineer at the same enterprise. His practical experience in the field of power engineering exceeds 20 years. A regular participant of investment projects for the poultry farm aimed at improvement of microclimatic conditions of poultry maintenance. His professional interests are also diversified by the problems of illumination impact on the poultry physiology.

Mykola Zablodskiy graduated in 1973, Faculty of Electrical Engineering of the Kommunarskiy Mining and Metallurgical Institute. He works at the Department of Electrical Machinery and Electrical Equipment Operating from 2015. In 2008 he defended my doctoral dissertation in the specialty 05.09.01 - Electric machines and apparatuses. Has over 280 scientific and educational works.

Index

U

V

www.ingramcontent.com/pod-product-compliance
Lightning Source LLC
LaVergne TN
LVHW062104300926
849161LV00010B/2374